IET CONTROL, ROBOTICS AND SENSORS SERIES 102

Recent Trends in Sliding Mode Control

Other volumes in this series:

Volume 8	**A History of Control Engineering, 1800–1930** S. Bennett
Volume 18	**Applied Control Theory, 2nd Edition** J.R. Leigh
Volume 20	**Design of Modern Control Systems** D.J. Bell, P.A. Cook and N. Munro (Editors)
Volume 28	**Robots and Automated Manufacture** J. Billingsley (Editor)
Volume 33	**Temperature Measurement and Control** J.R. Leigh
Volume 34	**Singular Perturbation Methodology in Control Systems** D.S. Naidu
Volume 35	**Implementation of Self-tuning Controllers** K. Warwick (Editor)
Volume 37	**Industrial Digital Control Systems, 2nd Edition** K. Warwick and D. Rees (Editors)
Volume 39	**Continuous Time Controller Design** R. Balasubramanian
Volume 40	**Deterministic Control of Uncertain Systems** A.S.I. Zinober (Editor)
Volume 41	**Computer Control of Real-time Processes** S. Bennett and G.S. Virk (Editors)
Volume 42	**Digital Signal Processing: Principles, devices and applications** N.B. Jones and J.D.McK. Watson (Editors)
Volume 44	**Knowledge-based Systems for Industrial Control** J. McGhee, M.J. Grimble and A. Mowforth (Editors)
Volume 47	**A History of Control Engineering, 1930–1956** S. Bennett
Volume 49	**Polynomial Methods in Optimal Control and Filtering** K.J. Hunt (Editor)
Volume 50	**Programming Industrial Control Systems Using IEC 1131-3** R.W. Lewis
Volume 51	**Advanced Robotics and Intelligent Machines** J.O. Gray and D.G. Caldwell (Editors)
Volume 52	**Adaptive Prediction and Predictive Control** P.P. Kanjilal
Volume 53	**Neural Network Applications in Control** G.W. Irwin, K. Warwick and K.J. Hunt (Editors)
Volume 54	**Control Engineering Solutions: A practical approach** P. Albertos, R. Strietzel and N. Mort (Editors)
Volume 55	**Genetic Algorithms in Engineering Systems** A.M.S. Zalzala and P.J. Fleming (Editors)
Volume 56	**Symbolic Methods in Control System Analysis and Design** N. Munro (Editor)
Volume 57	**Flight Control Systems** R.W. Pratt (Editor)
Volume 58	**Power-plant Control and Instrumentation: The control of boilers and HRSG systems** D. Lindsley
Volume 59	**Modelling Control Systems Using IEC 61499** R. Lewis
Volume 60	**People in Control: Human factors in control room design** J. Noyes and M. Bransby (Editors)
Volume 61	**Nonlinear Predictive Control: Theory and practice** B. Kouvaritakis and M. Cannon (Editors)
Volume 62	**Active Sound and Vibration Control** M.O. Tokhi and S.M. Veres
Volume 63	**Stepping Motors, 4th Edition** P.P. Acarnley
Volume 64	**Control Theory, 2nd Edition** J.R. Leigh
Volume 65	**Modelling and Parameter Estimation of Dynamic Systems** J.R. Raol, G. Girija and J. Singh
Volume 66	**Variable Structure Systems: From principles to implementation** A. Sabanovic, L. Fridman and S. Spurgeon (Editors)
Volume 67	**Motion Vision: Design of compact motion sensing solution for autonomous systems** J. Kolodko and L. Vlacic
Volume 68	**Flexible Robot Manipulators: Modelling, simulation and control** M.O. Tokhi and A.K.M. Azad (Editors)
Volume 69	**Advances in Unmanned Marine Vehicles** G. Roberts and R. Sutton (Editors)
Volume 70	**Intelligent Control Systems Using Computational Intelligence Techniques** A. Ruano (Editor)
Volume 71	**Advances in Cognitive Systems** S. Nefti and J. Gray (Editors)
Volume 72	**Control Theory: A guided tour, 3rd Edition** J. R. Leigh
Volume 73	**Adaptive Sampling with Mobile WSN** K. Sreenath, M.F. Mysorewala, D.O. Popa and F.L. Lewis
Volume 74	**Eigenstructure Control Algorithms: Applications to aircraft/rotorcraft handling qualities design** S. Srinathkumar
Volume 75	**Advanced Control for Constrained Processes and Systems** F. Garelli, R.J. Mantz and H. De Battista
Volume 76	**Developments in Control Theory towards Glocal Control** L. Qiu, J. Chen, T. Iwasaki and H. Fujioka (Editors)
Volume 77	**Further Advances in Unmanned Marine Vehicles** G.N. Roberts and R. Sutton (Editors)
Volume 78	**Frequency-Domain Control Design for High-Performance Systems** J. O'Brien
Volume 80	**Control-oriented Modelling and Identification: Theory and practice** M. Lovera (Editor)
Volume 81	**Optimal Adaptive Control and Differential Games by Reinforcement Learning Principles** D. Vrabie, K. Vamvoudakis and F. Lewis
Volume 83	**Robust and Adaptive Model Predictive Control of Nonlinear Systems** M. Guay, V. Adetola and D. DeHaan
Volume 84	**Nonlinear and Adaptive Control Systems** Z. Ding
Volume 88	**Distributed Control and Filtering for Industrial Systems** M. Mahmoud
Volume 89	**Control-based Operating System Design** A. Leva *et al.*
Volume 90	**Application of Dimensional Analysis in Systems Modelling and Control Design** P. Balaguer
Volume 91	**An Introduction to Fractional Control** D. Valério and J. Costa
Volume 92	**Handbook of Vehicle Suspension Control Systems** H. Liu, H. Gao and P. Li
Volume 93	**Design and Development of Multi-Lane Smart Electromechanical Actuators** F.Y. Annaz
Volume 94	**Analysis and Design of Reset Control Systems** Y.Guo, L. Xie and Y. Wang
Volume 95	**Modelling Control Systems Using IEC 61499, 2nd Edition** R. Lewis & A. Zoitl
Volume 96	**Cyber-Physical System Design with Sensor Networking Technologies** S. Zeadally and N. Jabeur (Editors)

Recent Trends in Sliding Mode Control

Edited by Leonid Fridman, Jean-Pierre Barbot and Franck Plestan

The Institution of Engineering and Technology

Published by The Institution of Engineering and Technology, London, United Kingdom

The Institution of Engineering and Technology is registered as a Charity in England & Wales (no. 211014) and Scotland (no. SC038698).

First published 2016

The Institution of Engineering and Technology
Michael Faraday House
Six Hills Way, Stevenage
Herts, SG1 2AY, United Kingdom

www.theiet.org

British Library Cataloguing in Publication Data
A catalogue record for this product is available from the British Library

ISBN 978-1-78561-076-9 (hardback)
ISBN 978-1-78561-077-6 (PDF)

Typeset in India by MPS Limited
Printed in the UK by CPI Group (UK) Ltd, Croydon

Contents

Preface

The editors dedicate this book to their colleague and now close to be their "young" retired friend, Alain Glumineau, Professor at *Ecole Centrale de Nantes* for his contributions to the field of sliding mode control. Early from 90's, Alain has been the initiator of the first works, in Nantes, on sliding mode control. By coupling these works with his activities on control of electrical machines, he has shown the great interest of sliding mode control for such applications. From 2000, with Franck Plestan who has just arrived at Nantes, he has proposed new results on high-order sliding mode control (based on LQ controllers, integral sliding mode concept, sliding mode observes). As previously, he has also focused his efforts to show the applicability of such approaches to the control of electrical machines. He has also been the co-chair of *VSS'14* (International Workshop on Variable Structure Systems, Nantes, France, July 2014).

To celebrate Professor Alain Glumineau the editors tried to justify the key contributions from defining the state of art in SMC, representing the new directions and highlight new applications.

This book, for the sake of clarity, is divided into four parts; each of them represents an important research domain for sliding mode community during these last recent years. As the parts are sufficiently self-content, this division allows reading each of them separately.

The first part is devoted to the **design of higher-order sliding-mode controllers**. The design of the higher-order reaching law can then be carried out using Lyapunov analysis. The Lyapunov approach enjoys many attractive features such as a systematic procedure for setting the controller gains and the assessment of input-to-state stability or passivity. Specific designs are presented in the context of disturbance rejection by means of observation and identification. As in the first-order scenario, the sliding surface for higher-order sliding mode controllers can be selected by optimizing a quadratic criterion or by assigning the eigenvalues of the sliding dynamics.

The second part offers **a set of tools for establishing different dynamic properties of systems with discontinuous right-hand sides**. An important property is homogeneity, from which global asymptotic stability and finite-time convergence can be concluded in specific cases. An useful property in the context of interconnected systems is L2 gain. It is shown that a sliding-mode controlled system can achieve an L2 gain not only with respect to matched perturbations, but also with respect to unmatched ones as well. The existence of transient oscillatory behaviors is approached from an extended principle of harmonic balance. The effect that measurement noise has on the type of convergence to the sliding surface is also studied.

Time discretization is addressed in the third part. First-order sliding modes are discretized using an implicit scheme. The resulting discrete-time system does not exhibit chattering and converges to the sliding surface in finite time. Higher-order sliding-mode differentiators, typically used in output-feedback schemes, are discretized in such a way that the optimal accuracy of their continuous-time counterparts is restored. Also, the use of discrete-time reaching laws that are functions of the sliding variables and the state, is analyzed.

The last part is dedicated to **applications**. In the context of energy conversion, sliding-mode control is applied to variable-speed wind turbines, fuel cell coupled to a power converter, rugged DC series motors and rectifiers with unity power factor, and electropneumatic actuator. Finally, an event-triggered sliding-mode scheme is proposed for networked control systems subject to packet loss, jitter and delayed transmissions.

Obviously, this book can present the only a part of the recent results developments in sliding mode control but we have done our best to present full picture.

Enjoy reading!

Editors

Acknowledgments

The editors wish to thank all the anonymous reviewers of the chapters of this book.

We wish to thank Professor Fernando Castaños and Xinming Yan for help in the book preparation.

The work of Leonid Fridman was partially supported by CONACYT grant 261737, PAPIIT under Grant IN113216, DGAPA PASPA Program.

List of contributors

V. Acary
INRIA Grenoble Rhône Alpes, France

L.T. Aguilar
Instituto Politecnico Nacional, CITEDI, Mesa de Otay, Tijuana, Mexico

L. Amet
Quartz EA 7393/ENSEA, Cergy-Pontoise, France

R.S. Ashok
Department of Electrical and Computer Engineering, The University of Alabama in Huntsville, Huntsville, AL, USA

B. Bandyopadhyay
IDP in Systems & Control, IIT Bombay, India

J.-P. Barbot
Quartz EA 7393/ENSEA, Cergy-Pontoise, France

E. Bernuau
Ecole Centrale de Nantes – IRCCyN, Nantes, France

D. Biel
Institute of Industrial and Control Engineering, Universitat Politechnica de Catalunya, Barcelona, Spain

I. Boiko
The Petroleum Institute, Abu Dhabi, United Arab Emirates

A. Boubakir
Faculty of Science and Technology, University of Jijel, Ouled Aissa, Jijel, Algeria

B. Brogliato
INRIA Grenoble Rhône Alpes, Saint-Ismier, France

F. Castaños
Departamento de Control Automático, Centro de Investigación y de Estudios Avanzados del IPN, Mexico

I. Castillo
División de Ingeniería Eléctrica, Departamento de Ingeniería de Control y Robótica, Facultad de Ingeniería, UNAM, Mexico

S. Chakraborty
IDP in Systems & Control, IIT Bombay, India

E. Cruz-Zavala
Instituto de Ingenierla, Universidad Nacional Autonoma de Mexico UNAM, Mexico D.F., Mexico

A. Doria-Cerezo
Institute of Industrial and Control Engineering, Universitat Politechnica de Catalunya, Barcelona, Spain

D. Efimov
INRIA Lille – Nord Europe, Villeneuve d'Ascq, France

A. Estrada
CONACYT Research Fellow – Center for Engineering and Industrial Development (CIDESI), Queretaro, Mexico

C. Evangelista
CONICET and LEICI, Facultad de Ingenieria, Universidad Nacional de La Plata, Argentina

A. Ferrara
Dipartimento di Ingegneria Industriale e dell'Informazione, University of Pavia, Pavia, Italy, Italy

A. Ferreira de Loza
Instituto Politecnico Nacional, CITEDI, Mesa de Otay, Tijuana, Mexico

E. Fossas
Institute of Industrial and Control Engineering, Universitat Politechnica de Catalunya, Barcelona, Spain

L. Fridman
Department of Control, Engineering Faculty, Universidad Nacional Autónoma de México (UNAM), México D.F., Mexico

M. Ghanes
Quartz EA 7393/ENSEA, Cergy-Pontoise, France

O. Huber
INRIA Grenoble Rhône Alpes, Saint-Ismier, France

G.P. Incermona
Dipartimento di Ingegneria Industriale e dell'Informazione, University of Pavia, Pavia, Italy, Italy

A. Levant
School of Mathematical Sciences, Tel-Aviv University, Ramat-Aviv, Tel-Aviv, Israel

M. Livne
School of Mathematical Sciences, Tel-Aviv University, Ramat-Aviv, Tel-Aviv, Israel

D. Lunz
School of Mathematical Sciences, Tel-Aviv University, Ramat-Aviv, Tel-Aviv, Israel

J. Moreno
Universidad Nacional Autónoma de México, Instituto de Ingenieria, Coordinación Eléctrica y Computación, Ciudad Universitaria, México, D.F., Mexico

Y. Orlov
Electronic and Telecommunication Department, CICESE Research Center, Carretera Tijuana-Ensenada, Mexico

T. Osuna
Electronic and Telecommunication Department, CICESE Research Center, Carretera Tijuana-Ensenada, Mexico

W. Perruquetti
INRIA Lille – Nord Europe, Villeneuve d'Ascq, France

A. Pisano
Department of Electrical and Electronic Engineering, University of Cagliari, Cagliari, Italy

F. Plestan
LUNAM Université, Ecole Centrale de Nantes, Nantes, France

A. Polyakov
INRIA Lille – Nord Europe,
Villeneuve d'Ascq, France

I. Ponce
Electronic and Telecommunication Department, CICESE Research Center, Carretera Tijuana-Ensenada, B.C. 22860, Mexico

P. Puleston
CONICET and LEICI, Facultad de Ingenieria, Universidad Nacional de La Plata, Argentina

T. Sanchez
Universidad Nacional Autónoma de México, Instituto de Ingenieria, Coordinación Eléctrica y Computación, Ciudad Universitaria, CP 04510 México, D.F., Mexico

G. Santomenna
GS Maintenance Co., France

Y.B. Shtessel
Department of Electrical and Computer Engineering, The University of Alabama in Huntsville, Huntsville, AL, USA

M. Taleb
ENSEM, Moulay Ismail University, Meknes, Morocco

E. Usai
Department of Electrical and Electronic Engineering, University of Cagliari, Cagliari, Italy

X. Yan
LUNAM Université, Ecole Centrale de Nantes, Nantes, France

S. Zhuk
Department of System Analysis and Decision Making Theory, Taras Shevchenko National University of Kyiv, Ukraine

Section 1

Novel sliding mode algorithms

Chapter 1.1

Lyapunov approach to higher-order sliding mode design

Emmanuel Cruz-Zavala and Jaime A. Moreno

Abstract

Analysis and design of the well-known higher-order sliding mode (HOSM) controllers are usually made by means of homogeneity and contraction properties. Up to now, HOSM algorithms have not been properly addressed within a Lyapunov framework, and there is no Lyapunov-based analysis and design, despite the fact that Lyapunov's methods are one of the most important analysis and design tools in modern control theory of nonlinear systems.

In this chapter, we design new families of homogeneous HOSM controllers for a class of single-input-single-output uncertain systems. In contrast to the well-known quasi-continuous and nested HOSM control (HOSMC) families, the proposed families of HOSM controllers are obtained by using the concept of control Lyapunov functions (CLFs). The CLFs are constructed explicitly by a modification of the well-known Backstepping technique and applying properties of homogeneous systems. These ingredients allow us to construct continuously differentiable CLFs recursively, avoiding the problem of using complex mathematical tools for non-continuously differentiable Lyapunov functions. A Lyapunov framework for the HOSM design leads to synthesize different families of homogeneous HOSM controllers and it will allow us to investigate important nonlinear properties of these algorithms which cannot be easily studied with the known techniques hitherto established in HOSM theory.

1.1.1 Introduction

A suitable robust control technique for nonlinear uncertain systems is the classical first-order sliding mode control (FOSMC). This technique provides strong insensitivity against uncertainties and disturbances satisfying the well-known *matching condition*, i.e., uncertainties and perturbations affecting the system through the same channels as the control input. FOSMC is easy to implement and reduces the order of the state equations. However, it also has some disadvantages. For instance, it is restricted to systems having relative degree equal to one and it yields the well-known chattering effect, a phenomenon of high-frequency and finite-amplitude oscillations, which degrades the performance of the controlled system.

The higher-order sliding mode (HOSM) technique has been developed to deal with these drawbacks [20, 22, 24]. Elimination of the restriction of relative degree equal to one, better accuracy with respect to discrete sampling time and chattering effect reduction are achieved by a HOSMC [21, 22]. In contrast to the conventional FOSMC design, where a Lyapunov framework is used, a HOSMC is designed applying homogeneity tools [22]. The analysis and design of these algorithms have not been properly addressed and linked with the classical nonlinear control theory. In this sense, it is well known that a Lyapunov framework is used extensively for nonlinear systems analysis providing a strong instrumental tool for synthesizing nonlinear feedback controllers [9, 32, 33].

Recently, the analysis and design of HOSMCs by means of Lyapunov functions have been investigated. In some works [25, 26, 28–31], Lyapunov's method has been successfully applied only for second-order discontinuous algorithms. For example, for the twisting and super-twisting algorithms, *weak* and *strong* differentiable and non-differentiable Lyapunov functions have been proposed. Only few attempts have been made to construct Lyapunov functions for HOSM algorithms but with limited success compared to second-order sliding modes. In Reference 35, a general method to construct a Lyapunov function for HOSMs was proposed. However, the weakness of the method lies in the necessity of having the explicit solutions of the differential inclusion (DI), what strongly restricts the applicability of the method to systems of order higher than two. Recently, some promising attempts have been initiated [34], which promise to extend the design of Lyapunov functions to arbitrary order systems (see also the Chapter 1.4 in this book by the same authors).

Other methods have been tried to design new families for HOSMC applying standard nonlinear control tools. For example, in Reference 11, a kind of Lyapunov redesign [17] has been applied to designing HOSMCs. Basically, it consists of using firstly a continuous controller to achieve finite-time stabilization of an unperturbed chain of integrators, and based on this a Lyapunov redesign is applied to enhance the robustness properties of the closed-loop system by means of a discontinuous control. The resulting controller rejects matched bounded uncertainties/disturbances, that the nominal continuous control law is not able to compensate. However, this approach needs a previously designed continuous finite-time controller, e.g., those proposed by the authors in References 13, 14 and 33, for which a Lyapunov function is already known, and the closed-loop system is not homogeneous. Therefore, the problem of designing a homogeneous HOSMC using a Lyapunov framework is not properly addressed. Homogeneity has been a crucial ingredient in the development of HOSMCs. It provides nice properties to the system and leads to conclude, in a simple form, the finite-time stability and the accuracy of the controlled system.

The present chapter focuses on how HOSMC design can be done by using the concept of control Lyapunov functions (CLFs). Originally, Lyapunov theory was developed for systems without inputs. For systems with control inputs, CLFs play a main role in the construction of feedback control laws [9, 18]. Particularly, we show that a modification of the Backstepping technique leads to construct – simultaneously – suitable homogeneous CLFs and homogeneous HOSMCs for a class of uncertain systems. This approach leads to design robust controllers with a simple structure, which in feedback with the system make the closed-loop system homogeneous. The robustness and the finite-time stability properties provided to the uncertain plant in feedback with the proposed controllers can be characterized by means of the CLFs. Also, the Lyapunov's framework leads to sufficient conditions for tuning the gains and to estimate the convergence time of the state trajectories.

In contrast to the classical families of HOSMC reported in References 10, 20–22, 37, (i) the proposed families of HOSMCs are obtained by means of an explicit construction of CLFs; (ii) they are also discontinuous on manifolds which defines continuous hyper-surfaces; and (iii) they improve transient performance with same control effort and provide good convergence rate.

1.1.2 Basic mathematical tools

We will use the following notation. $\mathbb{R}$ stands for the set of real numbers and $\mathbb{R}_{\geq 0} := \{x \in \mathbb{R} : x \geq 0\}$. For a given p belonging to the natural numbers $\mathbb{N}$, $\mathscr{C}^p$ is the set of continuous functions which are continuously differentiable up to the order p. Define an open ball centered at the origin with radius $R > 0$ by $B_R = \{x \in \mathbb{R}^n : \|x\| < R\}$. Finally, we adopt the notation $\lceil z \rfloor^m = |z|^m \operatorname{sign}(z)$, for any $z \in \mathbb{R}$ and any real number $m \geq 0$. This operator preserves the sign of the value of the functions. According to this, for any $m, n > 0$, we have

1. $\lceil z \rfloor^m = z^m$ for any odd integer m, $|z|^m = z^m$ for any even integer m, and $\lceil z \rfloor^0 = \operatorname{sign}(z)$.
2. $\lceil z \rfloor^m \lceil z \rfloor^n = |z|^{m+n}$, $\lceil z \rfloor^0 \lceil z \rfloor^m = |z|^m$, and $\lceil z \rfloor^0 |z|^p = \lceil z \rfloor^p$.
3. $\frac{\mathrm{d}}{\mathrm{d}z} \lceil z \rfloor^m = m|z|^{m-1}$ and $\frac{\mathrm{d}}{\mathrm{d}z} |z|^m = m\lceil z \rfloor^{m-1}$.

1.1.2.1 Stability concepts

Consider the following dynamical system:

$$\dot{x} = f(x), \; x_0 = x(0), \tag{1.1.1}$$

which can admit a discontinuous vector field $f(x)$. Solutions of (1.1.1) are understood in the sense of Filippov [8]. Now, we will present some notions related to finite-time stability. Assume that the origin $x = 0$ is an asymptotically stable (AS) equilibrium point.

Definition 1.1.1. *[2, 4, 31]. The origin of (1.1.1) is said to be finite-time stable (FTS) if it is AS and any solution $x(t, x_0)$ of (1.1.1) reaches $x = 0$ at some finite-time moment, i.e., $x(t, x_0) = 0$, $\forall t \geq T(x_0)$. $T(x_0) : B_R \setminus \{0\} \to (0, \infty)$ is called the settling-time function.*

The notion is global if Definition 1.1.1 is satisfied with $B_R = \mathbb{R}^n$.

Denote by $D^*\phi(t)$ the upper right-hand derivative of a function $\phi(t)$, $D^*\phi(t) = \lim_{h\to+0}\sup((\phi(t+h)-\phi(t))/h)$. The following lemma gives a characterization of the finite-time stability in terms of Lyapunov functions.

Lemma 1.1.1. *[4]. If there exists a continuous, positive definite and proper function $V:\mathbb{R}^n\to\mathbb{R}_{\geq 0}$ and real numbers $\alpha>0$, $p\in(0,1)$, such that any solution $x(t)$ of (1.1.1) satisfies the inequality $D^*V(x(t))\leq -\alpha V^p(x(t))$, then the origin of (1.1.1) is globally FTS (GFTS) and $T(x_0)\leq V^{1-p}(x_0)/\alpha(1-p)$, $\forall x_0\in\mathbb{R}^n$.*

In practice, it is more convenient to apply Lemma 1.1.1 with a Lipschitz continuous or a continuously differentiable function V [4]. In this case, since $V(x(t))$ is differentiable at t, the derivative $D^*V(x(t))$ yields the time derivative $\dot{V}(x(t)) = \frac{\partial V(x)}{\partial x}f(x)$.

1.1.2.2 Homogeneous systems

Finite-time stability analysis, robustness, and design of a HOSMC are usually attained by using homogeneity tools for DIs and contraction properties [22]. Particularly, homogeneity theory helps to establish global stability results and finite-time convergence.

For a given vector $x=(x_1,\dots,x_n)^T\in\mathbb{R}^n$, the dilation operator is defined as $\Delta_\varepsilon^{\mathbf{r}}x := (\varepsilon^{r_1}x_1,\dots,\varepsilon^{r_n}x_n)^T$, $\forall\varepsilon>0$, where $r_i>0$ are the weights of the coordinates $x_i, i=1,\dots,n$. Let $\mathbf{r}=(r_1,\dots,r_n)^T$ be the vector of weights. A function $V:\mathbb{R}^n\to\mathbb{R}$ (respectively, a vector field $f:\mathbb{R}^n\to\mathbb{R}^n$, or a vector-set field $F(x)\subset\mathbb{R}^n$) is called $\mathbf{r}$-homogeneous of degree $l\in\mathbb{R}$ if the identity $V(\Delta_\varepsilon^{\mathbf{r}}x)=\varepsilon^l V(x)$ holds (respectively, $f(\Delta_\varepsilon^{\mathbf{r}}x)=\varepsilon^l\Delta_\varepsilon^{r}f(x)$, or $F(\Delta_\varepsilon^{\mathbf{r}}x)=\varepsilon^l\Delta_\varepsilon^{\mathbf{r}}F(x)$) [2, 3, 22].

Definition 1.1.2. *A system is called homogeneous if its vector field (or vector-set field) is* **r***-homogeneous of some degree.*

Properties of homogeneous functions and vector fields are given in the sequel.

Definition 1.1.3. *[2]. The homogeneous norm is defined by $\|x\|_{\mathbf{r},p} := \left(\sum_{i=1}^n |x_i|^{\frac{p}{r_i}}\right)^{\frac{1}{p}}$, $\forall x\in\mathbb{R}^n$, for any $p\geq 1$. The set $S=\{x\in\mathbb{R}^n : \|x\|_{\mathbf{r},p}=1\}$ is the corresponding unit sphere.*

Lemma 1.1.2. *[2, 4, 36]. Let $\Delta_\varepsilon^{\mathbf{r}}x$ be any family of dilations on $\mathbb{R}^n$, and let V_1, V_2 (respectively, f) be continuous real-valued functions on $\mathbb{R}^n$ (respectively, a continuous vector field) and homogeneous of degrees m_1,m_2 (respectively, l) with respect to $\Delta_\varepsilon^{\mathbf{r}}x$. Then,*

- *(i)* *V_1V_2 is homogeneous of degree m_1+m_2.*
- *(ii)* *There exists a constant c_1 such that $V_1\leq c_1\|x\|_{\mathbf{r},p}^{m_1}$. Moreover, if V_1 is positive definite, there exists c_2 such that $c_2\|x\|_{\mathbf{r},p}^{m_1}\leq V_1$ and for every $x\in\mathbb{R}^n$, $\left[\min_{\{z:V_1(x)=1\}}V_2(z)\right]V_1^{m_2/m_1}\leq V_2\leq\left[\max_{\{z:V_1(x)=1\}}V_2(z)\right]V_1^{m_2/m_1}$.*

(iii) $\frac{\partial V_1(x)}{\partial x_i}$ *is homogeneous of degree* $m_1 - r_i$ *with* r_i *being the weight of* x_i.
(iiv) $\dot{V} = \frac{\partial V_1(x)}{\partial x} \cdot f(x)$ *is homogeneous of degree* $m_1 + l$.

The following lemma is a classical result for continuous homogeneous functions [12].

Lemma 1.1.3. *[1, 12]. Let* $\eta : \mathbb{R}^n \to \mathbb{R}$ *and* $\gamma : \mathbb{R}^n \to \mathbb{R}_{\geq 0}$ *be two continuous* **r***-homogeneous functions, with weights* $\mathbf{r} = (r_1, \ldots, r_n)^T$ *and degrees m, such that the following holds:*

$$\{x \in \mathbb{R}^n \setminus \{0\} : \gamma(x) = 0\} \subseteq \{x \in \mathbb{R}^n \setminus \{0\} : \eta(x) < 0\}. \tag{1.1.2}$$

Then, there exists a real number λ^* *such that, for all* $\lambda \leq \lambda^*$ *and for all* $x \in \mathbb{R}^n \setminus \{0\}$, $\eta(x) - \lambda\gamma(x) < 0$.

This lemma shows that the sum of two **r**-homogeneous functions, one that is non-positive, can be made negative definite by a large weighting of the non-positive one, if condition (1.1.2) is satisfied.

Any homogeneous system with an AS equilibrium point admits a homogeneous Lyapunov function (HLF) [2–4, 27]. For a homogeneous continuous vector field f of degree l with locally AS equilibrium point, a $\mathscr{C}^p$ HLF of degree m exists if $m > p \cdot \max_{1 \leq i \leq n}\{r_i\}$ for any $p \in \mathbb{N}$. Stability of homogeneous DIs can be also studied by means of a $\mathscr{C}^p$ HLF.

Theorem 1.1.1. *[3, 27]. Assume that the origin of a homogeneous Filippov DI,* $\dot{x} \in F(x)$, *is* strongly *globally AS (GAS). Then, there exists a* $\mathscr{C}^\infty$ strong *HLF.*

All solutions of the DI converge to the origin if it is *strongly* GAS. A DI $\dot{x} \in F(x)$ is called a Filippov DI if the vector set $F(x)$ is non-empty, closed, convex, locally bounded and upper-semicontinuous [8]. Finite-time stability of a homogeneous DI is characterized by the homogeneity degree l of its vector-set field.

Lemma 1.1.4. *[3, 22]. Assume that the origin of a Filippov DI,* $\dot{x} \in F(x)$, *is strongly GAS and the vector-set field F is* **r***-homogeneous of degree* $l < 0$. *Then,* $x = 0$ *is strongly GFTS and the settling time is continuous at zero and locally bounded.*

1.1.3 The standard HOSMC problem

Consider a single-input-single-output uncertain system affine in the input

$$\dot{x} = f(t,x) + g(t,x)u, \quad \sigma = h(t,x), \tag{1.1.3}$$

where $x \in \mathbb{R}^n$ defines the state vector, $u \in \mathbb{R}$ is the control input, and $h(t,x) : \mathbb{R}^n \times \mathbb{R} \to \mathbb{R}$ is a sufficiently smooth measurable output function. Functions $f(t,x)$ and

$g(t,x)$ are uncertain smooth vector fields on $\mathbb{R}^n$ and the dimension n can also be unknown. The control objective consists of making the output σ vanish in finite time and keep $\sigma = 0$ exactly by a bounded (discontinuous) feedback control. All differential equations are understood in Filippov's sense [8].

Assume that the relative degree of system (1.1.3) with respect to σ is known, well defined, time invariant and equal to ρ. The value of ρ defines the minimum order i of the total derivative $\sigma^{(i)} := \frac{d^i}{dt^i}h(x,t)$ at which the control u appears explicitly for the first time [15]. Also, suppose that the associated zero dynamics is weakly stable. Under these assumptions, the output satisfies the differential equation

$$\Sigma_T : \begin{cases} \dot{z}_i = z_{i+1}, & \forall i = 1,\dots,\rho - 1, \\ \dot{z}_\rho = w(x,t) + b(x,t)u, & z_0 = z(0), \end{cases} \tag{1.1.4}$$

where $z = (z_1,\dots,z_\rho)^T = (\sigma,\dots,\sigma^{(\rho-1)})^T$. The scalar functions $w(x,t)$ and $b(x,t)$ are unknown and the only information about them are their bounds, i.e., there exists suitable known positive constants C, K_m, and K_M such that

$$0 < K_m \le b(x,t) \le K_M, |w(x,t)| \le C. \tag{1.1.5}$$

In practice, conditions (1.1.5) are satisfied at least in compact operating regions. For simplicity, we will assume that (1.1.5) holds globally. Then, it implies the DI

$$\Sigma_{ID} : \begin{cases} \dot{z}_i = z_{i+1}, & \forall i = 1,\dots,\rho - 1, \\ \dot{z}_\rho \in [-C,C] + [K_m,K_M]u. \end{cases} \tag{1.1.6}$$

In Σ_{ID} the dependencies of the original system dynamics disappears and the DI only "remembers" the constants ρ, C, K_m, and K_M.

The stated problem can be solved by designing a discontinuous feedback controller

$$u = \vartheta_r(z_1,z_2,\dots,z_\rho), \tag{1.1.7}$$

that renders the origin

$$z_1 = z_2 = \cdots = z_\rho = 0, \tag{1.1.8}$$

FTS for Σ_T. The motion on the set (1.1.8) is called ρ-order sliding mode. Function ϑ_r has to be discontinuous at this set, otherwise the origin cannot be stabilized under the uncertainty conditions considered.

1.1.4 Homogeneous HOSMC design by using CLFs

Homogeneity properties to the closed-loop inclusion (1.1.6) can be imposed through the feedback (1.1.7) [22]. Due to the term $[-C,C]$, the right-hand side of (1.1.6) must have homogeneity degree 0 with $C \neq 0$ and the homogeneity degree of $z_{\rho-1}$ has to be opposite to the degree l of the whole system. For this case, the coordinates $(z_1, \ldots, z_\rho)$ have associated vector of weights $\mathbf{r} = (r_1, \ldots, r_\rho)^T$, where each weight r_i satisfies $(\rho + 1 - i)l + r_i = 0$, for all $i = 1, \ldots, \rho$. These equalities are fulfilled iff $l < 0$, since $r_i > 0$ and $\rho > 0$. Setting $l = -1$, the vector of weights becomes $\mathbf{r}^s = (\rho, \rho - 1, \ldots, 1)^T$ [22]. The controller (1.1.7) is called ρ-sliding homogeneous if the equality

$$\vartheta_r \left(\varepsilon^{r_1} z_1, \varepsilon^{r_2} z_2, \ldots, \varepsilon^{r_\rho} z_\rho\right) = \vartheta_r(z_1, z_2, \ldots, z_\rho), \tag{1.1.9}$$

holds for any $\varepsilon > 0$. Due to (1.1.9), a locally bounded control will be also globally bounded. Let us define $F(z) = f(z) + \Delta f(t, z)$ and $G(z) = g(z)\xi(z)$, with $f(z) = (z_2, z_3, \ldots, z_\rho, 0)^T$, $g(z) = (0, 0, \ldots, 0, 1)^T$, $\Delta f(t, z) = g(z)[-C, C]$ and $\xi(z) = [K_m, K_M]$. Therefore, system (1.1.6) can be rewritten as

$$\dot{z} = F(z) + G(z)u. \tag{1.1.10}$$

We assume that system (1.1.10) satisfies the following conditions.

Assumption 1.1.1. *For a given dilation $\Delta_\varepsilon^{\mathbf{r}} z$, with $\mathbf{r} = (r_1, \ldots, r_\rho)^T$,*

(i) system (1.1.10) is homogeneous of degree $l < 0$, where l satisfies $(\rho + 1 - i)l + r_i = 0$, $i = 1, \ldots, \rho$, for any $r_i > 0$.
(ii) There exists a homogeneous CLF $V(z)$ of degree m.

Conditions (i) and (ii) are necessary for the existence of a homogeneous discontinuous controller. The existence of a control law that renders the origin $z = 0$ AS is imposed on the system by using a CLF (this is, in general, a necessary and sufficient condition for stabilizability). A CLF is used to prove that a control system is feedback stabilizable. A function $V : \mathbb{R}^\rho \to \mathbb{R}_{\geq 0}$ is a CLF for system (1.1.10) if it is at least a $\mathscr{C}^1$ function which is positive-definite ($V(0) = 0$ and $V(z) > 0$ for any $z \neq 0$) and proper (for any $c \geq 0$, $V^{-1}(c)$ is a compact set) such that by selecting the feedback control values $u = \vartheta(z)$ the time derivative along the trajectories of (1.1.10) is imposed to be negative definite pointwise. Due to $\dot{V} < 0$ is imposed through the feedback control for system (1.1.10), it immediately implies that a CLF satisfies

$$0 \in L_G V \Rightarrow \sup\nolimits_{v \in F(z)} L_v V < 0, \quad \forall z \in \mathbb{R}^\rho \setminus \{0\}. \tag{1.1.11}$$

For a vector-set field $F(z)$, we denote by $L_v V(z) = \left\{ y \in \mathbb{R} | y = \frac{\partial V(z)}{\partial z} \cdot v, v \in F(z) \right\}$ the set of values taken by Lie's derivatives of V along all vector fields contained in $F(z)$. Condition (1.1.11) has been introduced in Reference 9 to define a robust CLF for systems with uncertainties. In general, $\vartheta_r(z)$ will be discontinuous. In view of the existing converse Lyapunov theorems for DIs [3, 5, 27], Assumption 1.1.1 seems to be reasonable.

Theorem 1.1.2. *Suppose that Assumption 1.1.1 holds for system (1.1.6). Then, the following* **r**-*homogeneous discontinuous feedback controllers*

$$u = -k\left\lceil L_{g(z)}V \right\rfloor^{0} \quad \text{and} \quad u = -k\frac{L_{g(z)}V}{\|z\|_{\mathbf{r},p}^{l+m}} \tag{1.1.12}$$

render GAS the origin $z = 0$ with sufficiently large gain k. Moreover, if $l < 0$ the origin is GFTS.

Remark 1.1.1. *The controllers (1.1.12) can be modified without changing their properties. For example, the controllers $u = -k\lceil\varphi(z)\rfloor^0$ and $u = -k\frac{\varphi(z)}{\|z\|_{\mathbf{r},p}^{l+m}}$ can be also applied for (1.1.6) if the scalar function $\varphi(z)$ is* **r**-*homogeneous of degree $m + l$ and satisfies $\varphi(z) = 0 \Leftrightarrow L_{g(z)}V = 0,\ \forall z \in \mathbb{R}^\rho \setminus \{0\}$. It means that functions $\varphi(z)$ and $L_{g(z)}V$ vanish at the same points.*

The design of a discontinuous state feedback control for (1.1.10) is simple and direct when the homogeneous CLF is known. Since the closed-loop system has negative degree, the controller guarantees (robust) finite-time stabilization at the origin $z = 0$. Moreover, the homogeneity of the closed-loop system is preserved under condition (1.1.5).

We have taken advantage of the homogeneity properties to obtain control laws with the most simple structure. It is well known that a complex structure of the control law can cause undesired transient state response of the controlled system [9].

Now, denote the closed-loop system dynamics in feedback with any of the controllers (1.1.12) by an **r**-homogeneous Filippov DI $\dot{z} = F_{cl}(z)$ of degree l. A robustness result can be stated for the closed-loop system perturbed by an additive term $\Xi(z)$.

Assumption 1.1.2. *The additive disturbance $\Xi(z)$ is* **r**-*homogeneous of degree l.*

Theorem 1.1.3. *Under Assumptions 1.1.1 and 1.1.2, the origin $z = 0$ of the perturbed closed-loop system*

$$\dot{z} = F_{cl}(z) + h\Xi(z) \tag{1.1.13}$$

is still a GAS (GFTS if $l < 0$) equilibrium point for $h \in \mathbb{R}$ sufficiently small.

Theorem 1.1.3 shows that the stability properties at the origin $z = 0$ are not modified by a (sufficiently small) matched (unmatched) perturbation that does not modify the homogeneity degree of the vector-set field $F_{cl}(z)$.

Remark 1.1.2. *Consider that each element of $\Xi(z) = (\xi_1, \ldots, \xi_\rho)^T$ is bounded such that $|\xi_i| \leq C_i\left(|z_1|^{(l+r_1)/r_1} + \cdots + |z_i|^{(l+r_i)/r_i}\right)$, for any arbitrary value of $C_i \geq 0$. It means that, for any $i = 1, \ldots, \rho$, the disturbances are dominated by* **r**-*homogeneous functions. Then, from Theorem 1.1.3, it follows that system (1.1.6), in feedback with any of the controllers (1.1.12) and perturbed by $\Xi(z)$, will be FTS for k sufficiently large.*

With help of a CLF, it is very easy to show that the convergence time is a bounded function of the initial states and can be estimated.

Proposition 1.1.1. *From any initial state $z_0 = z(0) \in \mathbb{R}^\rho$ and any $l < 0$, the controllers (1.1.12) enforce any state trajectory of system (1.1.10) to reach $z = 0$ in a finite time smaller than*

$$T(z_0) \leq \frac{m}{-l\kappa} V^{-\frac{l}{m}}(z_0), \quad i = 1, \ldots, \rho, \tag{1.1.14}$$

where κ is a positive constant.

Remark 1.1.3. *The convergence time estimation provided by (1.1.14) is, in general, very crude.*

For system (1.1.6), Assumption 1.1.1 is also fulfilled for a dilation $\Delta_\varepsilon^{\mathbf{r}^s} z$ with $\mathbf{r}^s = (\rho, \rho - 1, \ldots, 1)^T$. Therefore, the following result is still valid.

Corollary 1.1.1. *Let $V(z)$ be an $\mathbf{r}^s$-homogeneous CLF satisfying conditions in Assumption 1.1.1 for system (1.1.6). Then, the controllers given by (1.1.12) are ρ-sliding homogeneous for k sufficiently large.*

The homogeneous HOSMC design is essentially reduced to search for an appropriate **r**-homogeneous CLF. Note that for each different CLF, we can obtain different controllers.

1.1.5 Two r-homogeneous CLFs

In this section, two **r**-homogeneous CLFs are introduced for the standard HOSM model (1.1.6). From it and applying the results of Theorem 1.1.2, new families of homogeneous nested and quasi-continuous HOSM controllers are obtained.

Consider an integer $\rho \geq 2$. For a vector $\bar{z}_i = (z_1, \ldots, z_i)^T$, with $i = 1, \ldots, \rho$, we assign the vector of weights $\bar{\mathbf{r}}_i = (r_1, \ldots, r_i)^T$, where each weight r_i satisfies the equality $(\rho + 1 - i)l + r_i = 0$, for all $i = 1, \ldots, \rho$. The restriction on the weights and the homogeneity degree $l < 0$ is necessary to obtain a homogeneous HOSMC. It is worth pointing out that any weight r_i fulfills $r_1 > r_2 > \cdots > r_\rho$. This fact is proved by defining $l = -l'$. Since $l < 0$, the constant l' has to be positive. For the vector $\bar{z}_\rho : = z$, the condition on the weights yields $\bar{\mathbf{r}}_\rho := \mathbf{r} = (\rho l', (\rho - 1)l', (\rho - 2) l', \ldots, l')^T$. Hence, for any $\rho \in \mathbb{N}$,

$$\rho l' > (\rho - 1)l' > (\rho - 2)l' > \cdots > l' \Rightarrow r_1 > r_2 > \cdots > r_\rho.$$

To construct a homogeneous $\mathscr{C}^1$ CLF, its homogeneity degree m must satisfy $m > \max_{1 \leq i \leq \rho}\{r_i\}$. In our case, fixing $m \geq r_1 + r_2$ will suffice. We firstly define recursively the homogeneous and continuously differentiable functions

$$
\begin{aligned}
V_{i,1}(\bar{z}_i) &= \frac{r_i}{m}|z_i|^{\frac{m}{r_i}} - \lceil v_{i-1}(\bar{z}_{i-1})\rfloor^{\frac{m-r_i}{r_i}} z_i + \left(1 - \frac{r_i}{m}\right)|v_{i-1}(\bar{z}_{i-1})|^{\frac{m}{r_i}} \\
&\quad + \gamma_{i-1} k_{i-1}^{\frac{m}{r_i}} V_{i-1,1}(\bar{z}_{i-1}), \qquad (1.1.15) \\
v_i(\bar{z}_i) &= -k_i \lceil \sigma_i \rfloor^{r_{i+1}/r_1}, \quad \sigma_i = \lceil z_i \rfloor^{r_1/r_i} + k_{i-1}^{r_1/r_i} \sigma_{i-1}(\bar{z}_{i-1}),
\end{aligned}
$$

with $V_{1,1}(z_1) = \frac{r_1}{m}|z_1|^{m/r_1}$, $v_1(z_1) = -k_1 \lceil \sigma_1 \rfloor^{r_2/r_1}$, $\sigma_1 = z_1$, and positive constants γ_j. Note that $V_{i,1}(\bar{z}_i)$, $v_i(\bar{z}_i)$, and σ_i are **r**-homogeneous of degrees m, r_{i+1}, and r_1, respectively.

Now, we define the second set of homogeneous and continuously differentiable functions as

$$
\begin{aligned}
V_{i,2}(\bar{z}_i) &= \frac{r_i}{m}|z_i|^{\frac{m}{r_i}} - \lceil v_{i-1}(\bar{z}_{i-1})\rfloor^{\frac{m-r_i}{r_i}} z_i + \left(1 - \frac{r_i}{m}\right)|v_{i-1}(\bar{z}_{i-1})|^{\frac{m}{r_i}} \\
&\quad + \gamma_{i-1} k_{i-1}^{\frac{m}{r_i}} V_{i-1,2}(\bar{z}_{i-1}), \qquad (1.1.16) \\
v_i(\bar{z}_i) &= -k_i \lceil \sigma_{i,d} \rfloor^{\frac{r_{i+1}}{m-r_i}}, \quad \sigma_{i,d} = \lceil z_i \rfloor^{\frac{m-r_i}{r_i}} + k_{i-1}^{\frac{m-r_i}{r_i}} \lceil \sigma_{i-1,d}(\bar{z}_{i-1}) \rfloor^{\frac{m-r_i}{m-r_{i-1}}},
\end{aligned}
$$

with $V_{1,2}(z_1) = \frac{r_1}{m}|z_1|^{m/r_1}$, $v_1(z_1) = -k_1 \lceil \sigma_{1,d} \rfloor^{(m-r_2)/(m-r_1)}$, $\sigma_{1,d} = \lceil z_1 \rfloor^{(m-r_1)/r_1}$, and positive constants γ_j. Note that $V_i(\bar{z}_i)$, $v_i(\bar{z}_i)$, and σ_i are **r**-homogeneous of degrees m, r_{i+1}, and $m - r_i$, respectively.

Proposition 1.1.2. *The continuously differentiable functions $V_{\rho,1}(z)$ (1.1.15) and $V_{\rho,2}(z)$ (1.1.16) are **r**-homogeneous CLFs of degree m for system (1.1.6) for all $\rho \geq 2$, all $\gamma_j > 0$ and sufficiently large values of $k_j > 0, j = 1, 2, \ldots, \rho - 1$. In particular, if $r_1 = \rho$, the CLFs $V_{\rho,1}(z)$ and $V_{\rho,2}(z)$ are $\mathbf{r}^s$-homogeneous.*

Basically, the constants $(k_1, \ldots, k_{\rho-1})$ need to be chosen sufficiently large in the index order. The values of these parameters are obtained in the construction process of the CLF. The parameters γ_j add more flexibility to the gain tuning during the controllers design. Each CLF leads to design different families of discontinuous and quasi-continuous controllers. The following corollary is a direct consequence of Theorem 1.1.2.

Corollary 1.1.2. *Suppose that (1.1.5) holds globally. Then, for any $\rho \geq 2$, the discontinuous and the quasi-continuous controllers*

$$
u = -k_\rho \lceil \sigma_\rho \rfloor^0, \quad u = -k_\rho \frac{\sigma_\rho}{\|z\|_{\mathbf{r},p}^{\frac{m-r_\rho}{}}} = -k_\rho \frac{\lceil z_\rho \rfloor^{\frac{m-r_\rho}{r_\rho}} + k_{i-1}^{\frac{m-r_\rho}{r_\rho}} \lceil \sigma_{i-1} \rfloor^{\frac{m-r_\rho}{r_1}}}{\|z\|_{\mathbf{r},p}^{m-r_\rho}}, \qquad (1.1.17)
$$

$$
u = -k_\rho \lceil \sigma_{\rho,d} \rfloor^0, \quad u = -k_\rho \frac{\sigma_{\rho,d}}{\|z\|_{\mathbf{r},p}^{m-r_\rho}} = -k_\rho \frac{\lceil z_\rho \rfloor^{\frac{m-r_\rho}{r_\rho}} + k_{i-1}^{\frac{m-r_\rho}{r_\rho}} \lceil \sigma_{i-1,d} \rfloor^{\frac{m-r_\rho}{m-r_{\rho-1}}}}{\|z\|_{\mathbf{r},p}^{m-r_\rho}}, \qquad (1.1.18)
$$

are ρ-sliding homogeneous for k_ρ sufficiently large.

Remark 1.1.4. *The gain k_ρ satisfies $K_m k_\rho > \Lambda_{\rho-1} k_{\rho-1}^{\frac{r_{\rho-1}}{r_\rho}} + C$, where the value of $\Lambda_{\rho-1}$ can be computed by means of the CLFs (1.1.15) and (1.1.16).*

The controllers described by (1.1.17) are obtained from the CLF (1.1.15), while the controllers described by (1.1.18) are obtained from the CLF (1.1.16). Since l is related to the homogeneity degree m, we note that each different selection of the homogeneity degree l produces a particular family of CLFs derived from (1.1.15) and (1.1.16). Due to this, a lot of new families of homogeneous nested sliding controllers (NSCs) and quasi-continuous sliding controllers (QCSCs) can be synthesized, it includes the family of $\mathbf{r}^s$-homogeneous controllers.

According to Remark 1.1.1, the following controllers are also ρ-sliding homogeneous:

$$u = -k_\rho \lceil \varphi_\rho(z) \rfloor^0 \quad \text{and} \quad u = -k_\rho \lceil \varphi_{\rho,d}(z) \rfloor^0, \tag{1.1.19}$$

where $\varphi_\rho(z) = z_\rho + k_{\rho-1} \lceil \sigma_{\rho-1} \rfloor^{r_\rho/r_1}$ and $\varphi_{\rho,d}(z) = z_\rho + k_{\rho-1} \lceil \sigma_{\rho-1,d} \rfloor^{\frac{r_\rho}{m-r_{\rho-1}}}$, which are derived from the same proposed families of CLFs.

To end this section, we present an extension of the introduced controllers to the variable gain case. The use of variable gain in control algorithms with discontinuous terms leads to relax the condition (1.1.5); therefore, it makes possible controlling systems whose dynamics are not precisely bounded, and reduction of the chattering effect is achieved. By using geometrical arguments, the classical NCSs and QCSCs have been modified to admit a variable gain in Reference 24 too (see also Reference 37). In a Lyapunov framework, the analysis and design of HOSM algorithms with variable gain are straightforward.

Theorem 1.1.4. *Consider that the term $|w(x,t)| \leq C + \Xi(z,t)$, where the known function satisfies $\Xi(z,t) \geq 0$, $\forall t \geq 0$. Assume that $k_\rho = k'_\rho + K(z,t)$ for the controllers (1.1.17) and (1.1.18) with large enough gain k'_ρ and $K_m K(z,t) \geq \Xi(z,t)$. Then, the origin $z = 0$ of the closed-loop system (1.1.4) is GFTS.*

Other important issue which can be addressed is the improvement of the convergence rate of the HOSM controllers. Due to lack of space, we will not present these results but the reader can refer to References 6 and 7.

1.1.5.1 Behind the construction of the CLFs

The goal of this section is to show how the proposed CLFs were constructed for system (1.1.4). The design follows a recursive method inspired on the work [32]. For illustrative purposes, we focus on (1.1.16). In general, any of the CLFs introduced in this chapter was obtained following the same idea.

It is well known that Backstepping-like technique leads to construct systematically at the same time a CLF and a state feedback controller for stabilization of a cascade system [16, 18]. In Backstepping design, due to the need to calculate derivatives of various functions during the construction of the control law and the CLF, smoothness assumptions on the virtual controllers are commonly required. However, for our system such assumptions cannot be satisfied. To deal with this, we will use this recursive approach with a modification which integrates a desingularizing method proposed in Reference 32 and homogeneity properties to the standard Backstepping method. Applying these modifications, it is possible to construct a CLF with the following features:

1. The function is positive definite by construction.
2. The function is a $\mathscr{C}^1$ (or a $\mathscr{C}^p$) function. In general, it can be designed as many times differentiable as we want.
3. The time derivative of the function along system's trajectories is imposed to be negative definite by making the gain of the controller large enough.

In the sequel, we will present the construction of (1.1.16). We will start with the case $\rho = 2$. Now, instead of using the state variable $s_2 = z_2 - v_1(z_1) = z_2 + k_1\lceil z_1\rfloor^{1/2}$ to construct the CLF, we define the *desingularizing* function

$$\sigma_{2,d} = \lceil z_2\rfloor^2 + k_1^2 z_1. \tag{1.1.20}$$

Note that $s_2 = 0 \Leftrightarrow \sigma_{2,d} = 0$. In this case, both functions define the same curve $z_2 = v_1(z_1) = -k_1\lceil z_1\rfloor^{1/2}$. Also, they have the same sign, i.e., $s_2 s_2 > 0$ and $s_2\sigma_{2,d} > 0$. In contrast to s_2, the function $\sigma_{2,d}$ is at least once differentiable, i.e., it is a $\mathscr{C}^1$ function. The CLF is constructed as

$$V_{2,2}(z_2, z_1) = \Phi_2 + \gamma_1 k_1^3 V_{1,2}(z_1), \quad \Phi_2 = \int_{v_1}^{z_2} \left(\lceil \varsigma_2\rfloor^2 + k_1^2 z_1\right) d\varsigma_2, \tag{1.1.21}$$

where $V_{1,2}(z_1) = \frac{2}{3}|z_1|^{3/2}$ and $\gamma_1 > 0$. The explicit solution of the integral results in $\Phi_2 = \frac{1}{3}|z_2|^3 + k_1^2 z_1 z_2 + \frac{2}{3}k_1^3|z_1|^{3/2}$. Since Φ_2 is a positive-semidefinite function and $V_{1,2}(z_1)$ is a positive-definite function, $V_{2,2}(z_2, z_1)$ is a positive-definite function by construction. Taking the time derivative of $V_{2,2}(z_2, z_1)$ along the system's trajectories, we have

$$\dot{V}_{2,2}(z_2, z_1) = s_{2,d}[b(x,t)u + w(x,t)] + k_1^2 s_2 z_2 + \gamma_1 k_1^3 \lceil z_1\rfloor^{\frac{1}{2}} z_2.$$

Under Assumption 1.1.5 and applying $u = -k_2\lceil\sigma_2\rfloor^0$, $\sigma_2 = s_{2,d}$, we arrive at

$$\dot{V}_{2,2}(z_2, z_1) \leq -(k_m k_2 - C)|s_{2,d}| + k_1^2 s_2 z_2 + \gamma_1 k_1^3 \lceil z_1\rfloor^{\frac{1}{2}} z_2.$$

Note that when $s_{2,d} = 0$, $\dot{V}_{2,2}(z_2, z_1) \leq -k_1^4|z_1|$. Therefore, $V_{2,2}(z_2, z_1)$ is CLF and by Lemma 1.1.3 there exists gain k_2 large enough such that $V_{2,2}(z_2, z_1)$ is negative definite.

Now, we define the desingularizing functions of $s_i = z_i - v_{i-1}(\bar{z}_{i-1})$ as $\sigma_{i,d} = \lceil z_i\rfloor^{\frac{m-r_i}{r_i}} + k_{i-1}^{\frac{m-r_i}{r_i}}\lceil\sigma_{i-1,d}\rfloor^{\frac{m-r_i}{m-r_{i-1}}}$. These functions vanish at the same points, i.e., $s_i = 0 \Leftrightarrow \sigma_{i,d} = 0$. In contrast to s_i, $\sigma_{i,d}$ are $\mathscr{C}^1$ functions. The function (1.1.16) is constructed as

$$
\begin{aligned}
V_{i,2}(\bar{z}_i) &= W_i(\bar{z}_i) + \gamma_{i-1} k_{i-1}^{\frac{m}{r_i}} V_{i-1,2}(\bar{z}_{i-1}), \\
W_i(\bar{z}_i) &= \int_{v_{i-1}}^{z_i} \left(\lceil \varsigma_i \rfloor^{\frac{m-r_i}{r_i}} + k_{i-1}^{\frac{m-r_i}{r_i}} \lceil \sigma_{i-1,d} \rfloor^{\frac{m-r_i}{m-r_{i-1}}} \right) d\varsigma_i,
\end{aligned} \tag{1.1.22}
$$

where γ_{i-1} and k_{i-1} are appropriate positive constants. Also, (1.1.22) is a $\mathscr{C}^1$ and positive-definite function by construction. Since it is based on a modification of the Backstepping technique, we proceed by induction.

As induction hypothesis, we assume that $V_{i-1,2}\,(\bar{z}_{i-1})$ is a CLF for the $(i-1)$th-order system (1.1.4), and we show that $V_{i,2}(\bar{z}_i)$ is a CLF for the ith-order system (1.1.4). Therefore, taking the time derivative of the CLF $V_{i,2}(\bar{z}_i)$ along the system's trajectories of $\dot{z}_1 = z_2, \ldots, \dot{z}_i = z_{i+1}, \forall i = 1, \ldots, \rho - 1$, we obtain

$$
\dot{V}_{i,2}(\bar{z}_i) = \sigma_{i,d} v_i + \frac{m - r_i}{r_i} s_i |v_{i-1}|^{\frac{m-2r_i}{r_i}} \dot{v}_{i-1} + \gamma_{i-1} k_{i-1}^{\frac{m}{r_i}} \left(\sum_{j=2}^{i} \frac{\partial V_{i-1,2}(\bar{z}_{i-1})}{\partial z_{j-1}} z_j \right).
$$

By straightforward calculation, we can deduce that $\sum_{j=2}^{i} \frac{\partial V_{i-1,2}(\bar{z}_{i-1})}{\partial z_{j-1}} z_j = \dot{V}_{i-1,2}(\bar{z}_{i-1}) + \frac{\partial V_{i-1,2}(\bar{z}_{i-1})}{\partial z_{i-1}} s_i$, where $\frac{\partial V_{i-1,2}(\bar{z}_{i-1})}{\partial z_{i-1}} = \sigma_{i-1,d}$. Taking into account these facts, and applying the virtual controller $v_i(\bar{z}_i)$ as in (1.1.16), it follows that

$$
\begin{aligned}
\dot{V}_{i,2}(\bar{z}_i) &\le -k_i |\sigma_{i,d}|^{\frac{m-r_i+r_{i+1}}{m-r_i}} - \gamma_{i-1} k_{i-1}^{\frac{m}{r_i}} \mathscr{V}_{i-1}(\bar{z}_{i-1}) + \Phi_i, \\
\Phi_i &= \frac{m - r_i}{r_i} s_i |v_{i-1}|^{\frac{m-2r_i}{r_i}} \dot{v}_{i-1} + \gamma_{i-1} k_{i-1}^{\frac{m}{r_i}} s_i \sigma_{i-1,d},
\end{aligned}
$$

where $\mathscr{V}_{i-1}(\bar{z}_{i-1}) = -\dot{V}_{i-1,2}(\bar{z}_{i-1}) \ge 0$, since $\dot{V}_{i-1,2}\,(\bar{z}_{i-1}) \le 0$. The right-hand side of the previous differential inequality is homogeneous and should satisfy Lemma 1.1.3. Define

$$
\gamma(z) = -k_i |\sigma_{i,d}|^{\frac{m-r_i+r_{i+1}}{m-r_i}}, \quad \eta(z) = -\gamma_{i-1} k_{i-1}^{\frac{m}{r_i}} \mathscr{V}_{i-1}\,(\bar{z}_{i-1}) + \Phi_i.
$$

Note that at points $\{z \in \mathbb{R}^i \setminus \{0\} : \gamma(z) = 0\} \Rightarrow \eta(z) = \dot{V}_{i,2}(\bar{z}_i) = \gamma_{i-1} k_{i-1}^{\frac{m}{r_i}} \dot{V}_{i-1,2}(\bar{z}_{i-1}) < 0$. Then, when $s_i = \sigma_{i,d} = 0$ is established, for any $(z_1, \ldots, z_i) \ne 0$, the time derivative $\dot{V}_{i,2}(\bar{z}_i)$ is negative. Therefore, by invoking Lemma 1.1.3, for s_i and $\sigma_{i,d}$ different from zero, there exists k_i large enough such that $V_{i,2}(\bar{z}_i)$ is negative definite.

Finally, by induction argument, under condition (1.1.5) and by using the controller (1.1.18), we arrive at

$$
\dot{V}_{\rho,2}(z) \le -(K_m k_\rho - C)|\sigma_{\rho,d}| - \gamma_{\rho-1} k_{\rho-1}^{\frac{m}{r_\rho}} \mathscr{V}_{\rho-1}(\bar{z}_{\rho-1}) + \Phi_\rho,
$$

which is negative definite if k_ρ is chosen sufficiently large. The construction of (1.1.15) is similar. It proves Proposition 1.1.2.

Table 1.1.1 New NSCs derived from the CLFs (1.1.15) and (1.1.16)

ρ	**With CLF (1.1.15) and $l = -1$**	**With CLF (1.1.16) and $l = \frac{-1}{\rho - 1}$**
2	$u = -k_2 \left\lceil \lceil z_2 \rfloor^2 + k_1^2 z_1 \right\rfloor^0$	$u = -k_2 \left\lceil \lceil z_2 \rfloor^2 + k_1^2 z_1 \right\rfloor^0$
3	$u = -k_3 \left\lceil \lceil z_3 \rfloor^3 + k_2^3 \left(\lceil z_2 \rfloor^{\frac{3}{2}} + k_1^{\frac{3}{2}} z_1 \right) \right\rfloor^0$	$u = -k_3 \left\lceil \lceil z_3 \rfloor^4 + k_2^4 \left\lceil \lceil z_2 \rfloor^{\frac{3}{2}} + k_1^{\frac{3}{2}} z_1 \right\rfloor^{\frac{4}{3}} \right\rfloor^0$
4	$u = -k_4 \left\lceil \lceil z_4 \rfloor^4 + k_3^4 \left(\lceil z_3 \rfloor^2 + k_2^2 \left(\lceil z_2 \rfloor^{\frac{4}{3}} + k_1^{\frac{4}{3}} z_1 \right) \right) \right\rfloor^0$	$u = -k_4 \left\lceil \lceil z_4 \rfloor^6 + k_3^6 \left\lceil \lceil z_3 \rfloor^{\frac{5}{2}} + k_2^{\frac{5}{2}} \left\lceil \lceil z_2 \rfloor^{\frac{4}{3}} + k_1^{\frac{4}{3}} z_1 \right\rfloor^{\frac{5}{4}} \right\rfloor^{\frac{6}{5}} \right\rfloor^0$

1.1.6 Differences with the classical families of HOSMCs

There are two important classical families of controllers commonly used in HOSM: the NSCs and the QCSCs [22, 24]. The following controllers belong to the NSCs

(L2) $u_L = -k_2 \left\lceil z_2 + \beta_1 \lceil z_1 \rfloor^{\frac{1}{2}} \right\rfloor^0,$

(L3) $u_L = -k_3 \left\lceil z_3 + \beta_2 \left(|z_2|^3 + |z_1|^2 \right)^{\frac{1}{6}} \left\lceil z_2 + \beta_1 \lceil z_1 \rfloor^{\frac{2}{3}} \right\rfloor^0 \right\rfloor^0,$

while the following controllers belong to QCSCs,

(LQ2) $u_C = -k_2 \frac{z_2 + \beta_1 \lceil z_1 \rfloor^{1/2}}{|z_2| + \beta_1 |z_1|^{1/2}},$

(LQ3) $u_C = -k_3 \frac{z_3 + \beta_2 \left(|z_2| + \beta_1 |z_1|^{3/2} \right)^{-1/2} \left(z_2 + \beta_1 \lceil z_1 \rfloor^{3/2} \right)}{|z_3| + \beta_2 \left(|z_2| + \beta_1 |z_1|^{3/2} \right)^{1/2}}.$

Both families have a recursive structure and the complexity of the control structure increases with the relative degree. Finite-time stability applying these algorithms is usually proved by using homogeneity for DIs and contraction properties. The advantage of QCSCs, compared to NSCs, is the reduction of the high-frequency leaping of the transient response of the trajectories when they approach the HOSM [37]. The positive parameters $(\beta_1, \ldots, \beta_{\rho-1}, k_\rho)$ are properly chosen sufficiently large in the index order. In References 20–22, the gains $(\beta_1, \ldots, \beta_{\rho-1})$ are fixed *a priori* such that only the gain k_ρ needs to be adjusted sufficiently large, but no explicit conditions for the gain tuning are derived. In general, the gains adjustment of these algorithms is usually done by simulations [37].

New families of HOSMCs designed by means of a CLF are derived from Corollary 1.1.2. These controllers have a different and simpler structure than the classical

Table 1.1.2 New QCSCs derived from the CLFs (1.1.15) and (1.1.16)

ρ	**With CLF (1.1.15) and $l = -1$**	**With CLF (1.1.16) and $l = \frac{-1}{\rho - 1}$**
2	$u = -k_2 \dfrac{\lceil z_2 \rfloor^2 + k_1^2 z_1}{\left(\lvert z_1 \rvert^{\frac{p}{2}} + \lvert z_2 \rvert^{\frac{p}{1}}\right)^{\frac{2}{p}}}$	$u = -k_2 \dfrac{\lceil z_2 \rfloor^2 + k_1^2 z_1}{\left(\lvert z_1 \rvert^{\frac{p}{2}} + \lvert z_2 \rvert^{\frac{p}{1}}\right)^{\frac{2}{p}}}$
3	$u = -k_3 \dfrac{\lceil z_3 \rfloor^3 + k_2^3 \left(\lceil z_2 \rfloor^{\frac{3}{2}} + k_1^{\frac{3}{2}} z_1\right)}{\left(\lvert z_1 \rvert^{\frac{2p}{3}} + \lvert z_2 \rvert^{\frac{p}{1}} + \lvert z_3 \rvert^{\frac{2p}{1}}\right)^{\frac{3}{2p}}}$	$u = -k_3 \dfrac{\lceil z_3 \rfloor^4 + k_2^4 \left\lceil \lceil z_2 \rfloor^{\frac{3}{2}} + k_1^{\frac{3}{2}} z_1 \right\rfloor^{\frac{4}{3}}}{\left(\lvert z_1 \rvert^{\frac{2p}{3}} + \lvert z_2 \rvert^{\frac{p}{1}} + \lvert z_3 \rvert^{\frac{2p}{1}}\right)^{\frac{2}{p}}}$
4	$u = -k_4 \cdot \dfrac{\lceil z_4 \rfloor^4 + k_3^4 \left(\lceil z_3 \rfloor^2 + k_2^2 \left(\lceil z_2 \rfloor^{\frac{4}{3}} + k_1^{\frac{4}{3}} z_1\right)\right)}{\left(\lvert z_1 \rvert^{\frac{3p}{4}} + \lvert z_2 \rvert^{\frac{p}{1}} + \lvert z_3 \rvert^{\frac{3p}{2}} + \lvert z_4 \rvert^{\frac{3p}{1}}\right)^{\frac{4}{3p}}}$	$u = -k_4 \cdot \dfrac{\lceil z_4 \rfloor^6 + k_3^6 \left\lceil \lceil z_3 \rfloor^{\frac{5}{2}} + k_2^{\frac{5}{2}} \left\lceil \lceil z_2 \rfloor^{\frac{4}{3}} + k_1^{\frac{4}{3}} z_1 \right\rfloor^{\frac{5}{4}} \right\rfloor^{\frac{6}{5}}}{\left(\lvert z_1 \rvert^{\frac{3p}{4}} + \lvert z_2 \rvert^{\frac{p}{1}} + \lvert z_3 \rvert^{\frac{3p}{2}} + \lvert z_4 \rvert^{\frac{3p}{1}}\right)^{\frac{2}{p}}}$

NSCs and the QCSCs. Table 1.1.1 shows two new families of NSCs. While Table 1.1.2 shows two new families of QCSCs.

For relative degree one, we can apply the FOSMC $u = -k_1 \lceil z_1 \rfloor^0$. For relative degree $\rho = 2$, the proposed controllers are discontinuous on the set $\lceil z_2 \rfloor^2 + k_1^2 z_1 = 0$, which is the same for controller $u_L = -k_2 \left\lceil z_2 + \beta_1 \lceil z_1 \rfloor^{1/2} \right\rfloor^0$ [21]. Hence, both controllers are discontinuous on the same curve in the phase plane and described by $z_2 = -k_1 \lceil z_1 \rfloor^{1/2}$ (see Figure 1.1.1(a)). The stability of system (1.1.6) in feedback with $u_L = -k_2 \left\lceil z_2 + \beta_1 \lceil z_1 \rfloor^{1/2} \right\rfloor^0$ has been studied in detail in Reference 23. With this controller, the system's trajectories reach the origin in two different ways, depending on how the gains k_2 and β are chosen. Basically, if $K_m k_2 > 0.5\beta_1^2 + C$, the controllers provide for the establishment of the FTS second-order sliding mode (terminal sliding mode) and for $C < k_2 K_m \le 0.5\beta_1^2 + C$, the controller assures the twisting-like convergence to the FTS second-order sliding mode.

In the higher-order case, the controllers (1.1.17) and (1.1.18) are discontinuous in different manifolds from those commonly observed in the classical HOSMCs proposed in References 21 and 22. For instance, the controllers (1.1.17) are discontinuous on the manifold $\lceil z_\rho \rfloor^{r_1/r_\rho} + k_{\rho-1}^{r_1/r_\rho} \sigma_{\rho-1} = 0$, i.e., when $z_\rho = -k_{\rho-1} \lceil \sigma_{\rho-1} \rfloor^{r_\rho/r_1}$. This manifold defines a continuous hyper-surface (see Figure 1.1.1(b)). However, the family of NSCs is, in general, discontinuous on discontinuous hyper-surfaces. As an example, Figure 1.1.1(c) shows the discontinuity plane of NSCs with $\rho = 3$. While the third-order sliding mode determined by the set $\sigma_3 = \lceil z_3 \rfloor^3 + k_2^3 \left(\lceil z_2 \rfloor^{3/2} + k_1^{3/2} z_1\right) = 0$ defines a continuous hyper-surface, the manifold imposed by the dimensional condition $\sigma_{3L} = z_3 + 2(\lvert z_2 \rvert^3 + \lvert z_1 \rvert^2)^{1/6} \left\lceil z_2 + \lceil z_1 \rfloor^{2/3} \right\rfloor^0 = 0$ is itself a discontinuous hyper-surface. It shows that at those points where the controllers are discontinuous, the manifolds are quite different.

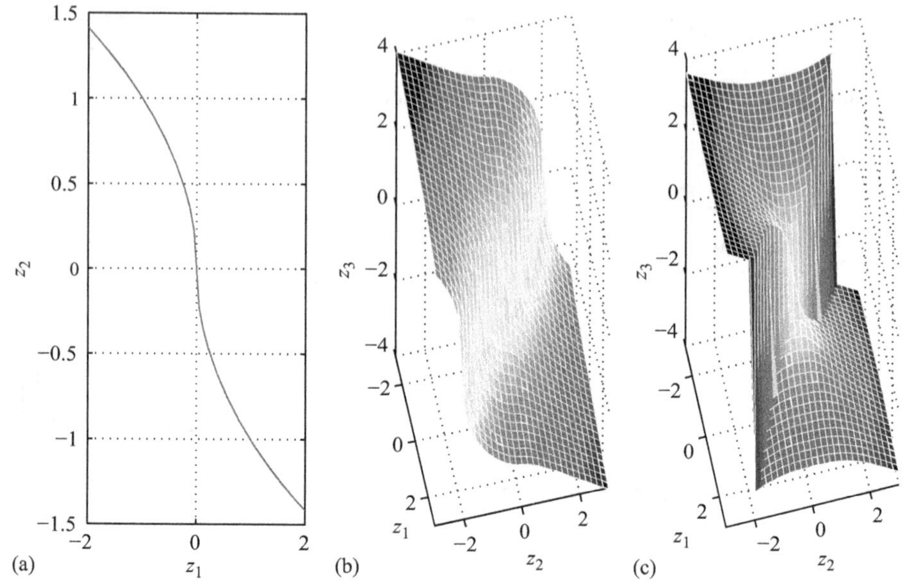

Figure 1.1.1 (a) The manifold $\sigma_2 = \lceil z_2 \rfloor^2 + k_2^2 z_1 = 0$; (b) the manifold $\sigma_3 = \lceil z_3 \rfloor^3 + k_2^3 \left(\lceil z_2 \rfloor^{3/2} + k_1^{3/2} z_1 \right) = 0$; and (c) the manifold $\sigma_{3L} = z_3 + 2\left(|z_2|^3 + |z_1|^2 \right)^{1/6} \left\lceil z_2 + \lceil z_1 \rfloor^{2/3} \right\rfloor^0 = 0$

1.1.7 Gain tuning

An important advantage of using CLFs is that they are useful to obtain conditions for designing the gains of the controllers. In general, obtaining an analytical expression of the gains for HOSM controllers is almost impossible for relative degree bigger than two. Thanks to the homogeneity properties of the proposed CLFs, search for a set of gains is turning into evaluating homogeneous functions and finding of them its maxima on a homogeneous sphere.

For practical purposes and for simplifying the gain design as much as possible, an appropriate way to compute the gains is to parameterize them in terms of the gain k_1, so only the gain k_ρ will have to be adjusted. For example, for the proposed CLFs (1.1.15) and the controller (1.1.17), the gains $(k_2, \ldots, k_\rho)$ can be parameterized in terms of the gain k_1 as follows:

$$k_i = \beta_{i-1} k_1^{\frac{\rho}{\rho-(i-1)}}, \quad \forall i = 2, \ldots, \rho - 1, \quad K_m k_\rho - C \geq \beta_{\rho-1} k_1^{\rho}, \tag{1.1.23}$$

where the parameters β_{i-1}, $i = 2, \ldots, \rho$, are suitable positive constants. In fact, these parameters result to be sufficiently large in the index order, i.e., they satisfy $\beta_1 < \beta_2 < \cdots < \beta_{\rho-1}$. The value of each β_{i-1} can be computed numerically with help of the CLF (1.1.15).

Another parametrization is obtained by introducing $\lambda = k_1^{\rho/(\rho-1)}$. In this case, (1.1.23) is reduced to

$$k_i = \beta_{i-1}\lambda^{\frac{\rho-1}{\rho-(i-1)}}, \quad \forall i = 2, \ldots, \rho-1, \quad K_m k_\rho - C \geq \beta_{\rho-1}\lambda_1^{\rho-1}. \tag{1.1.24}$$

The same k_1-parametrization and λ-parametrization arise for the CLF (1.1.16) and the controller (1.1.18). Expression (1.1.23) is almost an analytical expression that can be applied to tune the gains. It allows us to know how the structure of the gains is for any value of k_1. A similar parametrization has been proposed for the classical QCSCs in References 24, 37.

Now, let us show with a simple example how the parameters β_{i-1}, $i = 2, \ldots, \rho$, are computed from the CLF. Consider a relative degree $\rho = 2$ and the CLF (1.1.15), we know that applying a controller $u = -k_2\lceil\sigma_2\rfloor^0$ the time derivative of the CLF satisfies

$$\dot{V}_{2,1}(z_2, z_1) \leq -(K_m k_2 - C)|\sigma_2| + k_1^2 s_2 z_2 + \gamma_1 k_1^3 \lceil z_1\rfloor^{\frac{1}{2}} z_2$$

and it is imposed to be negative definite by doing the gain k_2 large enough. Therefore,

$$\dot{V}_{2,1}(z_2, z_1) < 0 \quad \Leftrightarrow -(K_m k_2 - C)|\sigma_2| + k_1^2 s_2 z_2 + \gamma_1 k_1^3 \lceil z_1\rfloor^{\frac{1}{2}} z_2 < 0,$$

or equivalently, if the following holds

$$\frac{K_m k_2 - C}{k_1^2} > \beta_1, \text{ where } \beta_1 > \chi := \frac{s_2 z_2 + \gamma_1 k_1 \lceil z_1\rfloor^{\frac{1}{2}} z_2}{|\sigma_2|}.$$

It is worth to note that function χ has two important properties: (i) it is homogeneous of degree zero and (ii) it tends to $-\infty$ when $\sigma_2 \to 0$. This last property is due to we have proposed a CLF. Besides, because of property (i), we only need to evaluate the function on a unit homogeneous sphere to find its maximum value. Then, once we have fixed a value for the parameter γ_1, β_1 is chosen such that

$$\beta_1 > \Lambda_1 := \max_{\{z:S\}}\{\chi\},$$

where $S = \left\{z \in \mathbb{R}^2 : \|z\|_{\mathbf{r},2} := |z_2|^{2/1} + |z_1|^{2/2} = 1\right\}$. In the case $\rho = 2$ and for a suitable $\gamma_1 > 0$, the gains of the proposed controllers satisfy $k_1 > 0$ and $k_2 K_m \geq \beta_1 k_1^2 + C$ (see Table 1.1.3). According to Levant [23], with this gain tuning, the trajectories inevitably hit the manifold $z_2 + k_1\lceil z_1\rfloor^{1/2} = 0$ and thereafter, they slide on it until the origin is reached in finite time, i.e., a terminal sliding mode is established.

Table 1.1.3 Parametrization of the gains by using CLF (1.1.15)

ρ	γ_{i-1}	**Without modification**	**Modified**
2	$\gamma_1 = 1/8$	$K_m k_2 \geq 1.26k_1^2 + C$	$K_m k_2 \geq k_1^2 + C$
3	$\gamma_2 = 0.966$	$K_m k_3 \geq 9.62k_1^3 + C$	$K_m k_3 \geq 3.23k_1^3 + C$
	$\gamma_1 = 1.485$	$k_2 = 1.5k_1^{3/2}$	$k_2 = 1.5k_1^{3/2}$
4	$\gamma_3 = 3.3$	$K_m k_4 \geq 739.5k_1^4 + C$	$K_m k_4 \geq 40k_1^4 + C$
	$\gamma_2 = 1.05$	$k_3 = 8.1k_1^2$	$k_3 = 8.1k_1^2$
	$\gamma_1 = 1.3$	$k_2 = 2k_1^{4/3}$	$k_2 = 2k_1^{4/3}$

For the arbitrary order cases, we have that by using the CLF (1.1.15), the corresponding value of Λ_{i-1}, $i = 2, \ldots, \rho$, is computed numerically with the following expression[1]:

$$\Lambda_{i-1} = \max_{\{z:S\}} \left\{ \frac{\frac{m-r_i}{r_i} k_{i-1}^{-\frac{r_{i-1}}{r_i}} s_i |v_{i-1}|^{\frac{m-2r_i}{r_i}} \dot{v}_{i-1} + \gamma_{i-1} k_{i-1}^{\frac{m-r_{i-1}}{r_i}} \left(\sum_{j=2}^{i} \frac{\partial V_{i-1,1}}{\partial z_{j-1}} z_j \right)}{|\sigma_{i,d}| |\sigma_i|^{\frac{r_{i+1}}{r_1}}} \right\}. \quad (1.1.25)$$

While with the CLF (1.1.16), the value of Λ_{i-1} is computed numerically with the following expression:

$$\Lambda_{i-1} := \max_{\{z:S\}} \left\{ \frac{\frac{m-r_i}{r_i} k_{i-1}^{-\frac{r_{i-1}}{r_i}} s_i |v_{i-1}|^{\frac{m-2r_i}{r_i}} \dot{v}_{i-1} + \gamma_{i-1} k_{i-1}^{\frac{m-r_{i-1}}{r_i}} \left(\sum_{j=2}^{i} \frac{\partial V_{i-1,2}}{\partial z_{j-1}} z_j \right)}{|\sigma_{i,d}|^{\frac{m-r_i+r_{i+1}}{m-r_i}}} \right\}, \quad (1.1.26)$$

where $S = \left\{ z \in \mathbb{R}^i : \|z\|_{\mathbf{r},2} = 1 \right\}$ is the unit homogeneous sphere, which is a compact set, and $s_i = z_i - v_{i-1}(\bar{z}_{i-1})$. Recall that for any $i = 1, \ldots, \rho$, each k_i needs to be replaced by the proposed k_1-parametrization (1.1.23). The maximization on S is feasible, since the function to be maximized has the following properties: (i) it is **r**-homogeneous of degree 0, so that it achieves all its values on the unit sphere S and (ii) it is continuous at the points where the denominator does not vanish, and when the denominator vanishes the numerator is negative.

In general, the value of Λ_{i-1} will depend on the selection of the parameters $(\gamma_{i-1}, \ldots, \gamma_1)$, $i = 2, \ldots, \rho$. By choosing a proper value of the degree of freedom

[1] Particular explicit expressions of Λ_{i-1}, $i = 2, \ldots, \rho$, are given in Section 1.1.10. They were utilized to obtain the parametrizations showed in the tables.

Table 1.1.4 Parametrization of the gains by using CLF (1.1.16)

ρ	γ_{i-1}	**Without modification**	**Modified**
2	$\gamma_1 = 1/8$	$K_m k_2 \geq 1.26k_1^2 + C$	$K_m k_2 \geq k_1^2 + C$
3	$\gamma_2 = 0.966$	$K_m k_3 \geq 9.62k_1^3 + C$	$K_m k_3 \geq 3.23k_1^3 + C$
	$\gamma_1 = 1.485$	$k_2 = 1.5k_1^{3/2}$	$k_2 = 1.5k_1^{3/2}$
4	$\gamma_3 = 2.5$	$K_m k_4 \geq 441k_1^4 + C$	$K_m k_4 \geq 45k_1^4 + C$
	$\gamma_2 = 1.05$	$k_3 = 8.1k_1^2$	$k_3 = 8.1k_1^2$
	$\gamma_1 = 1.3$	$k_2 = 2k_1^{4/3}$	$k_2 = 2k_1^{4/3}$

Table 1.1.5 New NSCs derived from the CLFs (1.1.15) and (1.1.16)

ρ	$\lambda > 0$	$\boldsymbol{\lambda > 0}$
2	$u = -k_2 \lceil z_2 + \lambda^{1/2}\lceil z_1 \rfloor^{1/2} \rfloor^0,$ $K_m k_2 \geq \lambda + C$	$u = -k_2 \lceil z_2 + \lambda^{1/2}\lceil z_1 \rfloor^{1/2} \rfloor^0$ $K_m k_2 \geq \lambda + C$
3	$u = -k_3 \lceil z_3$ $+\ 1.5\lambda \left\lceil \lceil z_2 \rfloor^{\frac{3}{2}} + \lambda z_1 \right\rfloor^{1/3} \Big\rfloor^0,$ $K_m k_3 \geq 3.23\lambda^2 + C$	$u = -k_3 \lceil z_3$ $+ 1.5\lambda \left\lceil \lceil z_2 \rfloor^{\frac{3}{2}} + \lambda z_1 \rfloor^{1/3} \right\rfloor^0,$ $K_m k_3 \geq 3.23\lambda^2 + C$
4	$u = -k_4 \lceil z_4 + 3.88\lambda^{3/2} \lceil \lceil z_3 \rfloor^2$ $+\ 4\lambda^2 \left(\lceil z_2 \rfloor^{\frac{4}{3}} + \lambda z_1 \right) \Big\rfloor^{1/4} \Big\rfloor^0,$ $K_m k_4 \geq 19.88\lambda^3 + C$	$u = -k_4 \Big\lceil z_4 + 3.8\lambda^{3/2} \Big\lceil \lceil z_3 \rfloor^{\frac{5}{2}}$ $+\ (2\lambda)^{\frac{5}{2}} \left\lceil \lceil z_2 \rfloor^{\frac{4}{3}} + \lambda z_1 \right\rfloor^{\frac{5}{4}} \Big\rfloor^{\frac{1}{5}} \Big\rfloor^0,$ $K_m k_4 \geq 30.25\lambda^3 + C$

γ_{i-1}, the value of Λ_{i-1} can be reduced. Once the value of each $(\gamma_{i-1}, \ldots, \gamma_1)$ has been chosen, we can fix any value for β_{i-1} such that it satisfies

$$\beta_{i-1} > \Lambda_{i-1}, \quad i = 2, \ldots, \rho.$$

Tables 1.1.3 and 1.1.4 show a particular k_1-parametrization of the gains for the controllers showed in Tables 1.1.1 and 1.1.2. The design of the gains from k_2 to $k_{\rho-1}$ is realized through the expressions (1.1.25) and (1.1.26). However, we observe that the restrictions over the gain k_ρ are relatively strong for $\rho \geq 4$, it cannot be applied in practice. The fourth column of Table 1.1.3 depicts an adjustment of gain k_ρ, whose values have been found by means of simulations.

Taking into account the previous observations, a λ-parametrization of two particular NSCs is shown in Table 1.1.5. This parametrization can be used along with the proposed QCSCs and the variable gain sliding controllers.

1.1.8 An academic example

This section is devoted to compare the classical HOSMCs with the new HOSMCs. For that, consider an academic kinematic model of a car [20], modeled in the state space by

$$\dot{x}_1 = v\cos(x_3), \quad \dot{x}_2 = v\sin(x_3), \quad \dot{x}_3 = \left(\frac{v}{L}\right)\tan(x_4), \quad \dot{x}_4 = u,$$

where x_1 and x_2 are Cartesian coordinates of the rear-axle middle point, x_3 is the orientation angle, x_4 is the steering angle, v is the longitudinal velocity ($v = 10\,\mathrm{m\,s^{-1}}$), L is the distance between the two axles ($L = 5$ m), and u is the control input. Note that actual control is x_4 and u is used as a new control input in order to avoid discontinuities on x_4. The control task is to steer the car from a given initial position to the trajectory $x_{2\mathrm{ref}} = 10\sin(0.05x_1) + 5$ in finite time. We turn on the controllers after 0.5 s assuming that there exists an observer providing the exact estimation of the states at most after 0.5 s [21]. Defining the sliding variable $\sigma = x_2 - x_{2\mathrm{ref}}$, the relative degree of system is three; therefore, the system can be rewritten as

$$\dot{z}_1 = z_2, \quad \dot{z}_2 = z_3, \quad \dot{z}_3 = \phi(\cdot) + \gamma(\cdot)u, \tag{1.1.27}$$

where the vector $z = \begin{bmatrix}\sigma & \dot{\sigma} & \ddot{\sigma}\end{bmatrix}^T$,

$$\begin{aligned}\phi(\cdot) = {} & \left[\frac{1}{800}\cos\left(\frac{x_1}{20}\right)\cos^2(x_3) - \frac{1}{4L}\sin\left(\frac{x_1}{20}\right)\sin(x_3)\tan(x_4)\right] \\ & \times v^3\cos(x_3) + \left[-\frac{1}{20}\sin\left(\frac{x_1}{20}\right)\cos(x_3)\sin(x_3)\right. \\ & \left. + \frac{1}{L}\left(\frac{1}{2}\cos\left(\frac{x_1}{20}\right)\cos(x_3) - \sin(x_3)\right)\tan(x_4)\right]\frac{v^3}{L}\tan(x_4), \\ \gamma(\cdot) = {} & \frac{v^2}{L}\left[\frac{1}{2}\cos\left(\frac{x_1}{20}\right)\cdot\sin x_3 + \cos(x_3)\right]\left[1 + \tan^2(x_4)\right].\end{aligned}$$

For simulation purposes, $x(0) = \begin{bmatrix}0 & 0 & 0 & 0\end{bmatrix}^T$ and sampling time $\tau = 0.0005$ with Euler's integration method were taken. It is assumed that the state variables x_3 and x_4 belong to $X = \{|x_3| \le \pi/4,\ |x_4| \le \pi/4\}$ and there is not more than 5% of uncertainty in the velocity. With this in mind, the functions ϕ and γ are bounded as $|\phi| \le C_0 = 49.63$ and $K_m = 6.38 \le \gamma \le K_M = 46.77$ [19].

1.1.8.1 Comparison between different NSCs

First, we compare the dynamical behavior of system (1.1.27) when we use the controllers

(L) $u = -k_{3L}\left\lceil [z_3 + 2\left(|z_2|^3 + |z_1|^2\right)^{1/6}\left\lceil z_2 + \lceil z_1 \rfloor^{2/3}\right\rfloor^0\right\rfloor^0$, where $k_{3L} = 20$.

(E) $u = -k_3\left\lceil z_3 + 1.5\lambda\left\lceil\lceil z_2\rfloor^{3/2} + \lambda z_1\right\rfloor^{1/3}\right\rfloor^0$, with $\lambda = 1$ and $k_3 = 20$.

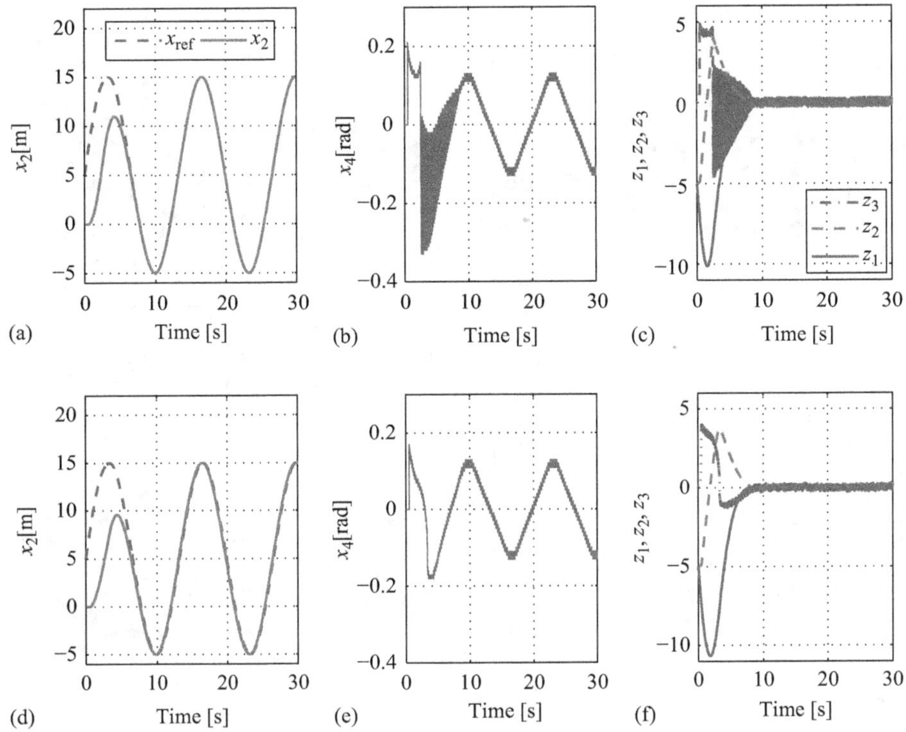

Figure 1.1.2 Controller ***(L)****: (a) the state x_2 (solid line); (b) the steering angle x_4; and (c) the sliding variable σ (solid line) and its time derivatives $\dot{\sigma}$ (dashed line), $\ddot{\sigma}$ (dashed-dot line). Controller* ***(E)****: (d) the state x_2; (e) the steering angle x_4; and (f) the sliding variable σ and $\dot{\sigma}$, $\ddot{\sigma}$*

The performance of the NSCs is displayed in Figure 1.1.2. Any controller enforces the state trajectory x_2 to reach the reference in finite time using the same control effort. The target is achieved in almost 9 s by using **(L)** and **(E)** (see Figure 1.1.2(a) and (d)).

Figure 1.1.2(b) shows oscillations of high frequency of steering angle x_4 during the transient response by using **(L)**. It may be caused by the set where the controller is discontinuous. Since it is also discontinuous. In contrast, by using **(E)** the steering angle x_4 is smoother than by using **(L)** (see Figure 1.1.2(e)). The same effect is observed in the dynamical behavior of z_3 (see Figure 1.1.2(c) and (f)). Figure 1.1.3(a) and (b) shows that the controllers **(L)** and **(E)** have almost the same accuracy.

The major practical achievement of controller **(E)** is that it guarantees good convergence rate and smooth transient response.

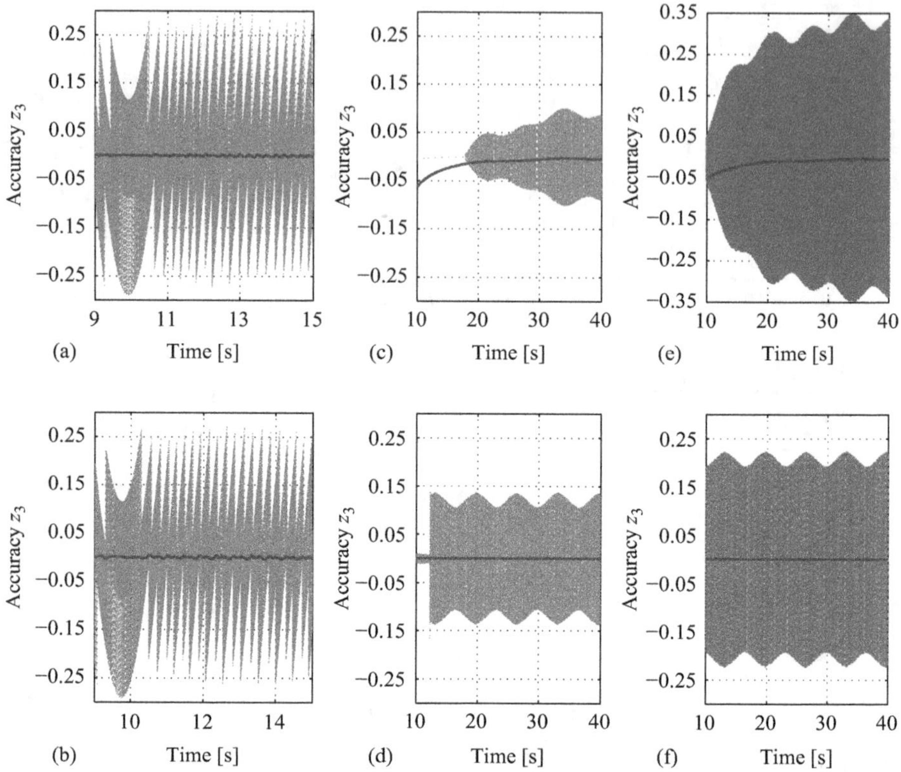

*Figure 1.1.3 Accuracy of z_3: (a) with **(L)**; (b) with **(E)**; (c) with **(LQ)**; (d) with **(EQ)**; (e) with **(LQ)** ($k_{3C} = 70$); and (f) with **(EQ)** ($k_3 = 40$)*

1.1.8.2 Comparison between different QCSCs

Finally, we compare the dynamical behavior of system (1.1.27) in feedback with the following quasi-continuous controllers:

(LQ) $u = -k_{3C}\frac{z_3+2\left(|z_2|+|z_1|^{3/2}\right)^{-1/2}\left(z_2+\lceil z_1\rfloor^{3/2}\right)}{|z_3|+2\left(|z_2|+|z_1|^{3/2}\right)^{1/2}}$, with $k_{3C} = 24.5$.

(EQ) $u = -k_3\frac{\lceil z_3\rfloor^3+(1.5\lambda)^3\left(\lceil z_2\rfloor^{3/2}+\lambda z_1\right)}{\left(|z_1|^{2p/3}+|z_2|^{p/1}+|z_3|^{2p/1}\right)^{3/2p}}$, with $\lambda = p = 1$ and $k_3 = 24.5$.

Figure 1.1.4 depicts the performance of the QCSCs. The controllers have been tuning to enforce the state trajectory x_2 to reach the reference in finite time using almost the same control effort. With the controller **(LQ)**, the target takes almost 18 s to be achieved (see Figure 1.1.3(c)). It takes almost the double of time than the previous NSCs. While with the controller **(EQ)**, the target takes almost 12 s to be achieved (see Figure 1.1.3(d)). Note that with the controller **(EQ)**, the dynamical behavior is similar to the NSC **(E)** (see Figure 1.1.4). Surprisingly, the state x_4, the sliding variable and its time derivatives present large transient response behaviors

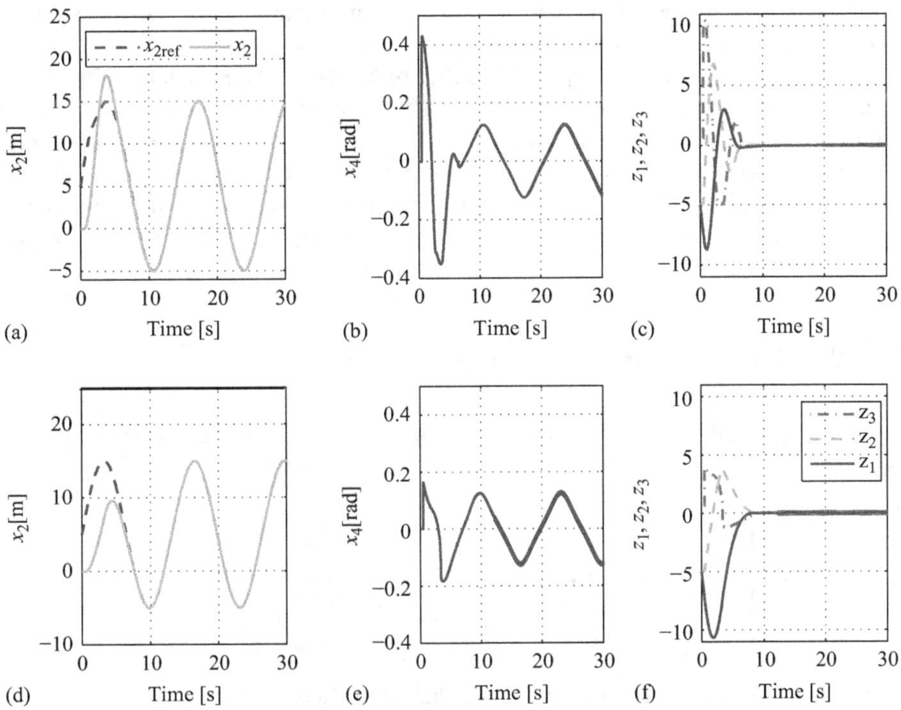

Figure 1.1.4 *Controller **(LQ)**: (a) the state x_2; (b) the steering angle x_4; and (c) the sliding variable σ and $\dot{\sigma}$, $\ddot{\sigma}$. Controller **(EQ)**: (d) the state x_2; (e) the steering angle x_4; and (f) the sliding variable σ and $\dot{\sigma}$, $\ddot{\sigma}$*

when the controller **(LQ)** is used (see Figure 1.1.4). However, the use of **(LQ)** and **(EQ)** provides better accuracy than the controllers **(L)** and **(E)**. An increment of the gains k_{3C} and k_3 in both controllers improves the convergence rate but it increases significantly the chattering effect and reduces the accuracy. Then, for the QCSCs exist a compromise between accuracy and convergence rate. Figure 1.1.3(e) and (f) shows that the accuracy of the QCSCs gets worse when it tries to have a convergence rate similar to **(L)** and **(E)**. The accuracy applying **(LQ)** is worse than if we use **(EQ)**.

1.1.9 Conclusions

This chapter focuses on designing HOSMCs. In contrast to other previously reported methods, the corresponding control law design developed in this chapter relies on the explicit construction of CLFs. Particularly, it is shown that a modification of the Backstepping method can be used to construct simultaneously the HOSM controllers as well as their corresponding CLFs. It establishes an important and novel methodological difference for designing homogeneous HOSM and connects these algorithms with standard techniques applied in nonlinear systems theory. It will lead to open new research directions in HOSM theory.

Notable features of the proposed families of NSCs are their capacity of reducing high-frequency transient response of the state trajectories of the controlled system and providing good convergence rate. Explicit conditions have been derived for gain tuning. For practical purposes, the gains are parameterized in advance, so only two parameters need to be properly adjusted. From many numerical and experimental results, rather than only theoretic studies, it was observed that the restrictions imposed on the gains by using the CLF are relatively strong.

1.1.10 Explicit expressions for Λ_{i-1}

Fix $\mathbf{r} = \left(\frac{\rho}{\rho-1}, 1, \frac{\rho-2}{\rho-1}, \ldots, \frac{1}{\rho-1}\right)$ and $m = r_1 + r_2$. In this case, expression (1.1.25) reduces to

$$\Lambda_{i-1} := \max_{\{z:S\}} \left\{ \frac{\frac{m-r_i}{r_1} s_i \left(k_{i-1}^{\frac{2}{r_i}} |\sigma_{i-1}|^{\frac{1}{r_1}} \right)^{(i-2)(1-p)} \dot{\sigma}_{i-1} + \gamma_{i-1} k_{i-1}^{\frac{m-r_{i-1}}{r_i}} \left(\sum_{j=2}^{i} \frac{\partial V_{i-1}}{\partial z_{j-1}} z_j \right)}{|\sigma_{i,d}| |\sigma_i|^{\frac{r_{i+1}}{r_1}}} \right\},$$

where $p = \frac{\rho-2}{\rho-1}$, and expression (1.1.26) reduces to this other

$$\Lambda_{i-1} := \max_{\{z:S\}} \left\{ \frac{\frac{m-r_i}{m-r_{i-1}} s_i \left(k_{i-1}^{\frac{2(i-2)(1-p)}{r_i}} |\sigma_{i-1,d}|^{\frac{1-p}{m-r_{i-1}}} \right) \dot{\sigma}_{i-1,d} + \gamma_{i-1} k_{i-1}^{\frac{m-r_{i-1}}{r_i}} \left(\sum_{j=2}^{i} \frac{\partial V_{i-1}}{\partial z_{j-1}} z_j \right)}{|\sigma_{i,d}|^{\frac{2}{m-r_i}}} \right\}.$$

Acknowledgments

The authors thank the financial support from PAPIIT-UNAM (Programa de Apoyo a Proyectos de Investigación e Innovación Tecnológica), Project IN113614; Fondo de Colaboración II-FI UNAM, Project IISGBAS-100-2015; CONACyT (Consejo Nacional de Ciencia y Tecnología), Project 241171, and CVU 267513.

References

[1] V. Andrieu, L. Praly, and A. Astolfi. Homogeneous approximation, recursive observer design and output feedback. *SIAM Journal on Control and Optimization*, 47(4):1814–1850, 2008.

[2] A. Bacciotti and L. Rosier. *Lyapunov Functions and Stability in Control Theory*, 2nd ed. Springer-Verlag, New York, NY, 2005.

[3] E. Bernuau, E. Efimov, W. Perruquetti, and A. Polyakov. On homogeneity and its application in sliding mode control. *Journal of the Franklin Institute*, 351(4):1866–1906, 2014.

[4] S.P. Bhat and D.S. Bernstein. Geometric homogeneity with applications to finite-time stability. *Mathematics of Control, Signals, and Systems*, 17(2): 101–127, 2005.

[5] F.H. Clarke, Y.S. Ledyaev, and R.J. Stern. Asymptotic stability and smooth Lyapunov functions. *Journal of Differential Equations*, 149:69–114, 1998.

[6] E. Cruz-Zavala and J.A. Moreno. Improved convergence rate of discontinuous finite-time controllers. In *19th IFAC World Congress*, pages 8636–8641, Cape Town, South Africa, 2014.

[7] E. Cruz-Zavala and J.A. Moreno. A new class of fast finite-time discontinuous controllers. In *Proceedings of the Variable Structure Systems Workshop*, pages 1–6, Nantes, France, 2014.

[8] A.F. Filippov. *Differential Equations with Discontinuous Righthand Side*. Kluwer, Dordrecht, The Netherlands, 1988.

[9] R.A. Freeman and P.V. Kokotovic. *Robust Nonlinear Control Design: State Space and Lyapunov Techniques*. Modern Birkhäuser Classics, Boston, MA, 1996.

[10] L. Fridman and A. Levant. Higher-Order Sliding Modes. In W. Perruquetti, and J.-P. Barbot, editors, *Sliding Mode Control in Engineering*, pages 53–102. Marcel Dekker Inc., New York, NY, 2002.

[11] M. Harmouche, S. Laghrouche, and Y. Chitour. Robust and adaptive higher order sliding mode. In *IEEE 51st Annual Conference on Decision and Control*, pages 6436–6441, Maui, HI, 10–13 December 2012.

[12] M.R. Hestenes. *Calculus of Variations and Optimal Control Theory*. John Wiley & Sons, New York, NY, 1966.

[13] Y. Hong. Finite-time stabilization and stabilizability of a class of controllable systems. *Systems & Control Letters*, 46:231–236, 2002.

[14] X. Huang and W. Lin. Global finite-time stabilization of a class of uncertain nonlinear systems. *Automatica*, 41:881–888, 2005.

[15] A. Isidori. *Nonlinear Control Systems*. Springer Verlag, Berlin, 1995.

[16] I. Kanellakopoulos, P.V. Kokotovic, and A.S. Morse. A toolkit for nonlinear feedback design. *Systems & Control Letters*, 18:83–92, 1992.

[17] H.K. Khalil. *Nonlinear Systems*, 3rd ed. Prentice-Hall, Englewood Cliffs, NJ, 2002.

[18] M. Krstic, I. Kanellakopoulos, and P.V. Kokotovic. *Nonlinear and Adaptive Control Design*. John Wiley & Sons, New York, NY, 1995.

[19] S. Laghrouche, F. Plestan, and A. Glumineau. Higher order sliding mode control based on integral sliding mode. *Automatica*, 43:531–537, 2007.

[20] A. Levant. Universal single-input single-output (SISO) sliding-mode controllers with finite-time. *IEEE Transactions on Automatic Control*, 46(9):1447–1451, 2001.

[21] A. Levant. High-order sliding modes: differentiation and output-feedback control. *International Journal of Control*, 76(9):924–941, 2003.

[22] A. Levant. Homogeneity approach to high-order sliding mode design. *Automatica*, 41:823–830, 2005.

[23] A. Levant. Construction principles of 2-sliding mode design. *Automatica*, 43:576–586, 2007.

[24] A. Levant. Adjustment of high-order sliding-mode controllers. *International Journal of Robust and Nonlinear Control*, 19:1657–1672, 2009.

[25] J.A. Moreno. Lyapunov approach for analysis and design of second-order sliding mode algorithms. In L. Fridman, J. Moreno, and R. Iriarte, editors, *Sliding Modes after the First Decade of the 21st Century*, LNCIS, 412, pages 113–150. Springer-Verlag, Berlin, 2011.

[26] J.A. Moreno and M. Osorio. Strict Lyapunov functions for the super-twisting algorithm. *IEEE Transactions on Automatic Control*, 57(4):1035–1040, 2012.

[27] H. Nakamura, Y. Yamashita, and H. Nishitani. Smooth Lyapunov functions for homogeneous differential inclusions. In *Proceedings of the 41st SICE Annual Conference*, pages 1974–1979, Osaka, Japan, 5–7 August 2002.

[28] Y.V. Orlov. *Discontinuous Systems: Lyapunov Analysis and Robust Synthesis under Uncertainty Conditions*. Springer-Verlag, Berlin, 2009.

[29] A. Polyakov and A. Poznyak. Lyapunov function design for finite-time convergence analysis: "twisting" controller for second-order sliding mode realization. *Automatica*, 45(2):444–448, 2009.

[30] A. Polyakov and A. Poznyak. Reaching time estimation for "super-twisting" second-order sliding mode controller via Lyapunov function designing. *IEEE Transactions on Automatic Control*, 54(8):1951–1955, 2009.

[31] A. Polyakov and A. Poznyak. Unified Lyapunov function for a finite-time stability analysis of relay second-order sliding mode control systems. *IMA Journal of Mathematical Control and Information*, 29(4):529–550, 2012.

[32] L. Praly, B. Andrea-Novel, and J. Coron. Lyapunov design of stabilizing controllers for cascaded systems. *IEEE Transactions on Automatic Control*, 36:1177–1181, 1991.

[33] C. Qian and W. Lin. A continuous feedback approach to global strong stabilization of nonlinear systems. *IEEE Transactions on Automatic Control*, 46(7):1061–1079, 2001.

[34] T. Sanchez and J.A. Moreno. A constructive Lyapunov function design method for a class of homogeneous systems. In *In IEEE 53rd Annual Conference on Decision and Control (CDC)*, pages 5500–5505, Los Angeles, CA, 15–17 December 2014.

[35] T. Sánchez and J.A. Moreno. Construction of Lyapunov functions for a class of higher order sliding mode algorithms. In *IEEE 51st Annual Conference on Decision and Control*, pages 6454–6459, Maui, HI, 10–13 December 2012.

[36] R. Sepulcre and D. Aeyels. Homogeneous Lyapunov functions and necessary conditions for stabilization. *Mathematics of Control, Signals, and Systems*, 9:34–58, 1996.

[37] Y. Shtessel, C. Edwards, L. Fridman, and A. Levant. *Sliding Mode Control and Observation*. Birkhäuser, New York, NY, 2014.

Chapter 1.2

Sliding surface design for higher-order sliding modes

Ismael Castillo, Fernando Castaños and Leonid Fridman

Abstract

Two main concepts for sliding surfaces design procedure: pole placement and optimal stabilization are generalized for the case of arbitrary order sliding modes. For the pole placement case, the formula of Ackermann-Utkin is extended allowing the design of sliding surfaces with arbitrary relative degree. The natural connection between order of singularity for singular optimal stabilization problem and order of sliding mode controller is shown and used in the design of the sliding surface and the sliding mode controller of corresponding order.

1.2.1 Introduction

One of the main advantages of sliding-mode control (SMC) is its ability to theoretically compensate matched uncertainties exactly. Another advantage is the reduction of the system dynamics to the sliding dynamics. The SMC design strategy consists of two steps [1–3]:

1. Design a sliding variable of relative degree one.
2. Enforce the finite-time convergence to the sliding surface.

The zero set of the sliding variable defines the so-called sliding surface. A prospective sliding surface should satisfy two aims. First, the system dynamics along it should exhibit a desired behaviour specified, e.g., by the locations of the poles or by a performance index to be optimised. Second, the sliding surface should be reachable by means of a suitable control action. In the context of classical SMC, this amounts to specifying a sliding surface of co-dimension one (equivalently, a sliding variable of relative degree one). In this chapter, we follow the same program, but we profit from the more general *higher-order sliding modes* (HOSM), with which we can effectively remove the constraint on the relative degree.

During the last 30 years, HOSM controllers [4, 5] have become very popular. A HOSM controller is capable of driving the system state to a sliding surface of co-dimension higher than one (the co-dimension of the sliding surface is called the order of the sliding mode). This concerns step 2 above, but regarding step 1, a reasonable question arises: how to design sliding dynamics of arbitrary order (see Section 1.2.3.3 for more details). To design the sliding surfaces, conventional SMC theory profits from two design methodologies for linear systems: pole placement and the linear quadratic regulator (LQR). The objective of this chapter is to extend both methodologies to the realm of higher-order sliding-mode control (HOSMC). In other words, we present two methodologies for designing stable sliding surfaces of higher relative degree. Afterwards, we apply HOSM controllers and compare the accuracy with the one obtained using a classical approaches.

1.2.1.1 Motivating example

Allow us to begin with a simple example of a sliding-mode surface of the first order. Consider a chain of integrators

$$\begin{aligned} \dot{x}_1 &= x_2, \\ \dot{x}_2 &= x_3, \\ \dot{x}_3 &= u + w, \end{aligned} \tag{1.2.1}$$

where $x = [x_1\ x_2\ x_3]^\top \in \mathbb{R}^3$ is the state and $u, w \in \mathbb{R}$ are the control input and a bounded unknown perturbation (we omit the time arguments for ease of notation). Suppose that we want to stabilize the origin. In the conventional approach to SMC, the last variable (i.e., x_3 in our example) is considered as a virtual control for the reduced order system

$$\begin{aligned} \dot{x}_1 &= x_2, \\ \dot{x}_2 &= x_3. \end{aligned} \tag{1.2.2}$$

Usually, the linear virtual control

$$x_3 = -Kx_{12}, \quad K = [k_1\ k_2], \quad x_{12} = [x_1\ x_2]^\top, \tag{1.2.3}$$

with free parameter K, is designed such that the origin of the reduced order system is exponentially stable. The virtual control leads to a sliding variable

$$\sigma = x_3 + Kx_{12}$$

of relative degree equal to one which, by construction, has a two-dimensional sliding dynamics (or zero dynamics associated to the output σ) given by (1.2.2) and (1.2.3). This is easier to see if one uses $\bar{x} = [x_1\ x_2\ \sigma]^\top$ as the state vector and writes down the equivalent system

$$\begin{aligned} \dot{x}_1 &= x_2, \\ \dot{x}_2 &= -Kx_{12} + \sigma, \\ \dot{\sigma} &= u + w + Kx_{23}, \quad x_{23} = [x_2\ x_3]^\top. \end{aligned} \tag{1.2.4}$$

Consider now the control law

$$u = -Kx_{23} - \bar{w}\,\mathrm{sign}(\sigma), \tag{1.2.5}$$

where $\bar{w}$ is the known upper bound for $|w|$. It is easy to see that the trajectories of the system will converge to the sliding surface $\{x \in \mathbb{R}^3 \mid \sigma = 0\}$ in finite time, regardless of w, and that the two-dimensional sliding dynamics,

$$\begin{aligned} \dot{x}_1 &= x_2, \\ \dot{x}_2 &= -Kx_{12}, \end{aligned} \tag{1.2.6}$$

converges exponentially to the origin, also independently of w.

1.2.1.1.1 Pole placement

In the pole-placement approach, one chooses a sliding variable such that the associated sliding dynamics has eigenvalues with negative real parts. Suppose, e.g., that we desire a sliding dynamics having an eigenvalue -1 with multiplicity two. We can use the well-known formula by Ackermann and Utkin [6] to obtain the sliding variable $\sigma = x_1 + 2x_2 + x_3$. Finally, we can apply the control law $u = -x_2 - 2x_3 - \bar{w}\,\mathrm{sign}(\sigma)$. It is not hard to see that the trajectories of the system converge globally exponentially to the origin at the desired exponential rate, regardless of w.

1.2.1.1.2 Linear quadratic regulator

Conventional SMC theory establishes a natural connection between a sliding surface and an optimal control problem having a singular performance index. For our simple example (1.2.1), we can show that the performance index

$$J = \frac{1}{2}\int_{t_1}^{\infty} \left(x^\top Q x\right) \mathrm{d}t, \quad Q = \begin{pmatrix} q_1 & 0 & 0 \\ 0 & q_2 & 0 \\ 0 & 0 & q_3 \end{pmatrix} \tag{1.2.7}$$

is optimal along a certain sliding surface. Recall that, since the performance index is singular with respect to the control, the corresponding optimal control is termed *free-cost control*. When x_3 is used as a virtual control, the performance index can be rewritten as

$$\bar{J} = \frac{1}{2}\int_{t_1}^{\infty} \left(x_{12}^\top \bar{Q} x_{12} + x_3 R x_3\right) \mathrm{d}t, \quad \bar{Q} = \begin{pmatrix} q_1 & 0 \\ 0 & q_2 \end{pmatrix}, \quad R = q_3, \tag{1.2.8}$$

which is no longer singular if $q_3 > 0$.

The solution of this reduced LQR problem is well known. The optimal virtual control is [7]

$$x_3 = -Kx_{12}, \quad K = R^{-1}B^\top P, \tag{1.2.9}$$

where $B = [0\ 1]^\top$ is the control matrix of (1.2.2) and P is the solution of the algebraic Riccati equation (ARE)

$$PA + A^\top P + \bar{Q} - PBR^{-1}B^\top P = 0$$

with system matrix

$$A = \begin{pmatrix} 0 & 1 \\ 0 & 0 \end{pmatrix}.$$

One can then select the sliding surface $\sigma = x_3 + R^{-1}B^\top P x_{12}$ and apply the control law $u = -R^{-1}B^\top P x_{23} - \bar{w}\,\mathrm{sign}(\sigma)$, $x_{23} = [x_2\ x_3]^\top$.

The performance index (1.2.7) achieves its minimum value along the sliding surface $\{x \mid \sigma = 0\}$ and the system trajectories converge to such surface in finite time. Finally, the trajectories of the system will converge globally exponentially to the origin with optimal dynamics, regardless of w.

1.2.1.1.3 Higher-order sliding-mode control

Consider now the case $\sigma = x_1$. The relative degree of σ is equal to the system's dimension, so there are no sliding dynamics to worry about. It is by now a standard result of HOSMC theory that the (substantially more complex) controller [8]

$$u = -\alpha \frac{\ddot{\sigma} + 2\left(|\dot{\sigma}| + |\sigma|^{2/3}\right)^{-1/2}\left(\dot{\sigma} + |\sigma|^{2/3}\,\mathrm{sign}(\sigma)\right)}{|\ddot{\sigma}| + 2\left(|\dot{\sigma}| + |\sigma|^{2/3}\right)^{1/2}}, \tag{1.2.10}$$

with $\alpha > 0$ high enough, drives the state to zero in finite time, regardless of w.

The computation of a sliding variable of relative degree equal to the dimension of the plant was simple because the system is in a canonical form. This suggests that, for a general linear controllable system, we first put it in controller canonical form and then take the state with highest relative degree as the sliding variable. In this way, the extreme case of relative degree equal to the system's dimension (no sliding dynamics) can be covered systematically.

Note, however, that there is no systematic method for constructing a sliding variable of intermediate relative degree (in our example, of relative degree two). To such a sliding variable there would correspond a sliding dynamics of dimension one. This dynamics can be enforced with a controller much simpler than (1.2.10), thus arriving at a fair compromise between order reduction and controller complexity.

1.2.1.2 Contribution

Our contributions concern single input single output (SISO) linear time-invariant (LTI) systems. The first one is a methodology which generalizes the formula of Ackermann and Utkin to the case of an arbitrary relative degree. That is, that allows the designer to construct a sliding surface with desired eigenvalues for the sliding dynamics. The second one is a methodology that allows the designer to construct a sliding surface along which the trajectories are optimal. A property of the performance index (to be defined later) is related to the relative degree of the sliding variable. Thus, by properly selecting the performance index, the relative degree of the sliding variable can be arbitrarily chosen.

1.2.2 Problem statement

Consider the perturbed LTI system

$$\dot{x} = Ax + B(u + w), \tag{1.2.11a}$$

where $A \in \mathbb{R}^{n\times n}$ is the system matrix, $B \in \mathbb{R}^n$ is the control matrix, $x \in \mathbb{R}^n$ is the state vector, $u \in \mathbb{R}$ is the control input and $w \in \mathbb{R}$ is a matched Lebesgue-measurable perturbation with a known bound $|w| \leq \bar{w}$. We assume that the pair (A, B)

is controllable (hence rank $(B) = 1$). The solutions of (1.2.11a) are taken in the sense of Filippov [9].

Our problem is to find a *virtual output*

$$\sigma = Cx, \quad \sigma \in \mathbb{R} \tag{1.2.11b}$$

of relative degree r and such that (1.2.11) has sliding dynamics that either:

1. Have eigenvalues at desired locations. That is, C is such that the zeros of the transfer function

   $$g(s) = C(sI - A)^{-1}B$$

   are located on the open left half-plane (the pole-placement problem).
2. Are optimal with respect to a singular performance index (the singular linear quadratic (LQ) problem).

1.2.3 Preliminaries

In this section, some of the necessary tools for the analysis and development of the HOSMC and surface design are presented. The classical controllable canonical form transformation is presented and applied to the original system and to its associated performance index for the singular LQR. Also, the iterative construction of the quasi-continuous controller as a particular case of HOSMCs to solve the robust stabilization problem.

1.2.3.1 Controllable canonical form

Controllability of the pair (A, B) ensures the existence of a non-singular transformation $z = Tx$ such that, in the new coordinates, the system

$$\dot{z} = \bar{A}z + \bar{B}u \tag{1.2.12}$$

takes the so-called *controllable canonical form*. That is,

$$\bar{A} = TAT^{-1} = \begin{pmatrix} 0 & 1 & 0 & \cdots & 0 \\ 0 & 0 & 1 & \cdots & 0 \\ \vdots & \vdots & \vdots & \ddots & 0 \\ 0 & 0 & 0 & \vdots & 1 \\ \alpha_1 & \alpha_2 & \cdots & \alpha_{n-1} & \alpha_n \end{pmatrix}$$

and

$$\bar{B} = TB = \begin{pmatrix} 0 \\ 0 \\ \vdots \\ 0 \\ 1 \end{pmatrix},$$

where $\{\alpha_1, \ldots, \alpha_n\}$ are the coefficients of the characteristic polynomial.

For the purposes of designing (1.2.11b), it is instructive to write the system in controller canonical form and then partition the state as $z^\top = \left(\bar{z}_1^\top \ \bar{z}_2^\top\right)$,

$$\begin{pmatrix} \dot{\bar{z}}_1 \\ \dot{\bar{z}}_2 \end{pmatrix} = \begin{pmatrix} \bar{A}_{11} & \bar{A}_{12} \\ \bar{A}_{21} & \bar{A}_{22} \end{pmatrix} \begin{pmatrix} \bar{z}_1 \\ \bar{z}_2 \end{pmatrix} + \begin{pmatrix} 0 \\ \bar{B}_2 \end{pmatrix} u. \tag{1.2.13}$$

In the conventional SMC scenario, we set $\bar{z}_1 \in \mathbb{R}^{n-1}$ and $\bar{z}_2 \in \mathbb{R}$. We can then assure that

$$\sigma = \bar{z}_2 + K\bar{z}_1 \tag{1.2.14}$$

is of relative degree one. It is not difficult to see that the sliding dynamics is given by

$$\dot{\bar{z}}_1 = \left(\bar{A}_{11} - \bar{A}_{12}K\right)\bar{z}_1.$$

Finally, since the controllability of the pair (A, B) implies the controllability of the pair $\left(\bar{A}_{11}, \bar{A}_{12}\right)$ [10], the designer is free to establish any desired eigenvalues for the sliding dynamics. (A simpler approach is of course to apply directly the formula by Ackermann and Utkin.)

Alternatively, one may choose a performance index

$$J(x) = \frac{1}{2}\int_{t_1}^{\infty} \left(x^\top Q x\right) \mathrm{d}t,$$

whose weighting matrix Q transforms as $\bar{Q} := (T^{-1})^\top Q T^{-1}$. We can then partition $\bar{Q}$ according to (1.2.13),

$$\bar{Q} = \begin{pmatrix} \tilde{Q}_{11} & \tilde{Q}_{12} \\ \tilde{Q}_{21} & \tilde{Q}_{22} \end{pmatrix}, \tag{1.2.15}$$

where $\tilde{Q}_{11} \in \mathbb{R}^{(n-1)\times(n-1)}$ and $\tilde{Q}_{22} \in \mathbb{R}$. Again, the variable $\bar{z}_2$ is used as a virtual control to stabilize the $\bar{z}_1$ subsystem. Matrix blocks $\tilde{Q}_{11} \geq 0$ and $\tilde{Q}_{22} > 0$ are then used to assign weights to $\bar{z}_1$ and $\bar{z}_2$, respectively. The resulting optimization problem leads to a regular ARE with matrices of reduced order and again to a sliding variable of the form (1.2.14) [1], i.e., of relative degree one.

1.2.3.2 Relative degree and zero dynamics

In the HOSM scenario, the virtual output may have relative degree larger than one. Then, recall that (1.2.11b) is said to have *relative degree r* if $CA^{i-1}B = 0$, $1 \leq i < r$ and $CA^{r-1}B \neq 0$. If (1.2.11b) has relative degree r, then it is possible to take σ and its successive $r-1$ time-derivatives as a partial set of coordinates $\xi_1, \ldots, \xi_r$. More precisely, there exists a full-rank matrix $B^{\perp} \in \mathbb{R}^{(n-r)\times n}$ such that $B^{\perp}B = 0$ and such that

$$T = \begin{bmatrix} B^{\perp} \\ \hline C \\ \vdots \\ CA^{r-1} \end{bmatrix}$$

is invertible. Hence,

$$\begin{bmatrix} \eta \\ \hline \xi \end{bmatrix} = Tx$$

is a coordinate transformation [11, Proposition 4.1.3]. It is straightforward to verify that, in the new coordinates, system (1.2.11) takes the *normal form*

$$\begin{bmatrix} \dot{\eta} \\ \hline \dot{\xi}_1 \\ \vdots \\ \dot{\xi}_{r-1} \\ \dot{\xi}_r \end{bmatrix} = \begin{bmatrix} A_0\eta + B_0\xi \\ \hline \xi_2 \\ \vdots \\ \xi_r \\ CA^r x \end{bmatrix} + \begin{bmatrix} 0 \\ \hline 0 \\ \vdots \\ 0 \\ CA^{r-1}B \end{bmatrix} (u + w) \tag{1.2.16a}$$

$$\sigma = \xi_1,\ \dot{\sigma} = \xi_2,\ \ldots,\ \sigma^{(r-1)} = \xi_r. \tag{1.2.16b}$$

The dynamics $\dot{\eta} = A_0\eta$, $\eta \in \mathbb{R}^{n-r}$, is the *zero dynamics*.

It is well known [12, Example 4.1.3] that the eigenvalues of A_0 coincide with the zeros of the transfer function

$$g(s) = C(sI - A)^{-1}B.$$

If the zeros of $g(s)$ have real part strictly less than zero, we say that the system is *minimum phase*.

Thus, the surface design for HOSM can be reformulated as the selection of the order and design of stable dynamics for the zero dynamics $\eta \in \mathbb{R}^{n-r}$, and consequently the order of $\xi \in \mathbb{R}^r$ (relative degree of (1.2.11b)).

1.2.3.3 Quasi-continuous higher-order sliding-mode controllers

The family of quasi-continuous higher-order sliding-mode controllers [8] guarantees the steering of a sliding variable σ to zero in finite time, even in the presence of matched bounded perturbations. Suppose that σ is of relative degree r and, furthermore, suppose that we have

$$\sigma^{(r)} = u + w. \tag{1.2.17}$$

The above-mentioned controller is defined recursively by

$$\begin{aligned}
\varphi_{0,r} &= \sigma, \quad N_{0,r} = |\sigma|, \quad \Psi_{0,r} = \tfrac{\varphi_{0,r}}{N_{0,r}} = \operatorname{sign}(\sigma),\\
\varphi_{j,r} &= \sigma^{(j)} + \phi_j N_{j-1,r}^{(r-j)/(r-j+1)} \Psi_{j-1,r},\\
N_{j,r} &= \left|\sigma^{(j)}\right| + \phi_j N_{j-1,r}^{(r-j)/(r-j+1)},\\
\Psi_{j,r} &= \tfrac{\varphi_{j,r}}{N_{j,r}},\\
v_r &= -\alpha_r \Psi_{r-1,r},
\end{aligned} \tag{1.2.18}$$

where $\phi_1, \ldots, \phi_{r-1}, \alpha_r$ are positive numbers, $j = 1, \ldots, r-1$, and r the order (relative degree) of the controller. For example, for the first order ($r = 1$), we have

$$\begin{aligned}
\varphi_{0,1} &= \sigma, \quad N_{0,1} = |\sigma|, \quad \Psi_{0,1} = \frac{\varphi_{0,i}}{N_{0,i}} = \operatorname{sign}(\sigma),\\
v_1 &= -\alpha_1 \operatorname{sign}(\sigma),
\end{aligned} \tag{1.2.19}$$

For the second order ($r = 2$), we have, setting $\phi_1 = 1$,

$$\begin{aligned}
&\varphi_{0,2} = \sigma, \quad N_{0,2} = |\sigma|, \quad \Psi_{0,2} = \frac{\varphi_{0,2}}{N_{0,2}} = \operatorname{sign}(\sigma),\\
&\varphi_{1,2} = \dot\sigma + |\sigma|^{\frac{1}{2}} \operatorname{sign}(\sigma), \quad N_{1,2} = |\dot\sigma| + |\sigma|^{\frac{1}{2}}, \quad \Psi_{1,2} = \frac{\dot\sigma + |\sigma|^{\frac{1}{2}} \operatorname{sign}(\sigma)}{|\dot\sigma| + |\sigma|^{\frac{1}{2}}},\\
&v_2 = -\alpha_2 \frac{\dot\sigma + |\sigma|^{\frac{1}{2}} \operatorname{sign}(\sigma)}{|\dot\sigma| + |\sigma|^{\frac{1}{2}}}.
\end{aligned} \tag{1.2.20}$$

For the third order ($r = 3$), we have, setting $\phi_1 = 1$ and $\phi_2 = 2$,

$$\begin{aligned}
&\varphi_{0,3} = \sigma, \quad N_{0,3} = |\sigma|, \quad \Psi_{0,3} = \frac{\varphi_{0,3}}{N_{0,3}} = \operatorname{sign}(\sigma),\\
&\varphi_{1,3} = \dot\sigma + |\sigma|^{\frac{2}{3}} \operatorname{sign}(\sigma), \quad N_{1,3} = |\dot\sigma| + |\sigma|^{\frac{2}{3}}, \quad \Psi_{1,3} = \frac{\dot\sigma + |\sigma|^{\frac{2}{3}} \operatorname{sign}(\sigma)}{|\dot\sigma| + |\sigma|^{\frac{2}{3}}},\\
&\varphi_{2,3} = \ddot\sigma + 2\frac{\dot\sigma + |\sigma|^{\frac{2}{3}} \operatorname{sign}(\sigma)}{\left(|\dot\sigma| + |\sigma|^{\frac{2}{3}}\right)^{\frac{1}{2}}}, \quad N_{2,3} = |\ddot\sigma| + 2\left(|\dot\sigma| + |\sigma|^{\frac{2}{3}}\right)^{\frac{1}{2}},\\
&\Psi_{2,3} = \frac{\ddot\sigma + 2\left(\dot\sigma + |\sigma|^{\frac{2}{3}} \operatorname{sign}(\sigma)\right)\left(|\dot\sigma| + |\sigma|^{\frac{2}{3}}\right)^{-\frac{1}{2}}}{|\ddot\sigma| + 2\left(|\dot\sigma| + |\sigma|^{\frac{2}{3}}\right)^{\frac{1}{2}}},\\
&v_3 = -\alpha_3 \frac{\ddot\sigma + 2\left(\dot\sigma + |\sigma|^{\frac{2}{3}} \operatorname{sign}(\sigma)\right)\left(|\dot\sigma| + |\sigma|^{\frac{2}{3}}\right)^{-\frac{1}{2}}}{|\ddot\sigma| + 2\left(|\dot\sigma| + |\sigma|^{\frac{2}{3}}\right)^{\frac{1}{2}}}.
\end{aligned} \tag{1.2.21}$$

The accuracy of the controllers is specified in the following theorem.

Theorem 1.2.1 ([8]). *Let the control value be updated at the moments* t_i, *with* $t_{i+1} - t_i = \tau = \text{const} > 0$; $t \in [t_i, t_{i+1})$ *(the discrete sampling case). Then, controller* (1.2.18) *provides in finite time for keeping the inequalities*

$$|\sigma| < \mu_0 \tau^r, \ |\dot{\sigma}| < \mu_1 \tau^{r-1}, \ \ldots, \ \left|\sigma^{(r-1)}\right| < \mu_{r-1}\tau \tag{1.2.22}$$

with some positive constants $\mu_0, \mu_1, \ldots, \mu_{r-1}$.

It is also shown in Reference 13 that in the presence of an actuator of the form $\tau\dot{z} = a(z, u)$, $v = v(z)$, $z \in \mathbb{R}^m$, $v \in \mathbb{R}$ with u the input of the actuator, v its output and μ the time constant, inequalities (1.2.22) also hold under reasonable assumptions.

It can be seen that the higher order of the controller, the higher accuracy it can be obtained, but also more complexity in the control law that requires more computational resources. Moreover, the maximal accuracy of a realization is restricted by the precision of the sensors and actuators. Then, it is reasonable to have the possibility to design a HOSMC of order higher than one but less than the order of the system in order to establish a compromise between a feasible accuracy and the complexity of the controller.

The development of the theory of HOSM allows to systematically find a family of controllers that can robustly stabilize in finite time a sliding surface of arbitrary degree; nevertheless, there is not a general design methodology to select a sliding surface with a desired relative degree. The need to design sliding surfaces of arbitrary relative degree arises in order to complete the two-step methodology of the HOSM theory.

1.2.4 Pole placement

1.2.4.1 Surface design

Problem (1) in page 33 can be easily solved by applying a generalization of the well-known formula of Ackermann and Utkin [14].

Theorem 1.2.2. *Let* $e_1 := (0 \ 0 \ \cdots \ 0 \ 1)$ *and let* F *be the system's controllability matrix. If*

$$C = e_1 F^{-1} \gamma(A), \tag{1.2.23}$$

with $\gamma(\lambda) = \lambda^{n-r} + \gamma_{n-r-1}\lambda^{n-r-1} + \cdots + \gamma_1\lambda + \gamma_0$, *then* σ *is of relative degree* r *and the roots of* $\gamma(\lambda)$ *are the eigenvalues of the sliding-mode dynamics in the intersection of the planes* $\sigma = \dot{\sigma} = \cdots = \sigma^{(r-1)} = 0$.

Proof. Let us assume first that the system is given in controller canonical form with system matrices $\bar{A}$ and $\bar{B}$. To verify (1.2.23), we will show that for $\bar{C} = e_1\bar{F}^{-1}\gamma\left(\bar{A}\right)$, the numerator of $g(s) = \bar{C}\left(sI - \bar{A}\right)^{-1}\bar{B}$ is equal to $\gamma(s)$.

It is a standard result that, for a system in controller canonical form, we have [15]

$$\begin{aligned} e_1\bar{F}^{-1} &= \begin{pmatrix}1 & 0 & \cdots & 0 & 0\end{pmatrix}, \\ \begin{pmatrix}1 & 0 & \cdots & 0 & 0\end{pmatrix}\bar{A} &= \begin{pmatrix}0 & 1 & \cdots & 0 & 0\end{pmatrix}, \\ &\ \ \vdots \\ \begin{pmatrix}1 & 0 & \cdots & 0 & 0\end{pmatrix}\bar{A}^{n-2} &= \begin{pmatrix}0 & 0 & \cdots & 1 & 0\end{pmatrix}, \\ \begin{pmatrix}1 & 0 & \cdots & 0 & 0\end{pmatrix}\bar{A}^{n-1} &= \begin{pmatrix}0 & 0 & \cdots & 0 & 1\end{pmatrix}. \end{aligned} \tag{1.2.24}$$

It then follows that

$$\bar{C} = \begin{pmatrix}\gamma_0 & \gamma_1 & \cdots & \gamma_{n-r-1} & 1 & 0 & \cdots & 0\end{pmatrix}.$$

Since $\bar{A}, \bar{B}$ and $\bar{C}$ are in controller canonical form, the transfer function is simply

$$g(s) = \frac{s^{n-r} + \gamma_{n-r-1}s^{n-r-1} + \cdots + \gamma_1 s + \gamma_0}{s^n + a_{n-1}s^{n-1} + \cdots + a_1 s + a_0},$$

which shows that the relative degree is r. Since the numerator is equal to $\gamma(s)$, the eigenvalues of the sliding-mode dynamics are equal to the roots of $\gamma(s)$.

Now, to address the general case, consider the transformation $T = F\bar{F}^{-1}$, which is such that $\bar{A} = T^{-1}AT$. We have $C = \bar{C}T^{-1}$, i.e., we have $C = e_1\bar{F}^{-1}\gamma\left(\bar{A}\right)T^{-1}$. Finally, from $\bar{F}^{-1} = F^{-1}T$ and $\gamma\left(\bar{A}\right) = T^{-1}\gamma(A)T$, we recover (1.2.23). □

1.2.4.2 Control design

Since $\sigma = Cx$ with C as in (1.2.23) is of relative degree r, we have

$$\sigma^{(r)} = CA^r x + CA^{r-1}B(u + w)$$

with $CA^{r-1}B \neq 0$. In view of the preceding discussion, we can take

$$u = \frac{v_r - CA^r x}{CA^{r-1}B} \tag{1.2.25}$$

and drive the system trajectories to the sliding surface in finite time.

1.2.4.3 Examples

Consider the linearized model of a real inverted pendulum on a cart [16]

$$\dot{x} = \begin{pmatrix} 0 & 1 & 0 & 0 \\ 0 & 0 & -1.56 & 0 \\ 0 & 0 & 0 & 1 \\ 0 & 0 & 46.87 & 0 \end{pmatrix} x + \begin{pmatrix} 0 \\ 0.97 \\ 0 \\ -3.98 \end{pmatrix} (u + w), \tag{1.2.26}$$

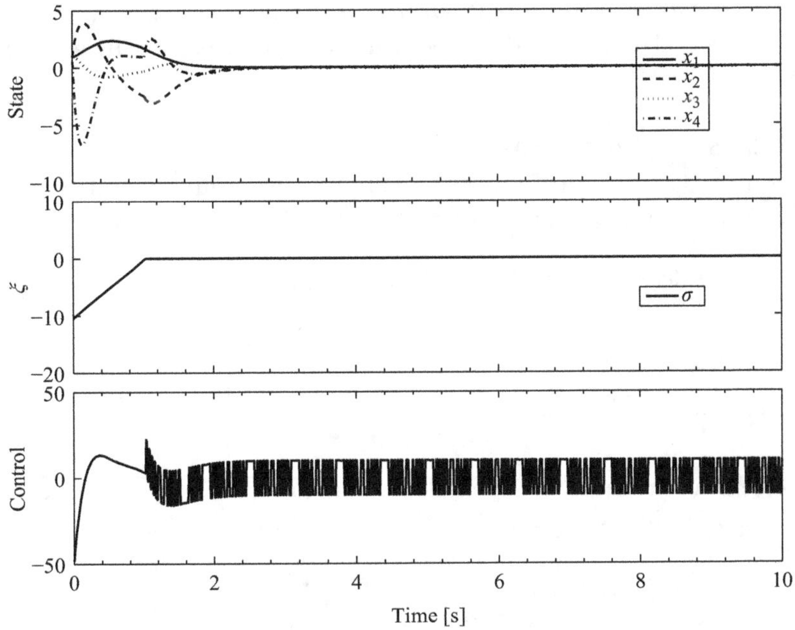

Figure 1.2.1 Simulation results for a first-order sliding-mode controller. The system is perturbed by $w = \sin(10\,t) + 3$. The control (1.2.25) is sampled and held every $\tau = 0.001$ s

where x_1, x_2, x_3 and x_4 are the position and velocity of the cart, and the angle and angular velocity of the pole, respectively. The system is controllable and the open-loop characteristic polynomial is $\lambda^2(\lambda + 6.85)(\lambda - 6.85)$. Suppose that we want to regulate the state to zero, in spite of any matched bounded perturbation.

1.2.4.3.1 First-order SMC

Consider the problem of designing a first-order sliding-mode controller with sliding-mode dynamics having eigenvalues $z_i = -5,\ i = 1, 2, 3$. Applying (1.2.23) with $\gamma(\lambda) = (\lambda + 5)^3$ gives

$$C = \begin{pmatrix} -3.2002 & -1.9201 & -4.5411 & -0.7166 \end{pmatrix},$$

which in turn yields the expected transfer function

$$g(s) = C(sI - A)^{-1}B = \frac{(s+5)^3}{s^2(s + 6.85)(s - 6.85)}.$$

To enforce a sliding motion on the surface $\sigma = 0$, we apply the control (1.2.25) with v_1 as in (1.2.19) and $\alpha_1 = 10$. Figure 1.2.1 shows the simulated response when

$$w = \sin(10t) + 3 \quad \text{and} \quad x_0^\top = \begin{pmatrix} 1 & 1 & 1 & 1 \end{pmatrix},$$

and the control law is sampled and held every $\tau = 0.001$ s. It can be seen that, once the state reaches the sliding surface, the state converges exponentially to the origin, despite w.

1.2.4.3.2 Second-order SMC

Suppose now that we desire a sliding-mode dynamics with eigenvalues $z_i = -5$, $i = 1, 2$. Applying (1.2.23) with $\gamma(\lambda) = (\lambda + 5)^2$ gives

$$C = \begin{pmatrix} -0.6400 & -0.2560 & -0.4062 & -0.0621 \end{pmatrix}$$

and

$$g(s) = \frac{(s+5)^2}{s^2(s+6.85)(s-6.85)}.$$

To enforce a second-order sliding motion on the surface $\sigma = \dot{\sigma} = 0$, we apply the control (1.2.25) with v_2 as in (1.2.20) and $\alpha_2 = 10$, i.e.,

$$u = -\frac{1}{CAB}\left(CA^2x + 10\frac{\dot{\sigma} + |\sigma|^{1/2}\,\mathrm{sign}(\sigma)}{|\dot{\sigma}| + |\sigma|^{1/2}}\right). \tag{1.2.27}$$

Figure 1.2.2 shows the simulated response for the same perturbation, initial conditions and sampling time as before. It can be seen that, once the state reaches the sliding surface, the state converges exponentially to the origin, again despite w.

1.2.4.3.3 Third-order SMC

Consider the problem of designing a third-order sliding-mode controller with sliding-mode dynamics having the eigenvalue $z_1 = -5$. Applying (1.2.23) with $\gamma(\lambda) = \lambda + 5$ gives

$$C = \begin{pmatrix} -0.1280 & -0.0256 & -0.0310 & -0.0062 \end{pmatrix},$$

which in turn yields the expected transfer function

$$g(s) = C(sI - A)^{-1}B = \frac{s+5}{s^2(s+6.85)(s-6.85)}.$$

To enforce a third-order sliding motion on the surface $\sigma = \dot{\sigma} = \ddot{\sigma} = 0$, we apply

$$u = -\frac{1}{CA^2B}\left(CA^3x + 10\frac{\ddot{\sigma} + 2\left(|\dot{\sigma}| + |\sigma|^{2/3}\right)^{-1/2}\left(\dot{\sigma} + |\sigma|^{2/3}\,\mathrm{sign}(\sigma)\right)}{|\ddot{\sigma}| + 2\left(|\dot{\sigma}| + |\sigma|^{2/3}\right)^{1/2}}\right). \tag{1.2.28}$$

Figure 1.2.3 shows the simulated response for the same perturbation, initial conditions and sampling time as before. Again, the state converges exponentially to the origin once the state reaches the sliding surface, despite w.

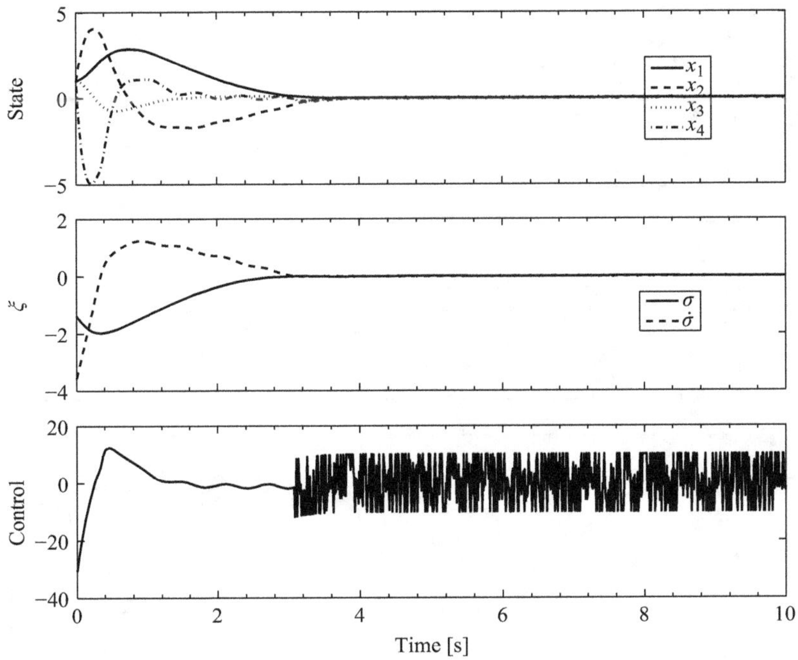

Figure 1.2.2 Simulation results for a second-order sliding-mode controller. The system is perturbed by $w = \sin(10\,t) + 3$. The control (1.2.27) is sampled and held every $\tau = 0.001$ s

1.2.4.3.4 Accuracy

To verify the order of accuracy established in (1.2.22), we take logarithms on both sides of the inequalities (the base is not important):

$$\log\left(\left|\sigma^{(i)}\right|\right) < \log(\mu_i) + (r - i)\log(\tau), \quad i = 0, \ldots, r - 1.$$

Notice that, on a logarithmic scale, the right-hand is a straight line with slope $r - i$ and ordinate at the origin $\log(\mu_i)$. To verify that the order of the error $\left|\sigma^{(i)}\right|$ as a function of τ is indeed $r - i$, the closed-loop system was simulated with a simple perturbation $w = \sin(10t) + 3$ for several values of τ, both for a zero order hold with sampling period τ and for a (previously neglected) actuator of the form $\tau\dot{v} = -v + u$. We recorded the maximum error after the transient, $\limsup_{t\to\infty}\left|\sigma^{(i)}\right|$. The best linear interpolation on a least square sense was then computed to recover an estimate of $\log(\mu_i)$ and $r - i$.

Figure 1.2.4 shows that the estimations agree well with (1.2.22) in the first-order control. Figures 1.2.5 and 1.2.6 show the system accuracy for several sampling times and several actuator time-constants in the case of second and third order sliding modes, respectively. Inequality (1.2.22) is again verified.

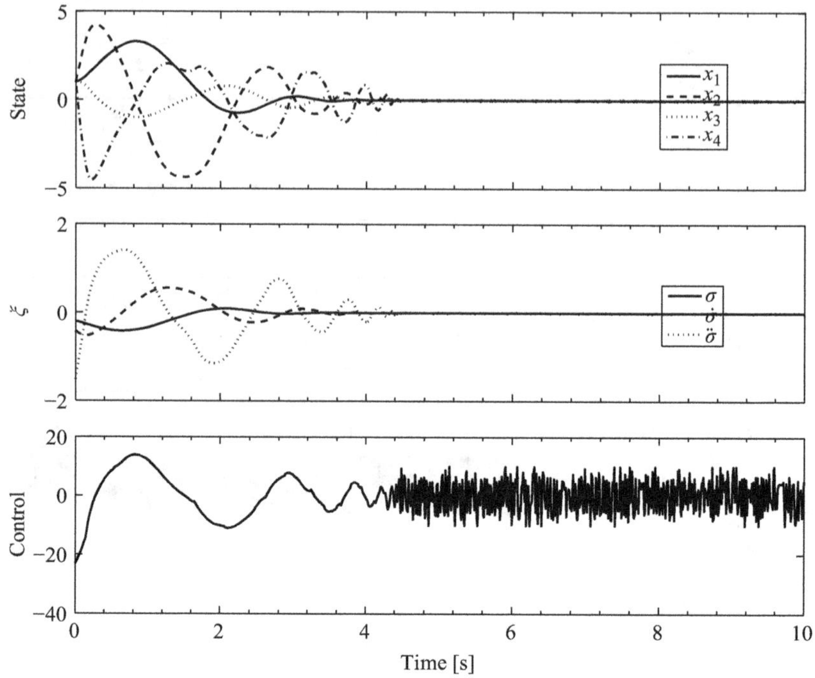

Figure 1.2.3 Simulation results for a third-order sliding-mode controller. The system is perturbed by $w = \sin(10\,t) + 3$. The control (1.2.28) is sampled and held every $\tau = 0.001$ s

1.2.5 Singular LQR

A possible choice of an $(n-1)$-dimensional sliding surface can be made through the use of a LQ performance index. With this method, an optimal sliding surface is designed as a solution of a singular optimal LQ problem with symmetric positive definite weighting matrix Q and *free-cost control* [1, 2, 17]. After that, the first-order sliding-mode controller can be designed ensuring the finite time convergence to the optimal sliding surface. It is well known from the theory of singular optimal control [18–20] that, for the case when the weighting matrix is semi-positive definite, the dimension of the optimal solution is less than $(n-1)$.

In this section, we show that the order of the sliding mode is in agreement with the singularity order of a given performance index. We use the optimal solution as a sliding surface and, as before, apply a quasi-continuous HOSM controller of the corresponding order.

Consider first the nominal system

$$\dot{x} = Ax + Bu \tag{1.2.29}$$

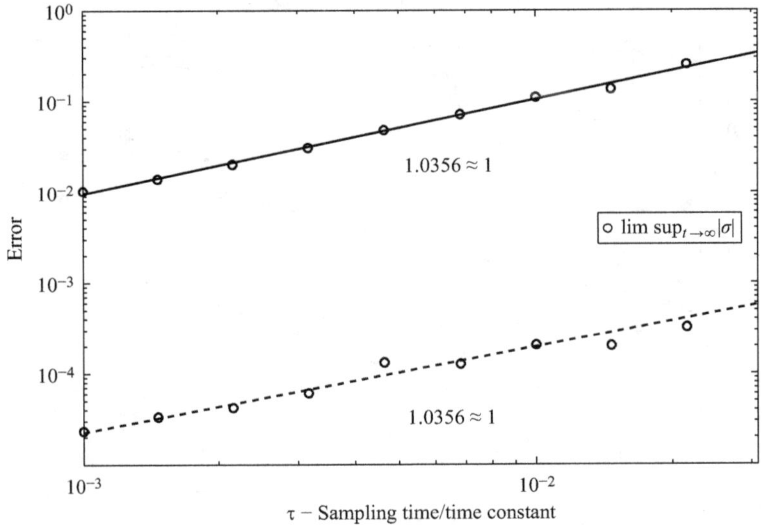

Figure 1.2.4 First-order sliding mode in the presence of a simple sinusoidal perturbation. Order of the errors as functions of the sampling period (solid) and the actuator time constant (dashed). The error order matches the sliding-mode order almost exactly

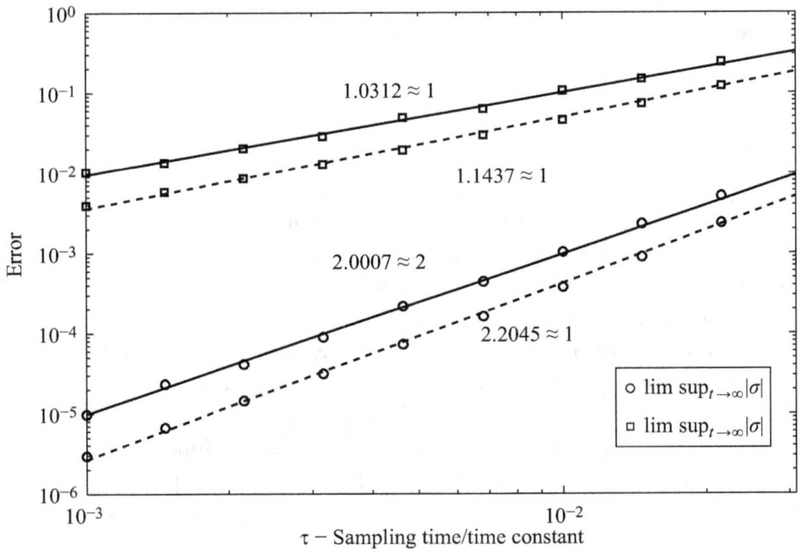

Figure 1.2.5 Second-order sliding mode in the presence of a simple sinusoidal perturbation. Order of the errors as functions of the sampling period (solid) and the actuator time constant (dashed). The error order for σ matches well with the sliding-mode order

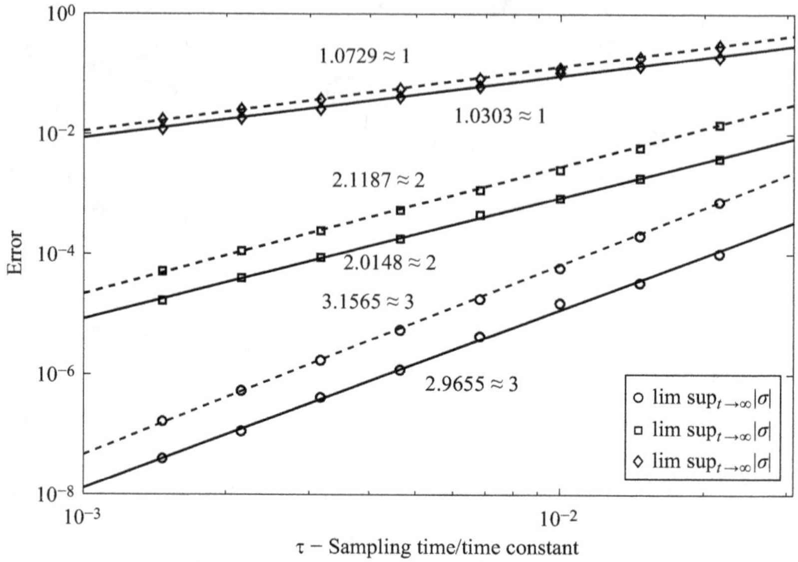

Figure 1.2.6 Third-order sliding mode in the presence of a simple sinusoidal perturbation. Order of the errors as functions of the sampling period (solid) and the actuator time constant (dashed). The error order for σ matches well with the sliding-mode order

together with the singular performance index

$$J(x) = \frac{1}{2}\int_{t_1}^{\infty} \left(x^\top Qx\right) \mathrm{d}t, \tag{1.2.30}$$

where $0 \leq Q = Q^\top \in \mathbb{R}^{n\times n}$. The criterion (1.2.30) is singular with respect to the scalar control input u. The solution of this singular optimal stabilization problems (1.2.29) and (1.2.30) lies on a manifold of smaller dimensionality.

The objective of this section is the HOSM surface design based on the reduced order manifold obtained by the solution of (1.2.29) and (1.2.30). In order to solve this singular optimal stabilization problem, the system (1.2.29) is transformed to a controllable canonical form. The criterion (1.2.30) is also transformed in order to keep the original weight of the states. Then, the connection between the *order of singularity* of the performance index and the order of the sliding mode is established. Finally, the sliding surface of the corresponding order as a solution of the singular optimal stabilization problem and the corresponding HOSMC ensuring the tracking on this surface is designed.

1.2.5.1 Regularization

The first step towards the construction of a sliding surface of higher relative degree is to set $\tilde{Q}_{22} = 0$.

Proposition 1.2.1. *If $\tilde{Q}_{22} = 0$ in (1.2.15), then $\tilde{Q}_{12} = 0$ and $\tilde{Q}_{21} = 0$.*

Proof. The result is a simple consequence of the symmetry and semi-positiveness of $\bar{Q}$. □

In other words, $\bar{Q}$ is necessarily of the form

$$\bar{Q} = \begin{pmatrix} \bar{Q}_{11} & \bar{Q}_{12} & 0 & \cdots & 0 \\ \bar{Q}_{21} & \bar{Q}_{22} & 0 & \cdots & 0 \\ 0 & 0 & 0 & \cdots & 0 \\ \vdots & \vdots & \vdots & \ddots & \vdots \\ \underbrace{0 \quad\;\; 0}_{n-k \text{ columns}} & & \underbrace{0 \quad \cdots \quad 0}_{k \text{ columns}} \end{pmatrix}, \tag{1.2.31}$$

where $0 \leq \bar{Q}_{11} = \bar{Q}_{11}^{\top} \in \mathbb{R}^{(n-k-1)\times(n-k-1)}$ and $0 < \bar{Q}_{22} \in \mathbb{R}$.

Definition 1.2.1. *The order of singularity is the natural number $s = k + 1$, where k is the maximal number of zero right-hand columns in $\bar{Q}$.*

Allow us to partition the state as $z^{\top} = \left[\bar{z}_1^{\top} \; \bar{z}_2 \; \bar{z}_3^{\top}\right]$, where $\bar{z}_1 \in \mathbb{R}^{n-k-1}$, $\bar{z}_2 \in \mathbb{R}$ and $\bar{z}_3 \in \mathbb{R}^k$. In view of (1.2.31), the performance index has the form

$$J = \frac{1}{2}\int_{t_0}^{\infty} \left(\bar{z}_1^{\top}\bar{Q}_{11}\bar{z}_1 + 2\bar{z}_1^{\top}\bar{Q}_{12}\bar{z}_2 + \bar{z}_2^{\top}\bar{Q}_{22}\bar{z}_2\right) \mathrm{d}t. \tag{1.2.32}$$

In order to eliminate cross terms, we introduce the change of variables

$$\nu = \bar{z}_2 + \bar{Q}_{22}^{-1}\bar{Q}_{12}^{\top}\bar{z}_1. \tag{1.2.33}$$

After the transformation T and the change of variables, the original singular performance index has become

$$J = \frac{1}{2}\int_{t_0}^{\infty} \left(\bar{z}_1^{\top}\hat{Q}\bar{z}_1 + \nu^{\top}\hat{R}\nu\right) \mathrm{d}t, \tag{1.2.34}$$

which is non-singular with ν as virtual control and $\hat{Q} := \bar{Q}_{11} - \bar{Q}_{12}\left(\bar{Q}_{22}\right)^{-1}\bar{Q}_{12}^{\top}$ and $\hat{R} := \bar{Q}_{22}$ as weighting matrices.

Note that, due to the form of $\bar{A}$, the partitions $\bar{A}_{11}$ and $\bar{A}_{12}$ are also in controller canonical form,

$$\bar{A}_{11} = \underbrace{\begin{pmatrix} 0 & 1 & 0 & \cdots & 0 \\ 0 & 0 & 1 & \cdots & 0 \\ 0 & 0 & 0 & \cdots & 0 \\ \vdots & \vdots & \vdots & \ddots & 1 \\ 0 & 0 & 0 & \cdots & 0 \end{pmatrix}}_{n-s \text{ columns}}, \quad \bar{A}_{12} = \left.\begin{pmatrix} 0 \\ 0 \\ 0 \\ 0 \\ \vdots \\ 1 \end{pmatrix}\right\} n-s \text{ rows}.$$

Finally, after the change of variables (1.2.33), we have

$$\dot{\bar{z}}_1 = \hat{A}\bar{z}_1 + \hat{B}\nu, \tag{1.2.35}$$

where

$$\hat{A} := \bar{A}_{11} + \bar{A}_{12}\left(\bar{Q}_{22}\right)^{-1}\bar{Q}_{12}^{\top} \in \mathbb{R}^{(n-s)\times(n-s)},$$
$$\hat{B} := \bar{A}_{12} \in \mathbb{R}^{(n-s)}.$$

1.2.5.2 *Surface design*

The singular optimal control problem given by (1.2.11a) and (1.2.30) has been transformed to a reduced-order, non-singular optimal control problem given by (1.2.35) and (1.2.34). Such problem can be solved via ARE

$$P\hat{A} + \hat{A}^{\top}P + \hat{Q} - P\hat{B}\hat{R}^{-1}\hat{B}^{\top}P = 0. \tag{1.2.36}$$

Recall that P is the only positive definite solution of the matrix Riccati equation if the pair $\left(\hat{A}, \bar{D}\right)$ is observable with any matrix $\bar{D}$ such that $\bar{D}^{\top}\bar{D} = \hat{Q}$ [7].

The reduced-order optimal control is

$$\nu = -\hat{R}^{-1}\hat{B}^{\top}P\bar{z}_1$$

which, in terms of $\bar{z}_2$ gives

$$\bar{z}_2 = -K\bar{z}_1 \tag{1.2.37}$$

with

$$K = \hat{R}^{-1}\left(\hat{B}^{\top}P + \bar{Q}_{12}^{\top}\right) = \bar{Q}_{22}^{-1}\left(\bar{A}_{12}^{\top}P + \bar{Q}_{12}^{\top}\right).$$

The virtual control (1.2.37) provides stability to the closed-loop sub-system (1.2.35). Also, the corresponding sliding variable,

$$\sigma = \bar{z}_2 + \left(\bar{Q}_{22}\right)^{-1}\left(\bar{A}_{12}^{\top}P + \bar{Q}_{12}^{\top}\right)\bar{z}_1, \tag{1.2.38}$$

is of relative degree higher than one.

Remark 1.2.1. *The* order of singularity *of the performance index is equal to the* relative degree *of the corresponding sliding variable. That is, $r = s$.*

1.2.5.3 Control design

The rth time-derivative of (1.2.38) can be generally written as

$$\sigma^{(r)} = \beta_1 z_1 + \beta_2 z_2 + \cdots + \beta_n z_n + u + w.$$

Similar to the pole-placement case, we propose the control law

$$u = -\beta_1 z_1 - \beta_2 z_2 - \cdots - \beta_n z_n + v_r. \tag{1.2.39}$$

If the system (1.2.12) is expressed in a new set of state variables as $\left[\bar{z}_1^{\top}, \sigma, \dot{\sigma}, \ldots, \sigma^{(r-1)}\right]^{\top}$, it takes the *normal form* (1.2.16a),

$$\begin{aligned} \dot{\bar{z}}_1 &= \left(\bar{A}_{11} - \bar{A}_{12}K\right)\bar{z}_1 + \bar{A}_{12}\sigma, \\ \sigma^{(r)} &= v_r + w, \end{aligned} \tag{1.2.40}$$

where $\eta = \bar{z}_1$, $\sigma = \xi_1$ and $K = \left(\bar{Q}_{22}\right)^{-1}\left[A_{12}^{\top}P + \bar{Q}_{12}^{\top}\right]$.

HOSMC is designed to restrict trajectories of the system to the manifold $\sigma = \dot{\sigma} = \cdots = \sigma^{r-1} = 0$, where the dynamics is described by the reduced order model

$$\dot{\bar{z}}_1 = \left(\bar{A}_{11} - \bar{A}_{12}K\right)\bar{z}_1. \tag{1.2.41}$$

The selection of the higher-order sliding-mode controller follows from the *singularity order* of the performance index (the relative degree of the sliding variable). It is well known that, in order to force all trajectories of the system to the manifold $\sigma = \dot{\sigma} = \cdots = \sigma^{r-1} = 0$, it is necessary to use a HOSMC of the same order as the relative degree of (1.2.38).

1.2.5.4 Examples

Consider the same linearized model of the inverted pendulum (1.2.26). Suppose that, again, we want to regulate the state to zero, in spite of any bounded matched perturbation.

First, the nominal system is transformed to its controllable canonical form,

$$\dot{z} = \bar{A}z + \bar{B}u, \quad \bar{A} = \begin{pmatrix} 0 & 1 & 0 & 0 \\ 0 & 0 & 1 & 0 \\ 0 & 0 & 0 & 1 \\ 0 & 0 & 46.87 & 0 \end{pmatrix}, \quad \bar{B} = \begin{pmatrix} 0 \\ 0 \\ 0 \\ 1 \end{pmatrix}. \tag{1.2.42}$$

The state z is obtained through the transformation $z = Tx$,

$$T = \begin{pmatrix} -0.0256 & 0 & -0.0062 & 0 \\ 0 & -0.0256 & 0 & -0.0062 \\ 0 & 0 & -0.2510 & 0 \\ 0 & 0 & 0 & -0.2510 \end{pmatrix}^{\top}.$$

1.2.5.4.1 First-order SMC

System (1.2.26) is associated with the singular performance index (1.2.30) with transformed weighting matrix

$$\bar{Q} = (T^{-1})QT^{-1} = \begin{pmatrix} 1 & 0 & 0 & 0 \\ 0 & 1 & 0 & 0 \\ 0 & 0 & 1 & 0 \\ 0 & 0 & 0 & 1 \end{pmatrix}.$$

The partition is

$$\bar{Q} = \begin{pmatrix} \bar{Q}_{11} & \bar{Q}_{12} \\ \bar{Q}_{21} & \bar{Q}_{22} \end{pmatrix} = \left(\begin{array}{ccc|c} 1 & 0 & 0 & 0 \\ 0 & 1 & 0 & 0 \\ 0 & 0 & 1 & 0 \\ \hline 0 & 0 & 0 & 1 \end{array}\right). \tag{1.2.43}$$

Then, the regular transformed performance index becomes

$$\bar{J}_1 = \frac{1}{2}\int_{t_1}^{\infty} \left(z_{13}^{\top}\bar{Q}_{11}z_{13} + z_4^{\top}\bar{Q}_{22}z_4\right) \mathrm{d}t,$$

where $z_{13} = \begin{pmatrix} z_1 & z_2 & z_3 \end{pmatrix}^{\top}$. System (1.2.42) is partitioned in the same way as $\bar{Q}$ in (1.2.43), i.e., $\bar{A}_{11}$ has the same dimensions of Q_{11} and we consider z_4 as virtual control,

$$\dot{z}_{13} = \bar{A}_{11}z_{13} + \bar{A}_{12}z_4, \quad \bar{A}_{11} = \begin{pmatrix} 0 & 1 & 0 \\ 0 & 0 & 1 \\ 0 & 0 & 0 \end{pmatrix}, \quad \bar{A}_{12} = \begin{pmatrix} 0 \\ 0 \\ 1 \end{pmatrix}.$$

The reduced-order optimal control is found through the ARE

$$P\bar{A}_{11} + \bar{A}_{11}^{\top}P + \bar{Q}_{11} - P\bar{A}_{12}\bar{Q}_{22}^{-1}\bar{A}_{12}^{\top}P = 0,$$

the solution of which is

$$P = \begin{pmatrix} 2.4142 & 2.4142 & 1.000 \\ 2.4142 & 4.8284 & 2.4142 \\ 1.000 & 2.4142 & 2.4142 \end{pmatrix}.$$

The virtual control and sliding surface of relative degree one are

$$z_4 = -\bar{Q}_{22}^{-1}\bar{A}_{12}Pz_{13} \quad \text{and} \quad \sigma = z_4 + \bar{Q}_{22}^{-1}\bar{A}_{12}Pz_{13}. \tag{1.2.44}$$

Finally, the control law is

$$u = \begin{pmatrix} 0 & -1 & -49.2859 & -2.4142 \end{pmatrix} z - \alpha_1 \operatorname{sign}(\sigma). \tag{1.2.45}$$

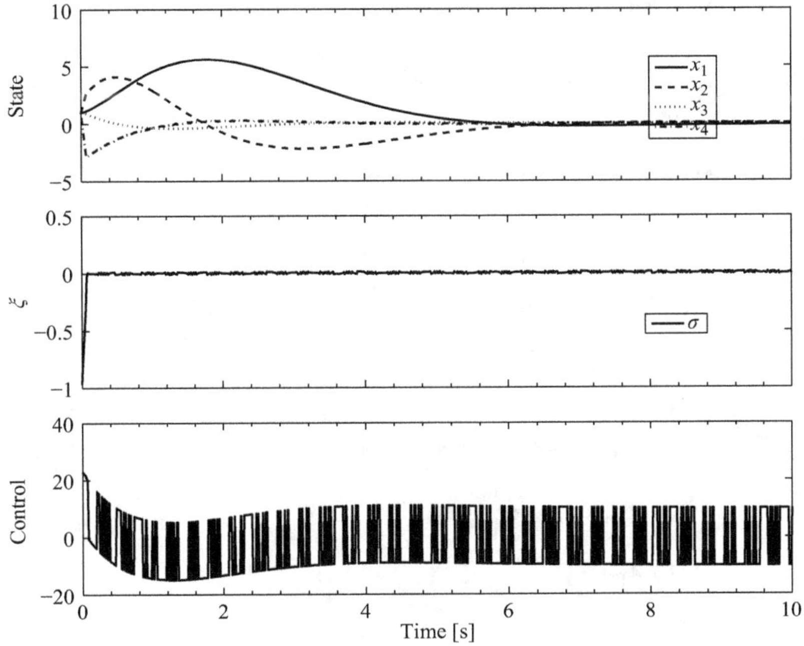

Figure 1.2.7 Simulation results for a first-order sliding-mode controller. The system is perturbed by $w = \sin(10\,t) + 3$. The control (1.2.45) is sampled and held every $\tau = 0.001$ s

Figure 1.2.7 shows the results of simulation of the system with the sliding surface (1.2.44) and the control law (1.2.45) with the same perturbation $w = \sin(10t) + 3$. It can be seen how the sliding surface of relative degree one is reached at the time t_1 (almost at the beginning of the simulation) and the sliding mode is kept during the rest of the simulation in spite of the perturbation w. When the sliding mode is reached, the three-dimensional sliding dynamics behaves optimally with respect to the performance index $\bar{J}_1$.

1.2.5.4.2 Second-order SMC

System (1.2.26) is associated with another singular performance index with transformed weighting matrix

$$\bar{Q} = \begin{pmatrix} \bar{Q}_{11} & \bar{Q}_{12} & \bar{0} \\ \bar{Q}_{21} & \bar{Q}_{22} & 0 \\ \bar{0} & 0 & 0 \end{pmatrix} = \left(\begin{array}{cc|cc} 1 & 0 & 0 & 0 \\ 0 & 1 & 0 & 0 \\ \hline 0 & 0 & 1 & 0 \\ 0 & 0 & 0 & 0 \end{array}\right). \tag{1.2.46}$$

Then, the regular transformed performance index takes the form

$$\bar{J}_2 = \frac{1}{2}\int_{t_1}^{\infty} \left(z_{12}^{\top}\bar{Q}_{11}z_{12} + z_3^{\top}\bar{Q}_{22}z_3\right) \mathrm{d}t, \tag{1.2.47}$$

where $z_{12} = (z_1\ z_2)^{\top}$. The reduced-order system is

$$\dot{z}_{12} = \bar{A}_{11}z_{12} + \bar{A}_{12}z_3, \quad \bar{A}_{11} = \begin{pmatrix} 0 & 1 \\ 0 & 0 \end{pmatrix}, \quad \bar{A}_{12} = \begin{pmatrix} 0 \\ 1 \end{pmatrix}.$$

The solution of the ARE is

$$P = \begin{pmatrix} 1.7321 & 1 \\ 1 & 1.7321 \end{pmatrix}.$$

The sliding surface of relative degree two is

$$\sigma = z_3 + \bar{Q}_{22}^{-1}\bar{A}_{12}Pz_{12}. \tag{1.2.48}$$

Finally, the control law is

$$u = \begin{pmatrix} 0 & 0 & -47.8717 & -1.7320 \end{pmatrix} z - \alpha_2 \frac{\dot{\sigma} + |\sigma|^{1/2}\,\mathrm{sign}(\sigma)}{|\dot{\sigma}| + |\sigma|^{1/2}}. \tag{1.2.49}$$

Figure 1.2.8 shows the results of simulation of the system with the sliding surface (1.2.48) and the control law (1.2.49). The sliding surface of relative degree two is reached at the time $t_1 = 1.2$ s, and the second-order sliding mode is kept during the rest of the simulation in spite of the perturbation w. The two-dimensional sliding dynamics behaves optimally with respect to $\bar{J}_2$ after the time t_1.

1.2.5.4.3 Third-order SMC

System (1.2.26) is associated with yet another singular performance index with weighting matrix

$$\bar{Q} = \begin{pmatrix} \bar{Q}_{11} & \bar{Q}_{12} & \bar{0} \\ \bar{Q}_{21} & \bar{Q}_{22} & \bar{0} \\ \bar{0} & \bar{0} & \bar{0} \end{pmatrix} = \left(\begin{array}{c|ccc} 1 & 0 & 0 & 0 \\ \hline 0 & 1 & 0 & 0 \\ 0 & 0 & 0 & 0 \\ 0 & 0 & 0 & 0 \end{array}\right). \tag{1.2.50}$$

The solution of the ARE is $P = 1$ and the sliding surface of relative degree three is

$$\sigma = z_2 + \bar{Q}_{22}^{-1}\bar{A}_{12}Pz_1. \tag{1.2.51}$$

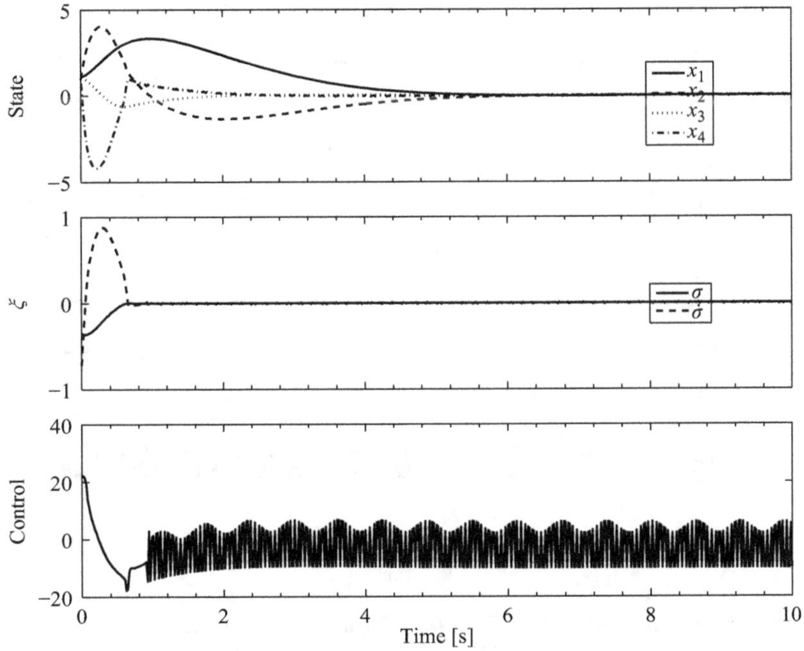

Figure 1.2.8 Simulation results for a second-order sliding-mode controller. The system is perturbed by w = sin(10 t) + 3. The control (1.2.49) is sampled and held every τ = 0.001 s

Finally, the control law is

$$u = \begin{pmatrix} 0 & 0 & -46.8717 & -1 \end{pmatrix} z - \alpha_3 \frac{\ddot{\sigma} + 2\left(|\dot{\sigma}| + |\sigma|^{2/3}\right)^{-1/2}\left(\dot{\sigma} + |\sigma|^{2/3}\,\mathrm{sign}(\sigma)\right)}{|\ddot{\sigma}| + 2\left(|\dot{\sigma}| + |\sigma|^{2/3}\right)^{1/2}}. \tag{1.2.52}$$

Figure 1.2.9 shows the results of simulation of the system with the sliding surface (1.2.51) and the control law (1.2.52). The sliding surface of relative degree three is reached at the time $t_1 = 2.2$ s, and the third-order sliding mode is kept during the rest of the simulation in spite of the perturbation w. The one-dimensional sliding dynamics behaves optimally with respect to $\bar{J}_3$ after the time t_1.

For the accuracy performance tests, we took the same methodology that in (1.2.4.3.4) with the same bounded matched perturbation $w = \sin(10\,t) + 3$. The results are shown in Figures 1.2.10, 1.2.11 and 1.2.12.

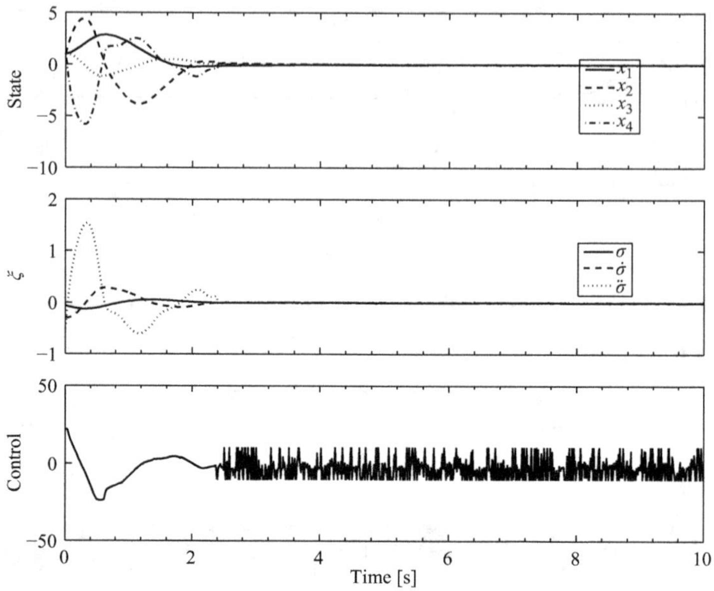

Figure 1.2.9 Simulation results for a third-order sliding-mode controller. The system is perturbed by $w = \sin(10\,t) + 3$. The control (1.2.52) is sampled and held every $\tau = 0.001$ s

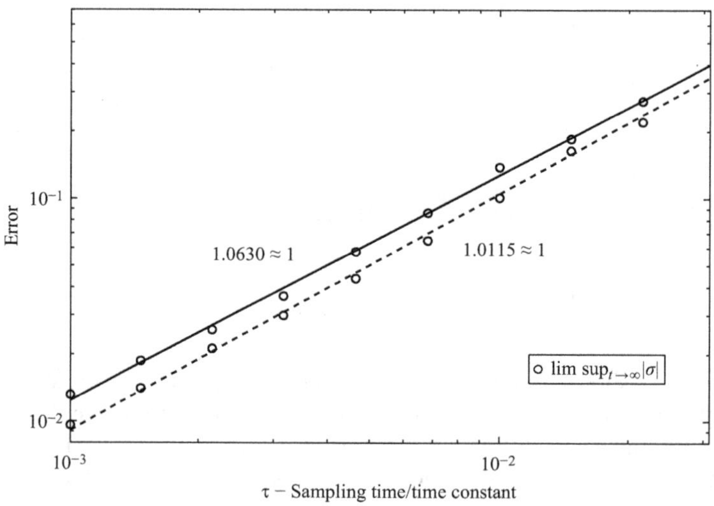

Figure 1.2.10 First-order sliding mode. Order of the errors as functions of the sampling period (solid) and the actuator time constant (dashed). The error order matches the sliding-mode order almost exactly

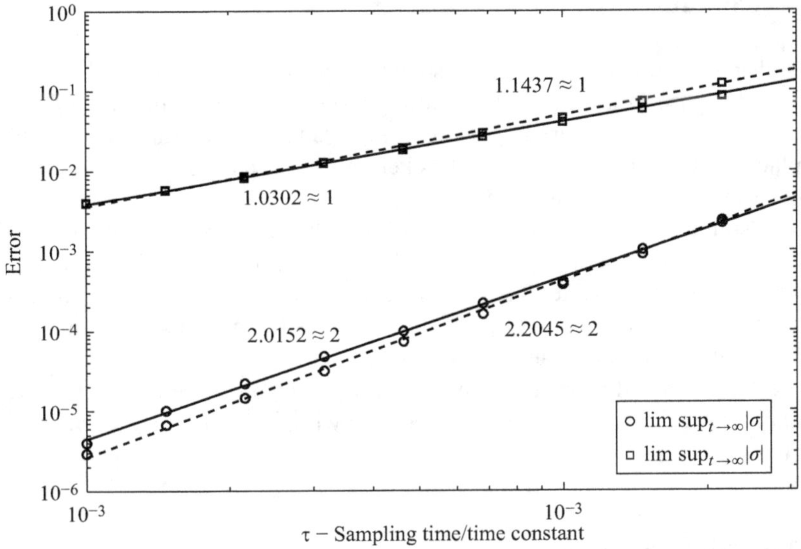

Figure 1.2.11 *Second-order sliding mode. Order of the errors as functions of the sampling period (solid) and the actuator time constant (dashed). The error order for σ matches well with the sliding-mode order*

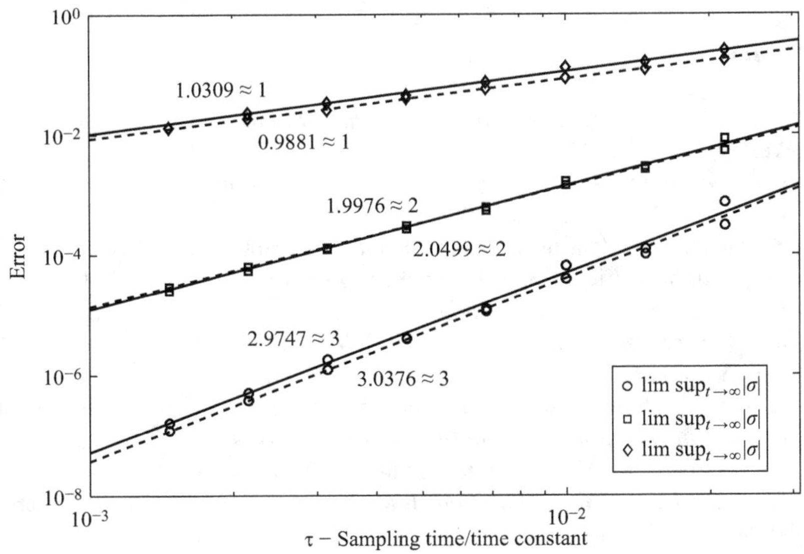

Figure 1.2.12 *Third-order sliding mode. Order of the errors as functions of the sampling period (solid) and the actuator time constant (dashed). The error order for σ matches well with the sliding-mode order*

1.2.6 Conclusions

In this chapter, the conventional two-step sliding-mode design procedure is generalized for the case of arbitrary order sliding modes. Two principal ideas for sliding surface design are considered: pole placement and optimal stabilization. For the pole-placement design, the formula of Ackermann–Utkin is generalized to design the sliding surfaces with arbitrary relative degree.

To design the sliding manifold of arbitrary relative degree ensuring optimal stabilization, the singular optimal control techniques are used. It is shown that there is a natural connection between order of singularity and order of sliding modes. The optimal surface for singular LQ problem is considered as the sliding manifold for HOSM of corresponding order.

The simulations illustrating output tracking in the presence of the fast actuators show that the precision of tracking rises together with the relative degree of the sliding dynamics.

Acknowledgment

The authors are grateful for the financial support of CONACyT (Consejo Nacional de Ciencia y Tecnología) CVU 425551; PAPIIT (Programa de Apoyo a Proyectos de Investigación e Innovación Tecnológica) 113216.

References

[1] V. I. Utkin. *Sliding Modes in Control and Optimization*, vol. 116. Springer-Verlag, Berlin, 1992.

[2] C. Edwards and S. K. Spurgeon. *Sliding Mode Control: Theory and Applications*. CRC Press, Padstow, 1998.

[3] Y. Shtessel, C. Edwards, L. Fridman, and A. Levant. *Sliding Mode Control and Observation*. Series: Control Engineering. Birkhäuser, Basel, 2013.

[4] L. Fridman and A. Levant. *Higher Order Sliding Modes in Sliding Mode Control in Engineering*. In J. P. Barbot and W. Perriguetti (Eds.), Marcel Dekker, New York, NY, 2002.

[5] A. Levant. Higher-order sliding modes, differentiation and output-feedback control. *International Journal of Control*, 76:924–941, 2003.

[6] J. Ackermann and V. Utkin. Sliding mode control design based on Ackermann's formula. *IEEE Transactions on Automatic Control*, 43:234–237, February 1998.

[7] B. D. O. Anderson and J. B. Moore. *Optimal Control: Linear Quadratic Methods*. Prentice-Hall, London, 1971.

[8] A. Levant. Quasi-continuous high-order sliding-mode controllers. *IEEE Transactions on Automatic Control*, 50:1812–1816, 2005.

[9] A. F. Filippov. *Differential Equations with Discontinuous Right Hand-Sides, Mathematics and Its Applications*. Kluwer Academic Publisher, Dordrecht, 1983.

[10] V. Utkin, J. Guldner, and J. Shi. *Sliding Modes in Electromechanical Systems*. Taylor & Francis, London, 1999.

[11] A. Isidori. *Nonlinear Control Systems*. Springer-Verlag, London, 1996.

[12] R. Marino and P. Tomei. *Nonlinear Control Design: Geometric, Adaptive, and Robust*. Prentice-Hall, Englewood, NJ, 1995.

[13] A. Levant. Chattering analysis. *IEEE Transactions on Automatic Control*, 55:1380–1389, 2010.

[14] D. Hernández-Zárate, F. Castaños and L. Fridman. Pole-placement in higher-order sliding-mode control, *Proc. IFAC World Congress*, Cape Town, South Africa, 1386–1391, Aug. 2014.

[15] R. L. Williams and D. A. Lawrence. *Linear State-Space Control Systems*. John Wiley & Sons, Inc., Hoboken, NJ, 2007.

[16] I. Fantoni and R. Lozano. *Non-Linear Control for Underactuated Mechanical Systems*. Springer-Verlag, London, 2002.

[17] A. G. Luk'yanov and V. I. Utkin. Design of optimal linear systems with a degenerate criterion. *Avtomatika i Telemekhanika*, (7):42–50, 1982.

[18] H. J. Kelley. A transformation approach to singular subarcs in optimal trajectory and control problems. *Journal of the Society for Industrial & Applied Mathematics, Series A: Control*, 2(2):234–240, 1964.

[19] D. H. Jacobson and J. L. Speyer. Necessary and sufficient conditions for optimality for singular control problems: A limit approach. *Journal of Mathematical Analysis and Applications*, 34(2):239–266, 1971.

[20] R. Gabasov and N. H. Choksy. *Singular Optimal Controls*. Kluwer Academic Publisher, Dordrecht, 2001.

Chapter 1.3

Robust output control of systems subjected to perturbations via high-order sliding modes observation and identification

Leonid Fridman, Alejandra Ferreira de Loza and Luis T. Aguilar

Abstract

The control of linear systems subjected to perturbations by means of high-order sliding modes observation and identification techniques is presented. First, the regulation control of a system affected by matched perturbations is addressed. Later on, the method is extended to deal with systems affected by matched and unmatched perturbations. The performance of the proposed controllers is estimated in terms of the deterministic noise upper bounds, sampling time and execution time. The feasibility of the approach is shown by means of experiments and numerical simulations.

1.3.1 Introduction

Systems suffer from uncertainties and disturbances which may affect and even destroy the systems performance. One of the most prospering control strategies insensitive with respect to uncertainties is sliding mode control (SMC) (see, e.g., References 1 and 2). This robust technique is well known for its ability to counteract external disturbances and model uncertainties which satisfy the matching condition. This condition is met when the perturbations or uncertainties are implicit at the input channels, for example in the case of completely actuated systems.

The SMC design methodology involves two stages: the design of a switching function which provides desirable system performance in the sliding mode and the design of the control law ensuring that the system states are driven to the sliding manifold and thus the aimed performance is attained and maintained in spite of the matched uncertainties/perturbations. Nevertheless, there are some disadvantages: first, the necessity to measure the whole state and second, the lack of robustness against unmatched uncertainties of the resulting controller.

With the advent of high-order sliding mode (HOSM) controllers [3] the chattering effect has been counteracted. Whereas the main solution to address the issue of robustness against unmatched perturbations has been the combination of sliding mode techniques with another robust strategies [4]. In the same direction, in References 5–7, the design of the sliding surface against unmatched perturbations considering only output information has been considered. The aforementioned methods deal with the unmatched perturbations in a robust way, i.e., the effects of the unmatched perturbations are attenuated at the expense of the nominal control performance. This is not the case of methods based on unknown inputs observers [8–11]. In these methods, the perturbation is estimated and then injected through the control action in order to compensate its effects. For instance, in Reference 8, a particular class of second-order systems affected by unmatched perturbations is tackled. On the other hand, in Reference 9, a wider class of systems is considered but it assumes full state measurements.

Concerning the identification and compensation approach, the development of finite-time exact observers based on HOSMs has allowed for the possibility of strike a wider class of uncertain systems considering only output information. Such observers provide theoretically exact estimation and finite-time convergence for the state and the unknown inputs. References 12 and 13 were shown that the finite time theoretically exact observation and unknown inputs identification for linear time invariant systems with unknown inputs is realizable when the system is strongly observable [14].

Here, compensation control methods to counteract the effect of the perturbations are proposed based on the aforementioned HOSM observers. The results of the papers [10, 11] are summarized in this chapter and the accuracy of the studied strategies is assessed in terms of deterministic noise upper bounds, sampling time and execution time issues.

This chapter is organized as follows. In Section 1.3.2, the notation is given. In Section 1.3.3, the problem formulation and control challenge are presented. Section 1.3.4 summarizes the HOSM observer design, additionally the precision of the observation and identification processes is discussed. Section 1.3.5 develops a strategy to cope with matched uncertainties. A comparison with integral sliding mode techniques based on the estimated values of the state is introduced. A guideline for selecting the best approach in terms of performance upon hardware characteristics is given. Experimental results in an inverted rotary pendulum are given at the end of the section. Section 1.3.6 presents a generalization of the method to deal with unmatched perturbations. A dynamic sliding surface is proposed to counteract the effect of the unmatched perturbations whereas a continuous SMC is considered to drive the state trajectories toward the sliding surface. Numerical simulations for a 3-Degrees of freedom (DOF) helicopter are given. Finally, some concluding remarks are provided in Section 1.3.7.

1.3.2 Notation

The following notation is used. For a matrix $X \in \mathbb{R}^{n\times m}$ with rank$(X) = r$, the matrix $X^{\perp} \in \mathbb{R}^{n-r\times n}$ with rank$(X^{\perp}) = n - r$ is defined such that $X^{\perp}X = 0$. If $X \in \mathbb{R}^{n\times m}$ with rank$(X) = m$, the matrix $X^{+} = (X^TX)^{-1}X^T$ is defined as the pseudo-inverse of X. For any $x \in \mathbb{R}$, the symbol $|x|$ denotes its absolute value. For an square matrix

$X \in \mathbb{R}^{n\times n}$, $\lambda_i X$, $i = \overline{1, n}$, is used to denote an eigenvalue of the matrix X, whereas $\lambda_{\min}(X)$ represents the smallest eigenvalue of X. If $x \in \mathbb{R}^n$, then $\|x\|$ states for the Euclidean norm. If $X \in \mathbb{R}^{n\times m}$ is a matrix, the symbol $\|X\|$ denotes the corresponding induced norm.

1.3.3 Problem statement

Consider a linear time invariant system affected by unknown inputs

$$\dot{x} = Ax + Bu + G\nu, \tag{1.3.1}$$

$$y = Cx, \tag{1.3.2}$$

where $x \in \mathbb{R}^n$, $y \in \mathbb{R}^p$ $(1 \leq p < n)$ represent the system's states and measured outputs, respectively. The vector $u \in \mathbb{R}^m$ represents the input and $\nu \in \mathbb{R}^q$ $(q \leq n - m)$ are the perturbations. The matrices A, B, G and C are of the corresponding dimensions.

The following is assumed henceforth.

Assumption 1.3.1. *The system is strongly observable, or equivalently the triplet (A, C, G) has no zeros.*

Assumption 1.3.2. *ν and its derivatives up to order ρ are bounded, i.e., $\|\nu\| \leq \breve{\nu}_0$ as well as $\|\nu^{(i)}\| \leq \breve{\nu}_i$ for $i = \overline{1, \rho}$, for all $t \geq 0$.*

Control aim. The goal is to design an output feedback continuous controller u which allows to achieve a prescribed task in spite of system perturbations ν. Hence, two cases are considered:

- In the first case, the output regulation control design for systems affected by matched perturbations (i.e., span$(G) \subset$ span(B)) is considered.
- The second approach deals with the more general case of designing an output tracking control for systems affected by unmatched perturbations. □

It is worth mentioning that, the aforementioned goals, rely on the availability of the state and the identified value of the perturbations. Thus, following the developments in Reference 12, a HOSM observer is introduced in order to reconstruct the state vector. Besides, such observer is able to provide for the perturbation vector as well as for the perturbation's successive derivatives up to the order $\rho - 1$.

1.3.4 HOSM observer

The observer design involves two stages. First, the boundedness of the estimation error is achieved by means of a Luenberger auxiliary system. In the second instance, the state vector is reconstructed via a differentiation scheme.

Thus, a dynamic auxiliary system is proposed to bound the observation error, i.e.,

$$\dot{\tilde{x}} = A\tilde{x} + Bu + L(y - C\tilde{x}), \tag{1.3.3}$$

where $\tilde{x} \in \mathbb{R}^n$ is the estimated state, the gain L is designed such that $\tilde{A} := A - LC$ be Hurwitz.

Let $e = x - \tilde{x}$ whose dynamics follows

$$\dot{e} = \tilde{A}e + Gv, \tag{1.3.4}$$

$$y_e = Ce. \tag{1.3.5}$$

Under the assumptions $A2$, the unknown inputs vector ν is bounded. Thus, it is well known that e will have a bounded norm, i.e., $\|e\| < \breve{e}$ for all $t > t_e$.

Now, for the second instance, the error vector will be represented as an algebraic expression of the output and its derivatives. To this aim, a decoupling algorithm is involved in order to get rid of the effects of the unknown input vector ν.

Starting with the definition of $M_1 := C$ and $J_1 := (M_1 G)^{\perp}$, let M_κ be defined in a recursive way in the following form:

$$M_\kappa = \begin{bmatrix} (M_{\kappa-1}G)^{\perp}M_{\kappa-1}\tilde{A} \\ M_1 \end{bmatrix} \quad J_{\kappa-1} = (M_{\kappa-1}G)^{\perp}\begin{bmatrix} J_{\kappa-2} & 0 \\ 0 & I_p \end{bmatrix}.$$

Therefore, the following algebraic expression can be constructed:

$$M_\kappa e = \frac{d^{\kappa-1}}{dt^{\kappa-1}}\begin{bmatrix} J_{\kappa-1} & 0 \\ 0 & I_p \end{bmatrix}\begin{bmatrix} y_e \\ \vdots \\ y_e^{[\kappa-1]} \end{bmatrix}, \tag{1.3.6}$$

where $y_e^{[i]}$ represents the ith anti-differentiator of y_e, i.e., $y_e^{[i]} = \int_0^t \cdots \int_0^{\tau_i} y_e d\tau_i \cdots dt$. Due to $A1$, there exists a unique positive integer $\kappa \leq n$ such that the constant matrix M_κ, generated recursively, satisfies the condition $\operatorname{rank}(M_\kappa) = n$ (see Reference 13). Thus, from the above equation a solution for e exists, i.e.,

$$e = \frac{d^{\kappa-1}}{dt^{\kappa-1}}M_\kappa^{+}\begin{bmatrix} J_{\kappa-1} & 0 \\ 0 & I_p \end{bmatrix}\begin{bmatrix} y_e \\ \vdots \\ y_e^{[\kappa-1]} \end{bmatrix} = \frac{d^{\kappa-1}}{dt^{\kappa-1}}Y. \tag{1.3.7}$$

Thus, a real-time HOSM differentiator can be used to provide exact and finite-time differentiation of Y [3]. It is given by

$$\begin{aligned} \dot{z}_{i,0} &= \lambda_0 \Upsilon^{\frac{1}{\ell+1}} \left\|z_{i,0} - Y_i\right\|^{\frac{\ell}{\ell+1}} \operatorname{sign}(z_{i,0} - Y_i) + z_{i,1}, \\ \dot{z}_{i,j} &= \lambda_j \Upsilon^{\frac{1}{\ell+1-j}} \left\|z_{i,j} - \dot{z}_{i,j-1}\right\|^{\frac{\ell-j}{\ell+1-j}} \operatorname{sign}\left(z_{i,j} - \dot{z}_{i,j-1}\right) + z_{i,j+1}, \\ \dot{z}_{i,\ell} &= \lambda_\ell \Upsilon \operatorname{sign}\left(z_{i,\ell} - \dot{z}_{i,\ell-1}\right), \end{aligned} \tag{1.3.8}$$

with $j = \overline{1, \ell - 1}$, ℓ is the differentiator order. The differentiator inputs Y_i, for $i = \overline{1, n}$, represent the components of vector Y in (1.3.7). A positive sequence of λ_j can be selected as in Reference 3. The gain Υ is a Lipschitz constant of $Y^{(\ell)}$, i.e., $\left\| Y^{(\ell+1)} \right\| < \Upsilon$. Due to assumption $A2$, the higher differentiator order may be $\ell = \kappa + \rho - 1$. Therefore, it can be shown that

$$\Upsilon \geq \left\| \tilde{A} \right\|^{\rho+1} \breve{e} + \|G\| \sum_{j=0}^{\rho} \left\| \tilde{A} \right\|^{\rho+1-j} \breve{\nu}_j. \tag{1.3.9}$$

In Reference 3, it was shown that with the proper choice of the constants λ_j, and Υ for all $j = \overline{0, \ell}$, there is a finite time $t \geq t_o$ such that $z_j = \frac{d^j}{dt^j} Y$ is fulfilled for all $j = \overline{0, \ell}$, $z_j = [\, z_{1,j} \ \cdots \ z_{n,j} \,]^T$.

1.3.4.1 State estimation

Hence, the vector e in (1.3.7) is recovered from the $(\kappa - 1)$th sliding dynamics, i.e., $e = z_{\kappa-1}$ holds for $t > t_o$. Consequently, the next expression holds

$$\hat{x} := z_{\kappa-1} + \tilde{x}, \tag{1.3.10}$$

where $\hat{x} \in \mathbb{R}^n$ is the estimated value of x for all $t > t_o$.

1.3.4.2 Perturbations identification

The perturbations function vector $\nu(t)$, as well as its successive derivatives up to the order $\rho - 1$, can be recovered from (1.3.4). Working out ν it yields to

$$\nu = G^{+}\left(\dot{e} - \tilde{A}e\right), \tag{1.3.11}$$

where $\dot{e}$, e can be obtained from the HOSM differentiator (1.3.8), i.e., the equalities $z_\kappa = \dot{e}$, $z_{\kappa-1} = e$ are accomplished for all $t > t_o$. Hence, taking the successive derivatives of (1.3.11) returns

$$\hat{\nu}^{(j)} = G^{+}\left(z_{\kappa+j} - \tilde{A} z_{\kappa-1+j}\right) \tag{1.3.12}$$

for $j = \overline{0, \rho - 1}$, where $z_{\kappa+j}, z_{\kappa-1+j}$ come from (1.3.8). Therefore, $\hat{\nu}^{(j)} \in \mathbb{R}^q$ represents the estimate of $\nu^{(j)}$ with $j = \overline{0, \rho - 1}$, for all $t > t_o$.

Theoretically, the HOSM observer provides exact state and perturbation estimation, however, the accuracy of the estimations may be spoiled by noise. In the following, the upper bounds of the estimation and identification errors are derived.

1.3.4.3 Precision of the observation and identification processes

Suppose that the observation is realized with a sampling step δ while a deterministic noise signal $\eta(t)$ (a Lebesgue measurable function of time with a maximal magnitude η) is presented in the system output. Let

$$Y(t) = Y_0(t) + \eta(t), \quad \left|Y_0^{\ell+1}(t)\right| < \Upsilon, \quad |\eta(t)| \leq \eta, \tag{1.3.13}$$

where $Y_0(t)$ is an unknown basic signal, $\eta(t)$ is a Lebesgue-measurable sampling noise.

The error provoked by the combined effect of the sampling time δ and the deterministic noise η follows from Reference [15, Theorem 3.1]. Let $\delta < k_\delta \Delta$ and $\eta < k_\eta \Upsilon \Delta^{\ell+1}$ with k_δ, k_η, Δ some positive constants and $\ell = \rho + \kappa - 1$. Then after a finite time, the following estimations are achieved after some finite time:

$$\left|z_j - Y^{(j)}\right| \leq O\left(\Delta^{\ell-j+1}\right), \quad j = \overline{0, \ell}. \tag{1.3.14}$$

Thus, in accordance with the aforementioned theorem, using the differentiator (1.3.8) of order $\ell = \kappa + \rho - 1$, the following estimations are achieved after some finite time.

State estimation error. It comes from (1.3.7) and (1.3.10):

$$\begin{aligned} \left\|x(t) - \hat{x}(t)\right\| &= \left\|Y^{(\kappa-1)}(t) - z_{\kappa-1}\right\| \\ &\leq O\left(\Delta^{\rho+1}\right). \end{aligned} \tag{1.3.15}$$

Perturbations identification error. It follows from (1.3.12) that for the perturbation and its successive derivatives, the error yields to

$$\begin{aligned} \left\|\hat{\upsilon}^{(j)} - \upsilon^{(j)}\right\| &= \left\|G^+\left(z_{\kappa+j} - \tilde{A}z_{\kappa-1+j}\right) - G^+\left(e^{(j+1)} - \tilde{A}e^{(j)}\right)\right\| \\ &\leq \left\|G^+\right\| \left\|z_{\kappa+j} - Y^{(\kappa+j)}(t)\right\| + \left\|G^+\tilde{A}\right\| \left\|z_{\kappa+j-1} - Y^{(\kappa+j-1)}(t)\right\| \\ &\leq O\left(\Delta^{\rho-j}\right) + O\left(\Delta^{\rho-j+1}\right) \leq O\left(\Delta^{\rho-j}\right) \end{aligned} \tag{1.3.16}$$

with $j = \overline{0, \rho - 1}$.

Below, the control of systems subjected to matched perturbation is developed. A compensation control method is proposed. Such control is designed based on the estimated states and the perturbation identification. As an alternative, an output integral sliding mode control (OISMC) is presented. The advantages/drawbacks of both methods are discussed in terms of deterministic noise, sampling step and execution time.

1.3.5 Control of systems affected by matched perturbations

Consider a linear time-invariant system affected by perturbations

$$\begin{aligned} \dot{x}(t) &= Ax(t) + B(u(t) + v(t)), \\ y(t) &= Cx(t), \end{aligned} \tag{1.3.17}$$

where $x(t) \in \mathbb{R}^n$, $u(t) \in \mathbb{R}^m$, $y(t) \in \mathbb{R}^p$ $(1 \le p < n)$ are the state vector, the control and the output of the system, respectively. The disturbances and system uncertainties are represented by the unknown inputs function vector $v(t) \in \mathbb{R}^m$.

The goal is to design a robust output control for system (1.3.17). The control has the following form:

$$u(t) = u_0(t) + u_1(t), \tag{1.3.18}$$

where $u_0(t)$, which may be designed following any control strategy, is the nominal control for the nominal system (i.e., $v(t) = 0$); $u_1(t)$ is a compensator of the unknown input vector $v(t)$.

In order to cope with the matched perturbation, two approaches are discussed: exact compensation and SMC. Later on, a performance comparison for both strategies is addressed.

The control design relies in the estimation of the state and, in the case of the compensation-based control, the identified value of the matched perturbation vector is also involved.

1.3.5.1 Exact output feedback stabilization

Here, a compensation-based control law is designed based on the estimated states and the unknown input identification. It is designated exact output feedback stabilization (EOFS).

Consider the nominal system

$$\dot{x}_0(t) = Ax_0(t) + Bu_0(t). \tag{1.3.19}$$

The signal $u_0(t)$ is a stabilization control law for the nominal system, $u_0(t) = -K_0x(t)$, the gain K_0 can be designed by any control strategy. The second part of the control input (1.3.18) is designed as $u_1(t) = -\hat{v}(t)$, where $\hat{v}(t)$ is the unknown input vector estimated in Section 1.3.4.2.

Theoretically, assuming exact observation and identification, the equalities $\hat{x}(t) \equiv x(t)$ and $\hat{v}(t) \equiv v(t)$ hold after a finite time t_o. Thus, the control law is given by

$$u(t) = -Kx(t) - \hat{v}(t) \tag{1.3.20}$$

applied to system (1.3.17) yields the following dynamic equation: $\dot{x}(t) = (A - BK)x(t)$. It means that, theoretically, the continuous control $u_1(t)$ exactly compensates the matched perturbations and the solutions for systems (1.3.17) and (1.3.19) coincide.

EOFS realization error. Theoretically, perturbations are exactly compensated in finite time. Nevertheless, discretization and deterministic output noise, present in the observation and identification processes, affect the compensation accuracy, as it was shown in Section 1.3.4.3. Furthermore, an additional error, due to the actuator time constant μ, will cause an error of order $O(\mu)$ (see References 1 and 16). Therefore, considering (1.3.15) and (1.3.16) with $j = 0$, the controller stabilization error may be estimated by

$$\varepsilon = O(\mu) + O(\Delta^{\rho+1}) + O(\Delta^{\rho}). \tag{1.3.21}$$

1.3.5.2 Output integral sliding mode control

In this section, an integral SMC is applied (see, e.g., Reference 17). Such controller is based on the estimated states obtained by the HOSM observer; thus, it is an OISMC.

Consider a control input of the form (1.3.18), where $u_0(t)$ is the nominal control for the system without uncertainties (1.3.19). Let the nominal control be $u_0(t) = -K_0 x(t)$. The compensator $u_1(t)$ should be designed to reject the disturbance $\nu(t)$ in the sliding mode on the manifold $\{x\colon s(x,t) = 0\}$ [18], i.e., the equivalent control $u_{1eq}(t) = -\nu(t)$ [1]. The switching function $s(x,t)$ is defined as

$$s(x,t) = B^{+}\left[x(t) - x(t_0) - \int_{t_i}^{t} [Ax(\tau) + Bu_0(x,\tau)]\, d\tau\right],$$

where $t_i \geq t_o$. The compensator $u_1(t)$ is designed as a discontinuous unitary control $u_1(t) = -\varphi \frac{s(x,t)}{\|s(x,t)\|}$. Thus, the sliding mode manifold $s(x,t)$ is attractive from $t_i \geq t_o$ if $\varphi > \breve{\nu}_0 \geq \|\nu\,(t)\|$. Finally, the control law (1.3.18) is designed as follows:

$$u(t) = -Kx(t) - \varphi \frac{s(x,t)}{\|s(x,t)\|}. \tag{1.3.22}$$

Again, in the ideal case, system (1.3.17) with $u(t)$ given by (1.3.22) takes the form of (1.3.19).

OISMC realization error. In the case of the OISMC, the stabilization error is the sum of the observation error plus the control error, i.e.,

$$\varepsilon = O(\mu) + O\left(\Delta^{\rho+1}\right). \tag{1.3.23}$$

1.3.5.3 EOFS and OISMC comparison

As it has been analyzed, in the ideal case both controllers, EOFS and OISCM, give identical results. However, in a practical case, their performance may be spoiled by discretization issues and the presence of deterministic noise in the output. Thus, here it is discussed the accuracy of the HOSM observer and the identification procedure, combined with both control methodologies, EOFS (1.3.21) and OISMC (1.3.23). Recall that $(\rho + \kappa - 1)$th-order HOSM differentiator is involved, moreover $(\kappa - 1)$th and κth derivatives for the state observation and unknown inputs identification are needed, respectively. Consider the following cases:

I. $O(\Delta^{\rho}) \ll O(\mu)$, i.e., the controller execution error is greater than the identification process error. In such a case, it would be suitable to use the EOFS strategy to avoid chattering.

II. $O(\Delta^{\rho+1}) \ll O(\mu) \ll O(\Delta^{\rho})$, i.e., the error related to the actuator time constant is less than the identification process error. Thus, the error in the EOFS control strategy is mainly determined by the identification error. In this case, OISMC strategy could be a better solution for systems tolerant to chattering with oscillation frequencies of order $O\left(\frac{1}{\mu}\right)$ [19].

III. $O(\mu) \ll O(\Delta^{\rho+1})$, i.e., the error provoked by the actuator time constant is less than the observation error. Once again, the precision of the EOFS controller is determined by the precision of the identification process $O(\Delta^{\rho})$, and the precision of the OISMC controller is determined by the accuracy of the observation process $O(\Delta^{\rho+1})$. However, it is necessary to remark that in this case the use of the OISMC controller could amplify the observer noise (see, e.g., Reference 19).

1.3.5.4 Experimental results: inverted rotary pendulum

Here, the experimental results of the inverted rotary pendulum system shown in Figure 1.3.1. An L-shaped arm, or hub, is connected to the DC motor shaft and pivots between ±180°. At the end of the arm, there is a suspended pendulum attached. The experimental setup includes a PC equipped with a real-time dSPACE acquisition platform with a minimum sample and integration time equal to 20 μs. The pendulum and hub angles $x_1(t)$, $x_2(t)$ are measured by two encoders with a resolution of 1024 pulses per rate and the motor driver bandwidth is 83 μs.

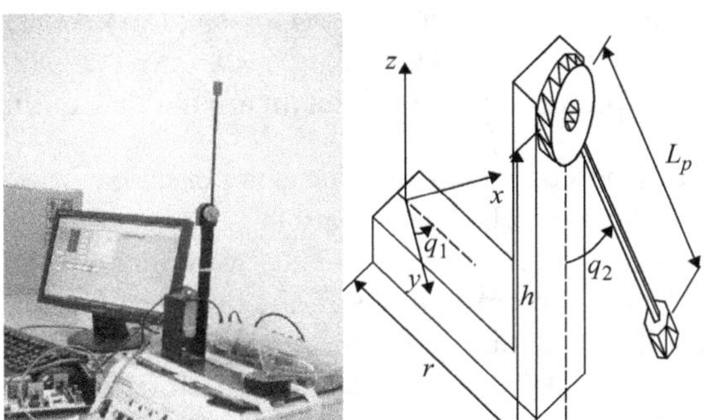

Figure 1.3.1 Inverted rotary pendulum

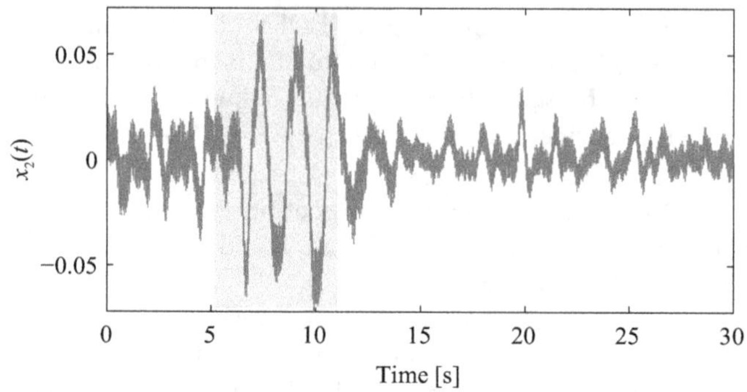

Figure 1.3.2 Pendulum position $x_2(t)$ [rad]

The system state equations with $x_3(t) = \dot{x}_1(t)$ and $x_4(t) = \dot{x}_2(t)$ linearized along the point $x = \begin{bmatrix} x_1 & \pi & 0 & 0 \end{bmatrix}$ are

$$\begin{aligned}
\dot{x}_1(t) &= x_3(t), \\
\dot{x}_2(t) &= x_4(t), \\
\dot{x}_3(t) &= 82.94x_2(t) - 1.31x_3(t) + 46.75(u(t) + \nu(t)), \\
\dot{x}_4(t) &= 56.81x_2(t) - 0.37x_3(t) + 46.75(u(t) + \nu(t)).
\end{aligned}$$

A second-order HOSM differentiator is designed (i.e., $\ell = 2$) with gain $\Upsilon = \begin{bmatrix} 50 & 50 & 500 & 200 \end{bmatrix}^T$. The state is recovered using the first derivative (i.e., $\kappa = 2$). The nominal control gain is given by $K_0 = \begin{bmatrix} 3.16 & -73.92 & 2.19 & -12.48 \end{bmatrix}$. First, an experiment was done to show the compensation effect using the control in (1.3.20) with $\delta = 20\,\mu\text{s}$ and $\mu = 400\,\mu\text{s}$. Recall that the state vector x and the disturbances ν are reconstructed using the HOSM observer. The pendulum position $x_2(t)$ is shown in Figure 1.3.2. The experiment stages are the following:

a. $0 \leq t < 5.1$, the nominal control signal is acting (i.e., $u(t) = -Kx(t)$);
b. $5.1 < t < 11$, a sinusoidal signal is added to the motor voltage acting as an exogenous matched perturbation (i.e., $\nu(t) = \sin(1.5t) + 0.3$);
c. $11 < t$, the compensator is added (i.e., $u(t) = -K_0x(t) - \hat{\nu}(t)$). It considerably improves the performance of the pendulum around the equilibrium point.

Therefore, it can be concluded that the compensation based control ensures robustness against smooth matched perturbations.

Furthermore, three different experiments were done considering different pairs of values for δ and μ corresponding to the three main cases aforementioned. Figure 1.3.3 shows the pendulum position $x_2(t)$ contrasting both methodologies: EOFS and OISMC. Table 1.3.1 summarizes the performance of the system with respect to the control philosophy as well as the sample and execution times.

In the following section, a generalization of the method to counteract not only the matched but also the unmatched perturbations is presented. To this aim, a dynamic

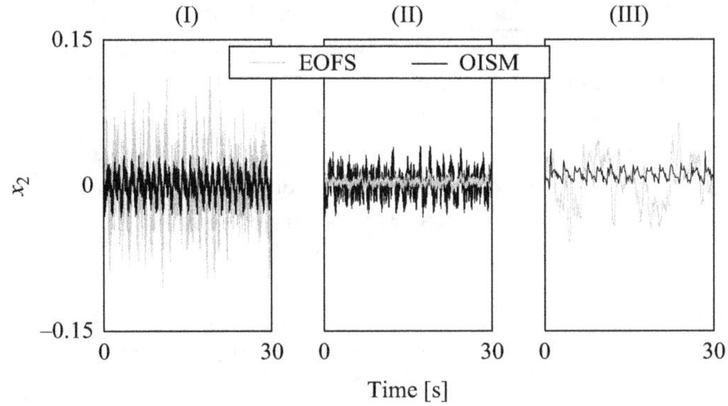

Figure 1.3.3 Comparison of the pendulum position $x_2(t)$ [rad] between EOFS and OISMC for three pairs of (δ, μ)

Table 1.3.1 Performance of the system for three sample and execution times

	μs	EOFS	OISMC
I	$\delta = 20$ $\mu = 400$	Compensates with better accuracy than OISMC	Less efficient than EOFS
II	$\delta = 500$ $\mu = 80$	Poor identification accuracy means high compensation error	Sensitive only to the observation accuracy
III	$\delta = 1000$ $\mu = 80$	Its efficiency is dominated by the identification accuracy	The chattering increases the stabilization error

sliding mode surface is proposed considering the estate estimations as well as the identified value of the perturbation and a certain number of its successive derivatives. Toward a better understanding of the method, the system (1.3.1) must be transformed into a convenient block-control form (see References 20 and 21).

1.3.6 Control of systems affected by unmatched perturbations

Consider a strict feedback form representation of system (1.3.1) (see References 20 and 21). Due to the pair (A, B) is controllable, system (1.3.1) can be always reduced to the block-controllable form, which is composed into r connected sub-systems, i.e.,

$$\dot{x}_1 = A_1 x_1 + B_1(x_2 + \Gamma_1 \nu), \tag{1.3.24}$$

$$\dot{x}_i = A_i \bar{x}_i + B_i(x_{i+1} + \Gamma_i \nu), \tag{1.3.25}$$

$$\dot{x}_r = A_r \bar{x}_r + B_r(u + \Gamma_r \nu) \tag{1.3.26}$$

for $i = \overline{2, r-1}$, where $\bar{x}_i = [x_1^T \; \cdots \; x_i^T]^T$, $x_i \in \mathbb{R}^{n_i}$, $n_i = \operatorname{rank}(B_i)$, $\sum_{i=1}^{r} n_i = n$. The state vector $x \in \mathbb{R}^n$ is given by $\bar{x} = [x_1^T \; x_i^T \; \cdots \; x_r^T]^T$. The measured output signal is given by (1.3.2) while the controlled output is given by $x_1 \in \mathbb{R}^{n_1}$. The sub-systems (1.3.24) and (1.3.25) for $i = \overline{1, r-1}$ represent the sub-actuated dynamics and (1.3.26) corresponds to the actuated dynamics thus, $x_r \in \mathbb{R}^m$. The matrices A_i, Γ_i are of the corresponding dimensions. Further details of the transformation can be found in Reference 22.

The goal is to design an output feedback continuous sliding mode controller u which allows the coordinate x_1 to track a smooth signal x_d in spite of system perturbations ν.

Comment 1.3.1. *It is worth mentioning that the tracking output x_1 may not be measurable. Nevertheless, it can be estimated using the HOSM observer.*

The control design relies on the availability, in finite time, of the exact estimation of the state and the identification of the perturbation and its successive derivatives until $r-2$th order. Thus, from (1.3.10), after $t > t_o$, the identities $\hat{x}_1 = x_1, \ldots, \hat{x}_r = x_r$ are certainly obtained.

In order to achieve the control design, first, the design of the sliding surface is realized in $r-1$ steps, it is carried out such that the sub-actuated dynamics (1.3.24) and (1.3.25) attains a desired behavior in spite of the perturbations. Then, at the rth step, the control law is conceived such that the state trajectories reach the sliding surface and there remain.

It is assumed, without loss of generality, that the successive derivatives of the desired trajectory x_d are available.

***Step* 1.** Starting with the sub-system (1.3.24), the coordinate x_2 can be exploited as a fictitious control input $x_2 := \phi_1$,

$$\phi_1 = -\Gamma_1 \hat{\nu} - B_1^{\dagger}\left(A_1 x_1 - \hat{A}_1(x_1 - x_d) - \dot{x}_d\right), \tag{1.3.27}$$

where $\hat{A}_1 \in \mathbb{R}^{n_1}$ is a Hurwitz matrix containing the desired convergence performance of x_1 toward x_d. Notice that since $\mathrm{rank}(B_i) = n_i$, then $B_i B_i^T$ is invertible, thus $B_i^{\dagger} = B_i^T (B_i B_i^T)^{-1}$ is the right-inverse matrix of B_i.

Step i. The coordinate x_{i+1} is a fictitious control input for (1.3.25), it is $x_{i+1} := \phi_i$

$$\phi_i = -\Gamma_i \hat{\nu} - B_i^{\dagger}\left(A_i \bar{x}_i - \hat{A}_i(x_i - \phi_{i-1}) + X_{i-1}(x_{i-1} - \phi_{i-2}) - \dot{\phi}_{i-1}\right) \tag{1.3.28}$$

for all $i = \overline{2, r-1}$, $\hat{A}_i \in \mathbb{R}^{n_i}$ is a Hurwitz matrix and $X_{i-1} = P_i^{-1} B_i^T P_{i-1}$, with P_i a positive definite matrix satisfying $P_i \hat{A}_i + \hat{A}_i^T P_i = -I$.

Finally, the sliding surface s is designed like

$$s = x_r - \phi_{r-1} \tag{1.3.29}$$

with $s \in \mathbb{R}^m$, $s = [\, s_1 \; \ldots \; s_m \,]^T$.

Step r. The control law takes the form

$$u = -B_r^{\dagger}\left(A_r \bar{x}_r - \dot{\tilde{\phi}}_r - B_{r-1}^T(x_{r-1} - \phi_{r-2}) + \hat{A}_r s + \vartheta\right) - \Gamma_r \hat{\nu}, \tag{1.3.30}$$

where $\dot{\tilde{\phi}}_r$ is a linear combination of the state x and $\hat{\nu}, \ldots, \hat{\nu}^{(r-2)}$, such that $\omega := \dot{\tilde{\phi}}_r - \dot{\phi}_{i-1} = B_{r-1}^{\dagger} \cdots B_2^{\dagger}\Gamma_1 \nu^{(r-1)}$. The control $\vartheta \in \mathbb{R}^m$ is conceived to steer the system trajectories to the sliding surface. Here, the super-twisting control (see Reference 23) is considered. It is given by

$$\vartheta = K_1\Psi_1 + K_2 \int_{t_o}^{t} \Psi_2 \, dt \tag{1.3.31}$$

with the matrices $K_j = \mathrm{diag}\big(k_{j1}, \ldots, k_{jm}\big)$, for $j = \overline{1,2}, \Psi_1 = \mathrm{diag}(\psi_{11}, \ldots, \psi_{1m})$ and $\Psi_2 = \mathrm{diag}(\psi_{21}, \ldots, \psi_{2m})$, where $\psi_{1j} = \|s_j\|^{\frac{1}{2}} \operatorname{sign}(s_j)$ and $\psi_{2j} = \operatorname{sign}(s_j)$ for $j = \overline{1,m}$.

The components of the matrices gains K_1 and K_2 are given as

$$k_{1j} > 0 \qquad k_{2j} > 3\breve{\omega} + 2\left(\frac{\breve{\omega}}{k_{j1}}\right)^2 \tag{1.3.32}$$

for $j = \overline{1,m}$, the constant $\left\|B_{r-1}^{\dagger} \cdots B_2^{\dagger}\Gamma_1 \nu^{(r)}\right\| \leq \breve{\omega}$. □

Now, the stability of the closed-loop system is shown. Beforehand, notice that x_{i+1} is exploited as a fictitious control for the coordinate x_i. Thus, the following error variables $\sigma_1 = x_1 - x_d, \sigma_i = x_i - \phi_{i-1}$ for $i = \overline{2, r-1}$ are introduced. Therefore, differentiating with respect to time every σ_i for $i = \overline{1, r-1}$, as well as the sliding surface s in (1.3.29) and taking into account (1.3.24)–(1.3.26), it results in

$$\begin{aligned}
\dot{\sigma}_1 &= A_1 x_1 + B_1(\phi_1 + \Gamma_1\nu) - \dot{x}_d + B_1\sigma_2, \\
\dot{\sigma}_i &= A_i\bar{x}_i + B_i(\phi_i + \Gamma_i\nu) - \dot{\phi}_{i-1} + B_i\sigma_{i+1}, \\
\dot{\sigma}_{r-1} &= A_{r-1}\bar{x}_{r-1} + B_{r-1}(\phi_r + \Gamma_{r-1}\nu) - \dot{\phi}_{r-1} + B_{r-1}s, \\
\dot{s} &= A_r\bar{x}_r + B_r(u + \Gamma_r\nu) - \dot{\phi}_r.
\end{aligned} \tag{1.3.33}$$

First, it is shown that the super-twisting controller drives the state trajectories to the sliding surface. For the sake of simplicity, the convergence of the individual components of the sliding surface is studied. Considering the last row in (1.3.33) and applying the control law (1.3.30), it yields to $\dot{s}_j = -k_{j1}\psi_{1j} - k_{j2}\int_{t_o}^{t}\psi_{2j}\,dt - \omega_j$ with $j = \overline{1,m}$; ω_j, s_j, represent the components of vectors s, ω.

The vector $\omega = -B_{r-1}^{\dagger} \cdots B_2^{\dagger}\Gamma_1\nu^{(r-1)}$ is an uncertainty due to the effects of the unmatched perturbation in the actuated dynamics. Defining $\mu_j = -k_{j2}\int_{t_o}^{t}\psi_{2j}dt + \omega_j$, yields to

$$\dot{s}_j = -k_{j1}\psi_{1j} + \mu_j, \tag{1.3.34}$$

$$\dot{\mu}_j = -k_{j2}\psi_{2j} + \dot{\omega}_j \tag{1.3.35}$$

with $\dot{\omega} = B_{r-1}^{\dagger} \cdots B_2^{\dagger}\Gamma_1\nu^{(r)}$ which under $A3$ is bounded, i.e., there exists a constant $\breve{\omega}$ such that $\|\dot{\omega}\| \leq \breve{\omega}$. In Reference 24, it was stated that the auxiliary system (1.3.35) can be stabilized by selecting the controller gains as in (1.3.32) achieving $s = 0$ in finite time.

In the second instance, given that $s = \dot{s} = 0$, the convergence of the reduced dynamics is exposed in the following. A composite Lyapunov function candidate, for the reduced order dynamics, is proposed as $V_\sigma = \sum_{i=1}^{r-1} \sigma_i^T P_i \sigma_i$. Thus, taking the Lyapunov function derivative with respect to time, it yields to

$$\begin{aligned}\dot{V}_\sigma &= 2\sigma_1^T \left(x_1^T \left(A_1 x_1 + B_1 \left(\phi_1 + \Gamma_1 \nu\right) - \dot{x}_1^d + B_1 \sigma_2\right)\right.\\ &\quad + 2\sum_{i=2}^{r-2} \sigma_i^T P_i \left(A_i \bar{x}_i + B_i(\phi_i + \Gamma_i \nu) - \dot{\phi}_{i-1} + B_i \sigma_{i+1}\right) \qquad (1.3.36)\\ &\quad + 2\sigma_{r-1}^T P_{r-1}\left(A_{r-1}\bar{x}_{r-1} + B_{r-1}(\phi_{r-1} + \Gamma_{r-1}\nu) - \dot{\phi}_{r-2}\right) + B_{r-1} s\right)\end{aligned}$$

due to $s \equiv 0$ was previously established, it follows from (1.3.29) that $x_r \equiv \phi_{r-1}$ and, in virtue of (1.3.27) and (1.3.28), the derivative of V_σ yields to

$$\dot{V}_\sigma = -\sum_{i=1}^{r-1} \|\sigma_i\|^2 \qquad (1.3.37)$$

as a result, every σ_i for $i = \overline{r-1, 1}$ converges asymptotically to zero. Consequently, $x_i \to \phi_{i-1}$ for $i = \overline{r-1, 2}$. Therefore, $x_1 \to x_d$.

Control realization error. The error provoked by discretization, deterministic output noise and actuator time constant may be estimated from (1.3.15) and (1.3.16), resulting in

$$\breve{\varepsilon} = O(\mu) + O\left(\Delta^{\rho+1}\right) + O(\Delta). \qquad (1.3.38)$$

Comment 1.3.2. *As a consequence, the closed-loop stability of the sub-actuated dynamics in (1.3.37) is impacted with the presence of the realization error. Thus, the Lyapunov derivative yields to*

$$\dot{V}_\sigma \le -\sum_{i=1}^{r-1} \|\sigma_i\| \left(\|\sigma_i\| - \breve{\varepsilon}\right). \qquad (1.3.39)$$

Let $D_{\breve{\varepsilon}} = \{\sigma_i : \|\sigma_i\| < \breve{\varepsilon}\}$ represent a disc centered at the origin. Then, from (1.3.39) it follows that for $\sigma_i \notin D_{\breve{\varepsilon}}$ $\dot{V}_\sigma < 0$. Therefore, there exists a $t_\sigma > t_o$ such that the error coordinates $\sigma_i \in D_\varepsilon$, for all $i = \overline{1, r-1}$ and $t > t_\sigma$. In other words, the tracking error coordinate σ is ultimately bounded with respect to D_ε (see Reference 2).

1.3.6.1 3-DOF helicopter example

Consider a 3-DOF helicopter (see Reference 25). The system is composed by two axles: the body and the arm. The body is a small arm with a propeller at each end. This is connected to a swinging axle connected in a fixed base (see Figure 1.3.4). The helicopter position is characterized by the elevation ε_1, pitch ε_2 and travel ε_3 movements. The system can rotate freely around three axes. The control variables are the input voltages of the propellers.

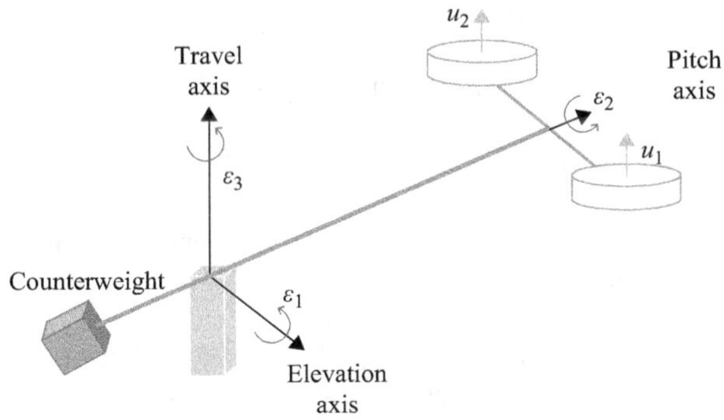

Figure 1.3.4 Schematic diagram of a 3-DOF helicopter

Table 1.3.2 3-DOF helicopter parameters

Symbol	Description	Unit
J_1	Moment of inertia around the elevation axis	kg m^2
J_2	Moment of inertia around the pitch axis	kg m^2
J_3	Moment of inertia around the travel axis	kg m^2
K_f	Force constant of the propeller	N
K_p	Proportionality constant of gravitational force	N
ℓ_a	Distance between the base and the helicopter body	m
ℓ_h	Distance from the elevation axis to either motor	m
T_g	Effective gravitational torque	N

The helicopter parameters are enlisted in Table 1.3.2.
A linearized model around $\varepsilon_2^* = 0$ is given by

$$\begin{aligned} \ddot{\varepsilon}_1 &= 0.45((u_1 + \nu_1) + (u_2 + \nu_2)), \\ \ddot{\varepsilon}_2 &= 3.05((u_1 + \nu_1) - (u_2 + \nu_2)), \\ \ddot{\varepsilon}_3 &= -0.49\varepsilon_2 + \nu_3. \end{aligned} \tag{1.3.40}$$

$$y = \begin{bmatrix} \varepsilon_1 & \varepsilon_2 & \varepsilon_3 \end{bmatrix}^T, \tag{1.3.41}$$

where $\varepsilon_1, \varepsilon_2, \varepsilon_3 \in \mathbb{R}$, $u \in \mathbb{R}^2$. The system is subjected to perturbations $\nu = [\nu_1 \quad \nu_2 \quad \nu_3]$. Where the perturbations ν_1, ν_2 could be considered as additive actuator faults. Whereas an unmatched disturbance signal given by ν_3 is acting on the travel dynamics. The measured outputs correspond to the angular positions (1.3.41) .

The goal is that the travel coordinate ε_3 tracks a desired trajectory $x_d = 5\sin(1.1t)$ in spite of the faults and disturbances affecting the system.

The simulation trial consists in four stages:

a. $0 \leq t < 60$, the system is working free of perturbations (i.e., $\nu_1, \nu_2, \nu_3 = 0$).
b. $60 \leq t$, an unmatched disturbance occurs $\nu_3 = 2\sin 3t + 1.5$.
c. $100 \leq t$, a type of liquid oscillatory fault occurs in the first actuator $\nu_1 = 5\cos(1.5t) + 5$.
d. $140 \leq t$, a drifting fault (i.e., $\nu_2 = \beta t : \beta > 0$) appears in the second actuator. The simulation sampling time is 100 μs.

It is worth noting to mention that according to aeronautics terminology, liquid failures correspond to additive oscillatory faults (see Reference 26).

Observer design. First, the state vector is selected as $x = [\varepsilon_1 \ \ \varepsilon_2 \ \ \varepsilon_3 \ \ \dot{\varepsilon}_1 \ \ \dot{\varepsilon}_2 \ \ \dot{\varepsilon}_3]^T$. Then, it can be easily shown that for the obtained state space representation, the triplet (A, C, G) is strongly observable. Thus, regarding (1.3.4), an auxiliary system matrix $\tilde{A}$ with eigenvalues $-\{1.6, 1.5, 1.8, 1.9, 1.7, 1.1\}$ is considered. Subsequently, for $\kappa = 2$ iterations, a full column rank matrix $M_2 = [-21 \ \ 0 \ \ 0 \ \ 1 \ \ 0 \ \ 0; \ 0 \ \ -13 \ \ 0 \ \ 0 \ \ 1 \ \ 0; \ C]$ is found.

The differentiator in (1.3.8) has an order $\ell = 3$, such that $\hat{x}$, $\hat{\nu}$, and $\hat{\nu}^{(1)}$ can be estimated. The bounds are $\breve{\nu} = 18$, $\breve{\nu}_1 = 6$, $\breve{\nu}_2 = 18$. Then, $\Upsilon = 150$, $\lambda_i = \{1.1, 1.5, 3, 5\}$.

Control design. The controller aim is that $\varepsilon_3 \to x_d$. Moreover, the elevation angle is kept at $\varepsilon_o = 15°$; so, the variables $\tilde{\varepsilon}_1 := \varepsilon_1 - \varepsilon_o$ and $\varepsilon_4 = \int_{t_0}^{t} \tilde{\varepsilon}_1 \, dt$ are introduced. As a consequence, an extended state vector is considered $x = [\tilde{\varepsilon}_1 \ \ \varepsilon_2 \ \ \varepsilon_3 \ \ \dot{\varepsilon}_1 \ \ \dot{\varepsilon}_2 \ \ \dot{\varepsilon}_3 \ \ \varepsilon_4]^T$. This extended system can be transformed into (1.3.24)–(1.3.26). Following the coordinate transformation given in Reference 22 with $x_1 = \varepsilon_3$, $x_2 = [\dot{\varepsilon}_3 \ \ \varepsilon_4]^T$, $x_3 = [0.98 \ \ 0.14; \ -0.14 \ \ -0.98][\tilde{\varepsilon}_1 \ \ \varepsilon_2]^T$ and $x_4 = [\dot{\varepsilon}_1 \ \ -\dot{\varepsilon}_2]^T$, it yields to

$$\dot{x}_1 = B_1 x_2, \tag{1.3.42}$$

$$\dot{x}_2 = B_2(x_3 + \Gamma_2 \nu), \tag{1.3.43}$$

$$\dot{x}_3 = B_3 x_4, \tag{1.3.44}$$

$$\dot{x}_4 = B_4(u + \Gamma_4 \nu), \tag{1.3.45}$$

where $B_1 = [1 \ \ 0]$, $B_2 = [0.49 \ \ 0.07; \ \ -0.14 \ \ 0.98]$, $\Gamma_2 = [1.99 \ \ -0.29]^T$, $B_3 = [0.14 \ \ 0.98; \ \ -0.98 \ \ 0.14]$ and $B_4 = [-0.45 \ \ -0.45; \ \ -3.05 \ \ 3.05]$.

The fictitious controls ϕ_i for $i = \overline{1, 3}$ are computed following (1.3.27) and (1.3.28). The sliding surface is thus designed as $s = x_4 - \phi_3$. The control signal (1.3.30) has $\hat{A}_1 = -1.8$, $\hat{A}_2 = \text{diag}(-2, -3)$, $\hat{A}_3 = \text{diag}(-4, -5)$ and $\breve{\omega} = 160$. The super-twisting gains (1.3.32) are $k_{1i} = 10$, $k_{2i} = 600$. For comparison purposes, a linear quadratic regulator (LQR) control law $u = -K_n x$ that involves full state measurements is considered. The LQR controller is designed for a nominal system without faults and disturbances (see Reference 25). The LQR control keeps the elevation at 15° while the travel coordinate tracks the desired trajectory x_d. Augmenting the state with $\tilde{\varepsilon}_3 = \varepsilon_3 - x_d$ and $\varepsilon_5 := \int \tilde{\varepsilon}_3$, i.e., $x = [\tilde{\varepsilon}_1 \ \ \varepsilon_2 \ \ \tilde{\varepsilon}_3 \ \ \dot{\varepsilon}_1 \ \ \dot{\varepsilon}_2 \ \ \dot{\tilde{\varepsilon}}_3 \ \ \varepsilon_4, \ \ \varepsilon_5]^T$.

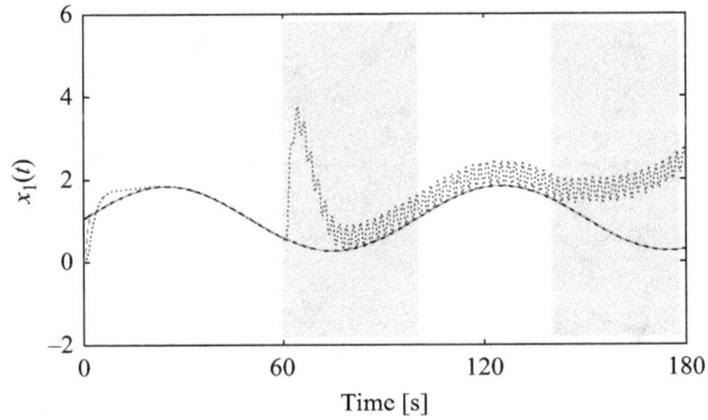

Figure 1.3.5 Travel tracking performance: reference signal x_{1d} (solid line), proposed control (dashed line) and LQR control (dotted line)

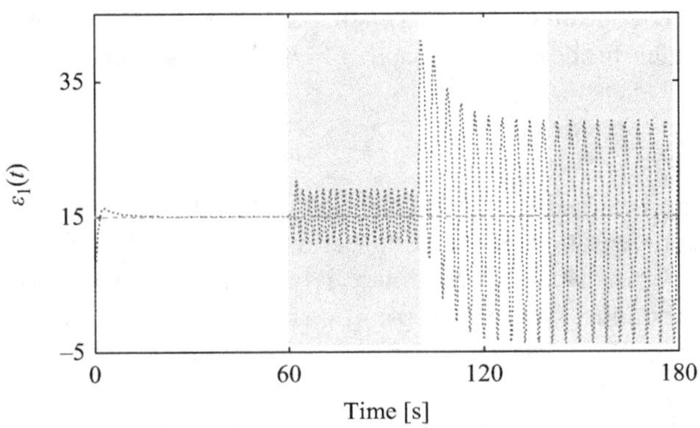

Figure 1.3.6 Elevation level: proposed control (dashed line) vs. LQR (dotted line)

The controller gain yields to $K_n = [15.6 \quad 9 \quad -7.8 \quad 9.7 \quad 3.6 \quad -20.5 \quad 2.2 \quad -1; 15.6 \quad -9 \quad 7.8 \quad -9.7 \quad -3.6 \quad 20.5 \quad 2.2 \quad 1]$.

Figure 1.3.5 shows the tracking performance with the four trial steps highlighted. It can be seen that both controllers achieve the desired travel goal during the nominal stage, see (a). However, when disturbances and/or faults act on the system, refer to the stages (b)–(d), the compensation-based controller (dashed line) is capable to attain the prescribed goal with a better performance than the LQR controller (dotted line).

Figure 1.3.6 shows the controllers' capability to maintain the prescribed elevation level (i.e., 15°).

Figure 1.3.7 shows the effects on the tracking accuracy provoked by the sampling time selection.

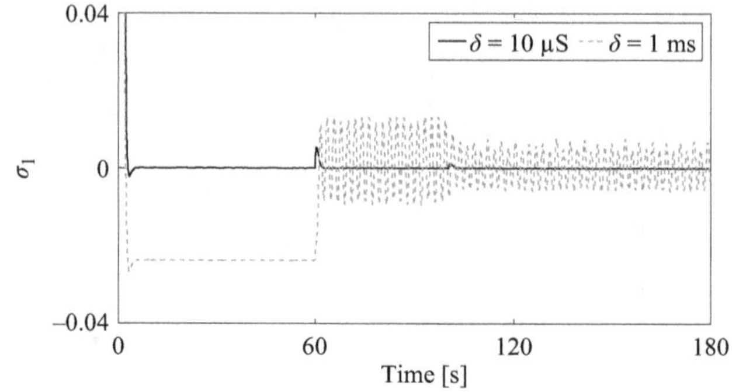

Figure 1.3.7 Tracking error accuracy considering different sampling steps: $\delta = 10\,\mu s$ and $\delta = 1\,ms$

1.3.7 Conclusions

The output-based control of linear systems affected by perturbations is studied. The methodologies proposed are based on HOSM observation and identification techniques.

Two problems are considered:

1. The regulation control of systems affected by matched perturbations by means of the exact compensation of the perturbations. Moreover, an OISMC is also proposed. The performance of both controllers is estimated in terms of deterministic noise upper bounds, sampling step and execution time. A guideline for the proper control philosophy selection based on hardware characteristics was proposed.
2. The tracking control problem of systems affected by matched and unmatched perturbations is proposed. A dynamic sliding mode surface copes with the effects of the unmatched perturbations while a continuous SMC steers the trajectories to the sliding surface. An estimated value of the tracking precision is given in a similar way that in the aforementioned matched case.

The feasibility of the methods was validated experimentally in an inverted rotary pendulum and in simulations for a 3-DOF helicopter.

References

[1] V. I. Utkin. *Sliding Modes in Control and Optimization.* Springer-Verlag, Berlin, 1992.
[2] C. Edwards and S. K. Spurgeon. *Sliding Mode Control.* Taylor & Francis, London, 1998.
[3] A. Levant. Higher-order sliding modes, differentiation and output-feedback control. *International Journal of Control*, 76:924–941, 2003.

[4] F. Castanos and L. Fridman. Analysis and design of integral sliding manifolds for systems with unmatched perturbations. *IEEE Transactions on Automatic Control*, 51(5):853–858, May 2006.

[5] X. G. Yan, C. Edwards, and S. K. Spurgeon. Dynamic sliding mode control for a class of systems with mismatched uncertainty. *European Journal of Control*, 11:1–10, 2005.

[6] H. H. Choi. Output feedback variable structure control design with an H_2 performance bound constraint. *Automatica*, 44:2403–2408, 2008.

[7] J.-L. Chang. Dynamic output integral sliding-mode control with disturbance attenuation. *IEEE Transactions on Automatic Control*, 54(11):2653–2658, 2009.

[8] X. Yu, J. Yan, and S. Li. Sliding-mode control for systems with mismatched uncertainties via a disturbance observer. *IEEE Transactions on Industrial Electronics*, 60:160–169, 2013.

[9] J. Davila. Exact tracking using backstepping control design and high-order sliding modes. *Transactions on Automatic Control*, 58:2077–2081, 2013.

[10] A. Ferreira, F. J. Bejarano, and L. Fridman. Robust control with exact uncertainties compensation: with or without chattering? *Transactions on Control Systems Technology*, 19(5):969–975, 2011.

[11] A. Ferreira de Loza, J. Cieslak, D. Henry, A. Zolghadri, and L. Fridman. Output tracking of systems subjected to perturbations and a class of actuator faults based on HOSM observation and identification. *Automatica*, 59:200–205, 2015.

[12] F. J. Bejarano and L. Fridman. High order sliding mode observer for linear systems with unbounded unknown inputs. *International Journal of Control*, 9:1920–1929, 2010.

[13] L. Fridman, J. Davila, and A. Levant. High-order sliding-mode observation for linear systems with unknown inputs. *Nonlinear Analysis: Hybrid Systems*, 5:337–347, 2011.

[14] B. P. Molinari. A strong controllability and observability in linear multivariable control. *IEEE Transactions on Automatic Control*, 21:761–764, 1976.

[15] A. Levant, M. T. Angulo, and L. Fridman. Robust exact finite-time output based control using high-order sliding modes. *International Journal of Systems Science*, 42:1847–1857, 2011.

[16] L. Fridman. Singularly perturbed analysis of chattering in relay control systems. *IEEE Transactions on Automatic Control*, 47:2079–2084, 2002.

[17] V. I. Utkin, J. Guldner, and J. Shi. *Sliding Modes in Electromechanical Systems*. Taylor & Francis, London, 1999.

[18] B. Drazenovic. The invariance conditions in variable structure systems. *Automatica*, 5:287–295, 1969.

[19] I. Boiko. *Discontinuous Control Systems: Frequency-Domain Analysis and Design*. Birkhäuser, Boston, MA, 2009.

[20] A. Louk'yanov and V. Utkin. Methods of reducing equations for dynamic systems to a regular form. *Automation and Remote Control*, 42(4):413–420, 1993.

[21] S. V. Drakunov, D. B. Izosimov, A. G. Luk'yanov, and V. I. Utkin. Block control principle I. *Automation and Remote Control*, 51:601–609, 1990.

[22] A. Polyakov. Nonlinear feedback design for fixed-time stabilization of linear control systems. *IEEE Transactions on Automatic Control*, 57:2106–2110, 2012.

[23] A. Levant. Sliding order and sliding accuracy in sliding mode control. *International Journal of Control*, 58:1247–1263, 1993.

[24] J. A. Moreno and M. Osorio. Strict Lyapunov functions for the super-twisting algorithm. *IEEE Transactions on Automatic Control*, 57:1035–1040, 2012.

[25] Quanser Inc. *3-DOF Helicopter Reference Manual, Document Number 644, Revision 2.1.*

[26] A. Zolghadri, D. Henry, J. Cieslak, D. Efimov, and P. Goupil. *Fault Diagnosis and Fault-Tolerant Control and Guidance for Aerospace Vehicles – From Theory to Application*. Springer-Verlag, London, 2013.

Chapter 1.4
Construction of Lyapunov functions for high-order sliding modes

Tonámetl Sánchez and Jaime A. Moreno

Abstract

High-order sliding modes (HOSM) are able to control highly uncertain systems effectively, providing very good properties to the closed loop system. They have been to date analyzed and designed using basically geometric and homogeneity tools. However, modern control theory for nonlinear systems use the Lyapunov and Lyapunov-like functions as a basic tool for analysis and control. Despite this fact, only recently the Lyapunov theory has been used for HOSM. The aim of this work is to present, in an informal manner, some recent methods to construct Lyapunov functions for HOSM. These constructive methods will be exemplified with some classical HOSM algorithms, such as twisting, terminal, and super-twisting.

1.4.1 Introduction

Lyapunov's direct method has become one of the main tools in the analysis and design of modern control systems [1, 8]. Beyond stability, it has also natural connections to optimality and robustness. Although originally used by Lyapunov as an stability analysis tool, it has become a standard to use it for design by means of control Lyapunov functions [6]. Recall the techniques of Lyapunov redesign or the backstepping methods that are now standard [8].

Sliding mode control [30] is a very effective control strategy to deal with uncertainties and perturbations in nonlinear systems, providing finite time convergence to a sliding manifold. By keeping the trajectories restricted to evolve on the sliding manifold by means of a discontinuous control signal, the dynamics of the reduced system is decoupled of the perturbation/uncertainty. Traditional sliding modes (first-order sliding modes (FOSMs)) require the sliding variable to have relative degree 1 with respect to the control variable. The design of the discontinuous control for the reaching phase uses the distance to the sliding surface as a (control) Lyapunov function. High-order sliding modes (HOSM) allow the sliding variable to have an arbitrary relative degree with respect to the control variable, they offer more flexibility in the design [7, 10–13, 29], and they provide also very nice properties to the controlled system, such as finite time stability and insensitivity to uncertainties/perturbations. In contrast to the FOSM Lyapunov functions are not the main tool for the design of HOSM. Basically, geometric methods and the theory of homogeneous systems [1] have been used in the analysis and design of HOSM [7, 10–13, 29]. Only recently, some attempts to use Lyapunov methods for HOSM have been reported in the literature [16, 18, 21, 22, 25, 27]. They are basically restricted to second-order systems due to the well-known difficulty of constructing Lyapunov functions for nonlinear systems, and in particular for discontinuous systems.

In the last 50 years, many methods to construct Lyapunov functions have been reported in the literature. Some classical results are given by Krasovskii's method [9], the variable gradient method [28] and Zubov's method [32]. They range between the very analytic methods (as Zubov's method in reference 32) and the more numerical-based ones (see, e.g., reference 2 and their references). For HOSM, there are however, very few constructive approaches. In references 21 and 22, the authors extend the idea of Zubov [32], solving a partial differential equation, to differential inclusions, and they are able to build Lyapunov functions for some second-order HOSM algorithms.

The objective of this chapter is to present some attempts performed by the authors to construct Lyapunov functions for HOSM that have been reported previously [14, 25, 26]. We provide an informal presentation of the basic ideas of these strategies. Instead of presenting the technical details, the techniques are illustrated by calculating explicitly some examples. The techniques are based on the three general ideas, which we describe (informally) in the following paragraphs.

Considering the dynamical system $\dot{x} = f(x), x \in \mathbb{R}^n$, whose origin is an asymptotically stable equilibrium point. Converse Lyapunov's theorem [8] assure the existence of a continuously differentiable positive definite function $V(x)$ and a positive definite function $W(x)$ such that the derivative of V along the trajectories of the system, i.e., $\dot{V}(x) = \frac{\partial V}{\partial x}(x) \cdot f(x) = -W(x)$. Finding a Lyapunov function amounts to solving the differential equation $\dot{V} = -W(x)$, where $V, W : \mathbb{R}^n \to \mathbb{R}$ are positive definite functions. In general, we can identify three main procedures to solve that differential equation:

Method 1. Trajectory integration method. Looking at the equation $\dot{V}(x) = -W(x)$ as a time differential equation, it is possible to obtain the function $V(x)$ by integrating $W(x)$ along the system's solution $\phi(t; t_0, x)$, where the pair (t_0, x) is the initial condition, and thus

$$V(x) = \int_{t_0}^{T} W(\phi(\tau; t_0, x)) \, d\tau.$$

Since this idea requires the knowledge of the trajectory of the system to be feasible, what is seldom the case, the method usually finds application in the proof of converse Lyapunov theorems [8]. For HOSM, this method is feasible, since many HOSM algorithms can be expressed as piecewise linear systems, for which it is easy to compute the trajectories.

Method 2. Partial differential equation method. After proposing a positive definite function W, a Lyapunov function V can be found by solving the partial differential equation:

$$\frac{\partial V(x)}{\partial x} f(x) = -W(x). \tag{1.4.1}$$

Method 3. Function parametrization method. Solving the PDE (1.4.1) is in general a difficult task, except for some particular cases. If the Lyapunov function $V(x)$, the vector field $f(x)$ and $W(x)$ can be expressed as a combination of some basis functions parametrized by some coefficients, then the PDE (1.4.1) becomes an algebraic equation in the coefficients that can be solved by algebraic methods. This is the case, e.g., for linear time invariant systems $\dot{x} = Ax$ and the family of quadratic functions $V(x) = x^T Px$ and $W(x) = x^T Qx$, which are parametrized by the coefficients matrices P, Q. Thus, a Lyapunov function is obtained by solving the algebraic Lyapunov equation $PA + A^T P = -Q$.

Thus, to design Lyapunov functions for HOSM, the *trajectory integration method* [25] uses **Method 1**, the *variable reduction method* [14] solves the problem in **Method 2**, and the idea of parametrizing a class of functions (**Method 3**) is used in the *generalized forms approach* [26].

In Section 1.4.2, we describe the trajectory integration method and apply it to the twisting algorithm. The variable reduction method is explained and applied to the terminal algorithm in Section 1.4.3. Finally, the generalized forms approach is used in Section 1.4.4 to design a Lyapunov function for the super-twisting algorithm.

1.4.1.1 Notation

We adopt the notation $\lceil z \rfloor^m = |z|^m \operatorname{sign}(z)$, for any $z \in \mathbb{R}$ and any real number $m \geq 0$. This operator preserves the sign of the value of the functions. According to this, for any $m, n \geq 0$, we have (i) $\lceil z \rfloor^m = z^m$ for any odd integer m, $|z|^m = z^m$ for any even integer m, and $\lceil z \rfloor^0 = \operatorname{sign}(z)$; (ii) $\lceil z \rfloor^m \lceil z \rfloor^n = |z|^{m+n}$, $\lceil z \rfloor^0 \lceil z \rfloor^m = |z|^m$, and $\lceil z \rfloor^0 |z|^p = \lceil z \rfloor^p$; and (iii) $\frac{d}{dz} \lceil z \rfloor^m = m|z|^{m-1}$ and $\frac{d}{dz}|z|^m = m\lceil z \rfloor^{m-1}$.

We need also to define homogeneous functions [1, 12, 3]. For a given vector $x = (x_1, ..., x_n)^T \in \mathbb{R}^n$, the dilation operator is defined as $\Delta_\varepsilon^{\mathbf{r}} x := (\varepsilon^{r_1} x_1, ..., \varepsilon^{r_n} x_n)$, $\forall \varepsilon > 0$, where $r_i > 0$ are the weights of the coordinates x_i, $i = 1, ..., n$. Let $\mathbf{r} = (r_1, ..., r_n)$ be the vector of weights. A function $V : \mathbb{R}^n \to \mathbb{R}$ (respectively, a vector field $f : \mathbb{R}^n \to \mathbb{R}^n$ or a vector-set field $F(x) \subset \mathbb{R}^n$) is called $\mathbf{r}$-homogeneous of degree $m \in \mathbb{R}$ if the identity $V(\Delta_\varepsilon^{\mathbf{r}} x) = \varepsilon^m V(x)$ holds (respectively, $f(\Delta_\varepsilon^{\mathbf{r}} x) = \varepsilon^l \Delta_\varepsilon^r f(x)$ or $F(\Delta_\varepsilon^{\mathbf{r}} x) = \varepsilon^l \Delta_\varepsilon^{\mathbf{r}} F(x)$).

1.4.2 Trajectory integration method

In this section, we describe briefly the method to build Lyapunov functions introduced in reference 25. Although such a method can be applied to a wider set of systems, here we restrict our attention to a class of HOSM. The idea of the method is based on the proofs of the converse Lyapunov theorems, sketched above in **Method 1**.

Consider the following nth order integrator:

$$\dot{x}_1 = x_2, \quad \ldots, \quad \dot{x}_{n-1} = x_n, \quad \dot{x}_n = u(x), \tag{1.4.2}$$

where $x \in \mathbb{R}^n$ is the state and $u(x) \in \mathbb{R}$ is the *discontinuous* control input generated by an HOSM controller. Let $\phi(t; t_0, x)$ be the solution of (1.4.2) with initial condition x at the instant $t_0 = 0$. Since one of the main characteristics of HOSM is its finite time convergence [12], we suppose that $x = 0$ is a finite time stable equilibrium point of (1.4.2). Therefore, there exists a time $T < \infty$ such that $\phi(t; t_0, x) = 0$, for all $t \geq T$. Moreover if $W : \mathbb{R}^n \to \mathbb{R}$ is a positive definite function, then, $W(\phi(t; t_0, x)) = 0$, for all $t \geq T$. Under these considerations, the integral given by

$$V(x) = \int_0^T W(\phi(\tau; 0, x))\,\mathrm{d}\tau \tag{1.4.3}$$

is well defined and V is a Lyapunov function for (1.4.2). The following are some characteristics of the method:

- The time derivative of (1.4.3) is given by $\dot{V} = -W(x)$ and therefore it is negative definite.
- The convergence time T is in general unknown, nevertheless, the finite time convergence allows us to set the upper integral limit to infinity.
- If the function W and the system are homogeneous, then V is also homogeneous.
- The method requires the knowledge of system's solutions.

The last point reveals a very serious disadvantage of the method. However, there are many HOSM systems whose solutions can be computed with relative simplicity. For example, those as (1.4.2) in closed loop with an HOSM controller that delivers a piecewise constant signal u. In this set of systems, we can find the controllers twisting [10], terminal [15], and the universal (nested) sliding mode controller [11]. Thus, the closed loop can be seen as a set of linear systems with constant inputs $\dot{x} = Ax + Bu_i$, where u_i is each of the constant values of the controller. Thus, the solutions of each subsystem are given by

$$\varphi_i(t; 0, x) = e^{At}x + \int_0^t e^{A(t-\tau)}\,\mathrm{d}\tau B u_i,$$

$$e^{At} = \begin{bmatrix} 1 & t & \frac{t^2}{2!} & \cdots & \frac{t^{n-1}}{(n-1)!} \\ 0 & 1 & t & & \vdots \\ \vdots & & \ddots & & \frac{t^2}{2!} \\ 0 & & & 1 & t \\ 0 & \cdots & \cdots & \cdots & 1 \end{bmatrix}, \quad \int_0^t e^{A(t-\tau)}\,\mathrm{d}\tau B = \begin{bmatrix} \frac{t^n}{n!} \\ \frac{t^{n-1}}{(n-1)!} \\ \vdots \\ \frac{t^2}{2!} \\ t \end{bmatrix}.$$

In the nominal case, i.e., for the system without matched perturbations or uncertainties, the construction method yields the necessary and sufficient conditions guaranteeing finite time stability of the origin. However, for the perturbed case, i.e., in the presence of perturbations or uncertainties, the stability of the origin can still be analyzed using the nominal Lyapunov function. However, in this case, in general, only sufficient conditions for finite time stability can be derived.

An important feature of the method is that for perturbed second-order systems, the method can be applied to obtain a robust Lyapunov function leading to necessary and sufficient conditions of stability under arbitrary (but bounded) perturbations. This is done by performing a disturbance analysis that consist in determining the *Majorizing curves*[1] for the system's trajectories.

Although homogeneity of the system and of the function W is not required for the application of the method, it greatly simplifies the calculation of the integral (1.4.3).

1.4.2.1 Example: twisting algorithm

Consider the following uncertain second-order system:

$$\dot{x}_1 = x_2, \quad \dot{x}_2 = f(t,x) + g(t,x)u, \tag{1.4.4}$$

where, $x^T = [x_1, x_2] \in \mathbb{R}^2$ is the state and $u \in \mathbb{R}$ is the control input. The uncertain functions $f(t,x)$, $g(t,x) \in \mathbb{R}$ satisfy $|f(t,x)| \leq F$ and $0 < G_m \leq g(t,x) \leq G_M$ for any $(t,x) \in \mathbb{R}_+ \times \mathbb{R}^n$ for some known real constants F, G_m and G_M. On the other hand, we have the second-order sliding mode controller known as twisting algorithm and is given by

$$u = -k_1 \operatorname{sign}(x_1) - k_2 \operatorname{sign}(x_2), \quad 0 < k_1, k_2 \in \mathbb{R}. \tag{1.4.5}$$

Note that (1.4.5) can be seen as a piecewise constant controller, this means that (defining $k = k_1 + k_2$ and $\bar{k} = k_1 - k_2$):

$$u = \begin{cases} u_1 = -k, & x_1 > 0,\ x_2 > 0 \\ u_2 = -\bar{k}, & x_1 > 0,\ x_2 < 0 \\ u_3 = k, & x_1 < 0,\ x_2 < 0 \\ u_4 = \bar{k}, & x_1 < 0,\ x_2 > 0 \end{cases}. \tag{1.4.6}$$

In order to apply the Lyapunov function design method described above, we have to set $f(t,x) \equiv 0$ and $g(t,x) \equiv 1$. Thus, we obtain the piecewise affine system

$$\dot{x}_1 = x_2, \quad \dot{x}_2 = u_i, \tag{1.4.7}$$

whose solution with initial condition x in the time $t_0 = 0$ is given by

$$\phi(t; 0, x) = \begin{bmatrix} x_1 + x_2 t + \frac{1}{2} u_i t^2 \\ x_2 + u_i t \end{bmatrix}.$$

[1] For this concept, see reference 5.

So, if the initial condition $x \in \{x_1 > 0,\ x_2 > 0\}$, then we have that

$$\phi(t;0,x) = \begin{bmatrix} x_1 + x_2 t - \frac{1}{2}kt^2 \\ x_2 - kt \end{bmatrix}.$$

This trajectory reaches the point $(x_{11},\ x_{21}) = \left(0,\ x_1 + x_2^2/2k\right)$, in a time $T_1 = x_2/k$. Thus, the trajectory in the quadrant $\{x_1 \geq 0,\ x_2 \leq 0\}$ is given by

$$\phi(t;T_1,(x_{11},x_{21})) = \begin{bmatrix} x_{11} + x_{21} t - \frac{1}{2}\bar{k}t^2 \\ x_{21} - \bar{k}t \end{bmatrix}$$

and so forth. Now, according to the method, we have to select a positive definite function $W : \mathbb{R}^2 \to \mathbb{R}$ to design a Lyapunov function for the closed loop (1.4.4), (1.4.5). Let us choose

$$W(x) = \begin{cases} 1, & x \neq 0 \\ 0, & x = 0 \end{cases}. \tag{1.4.8}$$

Observe that for this W, (1.4.3) must be rewritten as

$$V(x) = \sum_{j=1}^{\infty} T_j(x). \tag{1.4.9}$$

Each T_j is the transient time of the trajectory through each quadrant. If the starting point is such that $x \in \{x_1 > 0, x_2 > 0\}$, then we have from (1.4.9) that

$$V(x) = \sum_{j=1}^{\infty} T_j(x) \ = \ \tfrac{1}{k}x_2 + \tfrac{1}{k}\sqrt{x_2^2 + 2kx_1}\sum_{j=1}^{\infty}\left(p^j + p^{j-2}\right),$$

where $k = k_1 + k_2$, $\bar{k} = k_1 - k_2$ and $p = \sqrt{\bar{k}/k}$. Note that $V(x)$ is finite if and only if $\sum_{j=1}^{\infty}\left(p^j + p^{j-2}\right)$ is a geometric series. That occurs if and only if

$$p < 1 \quad \Longleftrightarrow \quad k_1 > k_2 > 0. \tag{1.4.10}$$

The same construction can be done for the remaining quadrants, so holding (1.4.10), $V(x)$ is expressed as follows:

$$V(x) = \begin{cases} \alpha_1 x_2 + \alpha_2\sqrt{x_2^2 + 2kx_1}, & x_1 > 0,\ x_2 > 0 \\ -\alpha_3 x_2 + \alpha_4\sqrt{x_2^2 + 2\bar{k}x_1}, & x_1 \geq 0,\ x_2 \leq 0 \\ -\alpha_1 x_2 + \alpha_2\sqrt{x_2^2 - 2kx_1}, & x_1 < 0,\ x_2 > 0 \\ \alpha_3 x_2 + \alpha_4\sqrt{x_2^2 - 2\bar{k}x_1}, & x_1 \leq 0,\ x_2 \geq 0 \end{cases},$$

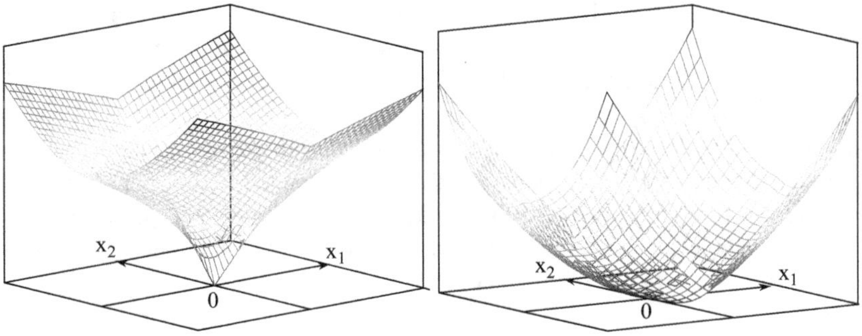

Figure 1.4.1 Example plot of (1.4.11) (left) and (1.4.12) (right)

$$\alpha_1 = \frac{1}{k}, \quad \alpha_2 = \frac{p^{-1}+p}{k(1-p)}, \quad \alpha_3 = -\frac{1}{\bar{k}}, \quad \alpha_4 = \alpha_1 + \alpha_2 - \alpha_3.$$

Observe that due to the symmetry with respect to the origin, this last function can be rewritten as

$$V(x) = \begin{cases} \alpha_1|x_2| + \alpha_2\sqrt{x_2^2 + 2k|x_1|}, & x_1x_2 > 0 \\ \alpha_3|x_2| + \alpha_4\sqrt{x_2^2 + 2\bar{k}|x_1|}, & x_1x_2 \le 0 \end{cases}. \tag{1.4.11}$$

Since $\dot{V} = -W(x)$, (1.4.11) is a Lyapunov function for the closed loop (1.4.4), (1.4.5), (in the nominal case) if and only if (1.4.10) holds. This function also ensures finite time stability of the system's origin; moreover, it is easy to see that $T = V(x(0))$ is the convergence time to the origin from the initial condition $x(0)$.

The closed loop (1.4.4), (1.4.5) is a homogeneous system of degree -1 with the weights $(r_1, r_2) = (2, 1)$. Note that (1.4.39) is a homogeneous function of degree 0 and (1.4.11) is homogeneous of degree 1 with the weights of the system. Figure 1.4.1 (left) shows a plot of (1.4.11). Note that (1.4.11) is not a Lipschitz function over the line $\{x_1 = 0\}$. However, the same procedure can be applied to design another Lyapunov function by considering $W(x) = |x_1| + x_2^2$ (this is a homogeneous function of degree 2), and by applying the design method we get the following Lyapunov function:

$$V(x) = \begin{cases} \beta_1|x_2|^3 + \beta_2x_1x_2 + \beta_3\left[x_2^2 + 2k|x_1|\right]^{\frac{3}{2}}, & x_1x_2 > 0 \\ \beta_4|x_2|^3 + \beta_5x_1x_2 + \beta_6\left[x_2^2 + 2\bar{k}|x_1|\right]^{\frac{3}{2}}, & x_1x_2 \le 0 \end{cases}; \tag{1.4.12}$$

$$\beta_1 = \frac{k+1}{3k^2}, \quad \beta_2 = \frac{1}{k}, \quad \beta_3 = \frac{p^{-1}(\bar{k}+1) + p^3(k+1)}{3k^2(1-p^3)},$$

$$\beta_4 = -\frac{\bar{k}+1}{3\bar{k}^2}, \quad \beta_5 = \frac{1}{\bar{k}}, \quad \beta_6 = \beta_1 + \beta_3 - \beta_4.$$

It is important to mention that (1.4.12) exists if and only if (1.4.10) holds. Its time derivative is $\dot{V} = -|x_1| - x_2^2$. Observe that (1.4.12) is homogeneous of degree three, and is a Lipschitz function over the whole $\mathbb{R}^2$. A plot of (1.4.12) is shown in Figure 1.4.1 (right).

Since (1.4.11) and (1.4.12) were designed considering the undisturbed case, it is necessary to perform a disturbance analysis by using these functions. This can be done by taking the derivative of V along the system's trajectories and verifying its negativeness. However, in the next section, we are going to show how the designing method can be applied taking into account the disturbances.

1.4.2.2 Perturbed case

Consider the closed loop (1.4.4), (1.4.5) that can be rewritten in integral form as follows:

$$x_1(t) = x_{10} + \int_0^t x_2(\tau)\,\mathrm{d}\tau, \tag{1.4.13}$$

$$x_2(t) = x_{20} + \int_0^t [f(\tau,x) + g(\tau,x)u]\,\mathrm{d}\tau. \tag{1.4.14}$$

The objective is to find, for each initial condition, the system's integral curve that bounds all the possible system's trajectories in the phase plane. Note that this is equivalent to find the *Majorizing curves* [5]. From (1.4.6), we have found that in the quadrant $\{x_1 > 0, x_2 > 0\}$ the control signal is $u = -k$. Thus, from (1.4.14), it is clear that

$$x_2(t) = x_{20} + \int_0^t [f(\tau,x) - g(\tau,x)k]\,\mathrm{d}\tau \le x_{20} - (kG_m - F)t. \tag{1.4.15}$$

So, in the quadrant $\{x_1 > 0, x_2 > 0\}$, we can build the Majorizing curve for (1.4.4) and (1.4.5) as follows: Define the parametric curve $\hat{x} : \mathbb{R} \to \mathbb{R}^2$, $\hat{x}(t) = [\hat{x}_1(t), \hat{x}_2(t)]^T$ such that $\dot{\hat{x}}_1 = \hat{x}_2$, and from (1.4.15), it is established that $\hat{x}_2 = \hat{x}_{20} - (kG_m - F)t$, then

$$\hat{x}(t) = \begin{bmatrix} \hat{x}_{10} + \hat{x}_{20}t - \frac{1}{2}(kG_m - F)t^2 \\ \hat{x}_{20} - (kG_m - F)t \end{bmatrix}.$$

Analogous analysis can be made for the remaining quadrants in $\mathbb{R}^2$. Then, recalling that $U = G_m k - F$ and $\bar{U} = G_M \bar{k} + F$, it is possible to write $\hat{x}(t)$ as follows:

$$\hat{x}(t) = \begin{bmatrix} \hat{x}_{10} + \hat{x}_{20}t + \frac{1}{2}\mathscr{U}t^2 \\ \hat{x}_{20} + \mathscr{U}t \end{bmatrix}, \quad \mathscr{U} = \begin{cases} -U, \ x_1 > 0, \ x_2 > 0 \\ -\bar{U}, \ x_1 > 0, \ x_2 < 0 \\ U, \ x_1 < 0, \ x_2 < 0 \\ \bar{U}, \ x_1 < 0, \ x_2 > 0 \end{cases}. \tag{1.4.16}$$

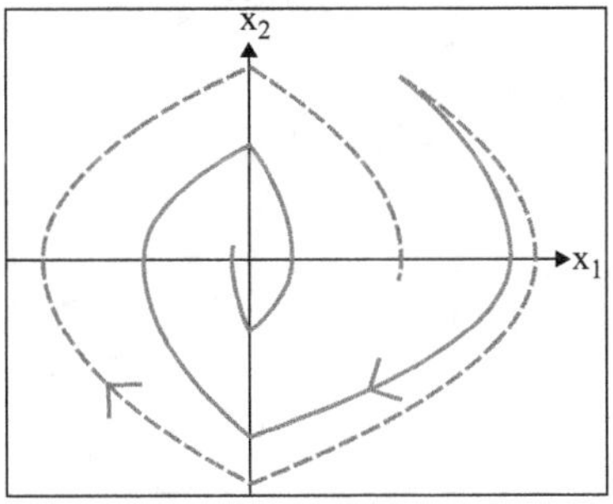

Figure 1.4.2 Trajectories of (1.4.4) and (1.4.5); Solid: nominal $x(t)$; Dashed: $\hat{x}(t)$

Thus, for each initial condition, (1.4.16) is the Majorizing curve for (1.4.4) and (1.4.5). In Figure 1.4.2, it is exemplified such a Majorizing curve. Now, observe that (1.4.16) can be seen as the solution of the following piecewise affine system:

$$\dot{x}_1 = x_2, \quad \dot{x}_2 = \mathscr{U}. \tag{1.4.17}$$

Therefore, it is possible to apply our design method to construct a Lyapunov function for (1.4.17). Let $\hat{x}(t;0,x)$ be the solution of (1.4.17) whose initial condition is x in the instant $t_0 = 0$, and T_i is each of the transient times of $\hat{x}$ through the different quadrants of the phase space. Suppose that the initial condition x is in the quadrant $\{x_1 > 0, x_2 > 0\}$. If we choose W as in (1.4.39), then from (1.4.9) we have that

$$\sum_{i=1}^{\infty} T_i = \tfrac{1}{U}x_2 + \tfrac{1}{U}\sqrt{x_2^2 + 2Ux_1}\sum_{j=1}^{\infty}\left(r^j + r^{j+1}\right). \tag{1.4.18}$$

Note that $V(x)$ exists if and only if $\sum_{j=1}^{\infty}\left(r^j + r^{j+1}\right)$ is a geometric series. This last occurs if and only if

$$r = \sqrt{\bar{U}/U} < 1 \quad \Leftrightarrow \quad G_m k - F > G_M \bar{k} + F, \quad G_m k - F > 0. \tag{1.4.19}$$

The procedure to compute V for other quadrants is similar. Let us underline that condition (1.4.19) emerges in the Lyapunov function's design in each quadrant. So, we have that

$$V(x) = \begin{cases} \alpha_1|x_2| + \alpha_2\sqrt{x_2^2 + 2U|x_1|}, & x_1x_2 > 0 \\ \alpha_3|x_2| + \alpha_4\sqrt{x_2^2 + 2\bar{U}|x_1|}, & x_1x_2 \le 0 \end{cases}; \tag{1.4.20}$$

$$\alpha_1 = \frac{1}{U}, \quad \alpha_2 = \frac{r^{-1} + r}{U(1-r)}, \quad \alpha_3 = -\frac{1}{\bar{U}}, \quad \alpha_4 = \alpha_1 + \alpha_2 - \alpha_3.$$

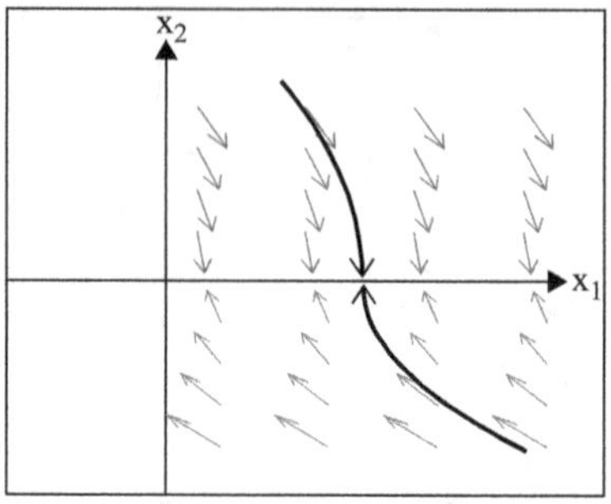

Figure 1.4.3 Vector field and trajectories of (1.4.4) and (1.4.5) supposing that $G_m\bar{k} \leq F$

is a Lyapunov function for (1.4.17). Now consider (1.4.20) as a Lyapunov function candidate for the disturbed closed loop (1.4.4), (1.4.5). Taking its derivative along system's trajectories, we obtain

$$\dot{V} \leq \begin{cases} -W(x), & x_1x_2 \geq 0 \\ -\dfrac{G_m\bar{k} - F}{G_M\bar{k} + F}, & x_1x_2 < 0 \end{cases}. \tag{1.4.21}$$

Hence, $\dot{V}$ is negative definite if $G_m\bar{k} > F$ holds. It is clear that the condition $G_m\bar{k} > F$ in (1.4.21) is not only sufficient but also necessary.

Consider from (1.4.4) and (1.4.5) the vector field $H(x) = [x_2, f(t,x) + g(t,x)u]^T$. Suppose that the perturbation acting over the system is given by $f(t,x) = F$, $g(t,x) = G_m$, and the controller's gains were chosen such that $G_m\bar{k} \leq F$. Under these conditions, it can be seen as an example plot of the vector field H in Figure 1.4.3. Note that if a trajectory hits the set $\{x_2 = 0\}$ in a time t_1, then a sliding motion will be produced due to the direction of the vector field H. Hence $x_2(t) = 0$ for all $t > t_1$, and from (1.4.13), it is easy to see that $x_1(t)$ will remain constant for all $t > t_1$. Therefore, in general, the trajectory is not able to reach the system's origin. So, the inequality $G_m\bar{k} > F$ is also a necessary condition to guarantee finite time stability of the origin of (1.4.4) and (1.4.5).

Now let us compute the convergence time estimation. Let $a \in \mathbb{R}$ be a constant and $\dot{z}(t) = a$ be a differential equation from which we have that

$$\int_0^T \frac{\mathrm{d}}{\mathrm{d}t} z(t)\,\mathrm{d}t = \int_0^T a\,\mathrm{d}t \quad \Rightarrow \quad z(T) - z(0) = aT.$$

Consider now a trajectory of (1.4.4) and (1.4.5) whose initial condition is $x(0)$ and T is its convergence time to the origin. Then, applying the comparison Lemma with (1.4.21), it is obtained that $V(x(T)) - V(x(0)) \leq -\gamma T$, and therefore, the convergence time T is such that

$$T \leq V(x(0))/\gamma, \quad \gamma = \frac{G_m\bar{k} - F}{G_M\bar{k} + F}. \tag{1.4.22}$$

The obtained results are summarized in the next theorem.

Theorem 1.4.1. *The trajectories of the uncertain closed loop (1.4.4), (1.4.5) converge in finite time to the system's origin if and only if the controller's gains* k_1 *and* k_2 *are chosen such that* (1.4.19) *and* $G_m\bar{k} > F$ *hold. Moreover,* (1.4.20) *is a Lyapunov function for (1.4.4) and (1.4.5), and the convergence time* T *can be estimated by (1.4.22).*

Remark 1.4.1. *It is worth to stress that (1.4.20) is similar to that presented in reference 22, and the convergence conditions are the same to those given in reference 10.*

Remark 1.4.2. *One important advantage of the method used here (compared, e.g., with reference 22) is that it is not difficult to design Lyapunov functions with higher homogeneous degree. This can be done by increasing the homogeneous degree of the function* W*, for instance, choosing* $W(x_1,x_2) = |x_1| + x_2^2$.

1.4.3 Variable reduction method

The Lyapunov function construction method that we are going to explain in this section was introduced in reference 14. The idea of the method is to transform the partial differential equation given by **Method 2** into an ordinary differential equation. This is done by taking advantage of the homogeneity property of the class of homogeneous second-order systems. A similar procedure was given in reference 32 for continuous homogeneous (in the classical sense) systems.

Consider the following second-order homogeneous dynamic system:

$$\dot{x}_1 = F_1(x_1,x_2), \quad \dot{x}_2 = F_2(x_1,x_2). \tag{1.4.23}$$

The system's homogeneous degree is $k \in \mathbb{R}$ with the weights $r = (r_1, r_2) \in \mathbb{R}^2$. On the other hand, the existence of a homogeneous Lyapunov function for homogeneous systems has been proved (see, e.g., reference 24 for continuous and references 3, 17 for discontinuous systems). Therefore, it is suitable to propose a homogeneous function $V : \mathbb{R}^2 \to \mathbb{R}$, of degree $m \in \mathbb{R}$, as a Lyapunov function candidate for (1.4.23). The homogeneity weights $\mathbf{r}$ of V are the same as the weights of the system. The derivative of V along the trajectories of (1.4.23) is

$$\dot{V} = \frac{\partial V(x)}{\partial x_1} F_1(x) + \frac{\partial V(x)}{\partial x_2} F_2(x) = -W(x), \tag{1.4.24}$$

where $W : \mathbb{R}^2 \to \mathbb{R}$ is a homogeneous function of degree $\bar{m} = m + k$ with the same weights $\mathbf{r}$. Note that, since V, W, and the vector field of (1.4.23) are homogeneous, they satisfy for any $x \in R^2$ and any positive $\varepsilon \in \mathbb{R}$ the following:

$$F_1(\varepsilon^{r_1}x_1, \varepsilon^{r_2}x_2) = \varepsilon^{k+r_1}F_1(x_1,x_2), \quad F_2(\varepsilon^{r_1}x_1, \varepsilon^{r_2}x_2) = \varepsilon^{k+r_2}F_2(x_1,x_2), \tag{1.4.25}$$

$$V(\varepsilon^{r_1}x_1, \varepsilon^{r_2}x_2) = \varepsilon^m V(x_1, x_2), \quad W(\varepsilon^{r_1}x_1, \varepsilon^{r_2}x_2) = \varepsilon^{m+k} W(x_1, x_2). \tag{1.4.26}$$

Now, defining $z = x_2|x_1|^{-\frac{r_2}{r_1}}$ and $\varepsilon = |x_1|^{-\frac{1}{r_1}}$ for any $x_1 \neq 0$, then from (1.4.25), it is obtained that

$$F_1(\lceil x_1 \rfloor^0, z) = |x_1|^{-\frac{k+r_1}{r_1}} F_1(x_1, x_2) \Leftrightarrow F_1(x_1, x_2) = |x_1|^{\frac{k+r_1}{r_1}} F_1(\lceil x_1 \rfloor^0, z).$$

Therefore, from the last equation, for $x_1 > 0$ the function F_1 can be expressed as

$$F_1(x_1, x_2) = |x_1|^{\frac{k+r_1}{r_1}} F_1(1, z) = |x_1|^{\frac{k+r_1}{r_1}} f_1^+(z),$$

and for $x_1 < 0$

$$F_1(x_1, x_2) = |x_1|^{\frac{k+r_1}{r_1}} F_1(-1, z) = |x_1|^{\frac{k+r_1}{r_1}} f_1^-(z).$$

For simplicity, we are going to use $f_1^\pm$ to refer f_1^+ and f_1^- simultaneously. Thus

$$F_1(x_1, x_2) = |x_1|^{\frac{k+r_1}{r_1}} f_1^\pm(z), \quad \forall x_1 \neq 0.$$

The same can be done for the functions F_2, V, and W; therefore,

$$F_1(x_1, x_2) = |x_1|^{\frac{k+r_1}{r_1}} f_1^\pm(z), \quad F_2(x_1, x_2) = |x_1|^{\frac{k+r_2}{r_1}} f_2^\pm(z), \tag{1.4.27}$$

$$W(x_1, x_2) = |x_1|^{\frac{\bar{m}}{r_1}} w^\pm(z), \quad V(x_1, x_2) = |x_1|^{\frac{m}{r_1}} v^\pm(z). \tag{1.4.28}$$

Note that the last equations give the relation between functions in variable x and their equivalents in variable z. Thus, the properties of a function in the variable $x \in \mathbb{R}^2$ can be determined through the properties of its equivalent in variable $z \in \mathbb{R}$. For example, the positive definiteness of V is guaranteed if $v^\pm$ satisfies the following:

A1. $v^\pm(0) \neq 0$ and $v^\pm(z) > 0, \forall z \neq 0$.
A2. The limits

$$\lim_{z\to\infty} \frac{v^\pm(z)}{|z|^{\frac{m}{r_2}}}, \quad \lim_{z\to-\infty} \frac{v^\pm(z)}{|z|^{\frac{m}{r_2}}},$$

exist and are constant.

The last condition means that the growth order of $v^\pm(z)$ and $|z|^{\frac{m}{r_2}}$ is the same. These conditions are the same for $w^\pm(z)$ but considering its homogeneous degree $\bar{m}$ instead of m.

Now, if we want V to be a strict Lyapunov function for (1.4.23), then in (1.4.24), functions V and W must be positive definite. This is equivalent that in variable z, $v^\pm$

and $w^{\pm}$ satisfy conditions A1 and A2. Thus, by substituting (1.4.27) and (1.4.28) in (1.4.24), it obtained the following ordinary differential equation

$$\frac{\mathrm{d}v^{\pm}(z)}{\mathrm{d}z} + a(z)v^{\pm}(z) = b(z); \tag{1.4.29}$$

$$a(z) = \pm\frac{mf_1^{\pm}(z)}{r_1 f_2^{\pm}(z) \mp r_2 z f_1^{\pm}(z)}, \quad b(z) = -\frac{r_1 w(z)}{r_1 f_2^{\pm}(z) \mp r_2 z f_1^{\pm}(z)}.$$

Observe that (1.4.29) indeed represents two different equations, one for v^+ and another for v^-. The solution for these equations are

$$v^{\pm}(z) = e^{-A(z)}\left[B(z) + c^{\pm}\right], \tag{1.4.30}$$

where $c^{\pm}$ is an integration constant and the functions A and B are given by

$$A(z) = \int a(z)\,\mathrm{d}z, \quad B(z) = \int e^{A(z)}b(z)\,\mathrm{d}z.$$

So, with the aim to come back to the domain in the variable x, from (1.4.28), we have

$$V(x_1, x_2) = \begin{cases} |x_1|^{\frac{m}{r_1}} v^+\left(x_2|x_1|^{-\frac{r_2}{r_1}}\right), & x_1 > 0 \\ |x_1|^{\frac{m}{r_1}} v^-\left(x_2|x_1|^{-\frac{r_2}{r_1}}\right), & x_1 < 0 \end{cases}. \tag{1.4.31}$$

Therefore, the above analysis has given us a procedure to compute a Lyapunov function V for (1.4.23), this method can described as follows:

Step 1. Given a system like (1.4.23), obtain the functions $f_1^{\pm}$, $f_2^{\pm}$, as stated in (1.4.27), by means of the change of variable $z = x_2|x_1|^{-\frac{r_2}{r_1}}$.
Step 2. Choose a function $w^{\pm}$ that satisfy A1 and A2.
Step 3. Compute $v^{\pm}$ as in (1.4.30).
Step 4. If $v^{\pm}$ satisfies A1 and A2, then (1.4.31) is a Lyapunov function for (1.4.23).

Observe that (1.4.31) is not defined for the set $x_1 = 0$. However, it is possible to extend it by a process of limit. Moreover, the integration constant $c^{\pm}$ can be used to tune the functions v^+ and v^- such that V be continuous over the set $\{x_1 = 0\}$.

1.4.3.1 Example: terminal algorithm

Consider the terminal algorithm given by the second-order system

$$\dot{x}_1 = x_2, \quad \dot{x}_2 = -\alpha\lceil\sigma\rfloor^0, \quad \sigma = x_2 + \beta\lceil x_1\rfloor^{\frac{1}{2}}, \quad 0 < \alpha, \beta \in \mathbb{R}. \tag{1.4.32}$$

This system is homogeneous of degree $k=-1$ with the weights $(r_1, r_2) = (2, 1)$. Thus, from the description in (1.4.23), $F_1(x) = x_2$ and $F_2(x) = -\alpha\lceil x_2 + \beta\lceil x_1 \rfloor^{\frac{1}{2}}\rfloor^0$. Observe that

$$F_2(x) = \begin{cases} -\alpha, & \sigma > 0 \\ \alpha, & \sigma < 0 \end{cases}. \tag{1.4.33}$$

With the aim to be more objective in the example, we are going to consider only the case $\sigma > 0$. Now, by applying the change of variable $z = x_2|x_1|^{-\frac{1}{2}}$, it is easy to see from (1.4.27) that

$$f_1^{\pm}(z) = z, \quad f_2^{\pm}(z) = -\alpha. \tag{1.4.34}$$

Note that $f_1^+ = f_1^-$ and $f_2^+ = f_2^-$. Let us choose $w^{\pm}(z) = 1$, this function satisfies A1 and A2 with $\bar{m} = 0$, and, since the homogeneous degree of (1.4.32) is $k = -1$, therefore $m = 1$. In order to obtain $v^{\pm}$ from (1.4.30), we have to compute the integrals

$$A(z) = \int a(z)\,\mathrm{d}z, \quad B(z) = \int e^{A(z)} b(z)\,\mathrm{d}z,$$

$$a(z) = \pm\frac{z}{-2\alpha \mp z^2} \quad \text{and} \quad b(z) = -\frac{2}{-2\alpha \mp z^2}.$$

Thus $A(z) = \ln\left(\frac{1}{\sqrt{z^2 \pm 2\alpha}}\right)$ and $B(z) = \frac{z}{\alpha\sqrt{z^2 \pm 2\alpha}}$. Therefore,

$$v^{\pm}(z) = e^{-A(z)}\left[B(z) + c^{\pm}\right] = \frac{z}{\alpha} + c^{\pm}\sqrt{z^2 \pm 2\alpha}. \tag{1.4.35}$$

It is clear that $v^{\pm}$ satisfies A2, and it is easy to proof that $v^{\pm}$ satisfies A1 if $c^{\pm} > \beta/(\alpha\sqrt{\beta^2 - 2\alpha})$, $\beta^2 > 2\alpha$. So, in variable x, we have from (1.4.31) that

$$V(x) = |x_1|^{\frac{1}{2}} v^{\pm}\left(x_2|x_1|^{-\frac{1}{2}}\right) = \frac{x_2}{\alpha} + c^{\pm}\sqrt{x_2^2 \pm 2\alpha|x_1|} = \frac{x_2}{\alpha} + c^{\pm}\sqrt{x_2^2 + 2\alpha x_1}.$$

An important observation is that the found restriction $\beta^2 > 2\alpha$ implies that the behavior of the system's trajectories is in "twisting-like mode." This means that $\sigma = 0$ is only a switching surface, therefore there is not an FOSM on such a surface. This can be seen in Figure 1.4.4. Analogously, for the case $\sigma < 0$, the process of designing $v^{\pm}$ yields, in variable x, the following function:

$$V(x) = -\frac{x_2}{\alpha} + c^{\pm}\sqrt{x_2^2 - 2\alpha x_1}.$$

In order to the function V be continuous and be defined for any x in $\mathbb{R}^2$, it is necessary and sufficient to choose $c^{\pm} = c$, where

$$c = \frac{2\beta}{\alpha\left(\sqrt{\beta^2 + 2\alpha} - \sqrt{\beta^2 - 2\alpha}\right)}.$$

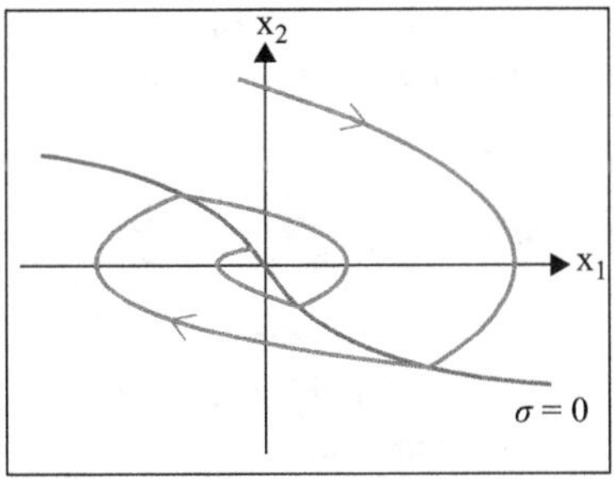

Figure 1.4.4 A trajectory of the terminal algorithm in "twisting-like mode"

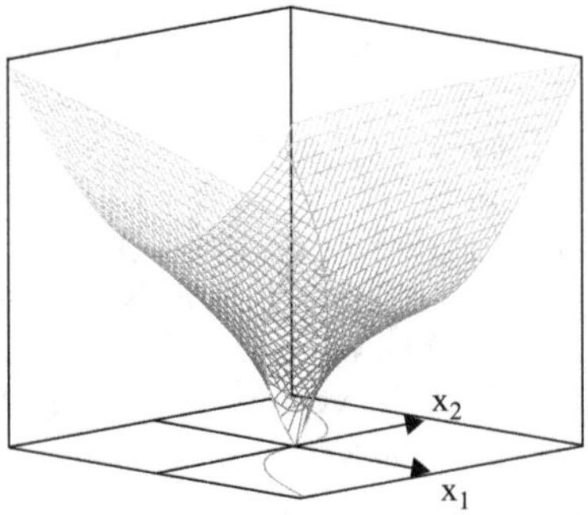

Figure 1.4.5 Example plot of the Lyapunov function for the terminal algorithm

Therefore, we have designed the following continuous Lyapunov function for (1.4.32):

$$V(x) = \begin{cases} \frac{1}{\alpha}x_2 + c\sqrt{x_2^2 + 2\alpha x_1}, & \sigma \geq 0 \\ -\frac{1}{\alpha}x_2 + c\sqrt{x_2^2 - 2\alpha x_1}, & \sigma < 0 \end{cases}.$$

This function coincides with that designed in reference 22, but here only for the nominal case. It is important to mention that for this example, $\dot{V} = -W(x)$ where the function $W : \mathbb{R}^2 \to \mathbb{R}$ is such that $W(x) = 1, \forall x \neq 0$ and $W(x) = 0, x = 0$. An example plot for $V(x)$ is shown in Figure 1.4.5.

1.4.4 Generalized forms approach

In this section, we show the procedure reported in reference 26. This method consists in proposing a class of parametrized functions as a family of candidate Lyapunov functions. Such parametrization allows us to choose a suitable candidate Lyapunov function and also to determine systematically its positive definiteness and the negative definiteness of its derivative.

With the aim of being brief but illustrative, we are going to explain the method through an example. But before that it is necessary to give the following definition:

Definition 1.4.1. *The function $f : \mathbb{R}^n \to \mathbb{R}$ is called a* generalized form *of degree m if:*

a) *it is a homogeneous function (of degree m for some vector of weights* **r***);*
b) *it consists of sums, products and sums of products of terms like:*

$$a\lceil x_k \rfloor^p, \quad b|x_k|^q, \quad a, b \in \mathbb{R},\ p, q \in \mathbb{Q}.$$

Consider the following dynamic system (super-twisting algorithm [10]):

$$\dot{x}_1 = -k_1\lceil x_1 \rfloor^{\frac{1}{2}} + x_2, \quad \dot{x}_2 = -k_2\lceil x_1 \rfloor^0. \tag{1.4.36}$$

This system is homogeneous of degree $k = -1$ with the weights $(r_1, r_2) = (2, 1)$. Note that each component of the vector field in (1.4.36) is a generalized form.

The first step in the designing of a Lyapunov function for (1.4.36) is to choose an adequate candidate function. Since the vector field of the system is made of generalized forms, it is reasonable to propose a generalized form as a candidate function. In this example, it is proposed

$$V(x) = \alpha_1|x_1|^{\frac{3}{2}} - \alpha_2 x_1 x_2 + \alpha_3|x_2|^3, \quad \alpha_1, \alpha_3 > 0. \tag{1.4.37}$$

This function is homogeneous of degree $m = 3$ with the same weights as the system's. The reason for the first and third terms in (1.4.37) seems to be obvious since the necessity of positive definiteness of V. The justification for the second term is clear from the derivative of (1.4.37) along the trajectories of (1.4.36). Thus, if it is defined $\dot{V} = -W(x)$, then

$$\begin{aligned} W(x) = {} & \tfrac{3}{2}\alpha_1 k_1|x_1| - \tfrac{3}{2}\alpha_1\lceil x_1 \rfloor^{\frac{1}{2}} x_2 - \alpha_2 k_1\lceil x_1 \rfloor^{\frac{1}{2}} x_2 \\ & + \alpha_2|x_2|^2 - \alpha_2 k_2|x_1| + 3\alpha_3 k_2\lceil x_1 \rfloor^0\lceil x_2 \rfloor^2. \end{aligned} \tag{1.4.38}$$

The first term in (1.4.38) is positive definite in x_1, while the fourth one is positive definite in x_2. Note that the fourth term only can be obtained from the second term in (1.4.37). Now, by simplifying (1.4.38), we obtain

$$W(x) = \beta_1|x_1| - \beta_2\lceil x_1 \rfloor^{\frac{1}{2}} x_2 + \beta_3 x_2^2, \tag{1.4.39}$$

where $\beta_1 = \frac{3}{2}\alpha_1 k_1 - \alpha_2 k_2$, and $\beta_2 = \frac{3}{2}\alpha_1 + \alpha_2 k_1$. The third coefficient is defined as follows: $\beta_3 = \beta_3^+ = \alpha_2 + 3\alpha_3 k_2$, for $\{x_1 x_2 > 0\}$, and $\beta_3 = \beta_3^- = \alpha_2 - 3\alpha_3 k_2$, for

$\{x_1x_2 < 0\}$. Note that $\beta_1, \beta_3 > 0$ are necessary conditions for positive definiteness of (1.4.39).

Generalized forms have two important properties: partial derivatives and products of generalized forms are also generalized forms. Thus, (1.4.39) is a generalized form of degree 2 with the same weights as those of the function V.

Now, if (1.4.37) is a strict Lyapunov function for (1.4.36), then it is sufficient to find the coefficients of (1.4.37) and (1.4.39) that make them positive definite functions. In general, this can be a difficult task; however, we can try to simplify it by using procedures to determine positiveness of polynomials. First of all, we have to transform (1.4.37) and (1.4.39) into polynomials, and this can be done by means of the following change of variables:

$$\begin{aligned} \sqrt{x_1} &= z_1, \quad x_2 = z_2, \quad \text{for} \quad x_1, x_2 \geq 0, \\ \sqrt{-x_1} &= z_1, \quad -x_2 = z_2, \quad \text{for} \quad x_1, x_2 \leq 0. \end{aligned}$$

Note that z_1, $z_2 \geq 0$ in any case. Thus, from (1.4.37) and (1.4.39), we obtain the following set of forms (homogeneous polynomials)

- For $\{x_1x_2 \geq 0\}$:

$$\begin{aligned} V(z) &= \alpha_1 z_1^3 - \alpha_2 z_1^2 z_2 + \alpha_3 z_2^3, \\ W(z) &= \beta_1 z_1^2 - \beta_2 z_1 z_2 + \beta_3^+ z_2^2. \end{aligned} \tag{1.4.40}$$

- For $\{x_1x_2 < 0\}$:

$$\begin{aligned} V(z) &= \alpha_1 z_1^3 + \alpha_2 z_1^2 z_2 + \alpha_3 z_2^3, \\ W(z) &= \beta_1 z_1^2 + \beta_2 z_1 z_2 + \beta_3^- z_2^2. \end{aligned} \tag{1.4.41}$$

Observe that the transformation only affects the functions V and W, and it does not produce any change in the system. Moreover, the positiveness analysis has been reduced to that of polynomials (with nonnegative variables). This analysis can be performed systematically, and here we are going to show two different procedures. The first one is by using Pólya's theorem [20], and the second one is by sum of squares (SOS) representation.

It is clear that V and W in (1.4.41) are positive definite, therefore, we have to analyze only the pair (1.4.40).

1.4.4.1 Pólya's theorem

Before explaining the procedure, we recall a strong version of Pólya's theorem [4, 20].

Theorem 1.4.2 (Pólya). *Let $\mathscr{P}_n$ be the set $\mathscr{P}_n = \{z \in \mathbb{R}^n \mid z \neq 0,\ z_i \geq 0, i = 1, 2, \ldots, n\}$, and let $F : \mathbb{R}^n \to \mathbb{R}$ be a form such that $F(z) > 0,\ \forall z \in \mathscr{P}_n$. Then, for any large enough $p \in \mathrm{N}$,*

$$(z_1 + z_2 + \cdots + z_n)^p F(z) = G(z), \quad \forall z \in \mathscr{P}_n,$$

for some form G whose coefficients are strictly positive.

So, the procedure to determine the positiveness of a form F consists in finding a p such that the coefficients of G are all strictly positive. Let us apply this to the function V from (1.4.40). For $p = 1$,

$$G_1(z) = (z_1 + z_2)V(z) = \alpha_1 z_1^4 + (\alpha_1 - \alpha_2)z_1^3 z_2 - \alpha_2 z_1^2 z_2^2 + \alpha_3 z_1 z_2^3 + \alpha_3 z_2^4,$$

notice that the third coefficient of G_1 cannot be positive since we have restricted $\alpha_2 > 0$. So, with $p = 1$ the conditions of Pólya's theorem cannot be satisfied. For $p = 2$,

$$\begin{aligned} G_2(z) = (z_1 + z_2)^2 V(z) &= \alpha_1 z_1^5 + (2\alpha_1 - \alpha_2)z_1^4 z_2 + (\alpha_3 - \alpha_2)z_1^2 z_2^3 \\ &\quad + (\alpha_1 - 2\alpha_2)z_1^3 z_2^2 + 2\alpha_3 z_1 z_2^4 + \alpha_3 z_2^5. \end{aligned}$$

Thus, all the coefficients of G_2 can be positive if α_1, α_2, and α_3 are chosen such that the following system of inequalities is satisfied:

$$\alpha_1 > 0, \quad \alpha_2 > 0, \quad \alpha_3 > 0, \quad 2\alpha_1 - \alpha_2 > 0, \quad \alpha_3 - \alpha_2 > 0, \quad \alpha_1 - 2\alpha_2 > 0.$$

This system is linear in the coefficients of V and can be rewritten as

$$A_v \alpha > 0, \quad A_v = \begin{bmatrix} 1 & 0 & 0 & 2 & 0 & 1 \\ 0 & 1 & 0 & -1 & -1 & -2 \\ 0 & 0 & 1 & 0 & 1 & 0 \end{bmatrix}^T, \quad \alpha = [\alpha_1 \quad \alpha_2 \quad \alpha_3]^T. \tag{1.4.42}$$

In (1.4.42) and henceforth, the inequality sign ">" is component-wise. The same procedure can be applied to W from (1.4.40), obtaining for each p an inequalities system $A_w \beta > 0$, $\beta = [\beta_1\ \beta_2\ \beta_3^+]^T$. Recall that each β_i depends (linearly) on α and (affinely) on $k = [k_1\ k_2]^T$, therefore the systems $A_v \alpha > 0$ and $A_w \beta > 0$ might be solved simultaneously. However, since $\beta = M(\alpha)[1\ k^T]^T$, where M is a matrix that depends linearly on α, we can perform the following procedure:

1. Solve the linear system of inequalities (1.4.42).
2. Pick an α from the set of solutions of (1.4.42) and substitute it in $M(\alpha)$.
3. Solve for k the linear system of inequalities $A_w \beta = A_w M(\alpha)[1\ k^T]^T > 0$.

An important note is that the procedure reduces to solve only linear systems of inequalities, and it is possible to find the *complete set of solutions* for this kind of systems by using available software.[2] For our example, the set of solutions of (1.4.42) is given by $\{\alpha \in \mathbb{R}^3 \mid \alpha = B_v\mu\}$, where

$$B_v = \begin{bmatrix} 2 & 0 & 2 \\ 0 & 0 & 1 \\ 0 & 2 & 1 \end{bmatrix}^T, \quad \mu = [\mu_1 \quad \mu_2 \quad \mu_3]^T,\ 0 < \mu_i \in \mathbb{R}.$$

By choosing $\mu = [0.05 \quad 0.05 \quad 1]^T$, it is obtained that $\alpha = [2.1 \quad 1 \quad 1.1]^T$. Thus the solution of $A_w\beta = A_wM(\alpha)[1 \quad k^T]^T > 0$ is given by $\{k \in \mathbb{R}^2 \mid k = B_w\gamma\}$, where $\gamma \in \mathbb{R}^q$, $\gamma_i > 0$, $\sum_{i=1}^q \gamma_i = 1$, and q being the number of columns of B_w. With $p = 6$,

$$B_w = \begin{bmatrix} 3.788 & 2.325 & 3.019 \\ 0.303 & 0.303 & 0.257 \end{bmatrix},$$

and by choosing $\gamma = (1/3)[1 \quad 1 \quad 1]^T$, it is obtained that $k_1 = 3.04$, $k_2 = 0.28$. If $p = 16$,

$$B_w = \begin{bmatrix} 0.922 & 10.107 & 4.019 & 2.271 & 1.534 & 1.138 \\ 0.303 & 0.303 & 0.114 & 0.107 & 0.148 & 0.216 \end{bmatrix},$$

and by choosing $\gamma = (1/15)[10 \quad 1 \quad 1 \quad 1 \quad 1 \quad 1]^T$, it is obtained that $k_1 = 1.88$, $k_2 = 0.26$.

1.4.4.2 SOS representation

Both polynomials in (1.4.40) are defined only on the positive quadrant, and the degree of V is not even. Therefore, if we want to apply SOS representation, we have to globalize such polynomials. This can be done by means of the transformation $(z_1, z_2) \mapsto (y_1^2, y_2^2)$, thus

$$V(y) = \alpha_1 y_1^6 - \alpha_2 y_1^4 y_2^2 + \alpha_3 y_2^6, \quad W(y) = \beta_1 y_1^4 - \beta_2 y_1^2 y_2^2 + \beta_3^+ y_2^4, \quad y \in \mathbb{R}^2.$$

As it was stated in reference 19, the problem of finding an SOS representation of a polynomial can be transformed into a problem of solving an LMI by writing the polynomial as a quadratic form. However, that quadratic form is in general only positive semidefinite. Therefore, instead of addressing the problem $V(y) > 0$, we are going to deal with $\bar{V}(y) = V(y) - \varepsilon(y_1^6 + y_2^6) > 0$, fore some $\varepsilon > 0$. Thus, we can rewrite $\bar{V}$ as the following quadratic form:

$$\bar{V}(y) = \psi^T(y) Q_v \psi(y), \quad \psi(y) = [y_1^3,\ y_1^2 y_2,\ y_1 y_2^2,\ y_2^3]^T.$$

[2] See, e.g., Skeleton [31], software available in http://www.uic.nnov.ru/~zny/skeleton.

So, the problem is reduced to find a positive definite matrix Q_v. There are several software packages that can be used to solve this problem, e.g., by using SOSTOOLS [23], it is obtained

$$Q_v = \begin{bmatrix} 10.17 & 0 & -5.30 & 0 \\ 0 & 6.19 & 0 & -3.81 \\ -5.30 & 0 & 7.63 & 0 \\ 0 & -3.81 & 0 & 9.10 \end{bmatrix},$$

for $\varepsilon = 0.1$. The eigenvalues of Q_v are $\{14.35,\ 3.60,\ 3.44,\ 11.80\}$, thus

$$V(y) = 10.27y_1^6 - 4.41y_1^4y_2^2 + 9.30y_2^6$$

is positive definite. Analogously for W, we have that $\bar{W}(y) = W(y) - \varepsilon(y_1^4 + y_2^4)$. Rewriting it as

$$\bar{W}(y) = \psi^T(y)Q_w\psi(y), \quad \psi(y) = [y_1^2,\ y_1y_2,\ y_2^2]^T,$$

we can find the positive definite matrix

$$Q_w = \begin{bmatrix} 148.15 & 0 & -102.64 \\ 0 & 8.56 & 0 \\ -102.64 & 0 & 71.74 \end{bmatrix},$$

for $k_1 = 1,\ k_2 = 0.1$. The eigenvalues of Q_w are $\{219.47,\ 0.42,\ 8.56\}$, thus

$$W(y) = 148.16y_1^4 - 196.72y_1^2y_2^2 + 71.75y_2^4$$

is a positive definite function. Although it is not necessary to show the SOS representation, we give it as follows:

$$\begin{aligned} \bar{V}(y) &= (3.04y_1^3 - 0.93y_1y_2^2)^2 + (-0.93y_1^3 + 2.59y_1y_2^2)^2 \\ &\quad + (2.38y_1^2y_2 - 0.71y_2^3)^2 + (-0.71y_1^2y_2 + 2.94y_2^3)^2, \end{aligned}$$

$$\bar{W}(y) = 8.56y_1^2y_2^2 + (10.20y_1^2 - 6.63y_2^2)^2 + (-6.63y_1^2 + 5.26y_2^2)^2.$$

In the last representation, there can be some residuals due to round-off error. The following is the SOS representation for V in original coordinates, recall that it is defined for the set $\{x_1x_2 > 0\}$:

$$\begin{aligned} V(x) - \varepsilon(|x_1|^{\frac{3}{2}} + |x_2|^3) &= (\alpha_1 - \varepsilon)|x_1|^{\frac{3}{2}} - \alpha_2x_1x_2 + (\alpha_3 - \varepsilon)|x_2|^3 \\ &= (3.04|x_1|^{\frac{3}{4}} - 0.93|x_1|^{\frac{1}{4}}|x_2|)^2 \\ &\quad + (-0.93|x_1|^{\frac{3}{4}} + 2.59|x_1|^{\frac{1}{4}}|x_2|)^2 \\ &\quad + (2.38|x_1|^{\frac{1}{2}}|x_2|^{\frac{1}{2}} - 0.71|x_2|^{\frac{3}{2}})^2 \\ &\quad + (-0.71|x_1|^{\frac{1}{2}}|x_2|^{\frac{1}{2}} + 2.94|x_2|^{\frac{3}{2}})^2. \end{aligned}$$

Method	Class of systems	Order	Advantages	Disadvantages
Integration	Piecewise affine	Arbitrary	• Necessary and sufficient conditions are obtained in nominal case • It can be extended to disturbed second-order systems • The existence of the function is guaranteed	• Computing system's solutions for orders greater than three can be very complicated • The extension for disturbed third-order systems is not clear
Variable reduction	Homogeneous	Second	• All the homogeneous Lyapunov functions are parameterized by the function W • The existence of the function is guaranteed	• The expression for the Lyapunov function is integral
Generalized forms	Generalized forms	Arbitrary	• Flexibility to analyze positiveness (available software)	• The existence of the function in the class is not guaranteed • Complexity in computation for high-order systems

Figure 1.4.6 Comparison of methods

We have described, through the above example, a process of designing a Lyapunov function by means of generalized forms. That procedure can be roughly summarized in the following steps:

Step 1. Consider the system $\dot{x} = f(x)$, $x \in \mathbb{R}^n$, whose vector field f is described by generalized forms. Obtain the homogeneous degree k and weights r of the system.

Step 2. Choose a generalized form $V : \mathbb{R}^n \to \mathbb{R}$ (of homogeneous degree m with the weights r) given by

$$V(x) = \sum_{i=1}^{n} \alpha_i |x_i|^{\frac{m}{r_i}} + P(\alpha_j, x), \quad j > n.$$

$P(\alpha_j, x)$ consisting of mixed-terms with coefficients α_j.

Step 3. Compute the generalized form $W(x) = -\dot{V}$.

Step 4. Apply an adequate change of variable $x \mapsto z$ in order to transform the generalized forms $V(x), W(x)$ into a set of classical forms $V(z)$, $W(z)$.

Step 5. Determine the coefficients of V and the parameters of the system such that the set of forms $V(z)$, $W(z)$ be positive definite. This can be done through Pólya's theorem or SOS representation.

1.4.5 Conclusions

To conclude this chapter, we summarize some characteristics of the methods in the chart of Figure 1.4.6.

Acknowledgment

The authors thank the financial support from PAPIIT-UNAM (Programa de Apoyo a Proyectos de Investigación e Innovación Tecnológica), project IN113614; Fondo de Colaboración II-FI UNAM, Project IISGBAS-100-2015; CONACyT (Consejo Nacional de Ciencia y Tecnología), project 241171; and CONACyT CVU:371652.

References

[1] A. Bacciotti and L. Rosier. *Lyapunov Functions and Stability in Control Theory*. Communications and Control Engineering. Springer, Berlin, 2nd edition, 2005.

[2] R. Baier, L. Grüne, and S. F. Hafstein. Linear programming based Lyapunov function computation for differential inclusions. *Discrete and Continuous Dynamical Systems—Series B*, 17(1):33–56, 2012.

[3] E. Bernuau, D. Efimov, W. Perruquetti, and A. Polyakov. On homogeneity and its application in sliding mode control. *Journal of the Franklin Institute*, 351(4):1866–1901, 2014.

[4] C. N. Delzell. Impossibility of extending Pólyas theorem to "forms" with arbitrary real exponents. *Journal of Pure and Applied Algebra*, 212:2612–2622, 2008.

[5] S. V. Emel'yanov, S. K. Korvin, and A. Levant. High-order sliding modes in control systems. *Computational Mathematics and Modeling*, 7(3):294–318, 1996.

[6] R. A. Freeman and P. V. Kokotovic. *Robust Nonlinear Control Design: State Space and Lyapunov Techniques*. Modern Birkhauser Classics, Boston, 1996.

[7] L. Fridman and A. Levant. *Sliding Mode Control in Engineering*, Chapter 3. Marcel Dekker Inc., NY, 2002.

[8] H. K. Khalil. *Nonlinear Systems*. Prentice Hall, 3rd edition, Upper Saddle River, (N.J.), 2002.

[9] N. N. Krasovskii. *Problems of the Theory of Stability of Motion*. Stanford University Press, Stanford, CA, 1963.

[10] A. Levant. Sliding order and sliding accuracy in sliding mode control. *International Journal of Control*, 58(6):1247–1263, 1993.

[11] A. Levant. Universal SISO sliding-mode controllers with finite-time convergence. *IEEE Transactions on Automatic Control*, 46(9), 2001.

[12] A. Levant. Homogeneity approach to high-order sliding mode design. *Automatica*, 41:823–830, 2005.

[13] A. Levant. Principles of 2 sliding mode design. *Automatica*, 43(4):576–586, 2007.

[14] F. Lopez-Ramirez, T. Sanchez, and J. A. Moreno. Construction of Lyapunov functions for homogeneous second-order systems. *IEEE 53rd Annual Conference on Decision and Control (CDC)*, 5494–5499, 2014.

[15] Z. Man, A. P. Paplinski, and H. R. Wu. A robust MIMO terminal sliding mode control for rigid robotic manipulators. *IEEE Transactions on Automatic Control*, 39(12):2464–2468, 1994.

[16] J. A. Moreno and M. Osorio. A Lyapunov approach to second-order sliding mode controllers and observers. *47th IEEE Conference on Decision and Control*, 2008.

[17] H. Nakamura, Y. Yamashita, and H. Nishitani. Smooth Lyapunov functions for homogeneous differential inclusions. In: *Proceedings of the 41st SICE Annual Conference*, 3:1974–1979, 2002.

[18] Y. Orlov. Finite time stability and robust control synthesis of uncertain switched systems. *SIAM Journal on Control and Optimization*, 43(4):1253–1271, 2005.

[19] P. A. Parrilo. *Structured Semidefinite Programs and Semialgebraic Geometry Methods in Robustness and Optimization*. PhD thesis, California Institute of Technology, Pasadena, California, 2000.

[20] G. Pólya. Über positive Darstellung von Polynomen. *Vierteljahrschrift Naturforschenden Ges*, 73:141–145, 1928.

[21] A. Polyakov and A. Poznyak. Lyapunov function design for finite-time convergence analysis: "twisting" controller for second-order sliding mode realization. *Automatica*, 45(2):444–448, 2009.

[22] A. Polyakov and A. Poznyak. Unified Lyapunov function for a finite-time stability analysis of relay second-order sliding mode control systems. *IMA Journal of Mathematical Control and Information*, 29(4):529–550, 2012.

[23] S. Prajna, A. Papachristodoulou, and P. A. Parrilo. SOSTOOLS: sum of squares optimization toolbox for MATLAB, 2002–2005. Available from: www.cds.caltech.edu/sostools and www.mit.edu/~parrilo/sostools.

[24] L. Rosier. Homogeneous Lyapunv function for homogeneous continuous vector field. *Systems & Control Letters*, 19(6):467–473, 1992.

[25] T. Sánchez and J. A. Moreno. Construction of Lyapunov functions for a class of higher order sliding modes algorithms. In: *51st IEEE Conference on Decision and Control (CDC)*, 6454–6459, 2012.

[26] T. Sanchez and J. A. Moreno. A constructive Lyapunov function design method for a class of homogeneous systems. In: *IEEE 53rd Annual Conference on Decision and Control (CDC)*, 5500–5505, 2014.

[27] R. Santiesteban, L. Fridman, and J. A. Moreno. Finite-time convergence analysis for "twisting" controller via a strict Lyapunov function. *11th International Workshop on Variable Structure Systems (VSS), Mexico City, Mexico*, 2010.

[28] D. G. Schultz and J. E. Gibson. The Variable Gradient Method for generating Lyapunov Functions. *Transactions of the American Institute of Electrical Engineers*, 81(4):203–210, 1962.

[29] Y. Shtessel, C. Edwards, L. Fridman, and A. Levant. *Sliding Mode Control and Observation*. Control Engineering. Birkháuser, New York, 2013.

[30] V. Utkin, J. Guldner, and J. Shi. *Sliding Mode Control in Electro–Mechanical Systems*. CRC Press, Taylor & Francis, London, UK, 2nd edition, 2009.

[31] N. Y. Zolotykh. New modification of the double description method for constructing the skeleton of a polyhedral cone. *Computational Mathematics and Mathematical Physics*, 52(1):146–156, 2012.

[32] V. I. Zubov. *Methods of A. M. Lyapunov and Their Applications*. P. Noordho Limited, Groningen, 1964.

Section 2

Properties of sliding mode algorithms

Chapter 2.1
Homogeneity of differential inclusions

Emmanuel Bernuau, Denis Efimov, Wilfrid Perruquetti, and Andrei Polyakov

Abstract

In this chapter, the notion of geometric homogeneity is extended for differential inclusions. This kind of homogeneity provides the most advanced coordinate-free framework for analysis and synthesis of nonlinear discontinuous systems. The main qualitative properties of continuous homogeneous systems are extended to the discontinuous setting: the equivalence of the global asymptotic stability and the existence of a homogeneous Lyapunov function; the link between finite-time stability and negative degree of homogeneity; the equivalence between attractivity and asymptotic stability are among the proved results.

2.1.1 Introduction

The *homogeneity* is an intrinsic property of an object on which the flow of a particular vector field operates as a scaling. This definition, rather simple, entails a lot of qualitative properties for a homogeneous object, and is of particular interest in view of stability purposes. The study of the stability or the asymptotic stability of a dynamical system is a central problem in the control theory. Given that the equations of a system are very often impossible to integrate explicitly, indirect methods have to be used for getting qualitative properties. Even though the results of Kurzweil [1] and Clarke *et al.* [2] prove the equivalence of the asymptotic stability and the existence of a smooth Lyapunov function, finding such a Lyapunov function may be a very difficult task. Qualitative results not involving the computation of a Lyapunov function are therefore of a great interest. This is why the homogeneity theory has been developed and used in control theory: the rigid properties of homogeneous systems simplify the study of the stability and give sufficient conditions for deriving it. The literature on the homogeneity theory is vast and detailed. A lot of theoretical and practical results have been proved in the last decades, and used in different context.

However, almost all the works done for homogeneity assume *continuity* [3–5]. Though, throughout the last decades, interests on discontinuous systems have been increasing. The theory of differential inclusions (DIs) is well established [6, 7]. Among others, it appears in optimal control theory or viability theory; when dealing with variable structure systems, systems with adaptive control, power electronic systems with switching devices or mechanical systems with friction, discontinuous right-hand sides appear naturally. Finally, the sliding mode control theory makes an important use of discontinuous controller to achieve finite-time stability (FTS) as well as robustness and the construction of a smooth homogeneous Lyapunov function is often used to prove the FTS of the system, especially when dealing with high-order sliding mode controls. Extensions of the homogeneity theory to discontinuous systems and to DIs constitute therefore a natural answer to theoretical and practical problems in control theory.

Only few extensions of the homogeneity theory to DI have yet been developed in the literature. First of all, Filippov [6] defined homogeneity for DI, but only in the context of classical homogeneity. Levant [8] and Orlov [9] studied also the subject in the context of weighted homogeneity, but important properties of continuous homogeneous systems were not extended. Finally, as far as we know, except our work [10] (this paper does not contain proofs), nothing has been done about geometric homogeneity for DI.

In this chapter, we shall define homogeneity for DIs, connect this definition with the usual definition for vector fields, and extend all the useful properties of homogeneous ordinary differential equation (ODE) to homogeneous DI. We shall particularly see how the *flow commutation property* can be extended, demonstrate a Rosier's theorem on the existence of a Lyapunov function for globally asymptotically stable (GAS) discontinuous systems and prove that the qualitative properties on homogeneous systems still apply with some slight changes. This chapter is organized as follows. Section 2.1.2 will recall the notions on DI that we will use in the sequel and the usual definitions of homogeneity for vector fields. Section 2.1.3 will be devoted to the definition and the first properties of homogeneous DIs. Section 2.1.4 will then present the main results on the qualitative properties of systems given by a homogeneous differential equation. A conclusion will finally sum up the chapter and give some directions of future works.

2.1.2 Preliminaries

2.1.2.1 Notations

We endow $\mathbb{R}^n$ with the Lebesgue measure and denote by $\mathcal{N}$ the set of zero-measure sets. If $x \in \mathbb{R}^n$ and $\varepsilon > 0$, the set $B(x, \varepsilon)$ denotes the Euclidean open ball centered at x and of radius ε. The set of locally essentially bounded vector fields on $\mathbb{R}$ is denoted by $\mathscr{L}^{\infty}_{\text{loc}}(\mathbb{R}^n, \mathbb{R}^n)$. Throughout the chapter, all considered vector fields belong to $\mathscr{L}^{\infty}_{\text{loc}}(\mathbb{R}^n, \mathbb{R}^n)$.

If $V : \mathbb{R}^n \to \mathbb{R}$ is a function (respectively, $\Phi : \mathbb{R}^n \to \mathbb{R}$ is a diffeomorphism), we denote $d_x V$ the differential of the function V at point $x \in \mathbb{R}^n$ (respectively, $d_x \Phi$ the differential of the diffeomorphism Φ at point $x \in \mathbb{R}^n$).

2.1.2.2 Differential inclusions

We refer to References 6, 7 for the basic definitions and the technical material on set-valued map and DIs. In this section, we will only recall the definitions and results that will be used hereafter, without any proof.

The Filippov's regularization procedure consists in the construction of a set-valued map F starting with a vector field $f \in \mathscr{L}^{\infty}_{\text{loc}}(\mathbb{R}^n, \mathbb{R}^n)$:

$$\mathscr{F}[f](x) = \bigcap_{\varepsilon > 0} \bigcap_{N \in \mathscr{N}} \overline{\text{conv}}(f(B(x, \varepsilon) \setminus N)). \tag{2.1.1}$$

By construction, for all $x \in \mathbb{R}^n$, the set $\mathscr{F}[f](x)$ is closed and convex. Moreover, the set-valued map $\mathscr{F}[f]$ is upper semi-continuous.

In many applications, the DI is given by the set-valued map coming from the Filippov's procedure. We will therefore focus on set-valued map with the properties inherited by this procedure.

Definition 2.1.1. *Let F be a set-valued map. We say that F verifies the* standard assumptions *if F is upper semi-continuous and if for any $x \in \mathbb{R}^n$, $F(x)$ is a nonempty compact convex set. The set of set-valued maps on $\mathbb{R}^n$ verifying the standard assumptions is denoted by* $\mathbf{F}(\mathbb{R}^n)$.

For DI, the standard assumptions suffice to get the existence of a solution to any Cauchy problem. That means that any differential equation with a locally essentially bounded right-hand side has a solution for any initial condition in the aforesaid sense.

In the classical setting of continuous vector fields with forward uniqueness of solutions, the flow or the semi-flow of the vector field provides a lot of qualitative information about the system. When the forward uniqueness is lost, a flow does not exist anymore. We shall in the sequel define a *generalized flow* and set its properties. The proofs of these results can be found in Reference 6 or 7.

Consider the following autonomous DI:

$$\dot{x} \in F(x), \quad x \in \mathbb{R}^n, \quad F \in \mathbf{F}(\mathbb{R}^n). \tag{2.1.2}$$

We say that a solution x of (2.1.2) *starts at* x_0 if x is defined on an interval containing 0 and $x(0) = x_0$. We will denote by $\mathscr{S}([0, T], A)$ the set of solutions of (2.1.2) defined on the interval $[0, T]$, $T > 0$, starting in $A \subset \mathbb{R}^n$. We also allow $T = +\infty$, and in this situation the interval $[0, T]$ has to be understood as $[0, +\infty]$. We will also denote $\mathscr{S}([0, T], x_0) = \mathscr{S}([0, T], \{x_0\})$.

Let $T \in]0, +\infty]$ be such that every solution of (2.1.2) starting in A is defined on $[0, T]$. We denote $\Psi^T(A) = \{x(T) : x \in \mathscr{S}([0, T], A)\}$. This set is the reachable set from A at time T, or the limit in case $T = +\infty$. Let us stress that with the assumption of uniqueness of solutions in forward time, Ψ^t corresponds to the semi-flow of F; this remark justifies that we call Ψ the *generalized flow* of F.

Theorem 2.1.1. *Let $K \subset \mathbb{R}^n$ be compact. There exists $T_0 > 0$ such that all the solutions of (2.1.2) starting in K are defined on the whole interval $[0, T_0]$. Moreover, for all $T > 0$ such that all the solutions of (2.1.2) starting in K are defined on the whole interval $[0, T]$, there exists a compact set $L \subset \mathbb{R}^n$ such that for all $t \in [0, T]$, $\Psi^t(K) \subset L$ and the set $\Psi^t(K)$ is compact.*

Corollary 2.1.1. *Let $T > 0$ be such that all solutions starting from the compact set $K \subset \mathbb{R}^n$ exist on $[0, T]$. Then $\mathscr{S}([0, T], K)$ is compact for the topology of uniform convergence on $[0, T]$.*

2.1.2.3 Homogeneity

We shall now present the geometric homogeneity, which needs first the notion of Euler vector field to be defined.

Definition 2.1.2 ([11]). *A vector field $\nu \in \mathscr{C}^1(\mathbb{R}^n, \mathbb{R}^n)$ is said to be* Euler *if it is complete and if the origin is a GAS equilibrium of $-\nu$. We will always write Φ the flow of ν, i.e., $\Phi^s(x)$ is the current state at time s of the trajectory of ν starting from x at $s = 0$.*

Definition 2.1.3. *Let ν be an Euler vector field.*

- *A function $V : \mathbb{R}^n \to \mathbb{R}$ is ν-homogeneous of degree $\kappa \in \mathbb{R}$ if:*

$$V(\Phi^s(x)) = e^{\kappa s} V(x) \quad \forall x \in \mathbb{R}^n, \quad \forall s \in \mathbb{R}.$$

- *A vector field $f : \mathbb{R}^n \to \mathbb{R}^n$ is ν-homogeneous of degree $\kappa \in \mathbb{R}$ if:*

$$f(\Phi^s(x)) = e^{\kappa s} d_x \Phi^s f(x) \quad \forall x \in \mathbb{R}^n, \quad \forall s \in \mathbb{R}. \tag{2.1.3}$$

The relation (2.1.3) can be recast under the more compact form $\mathscr{H}^s_\kappa(f) = f$, where the vector field $\mathscr{H}^s_\kappa(f)$ is defined by:

$$\mathscr{H}^s_\kappa(f) : x \mapsto e^{-\kappa s} (d_x \Phi^s)^{-1} f(\Phi^s(x)). \tag{2.1.4}$$

Consider now a homogeneous vector field f such that the following equation admits solutions:

$$\dot{x} = f(x). \tag{2.1.5}$$

Then the homogeneity property of f induces a very important property on the solutions of (2.1.5).

Proposition 2.1.1 ([12]). *Assume that the vector field $f : \mathbb{R}^n \to \mathbb{R}^n$ is ν-homogeneous of degree κ. For any solution $x(t)$ of (2.1.5) and for all $s \in \mathbb{R}$, the curve $t \mapsto \Phi^s(x(e^{\kappa s} t))$ is a solution of (2.1.5). If the system (2.1.5) admits a (semi-)flow $\Psi^t(x)$, we have*

$$\Phi^s \circ \Psi^{e^{\kappa s} t} = \Psi^t \circ \Phi^s. \tag{2.1.6}$$

2.1.3 Homogeneous DIs

In this section, we continue to consider the DI (2.1.2) for which the standard assumptions hold. Our aim is to define a homogeneity notion consistent with the conventional definition and see how the nice properties of the homogeneous continuous systems can be generalized. We adopt hereafter a natural definition, which is also a straightforward extension of the definition of Reference 8 (given only for weighted homogeneity).

Definition 2.1.4. *Let ν be an Euler vector field. A set-valued map $F : \mathbb{R}^n \rightrightarrows \mathbb{R}^n$ is ν-homogeneous of degree $\kappa \in \mathbb{R}$ if for all $s \in \mathbb{R}$ we have $\mathscr{H}_\kappa^s(F) = F$, where we extend the operator $\mathscr{H}_\kappa^s$ defined in (2.1.4) by $\mathscr{H}_\kappa^s(F) : x \mapsto e^{-\kappa s}\,(d_x\Phi^s)^{-1}\,F\,(\Phi^s(x))$.*

The following lemma is a result of an easy application of the definitions.

Lemma 2.1.1. *Let F be a set-valued map satisfying the standard assumptions. Then for all $\kappa \in \mathbb{R}$ and all $s \in \mathbb{R}$, the set-valued map $\mathscr{H}_\kappa^s(F)$ satisfies the standard assumptions.*

The following proposition is an extension of Proposition 2.1.1.

Proposition 2.1.2. *Let F be a ν-homogeneous set-valued map of degree κ, satisfying the standard assumptions. Then for all $x_0 \in \mathbb{R}^n$ and any solution x of the system (2.1.2) starting at x_0 and all $s \in \mathbb{R}$, the absolute continuous curve $t \mapsto \Phi^s\,(x\,(e^{\kappa s}t))$ is a solution of the system (2.1.2) starting at $\Phi^s(x_0)$.*

Proof. Consider a solution x of (2.1.2) starting at x_0. The curve $t \mapsto \Phi^s\,(x\,(e^{\kappa s}t))$ is clearly an absolute continuous curve for all $s \in \mathbb{R}$. Moreover, for almost all $t \in \mathbb{R}$, we have:

$$\begin{aligned} \frac{d}{dt}\Phi^s\,(x\,(e^{\kappa s}t)) &= e^{\kappa s} d_{x(e^{\kappa s}t)}\Phi^s \dot{x}\,(e^{\kappa s}t) \\ &\in e^{\kappa s} d_{x(e^{\kappa s}t)}\Phi^s F\,(x\,(e^{\kappa s}t))\,. \end{aligned}$$

Since F is ν-homogeneous of degree κ, we find that $\frac{d}{dt}\Phi^s(x(e^{\kappa s}t)) \in F(\Phi^s(x(e^{\kappa s}t)))$ and thus the curve $t \mapsto \Phi^s(x(e^{\kappa s}t))$ is a solution of (2.1.2) for all $s \in \mathbb{R}$. □

Remark 2.1.1. *This proposition may also be recast using the generalized flow, stating that, for all $t \geq 0$, for all $s \in \mathbb{R}$ and all compact sets $K \subset \mathbb{R}^n$:*

$$\Psi^t(\Phi^s(K)) = \Phi^s\big(\Psi^{e^{\kappa s}t}(K)\big)\,. \tag{2.1.7}$$

Now, similarly to the usual setting, a lot of properties can be extended from a sphere to everywhere outside the origin by homogeneity.

Proposition 2.1.3. *Let F be a ν-homogeneous set-valued map of degree κ, satisfying the standard assumptions. Then $F(x)$ is compact for all $x \in \mathbb{R}^n \setminus \{0\}$ iff $F(x)$ is compact for all $x \in \mathbb{S}$, where $\mathbb{S} = \{x \in \mathbb{R}^n : \|x\| = 1\}$. The same property holds for convexity or upper semi-continuity.*

Proof. The result about compactness or convexity is straightforward. Let us only prove that if $F(x)$ is upper semi-continuous (USC) on the sphere, so is F everywhere outside of the origin.

Set $y \neq 0$. There exists $s \in \mathbb{R}$ and $x \in \mathbb{S}$ such that $\Phi^s(x) = y$. Fix $\mathscr{V}$ a neighborhood of $F(y) = F(\Phi^s(x)) = e^{\kappa s} d_x \Phi^s F(x)$. Eventually, replacing $\mathscr{V}$ by a bounded neighborhood of $F(y)$ included in $\mathscr{V}$, we assume that $\mathscr{V}$ is bounded. Consider a bounded neighborhood $\mathscr{V}_0 \subset \mathscr{V}$ of $F(y)$ such that there exists $\alpha > 0$ with $\mathrm{d}(\mathscr{V}_0, \partial\mathscr{V}) \geq \alpha$, and denote by $\tilde{\mathscr{V}}_0 = e^{-\kappa s}(d_x\Phi^s)^{-1}\mathscr{V}_0$. $\tilde{\mathscr{V}}_0$ is a neighborhood of $F(x)$. Let us denote $M = \sup_{v \in \mathscr{V}_0} \|v\| > 0$. Let us also denote by $\sigma_{\max}(d_z\Phi^s(d_x\Phi^s)^{-1})$ the biggest singular value of the linear mapping $d_z\Phi^s(d_x\Phi^s)^{-1}$. The function φ : $z \mapsto \left|\sigma_{\max}(d_z\Phi^s(d_x\Phi^s)^{-1}) - 1\right|$ is continuous and vanishes at $z = x$. Therefore, there exists a neighborhood $\tilde{\mathscr{U}}$ of x on which $\varphi(z) < \frac{\alpha}{M}$. By upper semi-continuity of F at x, there exists $\tilde{\mathscr{U}}_0$ a neighborhood of x such that for all $\tilde{z} \in \tilde{\mathscr{U}}_0$, $F(\tilde{z}) \subset \tilde{\mathscr{V}}_0$. Set $\mathscr{U} = \Phi^s(\tilde{\mathscr{U}} \cap \tilde{\mathscr{U}}_0)$, then $\mathscr{U}$ is a neighborhood of y. Let z be an element of $\mathscr{U}$. Then there exists $\tilde{z} \in \tilde{\mathscr{U}} \cap \tilde{\mathscr{U}}_0$ such that $z = \Phi^s(\tilde{z})$. Therefore, $F(z) = F(\Phi^s(\tilde{z})) = e^{\kappa s} d_{\tilde{z}}\Phi^s F(\tilde{z}) \subset e^{\kappa s} d_{\tilde{z}}\Phi^s \tilde{\mathscr{V}}_0$ since $\tilde{z} \in \tilde{\mathscr{U}}_0$. But $\tilde{\mathscr{V}}_0 = e^{-\kappa s}(d_x\Phi^s)^{-1}\mathscr{V}_0$, thus $F(z) \subset d_{\tilde{z}}\Phi^s(d_x\Phi^s)^{-1}\mathscr{V}_0$. Let $v \in \mathscr{V}_0$ be fixed. We have:

$$\left\|d_{\tilde{z}}\Phi^s \left(d_x\Phi^s\right)^{-1} v - v\right\| \leq \varphi\left(\tilde{z}\right) M.$$

Since $\tilde{z} \in \tilde{\mathscr{U}}$, we find $\left\|d_{\tilde{z}}\Phi^s(d_x\Phi^s)^{-1} v - v\right\| \leq \alpha$, and hence $d_{\tilde{z}}\Phi^s(d_x\Phi^s)^{-1} v \in \mathscr{V}$. Finally, we conclude that $F(z) \subset \mathscr{V}$ and the proposition is proved. □

As we have seen, in many situations, the set-valued map F comes from the Filippov's regularization procedure of a discontinuous vector field f. Suppose that we have a vector field f, which is homogeneous in the sense of Definition 2.1.3. If we apply the regularization procedure, is the homogeneity property preserved in the sense of Definition 2.1.4? The answer is positive as shown in Corollary 2.1.2.

Proposition 2.1.4. *Let $f \in \mathscr{L}^{\infty}_{\mathrm{loc}}(\mathbb{R}^n, \mathbb{R}^n)$ be a vector field. Then $\mathscr{H}^s_\kappa(\mathscr{F}[f]) = \mathscr{F}[\mathscr{H}^s_\kappa(f)]$ holds for all $(s,\kappa) \in \mathbb{R} \times \mathbb{R}$.*

Proof. Since for all $\varepsilon > 0$ there exist $\varepsilon_- > 0$ and $\varepsilon_+ > 0$ such that $\Phi^s(B(x, \varepsilon_-)) \subset B(\Phi^s(x), \varepsilon) \subset \Phi^s(B(x, \varepsilon_+))$, we have

$$\begin{aligned}
\mathscr{F}[f](\Phi^s(x)) &= \textstyle\bigcap_{\varepsilon>0} \bigcap_{N\in\mathscr{N}} \overline{\operatorname{conv}}\,(f(y), y \in B(\Phi^s(x), \varepsilon) \setminus N) \\
&= \textstyle\bigcap_{\varepsilon>0} \bigcap_{N\in\mathscr{N}} \overline{\operatorname{conv}}\,(f(y), y \in \Phi^s(B(x, \varepsilon)) \setminus N) \\
&= \textstyle\bigcap_{\varepsilon>0} \bigcap_{N\in\mathscr{N}} \overline{\operatorname{conv}}\,(f(\Phi^s(z)), z \in B(x, \varepsilon) \setminus N)\,.
\end{aligned}$$

Hence, we find that

$$\mathscr{H}^s_\kappa(\mathscr{F}[f])(x) = \bigcap_{\varepsilon>0} \bigcap_{N\in\mathscr{N}} \overline{\operatorname{conv}}\left((d_x\Phi^s)^{-1} d_z\Phi^s \mathscr{H}^s_\kappa(f)(z), z \in B(x, \varepsilon) \setminus N\right).$$

Let us denote by $\sigma_{\max}\left((d_x\Phi^s)^{-1} d_z\Phi^s\right)$ the biggest singular value of the linear mapping $(d_x\Phi^s)^{-1} d_z\Phi^s$. The function $\phi : z \mapsto \left|\sigma_{\max}\left((d_x\Phi^s)^{-1} d_z\Phi^s\right) - 1\right|$ is continuous and therefore bounded on $B(x,\varepsilon)$ and moreover vanishes at $z = x$. For all $z \in B(x,\varepsilon)$, we have:

$$\left\|(d_x\Phi^s)^{-1}\, d_z\Phi^s \mathscr{H}_\kappa^s(f)(z) - \mathscr{H}_\kappa^s(f)(z)\right\| \le M(\varepsilon),$$

where the function M is defined by $M(0) = 0$ and for $\varepsilon \neq 0$ by $M(\varepsilon) = \sup_{z\in B(x,\varepsilon)}\varphi(z)\,\text{ess sup}_{z\in B(x,\varepsilon)}\left\|\mathscr{H}_\kappa^s(f)(z)\right\|$. The function M is continuous at zero. We have proved that $(d_x\Phi^s)^{-1}\, d_z\Phi^s\mathscr{H}_\kappa^s(f)(z) \in \mathscr{H}_\kappa^s(f)(z) + B(0, M(\varepsilon))$. It follows that

$$\begin{aligned}\mathscr{H}_\kappa^s(\mathscr{F}[f])(x) &\subset \textstyle\bigcap_{\varepsilon>0}\bigcap_{N\in\mathscr{N}}\overline{\text{conv}}\left(\mathscr{H}_\kappa^s(f)(z) + B(0,M(\varepsilon))\,, z\in B(x,\varepsilon)\setminus N\right)\\ &= \textstyle\bigcap_{\varepsilon>0}\left[\left(\bigcap_{N\in\mathscr{N}}\overline{\text{conv}}\left(\mathscr{H}_\kappa^s(f)(z), z\in B(x,\varepsilon)\setminus N\right)\right) + B(0,M(\varepsilon))\right]\\ &= \textstyle\bigcap_{\varepsilon>0}\bigcap_{N\in\mathscr{N}}\overline{\text{conv}}\left(\mathscr{H}_\kappa^s(f)(z), z\in B(x,\varepsilon)\setminus N\right)\mathscr{F}\left[\mathscr{H}_\kappa^s(f)\right](x).\end{aligned}$$

The proof of the converse inclusion is similar. Let K be a compact, convex neighborhood of $F(x)$. Following the previous consideration, there exists $\varepsilon_0 > 0$ such that for all $0 < \varepsilon < \varepsilon_0$ and for all $z \in B(x,\varepsilon)$, we have $F(z) + B(0, M(\varepsilon)) \subset \mathring{K}$, where $\mathring{K}$ denotes the interior of the set K. Therefore, $\overline{\text{conv}}\,\{F(z) + B(0, M(\varepsilon))\} \subset K$ and finally $F(\Phi^s(x)) \subset e^{\kappa s} d_x\Phi^s K$. Being compact and convex, $F(x)$ is equal to the intersection of all its compact convex neighborhood and hence $F(\Phi^s(x)) \subset e^{\kappa s} d_x\Phi^s F(x)$. Applying the same proof to $y = \Phi^s(x)$, we find $F(x) = F(\Phi^{-s}(y)) \subset e^{-\kappa s} d_y\Phi^{-s}F(y) = e^{-\kappa s}(d_x\Phi^s)^{-1}F(\Phi^s(x))$ and thus $F(\Phi^s(x)) = e^{\kappa s} d_x\Phi^s F(x)$. □

Corollary 2.1.2. *Let $f \in \mathscr{L}^\infty_{\text{loc}}(\mathbb{R}^n, \mathbb{R}^n)$ be a vector field. Suppose f is ν-homogeneous of degree κ. Then $\mathscr{F}(f)$ is ν-homogeneous of degree κ.*

Proof. Since f is ν-homogeneous of degree κ, we have $\mathscr{H}_\kappa^s(f) = f$. Hence, $\mathscr{F}(f) = \mathscr{F}\left(\mathscr{H}_\kappa^s(f)\right) = \mathscr{H}_\kappa^s(\mathscr{F}(f))$ by Proposition 2.1.4 and therefore $\mathscr{F}(f)$ is ν-homogeneous of degree κ. □

Example 2.1.1. *Consider the n-integrator with an input given by $u(x) = -\sum_i k_i \text{sign}(x_i)$, $k_i > 0$:*

$$\begin{cases}\dot{x}_1 &= x_2\\ \vdots & \quad\vdots\\ \dot{x}_{n-1} &= x_n\\ \dot{x}_n &= -\sum_i k_i sign(x_i)\end{cases}.$$

It is easy to check that this vector field is ν-homogeneous of degree -1 with $\nu(x) = (nx_1, \ldots, 2x_{n-1}, x_n)^T$. The associated DI is therefore ν-homogeneous of degree -1 as well.

2.1.4 Qualitative results on homogeneous discontinuous systems

In the previous section, we have seen how to define a homogeneous discontinuous system and the basic properties stemming from this definition. But the classical theory of homogeneity highlights a lot of very important and useful properties of homogeneous systems. Among those:

- the theorem of Rosier [13], which is a homogeneous converse Lyapunov theorem;
- the link between negative degree of homogeneity and FTS [14], and the properties of the settling-time function [15];
- the consequences of the existence of a strictly positively invariant (SPI) compact set [16];
- the equivalence of the notions of local attractivity and global stability for homogeneous systems [6, 16].

We shall generalize these properties in this section.

2.1.4.1 Converse Lyapunov theorem for homogeneous DIs

The following theorem asserts that a strongly GAS (i.e., asymptotic stability property holds for all solutions initiated from all initial conditions) homogeneous DI admits a homogeneous Lyapunov function. This result is a generalization of the theorem independently proved by Rosier [13] and Zubov [17] for continuous systems and of its extension, given in Reference 18, where weighted homogeneous locally essentially bounded vector fields were considered and the existence of a merely continuous Lyapunov function was proved.

Theorem 2.1.2. *Let F be a ν-homogeneous set-valued map of degree κ, satisfying the standard assumptions. Then the following statements are equivalent:*

- *The origin is (strongly) GAS for the system (2.1.2).*
- *For all $\mu > \max(-\kappa, 0)$, there exists a pair (V, W) of continuous functions, such that:*
  1. *V is of class $\mathscr{C}^\infty(\mathbb{R}^n, \mathbb{R})$, V is positive definite and ν-homogeneous of degree μ;*
  2. *W is of class $\mathscr{C}^\infty(\mathbb{R}^n \setminus \{0\}, \mathbb{R})$, W is strictly positive outside the origin and ν-homogeneous of degree $\mu + \kappa$;*
  3. *$\max_{v \in F(x)} d_x V v \leq -W(x)$ for all $x \neq 0$.*

Proof. By the result of Reference 2, the two following statements are equivalent:

- The system (2.1.2) is strongly GAS.
- There exist a pair (V_0, W_0) of continuous functions, such that:
  1. $V_0 \in \mathscr{C}^\infty(\mathbb{R}^n)$, V_0 is positive definite;
  2. $W_0 \in \mathscr{C}^\infty(\mathbb{R}^n \setminus \{0\})$, W_0 is strictly positive outside the origin;
  3. $\max_{v \in F(x)} d_x V_0 v \leq -W_0(x)$ for all $x \neq 0$.

Hence, it suffices to prove that the homogeneity condition allows us to build a homogeneous Lyapunov pair. The sequel of the proof is widely inspired by the proof from Reference 13. Let $a : [0, +\infty[\to [0, 1]$ be a $\mathscr{C}^\infty$ function such that for all $t \le 1$, $a(t) = 0$, for all $t \ge 2$, $a(t) = 1$ and for all $t \in]1, 2[$, $a'(t) > 0$. Set $\mu > \max(-\kappa, 0)$ and

$$V(x) = \int_{\mathbb{R}} e^{-\mu s} a(V_0(\Phi^s(x)))\, ds,$$

then $V(0) = 0$. For all $x \neq 0$, there exists s_1 such that for all $s \le s_1$, $V_0(\Phi^s(x)) \le 1$. Similarly, there exists s_2 such that for all $s \ge s_2$, $V_0(\Phi^s(x)) \ge 2$. Hence,

$$V(x) = \int_{s_1}^{s_2} e^{-\mu s} a(V_0(\Phi^s(x)))\, ds + \frac{e^{-\mu s_2}}{\mu},$$

and V is well defined.

The homogeneity of V is straightforward using a change of variable: $V(\Phi^\sigma(x)) = \int_{\mathbb{R}} e^{-\mu s} a\left(V_0(\Phi^s(\Phi^\sigma(x)))\right) ds = e^{\mu\sigma} \int_{\mathbb{R}} e^{-\mu u} a(V_0(\Phi^u(x))) du = e^{\mu\sigma} V(x)$.

On the other hand, for all $s \in \mathbb{R}$, $e^{-\mu s} a\left(V_0(\Phi^s(x))\right)$ is $\mathscr{C}^\infty$ and $\left|e^{-\mu s} a\left(V_0(\Phi^s(x))\right)\right| \le e^{-\mu s}$ which is integrable ($\mu > 0$). Thus V belongs to the class $\mathscr{C}^\infty$ on $\mathbb{R}^n$ and therefore proper [16]. Moreover, for all $v \in F(x)$:

$$d_x V v = \int_{\mathbb{R}} e^{-\mu s} a'\left(V_0(\Phi^s(x))\right) \left(d_{\Phi^s(x)} V_0\right) (d_x \Phi^s)\, v ds.$$

As F is homogeneous, there exists $\tilde{v} \in F(\Phi^s(x))$ such that $\tilde{v} = e^{\kappa s} d_x \Phi^s v$. Hence,

$$\begin{aligned} d_x V v &= \int_{\mathbb{R}} e^{-(\mu+\kappa)s} a'\left(V_0(\Phi^s(x))\right) \left(d_{\Phi^s(x)} V_0\right) \tilde{v} ds \\ &\le -\int_{\mathbb{R}} e^{-(\mu+\kappa)s} a'\left(V_0(\Phi^s(x))\right) W_0\left(\Phi^s(x)\right) ds. \end{aligned}$$

Let us denote

$$W(x) = \int_{\mathbb{R}} e^{-(\mu+\kappa)s} a'\left(V_0(\Phi^s(x))\right) W_0(\Phi^s(x))\, ds,$$

thus $\max_{v \in F(x)} d_x V v \le -W(x)$. It is clear that W is well defined and strictly positive. The function W is clearly homogeneous of degree $\kappa + \mu$ (this fact can be also proven using a simple change of variable). Moreover, for all $s \in \mathbb{R}$, the function $x \mapsto e^{-(\mu+\kappa)s} a'\left(V_0(\Phi^s(x))\right) W_0\left(\Phi^s(x)\right)$ is of class $\mathscr{C}^\infty(\mathbb{R}^n \setminus \{0\})$. Let us show that

$$\xi(x, s) = \left|e^{-(\mu+\kappa)s} a'\left(V_0(\Phi^s(x))\right) W_0(\Phi^s(x))\right|$$

is locally upper-bounded by an integrable function. Set $U_x = \bar{B}(x, |x|/2)$. For $x \neq 0$, U_x is a neighborhood of x. Since ν is Euler, there exists s_1, s_2 such that for all $y \in U_x \subset \mathbb{R}^n \setminus \{0\}$, for all $s \le s_1$, $V_0(\Phi^s(y)) \le 1$ and for all $s \ge s_2$, $V_0(\Phi^s(y)) \ge 2$. Hence, $a'(V_0(\Phi^s(y))) = 0$ for all $s \notin]s_1, s_2[$ and for all $y \in U_x$. Denote $c_1 = \sup_{y \in U_x} \sup_{s \in [s_1, s_2]} W_0(\Phi^s(y))$ and $c_2 = \sup_{t \in \mathbb{R}} a'(t)$. We get $\xi(x, s) \le e^{-(\mu+\kappa)s} \mathbf{1}_{U_x}(x) c_1 c_2$ which is clearly integrable (with $\mathbf{1}_A(x) = 1$ if $x \in A$ and 0 else). Therefore, W is $\mathscr{C}^\infty$ on a neighborhood of x for all $x \in \mathbb{R}^n \setminus \{0\}$, i.e., W is $\mathscr{C}^\infty$ on $\mathbb{R}^n \setminus \{0\}$.

The only point remaining to prove is the continuity of W at the origin. Let $\varepsilon > 0$ be fixed. There exists s_1 such that for all $s \leq s_1$, $V_0(\Phi^s(y)) \leq 1$ for all $y \in B(0, \varepsilon)$. Thus, introducing the sets $\mathscr{A} = \{V_0(\Phi^s(y)) > 2\}$ and $\mathscr{B} = \{V_0(\Phi^s(y)) \leq 2\}$, for all $y \in B(0, \varepsilon)$, we have

$$\begin{aligned} W(y) &= \int_{s_1}^{+\infty} e^{-(\mu+\kappa)s} a'(V_0(\Phi^s(y))) W_0(\Phi^s(y)) ds \\ &= \int_{s_1}^{+\infty} e^{-(\mu+\kappa)s} a'(V_0(\Phi^s(y))) W_0(\Phi^s(y)) \mathbf{1}_{\mathscr{A}} ds \\ &\quad + \int_{s_1}^{+\infty} e^{-(\mu+\kappa)s} a'(V_0(\Phi^s(y))) W_0(\Phi^s(y)) \mathbf{1}_{\mathscr{B}} ds. \end{aligned}$$

Since $a'(t) = 0$ for $t > 2$, the first part vanishes. But V_0 is proper, thus $\mathscr{B}$ is compact and W_0 is bounded by $c_3 > 0$ on this set. Therefore,

$$\begin{aligned} W(y) &= \int_{s_1}^{+\infty} e^{-(\mu+\kappa)s} a'(V_0(\Phi^s(y))) W_0(\Phi^s(y)) \mathbf{1}_{\mathscr{B}} ds \\ &\leq \int_{s_1}^{+\infty} e^{-(\mu+\kappa)s} c_2 c_3 ds < +\infty, \end{aligned}$$

since $\mu + \kappa > 0$. Finally, since W is homogeneous, it is also continuous at the origin and the proof is completed. □

2.1.4.2 Application to FTS

In this subsection, we aim at applying Theorem 2.1.2 to FTS. Indeed, the existence of a smooth homogeneous Lyapunov function provides information about the rate of convergence of such systems.

Definition 2.1.5. *Consider the system (2.1.2). The origin is said to be FTS if:*

1. *the origin is (strongly) stable;*
2. *there exists an open neighborhood $\mathscr{U}$ of the origin such that for all $x \in \mathscr{U}$, there exists $\tau \geq 0$ such that for all $t \geq \tau$, we have $\Psi^t(x) = \{0\}$ (strong finite-time convergence).*

The settling-time function is then defined for $x \in \mathscr{U}$ by $\mathsf{T}(x) = \inf\{\tau \geq 0 : \forall t \geq \tau, \Psi^t(x) = \{0\}\}$.

If the neighborhood $\mathscr{U}$ can be chosen to be $\mathbb{R}^n$, the system is said to be globally FTS (GFTS).

Corollary 2.1.3. *Let F be a ν-homogeneous set-valued map of degree $\kappa < 0$, satisfying the standard assumptions. Assume also that the origin is GAS for F. Then the* *origin is GFTS for F and the settling-time function is continuous at zero and locally bounded.*

Proof. The origin is GAS for F and F is homogeneous; thus by Theorem 2.1.2, F admits a homogeneous Lyapunov pair (V, W). Let us apply Lemma 4.2 of Reference 16 to the continuous functions V and W. We get that for all $x \in \mathbb{R}^n \setminus \{0\}$, and for all $v \in F(x)$:

$$d_x V v \leq -W(x) \leq -C(V(x))^{\frac{\kappa+\mu}{\mu}}, \tag{2.1.8}$$

where $C = \min_{\{V=1\}} W$. Since $\frac{\kappa+\mu}{\mu} < 1$, V converges to zero in a finite time, giving us the finite-time convergence of the system, which is therefore GFTS. Moreover, a direct integration of the inequation (2.1.8) gives $\mathsf{T}(x) \leq \frac{\mu V(x)^{\frac{-\kappa}{\mu}}}{-\kappa C}$, where T denotes the settling-time function. Since V is continuous, T is locally bounded and continuous at zero. □

It has been shown in Reference 16 that under the assumptions of homogeneity (of negative degree), continuity of the right-hand side and forward uniqueness of solutions, the settling-time function of a finite-time stable system is continuous. The two latter do obviously not hold in our context. We have seen that, however, under the standard assumptions, the settling-time function remains continuous at the origin and locally bounded. Let us emphasize that these conclusions are sharp and that the settling-time function is *not* continuous in general. See, for instance, Reference 19 or the following example.

Example 2.1.2. (A counterexample to the second statement of Theorem 1 from Reference 8.) *Consider the system defined on $\mathbb{R}^2$ by:*

$$\dot{x} = -(\text{sign}(x_1) + 2)\frac{x}{\|x\|}.$$

This system is clearly strongly (uniformly [8]) GFTS and ν-homogeneous of negative degree with $\nu = x_1 \frac{\partial}{\partial x_1} + x_2 \frac{\partial}{\partial x_2}$. A simple computation shows that the settling-time function is:

$$\mathsf{T}(x) = \begin{cases} \|x\| & x_1 \geq 0, \\ \dfrac{\|x\|}{3} & x_1 < 0, \end{cases}$$

which is discontinuous on $x_1 = 0$.

2.1.4.3 Sufficient conditions for global asymptotic stability

In this subsection, we focus on the qualitative properties of homogeneous discontinuous systems that can lead to GAS. The first result is a generalization of Theorem 6.1 of Reference 16.

Theorem 2.1.3. *Suppose that K is a strongly SPI compact subset of $\mathbb{R}^n$ for the homogeneous system (2.1.2). Then the origin is GAS for (2.1.2).*

Proof. Let us denote κ the degree of F. Since the solutions starting in K are bounded, they are defined for all $t \geq 0$, and thus $\Psi^t(K)$ is compact for all $t > 0$ by Theorem 2.1.1. From (2.1.7), we have ($\mathring{K}$ defines interior for a set $K \subset \mathbb{R}^n$)

$$\Psi^t(\Phi^s(K)) = \Phi^s\left(\Psi^{e^{\kappa s}t}(K)\right) \subset \Phi^s\left(\mathring{K}\right) = \overbrace{\Phi^s(K)}^{\circ}.$$

Therefore, the set $\Phi^s(K)$ is SPI for all $s \in \mathbb{R}$. We also note that $\Psi^{s+t}(K) = \Psi^t(\Psi^s(K)) \subset \Psi^t(K)$. Thus $\left(\Psi^t(K)\right)_{t\geq 0}$ is a nested family of compact sets and $K_\infty = \cap_{t\geq 0}\Psi^t(K)$ is a non-empty compact; it is also the biggest positively invariant compact subset of K. But for all $s \in \mathbb{R}$

$$\Phi^s(K_\infty) = \bigcap_{t\geq 0} \Phi^s(\Psi^t(K)) = \bigcap_{\tau\geq 0} \Psi^\tau(\Phi^s(K))$$

has the same property. Therefore, $K_\infty = \Phi^s(K_\infty)$, i.e., K_∞ is an invariant subset for Φ. Since ν is Euler, we conclude that $K_\infty = \{0\}$ and every solution starting in K converges to the origin, thus $0 \in K$. The stability follows from the SPI of the sets $\Phi^s(K)$ for all $s \in \mathbb{R}$. □

Let us illustrate how this theorem can be used to derive robustness properties for some homogeneous systems.

Example 2.1.3. *It is known that for $k_1 > k_2 > 0$, the following system is GAS:*

$$\begin{cases} \dot{x}_1 = x_2, \\ \dot{x}_2 = -k_1 \operatorname{sign}(x_1) - k_2 \operatorname{sign}(x_2). \end{cases}$$

As we have seen, this system is ν-homogeneous of degree -1 with $\nu(x) = (2x_1, x_2)^T$. Therefore, there exists a ν-homogeneous Lyapunov pair (V, W) of degrees $\kappa > 1$ and $\kappa - 1$. We denote F_0 the set-valued map associated to this vector field and for $\alpha \in \mathbb{R}^2$, we denote F_α the set-valued map associated to the vector field:

$$\begin{cases} \dot{x}_1 = x_2, \\ \dot{x}_2 = -(k_1 + \alpha_1)\operatorname{sign}(x_1) - (k_2 + \alpha_2)\operatorname{sign}(x_2). \end{cases}$$

We shall prove that the compact set $K = \{x \in \mathbb{R}^2 : V(x) \leq 1\}$ is SPI for F_α, for small values of α. Let $y \in K$ and $v \in F_\alpha(y)$. There exists $x \in S = \{x \in \mathbb{R}^n : V(x) = 1\}$ and $s \in \mathbb{R}$ such that $\Phi^s(x) = y$. By homogeneity, there also exists $w \in F_\alpha(x)$ such that $v = e^{-s} d_x \Phi^s w$. Therefore,

$$d_y V v = d_{\Phi^s(x)} V e^{-s}\, d_x \Phi^s w = e^{(\kappa-1)s}\, d_x V w.$$

Since $w \in F_\alpha(x)$, there exist $\sigma_1, \sigma_2 \in [-1, 1]$ such that $w = (x_2, -(k_1 + \alpha_1)\sigma_1 - (k_2 + \alpha_2)\sigma_2)^T$. Let us denote $\tilde{w} = (x_2, -k_1\sigma_1 - k_2\sigma_2)^T \in F_0(x)$. We have

$$
\begin{aligned}
d_y V v &= e^{(\kappa-1)s} d_x V w \\
&= e^{(\kappa-1)s} \left[d_x V \tilde{w} + d_x V (w - \tilde{w}) \right] \\
&\leq e^{(\kappa-1)s} \left[-W(x) + \sup_{x \in S} \|d_x V\| \cdot \|w - \tilde{w}\| \right] \\
&\leq e^{(\kappa-1)s} \left[-\inf_{x \in S} W(x) + \sup_{x \in S} \|d_x V\| \cdot \left\| \sum_{i=1}^{2} \alpha_i \sigma_i \right\| \right] \\
&\leq e^{(\kappa-1)s} \left[-\inf_{x \in S} W(x) + \sup_{x \in S} \|d_x V\| \cdot \sum_{i=1}^{2} \|\alpha_i\| \right].
\end{aligned}
$$

Therefore, if $|\alpha_1| + |\alpha_2| < \frac{\inf_{x \in S} W(x)}{\sup_{x \in S} \|d_x V\|}$, $d_y V v < 0$, which means that K is SPI. The set-valued map F_α being homogeneous, it is hence GAS for α small enough. Finally, any stabilizing control under the form $u(x) = -k_1 sign(x_1) - k_2 sign(x_2)$ is robust with respect to small time-varying errors on the gains k_i, like implementation errors.

As we have seen in Section 2.1.1, for continuous homogeneous systems (with forward uniqueness of solutions), the notions of local attractiveness and global asymptotic stability are merged. This fact admits a generalization in the discontinuous setting.

Theorem 2.1.4. *Let F be a ν-homogeneous set-valued map of degree κ with the standard assumptions. Assume moreover that all the solutions of the associated DI are defined for all $t \geq 0$ and tend to 0 when $t \to \infty$. Then the origin is strongly GAS.*

Proof. By contradiction, assume that the origin is unstable. Then there exists a neighborhood $\mathscr{U}$ of the origin such that for all neighborhoods of the origin $\mathscr{V} \subset \mathscr{U}$, there exists a solution starting in $\mathscr{V}$ which does not stay in $\mathscr{U}$. Taking $\mathscr{V} = B(0, 1/i)$, there exists a solution x_i such that $x_i(0) \in \mathscr{V}$, and there exists a real number t_i such that $x_i(t_i) \notin \mathscr{U}$. Therefore, when $i \to \infty$, $x_i(0) \to 0$ but $(x_i(t_i))$ does not converge to 0.

Let us denote by N a ν-homogeneous norm and denote $\delta_i = N(x_i(0))$. There exists $\varepsilon > 0$ such that $N(x_i(t_i)) \geq \varepsilon$. We can also assume that $\delta_i < \varepsilon$. Since $x_i(t_i) > 0$, we can finally assume that $\delta_i > 0$ by continuity.

Let us denote

$$
\begin{aligned}
a_i &= \sup\{t \in [0, t_i] : N(x_i(t)) = \delta_i\}, \\
b_i &= \inf\{t \in [0, t_i] : N(x_i(t)) = \varepsilon\}.
\end{aligned}
$$

We define $y_i(t) = x_i(t + a_i)$. The curves y_i are solutions of (2.1.2) defined on $[0, b_i - a_i]$ and we have $N(y_i(0)) = \delta_i$, $N(y_i(t)) \in (\delta_i, \varepsilon)$ for all $t \in (0, b_i - a_i)$ and $N(y_i(b_i - a_i)) = \varepsilon$.

By Proposition 2.1.2, for all $s \in \mathbb{R}$, the curve $t \mapsto \Phi^s(y_i(e^{\kappa s}t))$ is a solution. Set $s_i = -\ln \delta_i$, $z_i(t) = \Phi^{s_i}(y_i(e^{\kappa s_i}t))$ and $t_i^* = \delta_i^\kappa(b_i - a_i)$. We find $N(z_i(t)) = e^{s_i}N(y_i(e^{\kappa s_i}t)) = N(y_i(e^{\kappa_i}t))/\delta_i$. Hence, $N(z_i(0)) = 1$, for all $t \in (0, t_i^*)$, we have $N(z_i(t)) \geq 1$ and $N(z_i(t_i^*)) = \varepsilon/\delta_i$.

Assume that there exists a bounded subsequence of (t_i^*). By Theorem 2.1.1, the corresponding subsequence $(z_j(t_j^*))$ is bounded; however, $N(z_j(t_j^*)) = \frac{\varepsilon}{\delta_j} \to \infty$. Then the sequence (t_i) tends to $+\infty$.

By Corollary 2.1.1, let us now extract a subsequence $(z_{\varphi_1(i)})$ converging to a solution $\bar{z}$ on $[0, 1]$. Then we extract a sub-subsequence $(z_{\varphi_1 \circ \varphi_2(i)})$ converging to $\bar{z}$ on $[0, 2]$, etc. A diagonal extraction provides us the subsequence $(z_{\varphi_1 \circ \ldots \circ \varphi_i(i)})$ which is converging to $\bar{z}$ uniformly on $[0, j]$ for all $j \in \mathrm{N}$. For all $t > 0$, we have $N(\bar{z}(t)) = \lim_i N(z_{\varphi_1 \circ \ldots \circ \varphi_i(i)(t)})$. But for i large enough, we have $t \leq t^*_{\varphi_1 \circ \ldots \circ \varphi_i(i)}$ and thus $N(z_{\varphi_1 \circ \ldots \circ \varphi_i(i)(t)}) \geq 1$. Therefore, $N(\bar{z}(t)) \geq 1$ and $\bar{z}(t)$ does not tend to zero, which is a contradiction. □

Example 2.1.4. *Consider a double integrator endowed with an observer written with the error equation under the form*

$$\begin{cases} \dot{x}_1 = x_2, \\ \dot{x}_2 = -k_1 \operatorname{sign}(x_1) - k_2 \operatorname{sign}(x_2 - e_2), \\ \dot{e}_1 = e_2 - l_1 \lfloor e_1 \rceil^{1/2}, \\ \dot{e}_2 = -l_2 \operatorname{sign}(e_1), \end{cases} \tag{2.1.9}$$

with $k_1 > k_2 > 0$, $l_1, l_2 > 0$ *and* $\lfloor e_1 \rceil^{1/2} = |e_1|^{1/2}\operatorname{sign}(e_1)$. *This system is* ν*-homogeneous of degree* -1 *with* $\nu(x_1, x_2, e_1, e_2) = (2x_1, x_2, 2e_1, e_2)^T$.

Consider first the error subsystem:

$$\begin{cases} \dot{e}_1 = e_2 - l_1 \lfloor e_1 \rceil^{1/2}, \\ \dot{e}_2 = -l_2 \, sign(e_1). \end{cases} \tag{2.1.10}$$

Using the function $V(e) = l_2|e_1| + e_2^2/2$, *we see that the system (2.1.10) is FTS. Indeed, for* $e_1 \neq 0$, $\dot{V}(e) = -l_1 l_2 |e_1|^{1/2} < 0$. *Hence, the compact set* $K = \{V \leq 1\}$ *is SPI, since the line* $e_1 = 0$ *is not invariant. Being of negative degree, we find that the system (2.1.10) is FTS.*

The system (2.1.9) being clearly forward complete, it becomes equivalent in a finite time to the system:

$$\begin{cases} \dot{x}_1 = x_2, \\ \dot{x}_2 = -k_1 sign(x_1) - k_2 sign(x_2). \end{cases} \tag{2.1.11}$$

Using the Lyapunov function $V(x) = k_1|x_1| + x_2^2$ and $k_1 > k_2$, we find that the system (2.1.11) is FTS as well.

Finally, all the solutions of the system (2.1.9) converge in finite time to 0, but stability is not straightforward. However, Theorem 2.1.4 ensures us that the stability is a consequence of the attractiveness for homogeneous systems and we find that the system (2.1.9) is GFTS.

2.1.5 Conclusion

In this chapter, we have proposed a geometric definition of homogeneity for DI, consistent with the Filippov's regularization procedure. With this framework, we have been able to state extensions to the DI setting of results holding for continuous homogeneous systems: (i) A converse homogeneous Lyapunov theorem – if the origin is a GAS equilibrium, then there exists a homogeneous Lyapunov function (indeed, a homogeneous Lyapunov pair) for the DI. (ii) If the origin is a GAS equilibrium for a homogeneous DI of negative degree, then it is strongly FTS. (iii) The existence of a SPI compact set is equivalent to global asymptotic stability. (iv) The local attractiveness of the origin implies its global asymptotic stability.

Future works will include homogenization of a DI and study of the robustness of systems defined by homogeneous DI.

References

[1] J. Kurzweil, "On the inversion of Lyapunov's second theorem on stability of motion," *Translations of American Mathematical Society*, vol. 24, pp. 19–77, 1963.

[2] F. H. Clarke, Y. S. Ledyaev, and R. J. Stern, "Asymptotic stability and smooth Lyapunov functions," *Journal of Differential Equations*, vol. 149, pp. 69–114, 1998.

[3] M. Kawski, "Homogeneous stabilizing feedback laws," *Control Theory and Advanced Technology*, vol. 6, no. 4, pp. 497–516, 1990.

[4] H. Hermes, "Nilpotent approximations of control systems and distributions," *SIAM Journal on Control and Optimization*, vol. 24, no. 4, pp. 731–736, 1986.

[5] S. P. Bhat and D. S. Bernstein, "Finite-time stability of continuous autonomous systems," *SIAM Journal on Control and Optimization*, vol. 38, no. 3, pp. 751–766, 2000.

[6] A. Filippov, *Differential Equations with Discontinuous Righthand Sides*. Boston, MA: Kluwer Academic, 1988.

[7] J.-P. Aubin and A. Cellina, *Differential Inclusions*. Grundlehren der Mathematischen Wissenschaften 264. Berlin: Springer-Verlag, 1984.

[8] A. Levant, "Homogeneity approach to high-order sliding mode design," *Automatica*, vol. 41, no. 5, pp. 823–830, 2005.

[9] Y. Orlov, "Finite time stability and robust control synthesis of uncertain switched systems," *SIAM Journal of Control on Optimization*, vol. 43, no. 4, pp. 1253–1271, 2005.

[10] E. Bernuau, A. Polyakov, D. Efimov, and W. Perruquetti, "On extension of homogeneity notion for differential inclusions," in *Proceedings of ECC 2013*, Zurich, 2013.

[11] M. Kawski, "Families of dilations and asymptotic stability," in *Analysis of Controlled Dynamical Systems (Lyon, 1990)*, vol. 8 of *Progress in Systems Control Theory*, pp. 285–294. Boston, MA: Birkhäuser Boston, 1991.

[12] M. Kawski, "Geometric homogeneity and stabilization," in *Proceedings of the IFAC Nonlinear Control Symposium* (A. Krener and D. Mayne, eds.), (Lake Tahoe, CA), pp. 164–169, 1995.

[13] L. Rosier, "Homogeneous Lyapunov function for homogeneous continuous vector field," *Systems & Control Letters*, vol. 19, pp. 467–473, 1992.

[14] S. P. Bhat and D. S. Bernstein, "Continuous finite-time stabilization of the translational and rotational double integrator," *IEEE Transactions on Automatic Control*, vol. 43, pp. 678–682, May 1998.

[15] S. P. Bhat and D. S. Bernstein, "Finite-time stability of homogeneous systems," in *Proceedings of the American Control Conference*, Albuquerque, NM, pp. 2513–2514, 1997.

[16] S. P. Bhat and D. S. Bernstein, "Geometric homogeneity with applications to finite-time stability," *Mathematics of Control, Signals and Systems*, vol. 17, pp. 101–127, 2005.

[17] V. Zubov, "On ordinary differential equations with generalized homogeneous right-hand sides," *Izvestiya Vuzov. Matematika*, vol. 1, no. 2, pp. 80–88, 1958.

[18] L. Rosier, "Inverse of Lyapunov's second theorem for measurable functions," in *Proceedings of the NOLCOS 92*, Bordeaux, France, pp. 655–660, 1992.

[19] A. Polyakov, "On settling time function and stability of vector relay systems," in *International Workshop on Variable Structure Systems*, pp. 149–154, IEEE, Mumbai, 2012.

Chapter 2.2
Minimax observer for sliding mode control design

Sergiy Zhuk and Andrey Polyakov

Abstract

We consider the classical reaching problem of sliding mode control design, that is to find a control law which steers the state of a linear time-invariant (LTI) system toward a given hyperplane in a finite time. Since the LTI system is subject to unknown but bounded disturbances, we apply the minimax observer which provides the best possible estimate of the system's state. The reaching problem is then solved in observer's state space by constructing a feedback control law. The cases of discontinuous and continuous admissible feedbacks are studied. The theoretical results are illustrated by numerical simulations.

2.2.1 Introduction

Sliding mode is the oldest robust control technique introduced more than 50 years ago (see, e.g., Reference 14 and references therein). This method had opened new research areas from purely theoretical domains to practical applications. The key theoretical advantage of sliding mode control is that it is insensitive to the so-called matched disturbances and uncertainties, see References 5, 13 and 15.

We stress that in control practice, it is often required to design sliding modes for systems with mismatched uncertainties [1, 10, 16]. The same holds true for output-based feedback control application [4, 13]. These practical issues prompt for new developments in sliding mode control methodology.

This chapter treats the problem of output-based sliding mode control design for an LTI system with additive exogenous disturbances and bounded deterministic measurement noises. In this case, ensuring of the ideal sliding mode in the state space of the original system is impossible due to incomplete and noised measurements. The control law, which provides the motion of the closed-loop system as close as possible to the selected sliding surface, can only be designed. The conventional (first order) sliding mode control design principles are studied in the chapter for the case of $\mathbb{L}^2$-bounded noises and disturbances. It is known (see, e.g., Reference 2) that realization of the reaching phase of the sliding surface can be formulated as a special optimization problem. This chapter shows that for the case of incomplete and noisy output measurements, the sliding mode control algorithm is just one possible solution of the corresponding optimization problem, which, in fact, admits both continuous and discontinuous optimal control laws.

The control design relies upon minimax state estimation framework [3, 7, 8] and duality argument [17, 19, 20]. The minimax state estimator or observer constructs the best linear estimate of the system's state provided the uncertain parameters (model disturbance, observation error, error in the initial condition) belong to a given bounding set. Statistically, the latter assumption implies uniform distributions for uncertain parameters and, under these assumptions, the minimax filter is designed so that for any realization of uncertain parameters the estimation error is minimal. Given the best linear estimate of the state, we apply the linear separation principle and transform the problem of sliding mode control design in the state space of the original system to the optimal control problem for the observer's variables.

It is worth noting that the minimax observer (in the form of a linear functional of observations), used in this paper, is optimal among all observers represented by measurable functionals of observations [22]. Thus, at least theoretically, the proposed control design cannot be further improved by using observers realized by nonlinear functionals of the observations. We refer the reader to References 5, 11, 13 and 15 for more information about nonlinear sliding mode observers. Numerical methods designed for minimax observers may be found in References 6 and 21.

The chapter is organized as follows. The next section presents the notations used in this chapter. Then the problem statement and basic assumptions are considered. The minimax observed design is given in Section 2.2.4. Next, the control design algorithms are discussed. Finally, the numerical simulations and conclusions are presented.

2.2.2 Notation

Throughout the paper, the following notations will be used:

- $\mathbb{R}_+ = \{x \in \mathbb{R} : x > 0\}, \mathbb{R}_- = \{x \in \mathbb{R} : x < 0\}$, where $\mathbb{R}$ is the set of real number;
- $\|\cdot\|$ is the Euclidian norm in $\mathbb{R}^n$, i.e., $\|x\| = \sqrt{x_1^2 + \cdots + x_n^2}$ for $x = (x_1, \ldots, x_n)^T \in \mathbb{R}^n$;
- if $P \in \mathbb{R}^{n\times n}$, then the inequality $P > 0$ ($P \geq 0$, $P < 0$, $P \leq 0$) means that P is symmetric and positive definite (positive semidefinite, negative definite and negative semidefinite);
- $\mathbb{L}^2_{[a,b]}$ is a set of Lebesgue quadratically integrable functions defined on $[a, b]$.

2.2.3 Problem statement

Consider the linear output control system

$$\dot{x} = Ax + Bu + Dg(t), \tag{2.2.1}$$

$$y = Cx + w(t), \tag{2.2.2}$$

$$t \in [0, T), \quad x(0) = x_0 \in \mathbb{R}^n,$$

where

- $T \in \mathbb{R}_+$ is a finite instant of time or $T = +\infty$,
- $x \in \mathbb{R}^n$ is the vector of system state,
- $u \in \mathbb{R}$ is the scalar control input,
- $y \in \mathbb{R}^k$ is the measured output,
- the function $g : \mathbb{R}_+ \to \mathbb{R}^p, p \leq n$ describes the matched external disturbances,
- the function $w : \mathbb{R}_+ \to \mathbb{R}^k$ is a deterministic measurement noise,
- the system parameters $A \in \mathbb{R}^{n\times n}, B \in \mathbb{R}^{n\times m}, C \in \mathbb{R}^{k\times n}, D \in \mathbb{R}^{n\times p}$ are assumed to be known and time invariant.

We study this system under the standard assumptions (see References 5 and 15).

Assumption 2.2.1. *The pair (A, C) is observable, and the pair (A, B) is controllable.*

In general, we do not assume that the disturbances are matched, but the case range $D \subseteq$ range(B) is also possible.

The noise measurements $w \in \mathbb{L}^2_{[0,T]}$ and exogenous disturbances $g \in \mathbb{L}^2_{[0,T]}$ are assumed to be deterministic and satisfy the following inequality:

$$x_0^T P_0^{-1} x_0 + \int_0^T \left(w^T(\tau) R w(\tau) + g^T(\tau) Q g(\tau) \right) d\tau \leq 1, \tag{2.2.3}$$

where $P_0 \in \mathbb{R}^{n\times n}$, $R \in \mathbb{R}^{k\times k}$ and $Q \in \mathbb{R}^{p\times p}$ are symmetric positive definite matrices. The above inequality defines an ellipsoid in the corresponding space. We note that each uncertain parameter above (x_0, w and g) may belong to a separate ellipsoid and, in this case, the above ellipsoid represents an approximation of three independent ellipsoids, provided it has been rescaled appropriately.

The admissible control law is assumed to belong to $\mathbb{L}_{[0,T]}$-space, which contains both continuous and discontinuous functions.

The **classical control problem** (see References 5 and 15) for the system (2.2.1) is to design the control algorithm, which realizes finite-time reaching of a given linear plane like

$$F^T x = 0, \quad F \in \mathbb{R}^{m\times n}, \quad \det(FB) \neq 0$$

and further sliding on this plane. It is worth to stress that the condition $\det(FB) \neq 0$ is necessary for realization of the *first-order* sliding mode control principles [14].

Let us consider the sliding mode control design problem for the reaching phase, i.e., we need to find the control law u such that $Fx(T) = 0$. The considered problem can be equivalently rewritten

$$\begin{aligned} &\|Fx(T)\| \to \min \\ &\text{s.t. (2.2.1)–(2.2.3).} \end{aligned} \tag{2.2.4}$$

Indeed, obtaining a solution of this optimization problem with zero value of the cost functional guarantees the successful realization of the reaching phase. Due to measurement noises and system disturbances, the sliding mode of the given surface $Fx = 0$ may not be guaranteed. In this case, it is important to know, which sort of feedback control will be optimal in order to provide the system motion as close as possible to the surface.

In what follows, we only study the observer-based feedback design assuming that static output-based sliding mode control (see, e.g., Reference 13) cannot be applied, i.e., $\text{range}(F^T) \not\subseteq \text{range}(C^T)$.

2.2.4 Min-max optimal state observer design

According the classical methodology of the sliding mode control design, the precise knowledge of the so-called sliding variable $s(t) := Fx(t)$ is required in order to ensure the motion of the system (2.2.1) on the surface $Fx = 0$. We stress that this information is not available as the given output $y(t)$ is incomplete and noisy. In this situation, the best available information about the value of $Fx(t)$ is represented by the minimax estimate of the state.

Let x_u, x_g denote the solutions of the following ordinary differential equations (ODEs):

$$\begin{aligned} \frac{dx_u}{dt} &= Ax_u + Bu, \quad x_u(0) = 0; \\ \frac{dx_g}{dt} &= Ax_g + Dg, \quad x_g(0) = x_0. \end{aligned} \tag{2.2.5}$$

Then, clearly, $x(t) = x_u(t) + x_g(t)$ and

$$y_g(t) := y(t) - Cx_u(t) = Cx_g(t) + w(t). \tag{2.2.6}$$

The function x_g may be considered as a noisy part of x corresponding to disturbances from the ellipsoid (2.2.3) and x_u represents its "mean" value corresponding to the case of zero disturbances $x_0 = 0$ and $g = 0$, which forms (together with $w = 0$) the central point of the ellipsoid (2.2.3). Since $x_g(t)$ does not depend on the control parameter u, we may first construct an estimate of the noisy part. Let us introduce the following definition.

Definition 2.2.1. *Assume that $l \in \mathbb{R}^n$ and $\widehat{U} \in L^2(0, t^*)$. A linear functional*

$$\widehat{\mathscr{U}_l}(y) := \int_0^{t^*} \widehat{U}^T(\tau)y(\tau)d\tau$$

is called a ***minimax estimate*** *of $l^T x(t)$ iff*

$$\begin{aligned} \sigma\left(\widehat{U}, l, t^*\right) &:= \sup\nolimits_{(x_0, g, w) \in \Omega^*} \left(l^T x(t^*) - \widehat{\mathscr{U}_l}(y)\right)^2 \\ &\leq \sigma(U, l, t^*), \quad \forall U \in L^2(0, t^*), \end{aligned}$$

where Ω^* is defined by (2.2.3) *with $T = t^*$.*

The number $\hat{\sigma}(l, t^*) := \sigma\left(\widehat{U}, l, t^*\right)$ is called the ***minimax estimation error****.*

Let $\mathscr{G}$ denote a set of all continuous mappings of $\mathbb{L}_2$ into $\mathbb{R}$ and let $g \in \mathscr{G}$. Then it can be proven that

$$\sigma\left(\widehat{U}, l, t^*\right) = \inf_{g \in \mathscr{G}} \sup_{(x_0, d_i, w) \in \Omega^*} \left|l^\top x(t^*) - g(y)\right|.$$

In other words, the worst-case estimation error of any continuous mapping g defined by the right-hand side of the above formula is greater or equal to the minimax error of the linear functional $\widehat{\mathscr{U}_l}(y)$ which has minimal worst-case estimation error $\hat{\sigma}$.

Proposition 2.2.1. *Let $\hat{x}_g(t) \in \mathbb{R}^n$ be the solution of the following ODE:*

$$\begin{cases} \dfrac{d\hat{x}_g(t)}{dt} = A\hat{x}_g + P(t)C^T R\big(y_g(t) - C\hat{x}_g(t)\big), \\ \hat{x}_g(0) = 0, \end{cases}$$

where $P(t) \in \mathbb{R}^{n \times n}$ is the solution of the following differential Riccati equation:

$$\begin{cases} \dot{P}(t) = AP(t) + P(t)A^T + DQ^{-1}D^T - P(t)C^T RCP(t), \\ P(0) = P_0. \end{cases} \tag{2.2.7}$$

Then $\widehat{\mathscr{U}_l}(y_g) = l^T \hat{x}_g(t^*)$ and $\hat{\sigma}(l, t^*) = \left(l^T P(t^*) l\right)^{\frac{1}{2}}$, where y_g is defined by (2.2.6).

The detailed proof of this proposition is available in the literature (see for instance [9, 19]).

Let us stress that the proposed observer is *stable*, i.e., $A - P^\infty C^T RC$ is a Hurwitz matrix, where P^∞ solves the following algebraic Riccati equation:

$$0 = AP + PA^T + DQ^{-1}D^T - PC^T RCP.$$

The latter has the unique symmetric non-negative definite solution provided $\{A, C\}$ is observable or, more generally, detectable and $\{A, D\}$ is controllable or, more generally, stabilizable [12].

The definition of the minimax estimate $\widehat{U}$ implies that

$$\left(l^T x_g(t^*) - l^T \hat{x}_g(t^*)\right)^2 \leq l^T P(t^*) l.$$

Now we recall that $x(t^*) = x_u(t^*) + x_g(t^*)$ and so

$$x(t^*) = x_u(t^*) + \hat{x}_g(t^*) + e(t^*),$$

where $e(t^*) = x_g(t^*) - \hat{x}_g(t^*)$ and $l^T e(t^*) \leq \left(l^T P(t^*) l\right)^{\frac{1}{2}}$ **does not depend** on u. Define

$$\hat{x} := x_u + \hat{x}_g.$$

Then it is straightforward to check that

$$\begin{cases} \dfrac{d\hat{x}(t)}{dt} = A\hat{x} + P(t)C^T R(y(t) - C\hat{x}(t) + Bu(t), \\ \hat{x}(0) = 0. \end{cases} \tag{2.2.8}$$

Since the calculations above hold true for any $0 < t < t^*$, we obtain

$$x(t) = \hat{x}(t) + e(t),$$

where the estimation error satisfies the inequality

$$l^T e(t) \leq \left(l^T P(t) l\right)^{\frac{1}{2}}, \tag{2.2.9}$$

and the latter estimate does not depend on the control u. The inequality (2.2.9) holds for all $l \in \mathbb{R}^n$ proving the following optimal (in the minimax sense) guaranteed estimate of the system state:

$$x(t) \in \left\{z \in \mathbb{R}^n : \quad z = \hat{x}(t) + e, \quad e^T P^{-1}(t) e \leq 1\right\}, \tag{2.2.10}$$

i.e., the state vector $x(t)$ belongs to the ellipsoid centered at $\hat{x}(t)$ with the shape matrix $P(t)$. Recall that the formula (2.2.9) with $\forall l \in \mathbb{R}^n$ is just a way to define the ellipsoid (see, e.g., Reference 7).

It is worth to stress that the minimax approach to observer design provides the "exact" estimate of all admissible system's states, namely, for any $e_* \in \mathbb{R}^n$ belonging to the estimating ellipsoid (i.e., $e_*^T P^{-1}(t) e_* \leq 1$) and for any $t \in [0, T]$, there exist $x_0^* \in \mathbb{R}^n$, $w^* \in \mathbb{L}^2$ and $g^* \in \mathbb{L}^2$ satisfying (2.2.3) such that the equality $x(t) = \hat{x}_u(t) + e^*$ holds. In addition, one may further improve the aforementioned estimate by filtering out the states which are incompatible with observations [18], provided the realization of y is available beforehand.

Note that P does not depend on the control parameter explicitly. This suggests to design a controller as a function of the center of the ellipsoid, i.e., $\hat{x}_u$. The next section presents the controller design.

2.2.5 Control design

Now let us consider the problem of the reaching phase realization of the sliding mode control for the systems (2.2.1)–(2.2.3), which is equivalently rewritten in the form (2.2.4).

Denote the sliding variable by

$$\sigma = Fx.$$

Using (2.2.10), we derive

$$\sigma(T) = Fx(T) = F\hat{x}(T) + Fe(T),$$

where the state estimate $\hat{x}$ satisfies (2.2.8) and e is the observation error, which is not depended on the control input u. Recall that the ellipsoid $\{e \in \mathbb{R}^n : e^T P(T) e \leq 1\}$ describes the best possible estimate of the observation error at time instant $t = T$. Since $Fe(T)$ also belongs to some ellipsoid in $\mathbb{R}^m$ with the center at the origin, then any control u, which guarantees

$$F\hat{x}(T) = 0 \tag{2.2.11}$$

is the solution of the optimization problem (2.2.4).

2.2.5.1 The conventional sliding mode feedback

Following the classical methodology of the sliding mode control design for linear plants [5, 15], we should define the control law in the form

$$u(t) = u_{\text{eq}}(t) + u_{\text{d}}(t), \tag{2.2.12}$$

where

$$u_{\text{eq}}(t) = -(FB)^{-1} FAx(t)$$

is the so-called equivalent control part and

$$u_{\text{d}}(t) = -(FB)^{-1} K(t) \operatorname{sign}[\sigma(t)]$$

is the discontinuous (relay) term with sufficiently large gain $K(t) > 0$.

Since the only observed state is admissible, then the following laws must be applied:

$$\begin{aligned} u_{\text{eq}}(t) &= -(FB)^{-1} FA\hat{x}(t), \\ u_{\text{d}}(t) &= -(FB)^{-1} K(t) \operatorname{sign}\left[\hat{\sigma}(t)\right], \end{aligned} \tag{2.2.13}$$

where

$$\hat{\sigma}(t) := F\hat{x}(t). \tag{2.2.14}$$

Multiplying both sides of the system (2.2.8) by F, we obtain the following equation:

$$\frac{d\hat{\sigma}(t)}{dt} = FA\hat{x}(t) + FP(t)C^T R\left(y(t) - C\hat{x}(t)\right) + FBu(t), \tag{2.2.15}$$

which defines the dynamic of sliding variable $\hat{\sigma}$ for the observer state space. Substituting (2.2.12) for the control law, we derive

$$\frac{d\hat{\sigma}(t)}{dt} = FP(t)C^T R\left(y(t) - C\hat{x}(t)\right) - K(t)\,\mathrm{sign}\left[\hat{\sigma}(t)\right].$$

Taking into account $\hat{x}(0) = 0$, we derive that for any

$$K(t) > \left|FP(t)C^T R(y(t) - C\hat{x}(t))\right| \tag{2.2.16}$$

the control (2.2.12) guarantees achievement of the aim (2.2.11).

Therefore, *the convectional sliding mode control is the solution of the optimization problem* (2.2.4) *if the design technique is based on minimax observer application*. Equation (2.2.16) represents the rule for selection of the relay feedback gain.

2.2.5.2 The optimal continuous control

Evidently, the continuous control can also be designed in order to guarantee the condition (2.2.11). For instance, the continuous feedback law

$$u_c(t) = -(FB)^{-1}F\left[A\hat{x}(t) + P(t)C^T R(y(t) - C\hat{x}(t))\right] \tag{2.2.17}$$

is also optimal for the problem (2.2.4). Indeed, this feedback provides

$$F\frac{d\hat{x}}{dt} = 0.$$

So, taking into account $\hat{x}(0) = 0$, we obtain $\hat{x}(T) = 0$.

In fact, *the optimal continuous control law is the equivalent control designed for the observer equation* (2.2.8).

2.2.6 Numerical simulations

Let us consider the model of the linear oscillator with

$$A = \begin{pmatrix} 0 & 1 \\ -1 & 0 \end{pmatrix}, \quad B = D = \begin{pmatrix} 0 \\ 1 \end{pmatrix}, \quad C = \begin{pmatrix} 1 & 0 \end{pmatrix},$$

and select the desired sliding surface $Fx = 0$ with

$$F = \begin{pmatrix} 1 & 1 \end{pmatrix}.$$

The restrictions to energy measurement noises, exogenous disturbances and uncertainty of initial conditions are represented by the inequality (2.2.3) with the following parameters:

$$P_0 = \frac{3\pi^2}{4}\begin{pmatrix} 1 & 0 \\ 0 & 1 \end{pmatrix}, \quad Q = R = 40/3, \quad T = 10.$$

We compare the *linear* continuous feedback (2.2.17) with the *sliding mode* control of the form (2.2.12) with $K(t) = 10$.

The initial condition selected for simulation is

$$x_0 = \begin{pmatrix} 0 & \frac{\pi}{2} \end{pmatrix}^T.$$

The deterministic noise and disturbance functions are defined by

$$w(t) = 0.05\,\mathrm{sign}[\sin(10t)] \quad \text{and} \quad g(t) = 0.05\,\mathrm{sign}[\cos(10t)].$$

The numerical simulations have been made using explicit Euler method with fixed step size $h = 0.01$.

Figures 2.2.1–2.2.3 present the simulation results. The sliding mode control is subjected to chattering during simulation. Due to this, the continuous control provides better results.

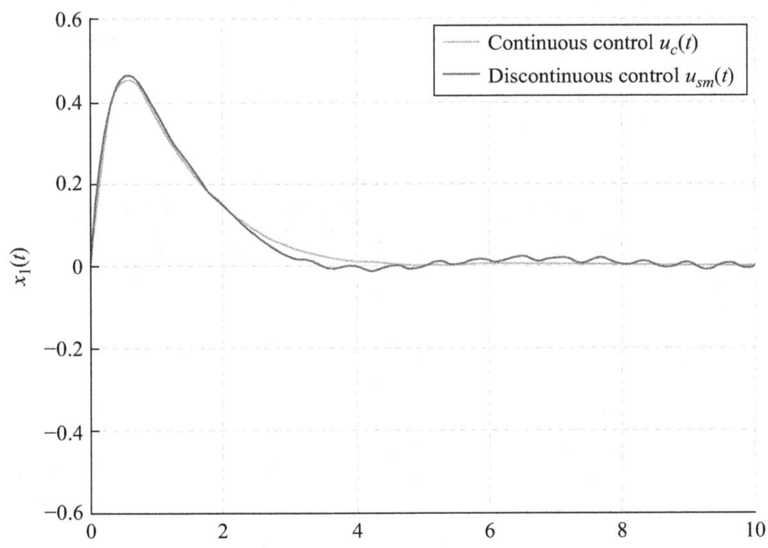

Figure 2.2.1 Evolution of the state x_1

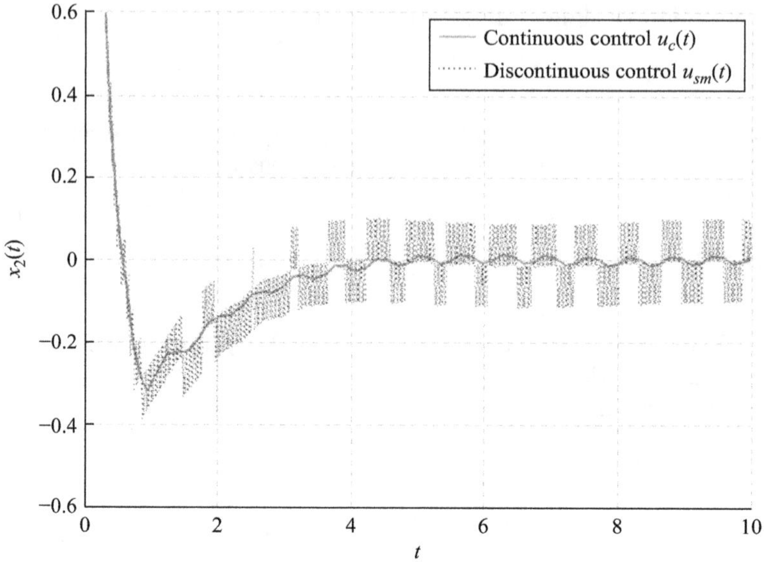

Figure 2.2.2 Evolution of the state x_2

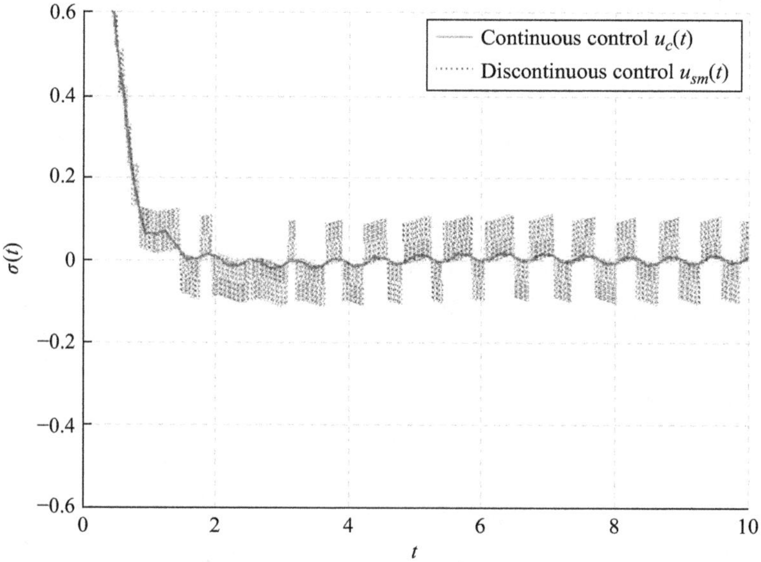

Figure 2.2.3 Evolution of the sliding variable σ

2.2.7 Conclusion

The problem of the output-based sliding mode control design in the reaching phase is studied for the LTI system with $\mathbb{L}^2_{[0,T]}$-bounded additive exogenous disturbances and the noised measurements of the output. The control providing optimal reaching (as close as possible) of the selected linear sliding surface is obtained using minimax observation approach.

The following important facts were discovered:

- The conventional sliding mode control law is an optimal solution to the considered problem.
- The optimal control law is not unique. The continuous optimal feedback is designed in the form of equivalent control for the observer equation. Therefore, discontinuity of sliding mode control may be unnecessary in the case of noised measurements.

The last fact poses the question on consistency of existing discontinuous sliding mode control design methodology for LTI systems with noised measurements.

References

[1] J. M. Andrade-Da Silva, C. Edwards, and S. K. Spurgeon. Sliding-mode output-feedback control based on LMIs for plants with mismatched uncertainties. *IEEE Transaction on Industrial Electronics*, 56(9):3675–3683, 2009.

[2] V. Azhmyakov, A. Polyakov, and A. Poznyak. Consistent approximations and variational description of some classes of sliding mode control processes. *Journal of the Franklin Institute*, 351(4): 1964–1981, http://dx.doi.org/10.1016/j.jfranklin.2013.01.011, 2014.

[3] F. L. Chernousko. *State Estimation for Dynamic Systems*. Boca Raton, FL: CRC Press, 1994.

[4] C. Edwards, A. Akoachere, and S. K. Spurgeon. Sliding-mode output feedback controller design using linear matrix inequalities. *IEEE Transactions on Automatic Control*, 46(1):115–119, 2001.

[5] C. Edwards and S. Spurgeon. *Sliding Mode Control: Theory and Applications*. Boca Raton, FL: CRC Press, 1998.

[6] J. Frank and S. Zhuk. Symplectic Möbius integrators for LQ optimal control problems. In *Proceedings of the IEEE Conference on Decision and Control*, Los Angeles, California, U.S.A., 2014.

[7] A. Kurzhanski and I. Valyi. *Ellipsoidal Calculus for Estimation and Control*. Boston, MA: Birkhäuser Boston Inc., 1997.

[8] M. Milanese and R. Tempo. Optimal algorithms theory for robust estimation and prediction. *IEEE Transactions on Automatic Control*, 30(8):730–738, 1985.

[9] A. Nakonechny. A minimax estimate for functionals of the solutions of operator equations. *Archivum Mathematicum (Brno)*, 14(1):55–59, 1978.

[10] A. Polyakov and A. Poznyak. Invariant ellipsoid method for minimization of unmatched disturbance effects in sliding mode control. *Automatica*, 47(7):1450–1454, 2011.

[11] A. S. Poznyak. *Variable Structure Systems: From Principles to Implementation*, volume 66 of *IEE Control Series*. Chapter: Deterministic output noise effects in sliding mode observation, pages 45–78. The IET, London, UK, 2004.

[12] A. S. Poznyak. *Advanced Mathematical Tools for Automatic Control Engineers: Deterministic Techniques*. Amsterdam: Elsevier, 2008.

[13] Y. Shtessel, C. Edwards, L. Fridman, and A. Levant. *Sliding Mode Control and Observation*. Boston, MA: Birkhäuser, 2014.

[14] V. I. Utkin. Variable structure systems with sliding modes. *IEEE Transactions on Automatic Control*, 22(2):212–222, 1977.

[15] V. I. Utkin, J. Guldner, and J. Shi. *Sliding Mode Control in Electro-Mechanical Systems*. Boca Raton, FL: CRC Press, 2009.

[16] J. Yang, S. Li, and X. Yu. Sliding-mode control for systems with mismatched uncertainties via a disturbance observer. *IEEE Transactions on Automatic Control*, 60(1):160–169, 2013.

[17] S. Zhuk. Estimation of the states of a dynamical system described by linear equations with unknown parameters. *Ukrainian Mathematical Journal*, 61(2):214–235, 2009.

[18] S. Zhuk. Minimax state estimation for linear discrete-time differential-algebraic equations. *Automatica Journal IFAC*, 46(11):1785–1789, 2010.

[19] S. Zhuk. Minimax state estimation for linear stationary differential-algebraic equations. In *Proceedings of the 16th IFAC Symposium on System Identification*, Brussels, Belgium, 2012.

[20] S. Zhuk. Kalman duality principle for a class of ill-posed minimax control problems with linear differential-algebraic constraints. *Applied Mathematics and Optimisation*, 68(2):289–309, 2013.

[21] S. Zhuk, J. Frank, I. Herlin, and R. Shorten. Data assimilation for linear parabolic equations: minimax projection method. *SIAM Journal on Scientific Computing*, 37(3), A1174–A1196, 2015.

[22] S. Zhuk, O. Nakonechnyi, and A. Polyakov. Note on observer-based sliding mode control design. 2015 (submitted).

Chapter 2.3

$\mathscr{L}_2$-Gain analysis of sliding mode dynamics

Topacio Osuna, Israel U. Ponce, Yury Orlov and Luis T. Aguilar

Abstract

This chapter extends the $\mathscr{L}_2$-gain analysis toward sliding mode dynamic systems. The developed analysis is applied to standard sliding mode algorithms of the first order and to the popular twisting and supertwisting algorithms of the second order. The above algorithms are shown to be capable of not only rejecting matched uniformly bounded disturbances but also attenuating unbounded ones, including unmatched disturbances. Numerical simulations are involved to support the theory. Experimental results on a DC motor with friction complement the chapter.

2.3.1 Introduction

Sliding mode (SM) control algorithms are well recognized for their useful robustness features against matching disturbances with *a priori* known bounds on their magnitudes. Their capability of attenuating disturbances with *a priori* unknown bounds on their magnitudes, which remains unattended in the literature, constitutes the topic of this chapter.

Typically, the works related to SM control (see the milestone monograph [25] and references therein) assume that only bounded and matched disturbances affect the underlying system. In relevant works [2, 7, 12, 13, 16, 21] to name a few, analysis of SM control was made within the framework of nonsmooth Lyapunov functions where certain conditions were imposed on the controller parameters and the upper bound of the magnitude of matched disturbances to guarantee finite-time stability, whereas general analysis, capturing both matched and unmatched disturbances with unknown upper bounds on their magnitudes, was not provided. The Lyapunov characterization of integral Input-to-State Stability of SM systems (see reference 3 and references therein) relied on the weighted homogeneity property, thus confining applications to matched disturbances and leaving nonhomogeneous systems unattended. Recently, reference 4 demonstrated that the closed-loop system, driven in the SM, is capable of presenting good performance in the presence of unmatched disturbances as well; however, the $\mathscr{L}_2$-gain analysis of such systems has not been addressed yet. As opposed to the above works, reference 23 proposed the $\mathcal{L}_2$-gain approach for analysis and synthesis of feedback controllers of discontinuous time-delay systems where specific SM analysis was not, however, made. Apart from this, reference 24 considered the finite-time stability and finite-time boundedness problems for switched linear systems subject to $\mathcal{L}_2$ disturbances where SMs were ruled out. The recent works [19, 20] suffered from the same shortcoming of the absence of the SM analysis.

This chapter follows the line of investigation, initiated in references 17–20, and its primary concern is to extend the $\mathcal{L}_2$-gain analysis toward SM dynamic systems. The extended analysis is then tested on a preselected SM control algorithm of the first order. Apart from this, a second order SM-based output feedback synthesis is analyzed where the twisting state feedback controller is used along with the supertwisting algorithm which is implemented to design a velocity observer, running in parallel to feed the twisting controller. These tests demonstrate that SMs are capable of not only rejecting matched bounded disturbances but also of attenuating disturbances of class $\mathcal{L}_2$, including the mismatched ones.

2.3.2 Generic $\mathcal{L}_2$-gain analysis

The $\mathcal{L}_2$-gain analysis, presented here, is based on the game-theoretic approach from reference 1 and extends the results from references 10, 15 and 22, where investigations were confined to smooth autonomous systems, toward piecewise continuous autonomous systems.

2.3.2.1 Basic assumptions and definitions

The $\mathcal{L}_2$-gain analysis is developed for an autonomous system of the form

$$\dot{x} = \varphi(x) + \psi(x)\omega(t) \tag{2.3.1}$$

and is made with respect to the output

$$z = h(x). \tag{2.3.2}$$

Hereinafter, $x(t) \in \mathbb{R}^n$ is the state vector, $t \in \mathbb{R}$ is the time variable, $\omega(t) \in \mathbb{R}^r$ are the unknown disturbances, $z(t) \in \mathbb{R}^p$, $\varphi(x) : \mathbb{R}^n \mapsto \mathbb{R}^n$, and $h(x) : \mathbb{R}^n \mapsto \mathbb{R}^p$ are vector functions, and $\psi(x) : \mathbb{R}^n \mapsto \mathbb{R}^{n\times r}$ is a matrix function. The following *assumptions* are imposed on the system.

A1. The functions $\psi(x)$ and $h(x)$ are locally Lipschitz continuous, whereas $\varphi(x)$ is piecewise locally Lipschitz continuous.
A2. $\varphi(0) = 0$ and $h(0) = 0$ for almost all t.

Recall that the function $\varphi(x) : \mathbb{R}^n \mapsto \mathbb{R}^n$ is piecewise (locally Lipschitz) continuous iff $\mathbb{R}^n$ is partitioned into a finite number of domains $G_j \subset \mathbb{R}^n$, $j = 1, \ldots, N$, with disjoint interiors and boundaries ∂G_j of measure zero such that $\varphi(x)$ is (locally Lipschitz) continuous within each of these domains and for all $j = 1, \ldots, N$ it has a finite limit $\varphi^j(x)$ as the argument $x^* \in G_j$ approaches a boundary point $x \in \partial G_j$.

Assumption **A2** is made to ensure that the origin is an equilibrium point of the nominal (i.e., disturbance-free) system, whereas assumption **A1** admits the underlying system to undergo discontinuities on the boundaries ∂G_j of measure zero, which is why the precise meaning of the differential equation (2.3.3) with a piecewise continuous right-hand side is throughout defined in the sense of Filippov. For convenience of the reader, the following definition is recalled from reference 8.

Definition 2.3.1. *Given the differential equation*

$$\dot{x} = \varphi(x), \tag{2.3.3}$$

let us introduce for each point $x \in \mathbb{R}^n$ *the smallest convex closed set* $\Phi(x)$ *which contains all the limit points of* $\varphi(x^*)$ *as* $x^* \to x$, *and* $x^* \in \mathbb{R}^n \setminus (\cup_{j=1}^N \partial G_j)$. *An absolutely continuous function x, is said to be a solution of (2.3.3) if it satisfies the differential inclusion*

$$\dot{x} \in \Phi(x). \tag{2.3.4}$$

Note that the extension of the Filippov solution concept to the perturbed differential equation (2.3.1) is straightforward.

At any continuity point, $x \in \cup_{i=1}^N G_i$ of the function φ the Filippov set $\Phi(x)$ consists of the only point $\varphi(x)$, and the Filippov solution satisfies (2.3.3) in the conventional sense. If the function $\varphi(x)$ undergoes discontinuities on a smooth surface S, governed by the scalar equation $s(x) = 0$, then the discontinuity set S separates the x space into domains $G^- = \{x \in \mathbb{R}^n : s(x) < 0\}$ and $G^+ = \{x \in \mathbb{R}^n : s(x) > 0\}$, and the Filippov set $\Phi(x)$ is a linear segment joining the endpoints of the vectors

$$\varphi^-(x) = \lim_{\xi \in G^-,\, \xi \to x} \varphi(\xi), \qquad \varphi^+(x) = \lim_{\xi \in G^+,\, \xi \to x} \varphi(\xi).$$

In the present work, an extra assumption is imposed on the piecewise continuous function $\varphi(x)$.

A3. The function $\varphi(x)$ undergoes discontinuities on a smooth surface S, governed by the scalar equation

$$s(x) = 0. \tag{2.3.5}$$

According to Definition 2.3.1, an SM on the discontinuity set $s(x) = 0$, if any, is governed by

$$\dot{x} = \varphi^0(x) + \psi^0(x)\omega \tag{2.3.6}$$

where the intersection of the Filippov segment $\Phi(x)$ and the plane T, tangential to S, determines the endpoint of the vector $\varphi^0(x)$. Analytically, this vector is expressed in the form

$$\varphi^0(x) + \psi^0(x)\omega = \mu(x)\varphi^+(x) + [1 - \mu(x)]\varphi^-(x) + \psi(x)\omega \tag{2.3.7}$$

where

$$\mu(x) = \frac{\text{grad}^T s(x)[\varphi^-(x) + \psi(x)\omega]}{\text{grad}^T s(x)[\varphi^-(x) - \varphi^+(x)]} \tag{2.3.8}$$

is found from the condition

$$\text{grad}^T s(x)\left\{\mu(x)\varphi^+(x) + [1 - \mu(x)]\varphi^-(x) + \psi(x)\omega\right\} = 0 \tag{2.3.9}$$

that the velocity vector (2.3.7) is in the plane T, tangential to S. Summarizing, the following relations

$$\varphi^0(x) = \frac{\text{grad}^T s(x)\varphi^-(x)}{\text{grad}^T s(x)[\varphi^-(x) - \varphi^+(x)]}\varphi^+(x) - \frac{\text{grad}^T s(x)\varphi^+(x)}{\text{grad}^T s(x)[\varphi^-(x) - \varphi^+(x)]}\varphi^-(x) \tag{2.3.10}$$

$$\begin{aligned}\psi^0(x)\omega &= \psi(x)\omega + \frac{\text{grad}^T s(x)\psi(x)\omega}{\text{grad}^T s(x)[\varphi^-(x) - \varphi^+(x)]}\varphi^+(x) \\ &\quad + \left[1 - \frac{\text{grad}^T s(x)\psi(x)\omega}{\text{grad}^T s(x)[\varphi^-(x) - \varphi^+(x)]}\right]\varphi^-(x)\end{aligned} \tag{2.3.11}$$

are derived to determine the functions $\varphi^0(x)$ and $\psi^0(x)$.

We are now in a position to introduce the finite $\mathcal{L}_2$-gain concept for the underlying discontinuous system (2.3.1).

Definition 2.3.2. *Given a real number $\gamma > 0$, further referred to as a disturbance attenuation level, it is said that the system (2.3.1) (locally) possesses $\mathcal{L}_2$-gain less than γ with respect to output (2.3.2) if the response $z(t)$, resulting from $\omega(t)$ for initial state $x(t_0) = 0$, satisfies*

$$\int_{t_0}^{t_1} \|z(t)\|^2 dt < \gamma^2 \int_{t_0}^{t_1} \|\omega(t)\|^2 dt \tag{2.3.12}$$

for all $t_1 > t_0$ and all piecewise continuous functions $\omega(t)$ (locally around the origin).

A standard notation

$$DV(x; \nu) = \lim_{\tau \to 0} \frac{V(x + \tau\nu) - V(x)}{\tau} \tag{2.3.13}$$

stands throughout for a Dini derivative (if any) of a scalar function $V(x)$, computed in the direction $\nu \in \mathbb{R}^n$ at $x \in \mathbb{R}^n$.

The following notion is due to reference 5. A vector $\zeta(\hat{x}) \in \mathbb{R}^n$ is a supergradient of a scalar function $f(x)$ at $\hat{x} \in \mathbb{R}^n$ if there exists some $\sigma(\hat{x}) > 0$ such that

$$f(x) \leq f(\hat{x}) + \zeta^T(\hat{x})(x - \hat{x}) + \sigma(\hat{x})\|x - \hat{x}\|^2 \tag{2.3.14}$$

for all x in some neighborhood $U(\hat{x})$ of $\hat{x}$.

The set of supergradients at x is denoted $\partial f(x)$, and is referred to as the *superdifferential.*

In addition, technical lemmas are extracted from reference 14 to be used in the subsequent $\mathscr{L}_2$-gain analysis.

Lemma 2.3.1. *Let $x \in \mathbb{R}^n$ be an absolutely continuous function of time variable t and let $V(x)$ be a scalar locally Lipschitz function around $x \in \mathbb{R}^n$. Then the composite function $V(x)$ is absolutely continuous, and its time derivative is given by*

$$\frac{d}{dt}V\,(x(t)) = DV\,(x(t), \dot{x}(t)) \tag{2.3.15}$$

almost everywhere. Furthermore,

$$DV\,(x(t), \dot{x}(t)) \leq \frac{\partial V}{\partial x}\dot{x}(t) \tag{2.3.16}$$

for almost all t and for all supergradients, $\left(\frac{\partial V}{\partial x}\right)^T \in \partial V(x)$, if any.

Lemma 2.3.2. *Let system (2.3.3) possess a Lyapunov function $V(x)$ with a nonpositive definite temporal derivative $\dot{V}(x(t))$ of the composite function $V(x(t))$ along the system trajectories. Then system (2.3.3) is stable. If in addition, the function $V(x)$ is a strict Lyapunov function with a negative definite temporal derivative $\dot{V}(x(t))$ (and it is radially unbounded), then system (2.3.3) is (globally) asymptotically stable.*

2.3.2.2 Hamilton–Jacobi inequality and its proximal solutions

The Hamilton–Jacobi inequality

$$\frac{\partial V}{\partial x}\varphi(x) + \frac{1}{4\gamma^2}\frac{\partial V}{\partial x}\psi(x)\psi^T(x)\left(\frac{\partial V}{\partial x}\right)^T + h^T(x)h(x) \leq -\nu(x) \tag{2.3.17}$$

with some positive γ and some positive definite function $\nu(x)$ is introduced in a standard manner within the continuity regions G^- and G^+ (i.e., outside the discontinuity surface S), whereas on the discontinuity surface (2.3.5), inequality (2.3.17) is specified according to (2.3.7) with

$$\varphi(x) = \varphi^0(x) \quad \text{provided that} \quad s(x) = 0. \tag{2.3.18}$$

In other words, the Hamilton–Jacobi inequality, if confined to the discontinuity surface (2.3.5), takes the form

$$\frac{\partial V}{\partial x}\varphi^0(x) + \frac{1}{4\gamma^2}\frac{\partial V}{\partial x}\psi(x)\psi^T(x)\left(\frac{\partial V}{\partial x}\right)^T + h^T(x)h(x) \leq -v(x). \tag{2.3.19}$$

Definition 2.3.3. *A locally Lipschitz continuous function $V(x)$ is said to be a (local) proximal solution of the partial differential inequality* (2.3.17) *subject to (2.3.18), specified according to (2.3.7) iff its proximal superdifferential $\partial^P V(x)$ is everywhere nonempty, and* (2.3.17) *holds with $V(x)$ beyond the discontinuity surface (2.3.5) (locally around the origin) for all proximal supergradients $\frac{\partial V}{\partial x} \in \partial^P V(x)$, whereas the SM Hamilton–Jacobi inequality (2.3.19) is satisfied on the discontinuity surface (2.3.5) (locally around the origin) for all $\frac{\partial V}{\partial x} \in \partial^P V(x)$.*

The interested reader may refer to reference 5 for the proximal superdifferential concept for continuous vector fields.

System (2.3.1) and (2.3.2) is subsequently analyzed under the hypothesis that

H. The Hamilton–Jacobi inequality, given by (2.3.17) beyond the discontinuity surface (2.3.5) and specified with (2.3.7) and (2.3.18) along this surface, possesses a (local) positive definite, radially unbounded proximal solution $V(x)$ under some positive γ and some positive definite function $v(x)$.

2.3.2.3 Main result

The following result presents sufficient conditions of the discontinuous system (2.3.1), (2.3.2) to be internally asymptotically stable and to possess $\mathcal{L}_2$-gain less than γ.

Theorem 2.3.1. *Let Assumptions* **A1**, **A2**, *and* **A3** *be in force, and let Hypothesis* **H** *be satisfied (locally). Then the nominal system (2.3.3) is globally (locally) asymptotically stable, whereas its disturbed version (2.3.1) possesses (locally) $\mathcal{L}_2$-gain less than γ with respect to output (2.3.2).*

Proof. The proof is rather technical and since it follows the standard arguments used in the nonlinear $\mathcal{L}_2$-gain analysis [10, 22], here we provide only a sketch. To begin with, the validity of the theorem is justified for the trajectories of the closed-loop system that remain beyond the discontinuity set $s = 0$ during the motion.

It is clear that Lemma 2.3.1 is applicable to a proximal solution $V(x)$ of the Hamilton–Jacobi inequality (2.3.17) viewed on the solutions $x(t)$ of the disturbance-free system (2.3.3). Then relations (2.3.15)–(2.3.17), coupled together, result in

$$\frac{d}{dt}V(x) = DV(x,\dot{x}) \leq \frac{\partial V}{\partial x}\dot{x} = \frac{\partial V}{\partial x}\varphi(x) \leq -v(x), \tag{2.3.20}$$

where $v(x)$ is some positive definite function.

Taking into account that (2.3.20) holds almost everywhere, hypothesis **H** thus ensures that $V(x)$ is a strict radially unbounded Lyapunov function of the nominal system (2.3.3). By Lemma 2.3.2, system (2.3.3) is globally (locally) asymptotically stable.

It remains to show that the disturbed system (2.3.1) (locally) possesses $\mathscr{L}_2$-gain less than γ with respect to output (2.3.2). For this purpose, let us introduce the multivalued function

$$H(x,\omega) = \frac{\partial V(x)}{\partial x}\left[\varphi(x) + \psi(x)\omega\right] + h^T(x)h(x) - \gamma^2\omega^T\omega, \tag{2.3.21}$$

where $\frac{\partial V}{\partial x} \in \partial^P V(x)$. Clearly, the multivalued function (2.3.21) is quadratic in ω. Then,

$$\left.\frac{\partial H(x,\omega)}{\partial \omega}\right|_{\omega=\alpha(x)} = \frac{\partial V(x)}{\partial x}\psi(x) - 2\gamma^2\alpha^T(x) = 0 \tag{2.3.22}$$

for $\alpha(x) = \frac{1}{2\gamma^2}\psi^T(x)\left(\frac{\partial V(x)}{\partial x}\right)^T$ and $\frac{\partial V}{\partial x} \in \partial^P V(x)$. Expanding the quadratic function $H(x,\omega)$ in Taylor series, we derive that

$$H(x,\omega) = H\left(x,\alpha(x)\right) - \gamma^2\|\omega - \alpha(x)\|^2, \tag{2.3.23}$$

where $H(x,\alpha(x)) \leq -\nu(x)$ due to (2.3.17). Hence,

$$H(x,\omega) \leq -\gamma^2\|\omega - \alpha(x)\|^2 - \nu(x) \tag{2.3.24}$$

and employing (2.3.21) and (2.3.23), we arrive at

$$\frac{\partial V(x)}{\partial x}\left[\varphi(x) + \psi(x)\omega\right] \leq -\gamma^2\|\omega - \alpha(x)\|^2 - \nu(x) - \|h(x)\|^2 + \gamma^2\|\omega\|^2. \tag{2.3.25}$$

By applying Lemma 2.3.1 and taking (2.3.25) into account, the time derivative of the solution $V(x)$ of the Hamilton–Jacobi inequality (2.3.17) on the trajectories of (2.3.1) is estimated as follows:

$$\frac{d}{dt}V(x) \leq -\gamma^2\|\omega - \alpha(x)\|^2 - \nu(x) - \|z\|^2 + \gamma^2\|\omega\|^2. \tag{2.3.26}$$

As a matter of fact, the latter inequality ensures that

$$\int_{t_0}^{t_1}(\gamma^2\|\omega(t)\|^2 - \|z(t)\|^2)dt \geq V(x(t_1)) - V(x(t_0))$$
$$+\gamma^2\int_{t_0}^{t_1}\left[\|\omega(t) - \alpha(x(t))\|^2 + \nu(x(t))\right]dt > 0 \tag{2.3.27}$$

for any trajectory of (2.3.1) and (2.3.2), initialized with $x(t_0) = 0$. Thus, inequality (2.3.12) is established thereby completing the proof of Theorem 2.3.1. To complete the proof it remains to note that the validity of the theorem for the SMs that possibly occur in the closed-loop system is justified following the same line of reasoning used in the proof of the result when confined to the dynamics with no SMs. Technical details of the complete proof are therefore omitted and left to the reader. □

2.3.3 A case study: first order SM dynamics

To support our theoretical results, system (2.3.1) is specified in the state space $x \in \mathbb{R}^2$ with

$$x = \begin{bmatrix} x_1 & x_2 \end{bmatrix}^T, \tag{2.3.28}$$

$$\omega = \begin{bmatrix} \omega_1 & \omega_2 \end{bmatrix}^T, \tag{2.3.29}$$

$$\varphi(x) = \begin{bmatrix} x_2 & -M\,\mathrm{sign}(cx_1 + x_2) \end{bmatrix}^T, \tag{2.3.30}$$

$$\psi(x) = \begin{bmatrix} d_1 & 0 \\ 0 & d_2 \end{bmatrix}, \tag{2.3.31}$$

where $M > 0$ and $c > 0$ are the parameters of the switched input $M\,\mathrm{sign}(cx_1 + x_2)$. The above system represents a controlled double integrator that can readily be interpreted in terms of dimensionless motor dynamics (to experimentally be studied in Section 2.3.4.4), where x_1 stands for a position deviation and x_2 is for its velocity. Both a piecewise continuous unmatched disturbance $\omega_1 \in \mathbb{R}$ and a matched disturbance $\omega_2 \in \mathbb{R}$ of the same class affect the system. It is well known that the control input $M\,\mathrm{sign}(cx_1 + x_2)$ imposes disturbance-independent sliding motions on system (2.3.1) thus specified to slide along the linear surface $cx_1 + x_2 = 0$, provided that only matched disturbances are admitted with an upper bound on their magnitude not exceeding the control gain M.

The aim of this section is to demonstrate that the so-called first order SM controller $M\,\mathrm{sign}(cx_1 + x_2)$, while driving the above system, is capable of not only rejecting matched uniformly bounded disturbances but also attenuating unbounded ones, including mismatched disturbances. For this purpose, we further specify the closed-loop system output (2.3.2) to be controlled to consist of the position deviation and the sliding variable $s = cx_1 + x_2$:

$$z = h(x) = \begin{bmatrix} x_1 & s \end{bmatrix}^T. \tag{2.3.32}$$

Thus, the underlying system is represented as follows:

$$\dot{x} = \begin{bmatrix} x_2 + d_1\omega_1(t) \\ u + d_2\omega_2(t) \end{bmatrix} \tag{2.3.33}$$

and it is driven by

$$u = -M\,sign(s), \tag{2.3.34}$$

where the switching surface is governed by

$$s = cx_1 + x_2 = 0. \tag{2.3.35}$$

It is well known [25] that the closed-loop system (2.3.33) and (2.3.34) is globally asymptotically stable whenever $d_1 = 0$ (no unmatched disturbances ω_1 affect the system), $c > 0$, and the state-dependent controller gain $M = M(x) > c|x_2| + d_2 \sup_{t\geq 0} |\omega_2(t)|$ is available.

2.3.3.1 Verification of the Hamilton–Jacobi inequality beyond the switching surface

Let us first demonstrate that beyond the switching manifold (2.3.35) the positive definite function

$$V = x_1^2 + |s| \tag{2.3.36}$$

satisfies the Hamilton–Jacobi inequality (2.3.17). Indeed, substituting (2.3.30)–(2.3.32), (2.3.36) into the left-hand side of (2.3.17), one has

$$\mathscr{H} = 2x_1x_2 + cx_2\,\text{sign}(s) - M + \frac{d_1^2}{4\gamma^2}(2x_1 + c\,\text{sign}(s))^2 + \frac{d_2^2}{4\gamma^2} + x_1^2 + s^2, \tag{2.3.37}$$

where the Hamiltonian $\mathscr{H}$ stands for the left-hand side of the Hamilton–Jacobi inequality (2.3.17). Then taking into account that $s = cx_1 + x_2$ and using straightforward manipulations, involving the well-known inequality $2x_1x_2 \leq x_1^2 + x_2^2$, it follows that

$$\mathscr{H} \leq -M + \left(2 + 2c^2 + \frac{2d_1^2}{\gamma^2}\right)x_1^2 + \frac{7}{2}x_2^2 + \frac{c^2}{2} + \frac{c^2 d_1^2}{2\gamma^2} + \frac{d_2^2}{4\gamma^2}. \tag{2.3.38}$$

Within the ball

$$\mathscr{B}_R = \{x : \|x\| \leq R\} \tag{2.3.39}$$

of radius R the above inequality (2.3.38) is simplified to

$$\mathscr{H} \leq -M + \eta_1 R^2 + \eta_2, \tag{2.3.40}$$

where

$$\eta_1 = \max\left\{2 + 2c^2 + \frac{2d_1^2}{\gamma^2}, \frac{7}{2}\right\}, \tag{2.3.41}$$

$$\eta_2 = \frac{c^2}{2} + \frac{c^2 d_1^2}{2\gamma^2} + \frac{d_2^2}{4\gamma^2}. \tag{2.3.42}$$

So, the Hamiltonian $\mathscr{H}$ proves to be negative definite within the ball (2.3.39) provided that the controller gain M is chosen according to

$$M > \eta_1 R^2 + \eta_2. \tag{2.3.43}$$

Thus, the Hamilton–Jacobi inequality (2.3.17) is shown to hold outside the switching surface (2.3.35), locally within the region $\mathscr{B}_R$.

2.3.3.2 Verification of the Hamilton–Jacobi inequality on the switching surface

The SM equation, governing the system dynamics on the switching surface (2.3.35), is obtained by applying the equivalent control method. Thus, if confined to the switching manifold (2.3.35), the double integrator (2.3.33) reduces to the first order system

$$\dot{x}_1 = -cx_1 + d_1\omega_1, \tag{2.3.44}$$

and its output (2.3.32) is then specified to

$$z = \begin{bmatrix} x_1 & 0 \end{bmatrix}^T. \tag{2.3.45}$$

In turn, the positive definite function (2.3.36) on the SMs $s(x) = 0$ is simplified to

$$V = x_1^2. \tag{2.3.46}$$

Let us now demonstrate that the Hamilton–Jacobi inequality (2.3.19), while being specified for the SM equation (2.3.44), is solved with the positive definite function (2.3.46). By substituting (2.3.46) into the Hamilton–Jacobi inequality (2.3.19), thus specified, one derives

$$\mathscr{H} = -\left(2c - \frac{1}{\gamma^2}d_1^2 - 1\right)x_1^2 < 0 \tag{2.3.47}$$

provided the surface parameter is chosen according to

$$c > \frac{1}{2}\left(\frac{1}{\gamma^2}d_1^2 + 1\right). \tag{2.3.48}$$

The validity of the Hamilton–Jacobi inequality (2.3.19) is thus straightforwardly verified on the switching surface (2.3.35) subject to parameter choice (2.3.48).

2.3.3.3 $\mathscr{L}_2$-Gain analysis of the overall system

Summarizing derivations of Subsections 1.3.1 and 1.3.2, and applying Theorem 1 to the system in question, the following result is obtained.

Theorem 2.3.2. *Given arbitrary $\gamma > 0$ and radius $R > 0$, let the controller parameters M and c be chosen to ensure that inequalities* (2.3.43) *and* (2.3.48) *be satisfied. Then the closed-loop system* (2.3.33)–(2.3.35) *is internally asymptotically stable and it possesses $\mathscr{L}_2$-gain less than γ with respect to output* (2.3.32), *locally within the ball $\mathscr{B}_R$ of radius R.*

2.3.3.4 Numerical performance analysis

The performance of the closed-loop system (2.3.33)–(2.3.35) was numerically tested under the parameter values $d_1 = d_2 = 1$ in the presence of the unmatched and matched harmonic disturbances

$$\omega_1 = 0.3\sin(3t), \quad \omega_2 = 0.3\sin(3t). \tag{2.3.49}$$

In the simulation runs, performed with Simulink. using the Runge–Kutta numerical integration method *ode*4 with the fixed integration step 1×10^{-3} s, the initial conditions were set to $x_1(0) = 0$ and $x_2(0) = 0$, and the numerical study was confined to the ball $\mathscr{B}_R$ of radius $R = 0.02$. Specifying the desired attenuation level with $\gamma = 0.5$, the controller parameters were then set to $c = 2.6$ and $M = 19$ to ensure that conditions (2.3.43) and (2.3.48) of Theorem 2.3.1 were satisfied.

The corresponding closed-loop dynamics with unmatched and matched disturbance and $\mathscr{L}_2$-gain performance are depicted in Figures 2.3.1 and 2.3.2, respectively. Good performance, concluded from these figures, supports the theory in that SM controllers are capable of not only rejecting matched disturbances but also attenuating unmatched disturbances.

Also to demonstrate the attenuation of matched disturbances whose magnitude is larger than that of the controller, additional simulation runs were performed for the controller of the smaller magnitude $M = 0.25$ than that of the matched disturbance. To respect conditions (2.3.43) and (2.3.48) the parameters were tuned to $R = 0.05$, $c = 0.6$, and $\gamma = 5$. Figure 2.3.3 depicts the closed-loop dynamics, thus specified. The $\mathscr{L}_2$-gain performance, shown in Figure 2.3.4, demonstrates that the matched disturbance, exceeding the controller gain, is actually attenuated in the closed-loop.

2.3.4 A case study: second order SM dynamics

2.3.4.1 Twisting controller

The $\mathscr{L}_2$-gain analysis is now applied to the popular twisting controller [9] to demonstrate that apart from rejecting bounded disturbances, it is also capable of attenuating unbounded ones of class $\mathscr{L}_2$. In contrast to (2.3.33), the double integrator

$$\dot{x}_1 = x_2, \quad \dot{x}_2 = u + \omega_u \tag{2.3.50}$$

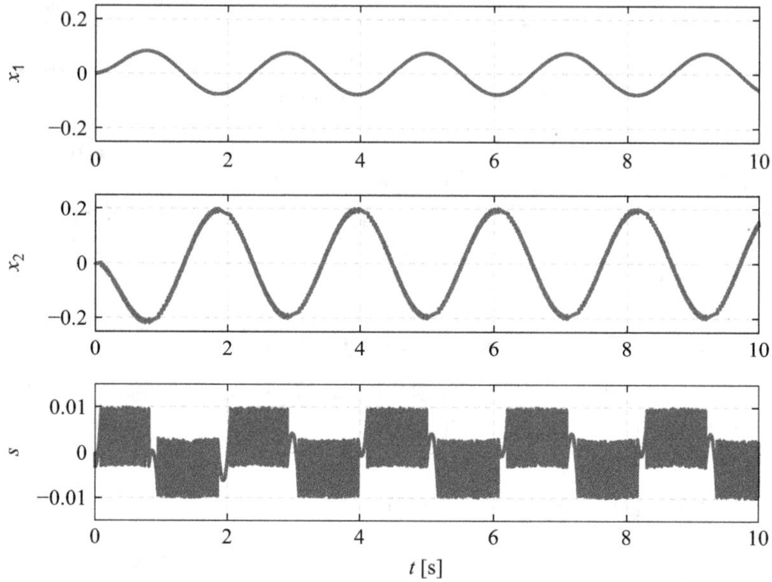

Figure 2.3.1 Dynamics of the closed-loop system (2.3.33)–(2.3.35) affected by unmatched and matched harmonic disturbance (2.3.49)

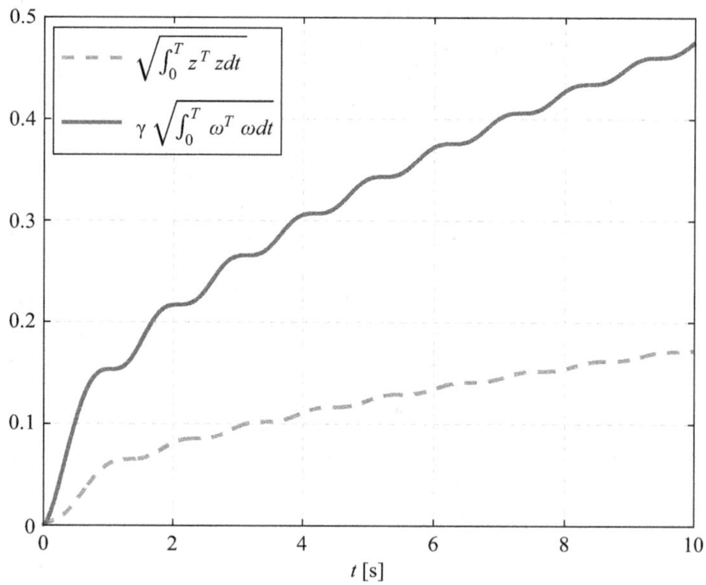

Figure 2.3.2 $\mathscr{L}_2$ gain performance of the closed-loop system (2.3.33)–(2.3.35) affected by unmatched and matched harmonic disturbances (2.3.49)

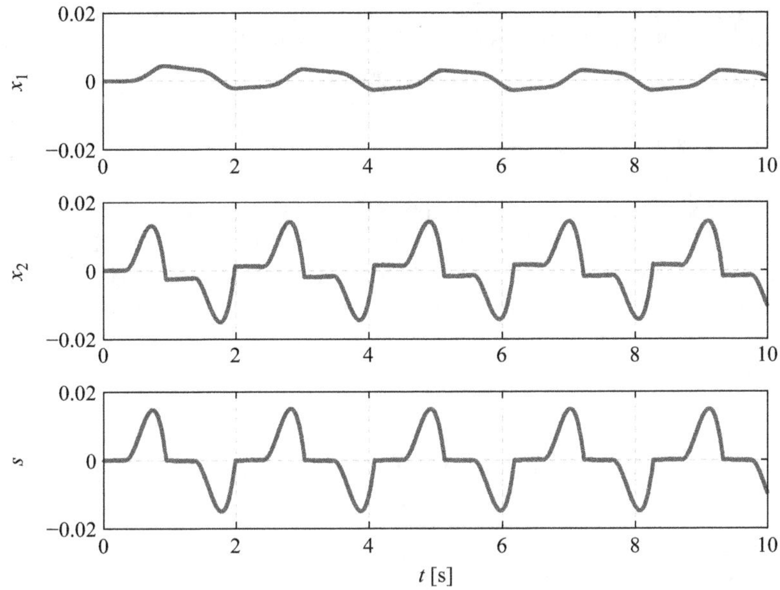

Figure 2.3.3 *Dynamics of the closed-loop system (2.3.33)–(2.3.35) affected only by matched harmonic disturbance of (2.3.49)*

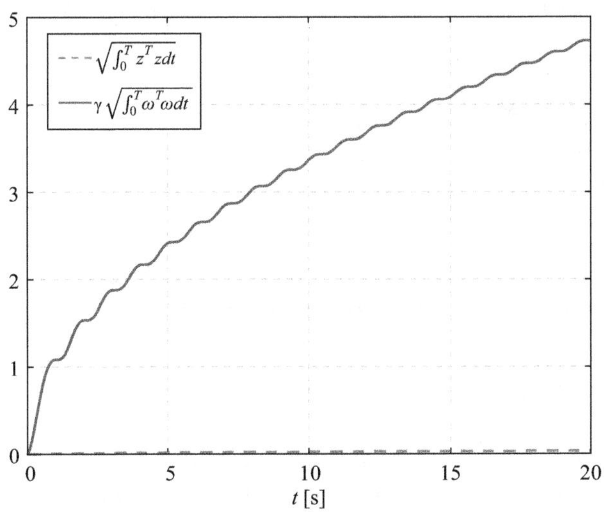

Figure 2.3.4 *$\mathscr{L}_2$ gain performance of the closed-loop system (2.3.33)–(2.3.35) affected only by matched harmonic disturbance of (2.3.49)*

is presently affected by a piecewise continuous matched disturbance $\omega_u \in \mathbb{R}$ only what is actually the case of fully actuated electromechanical plants. The control input is subsequently specified with the state feedback

$$u = -k_1 \operatorname{sign}(x_1) - k_2 \operatorname{sign}(x_2), \tag{2.3.51}$$

typically referred to as the twisting controller, whose gains $k_1, k_2 \in \mathbb{R}^+$ are to be selected to ensure a certain disturbance attenuation level. It should be pointed out that the closed-loop system (2.3.50), (2.3.51) is well known [13] to be globally asymptotically stable if

$$\sup_{t \geq 0} |\omega_u(t)| < k_2 < k_1 - k_2.$$

Setting $x = \begin{bmatrix} x_1 & x_2 \end{bmatrix}^T$, the following state-space representation of (2.3.50)–(2.3.51)

$$\dot{x} = \begin{bmatrix} x_2 \\ -k_1 \operatorname{sign}(x_1) - k_2 \operatorname{sign}(x_2) + \omega_u \end{bmatrix} \tag{2.3.52}$$

is obtained and coupled to the preselected output

$$z = x \tag{2.3.53}$$

to arrive at the generic representation (2.3.1)–(2.3.2) specified with

$$\varphi(x) = \begin{bmatrix} x_2 \\ -k_1 \operatorname{sign}(x_1) - k_2 \operatorname{sign}(x_2) \end{bmatrix}, \tag{2.3.54}$$

$$\psi(x) = \begin{bmatrix} 0 & 1 \end{bmatrix}^T, \tag{2.3.55}$$

$$h(x) = \begin{bmatrix} x_1 & x_2 \end{bmatrix}^T. \tag{2.3.56}$$

To meet conditions of Theorem 2.3.1 the Hamilton–Jacobi inequality (2.3.17) is next verified with the following positive definite function

$$V = \frac{1}{2}(x_1 + x_2)^2 + k_1 |x_1|. \tag{2.3.57}$$

Remark 2.3.1. *It is clear that the closed-loop system (2.3.52) possesses no SMs on the discontinuity surface $x_2 = 0$ but the origin whereas on the discontinuity surface $x_2 = 0$, SMs, if any, turn out to be disturbance-free because only matched disturbances are admitted. Thus, the $\mathcal{L}_2$-gain inequality (2.3.12) is trivially satisfied along SMs so that no need to verify the Hamilton–Jacobi inequality (2.3.19) on the discontinuity surfaces.*

Substituting (2.3.54)–(2.3.57) in the left-hand side of (2.3.17) yields

$$\mathscr{H} = x_1x_2 + x_2^2 - k_1|x_1| - k_2x_1\,\mathrm{sign}(x_2) - k_2|x_2| + \frac{1}{4\gamma^2}(x_1 + x_2)^2 + x_1^2 + x_2^2. \tag{2.3.58}$$

Then letting

$$k_1 > k_2 \tag{2.3.59}$$

and setting

$$\lambda_1 = \min\{k_1 - k_2, k_2\}, \quad \lambda_2 = \frac{5}{2} + \frac{1}{2\gamma^2}, \tag{2.3.60}$$

one derives the inequality

$$\mathscr{H} \leq -(\lambda_1 - \lambda_2\|x\|)\,\|x\| \tag{2.3.61}$$

which validates the negative definiteness of the Hamiltonian $\mathscr{H}$ locally within the ball

$$\mathscr{B}_{R_c} = \left\{x \in \mathbb{R}^2 : \|x\| < R_c\right\} \tag{2.3.62}$$

of a radius R_c, such that

$$\lambda_1 > \lambda_2 R_c. \tag{2.3.63}$$

Hypothesis **H** is thus locally verified for all $x \in \mathscr{B}_c \subset \mathbb{R}^2$ and by applying Theorem 2.3.1, the following result is concluded.

Theorem 2.3.3. *Given arbitrary $\gamma > 0$ and radius $R_c > 0$, let the controller gains k_1 and k_2 be chosen to ensure that inequalities (2.3.59), (2.3.63), coupled to (2.3.60), hold true. Then the closed-loop system (2.3.52) is internally asymptotically stable and it possesses $\mathscr{L}_2$-gain less than γ with respect to the output (2.3.53), locally within the ball $\mathscr{B}_{R_c}$ of radius R_c.*

2.3.4.2 Super-twisting observer

The investigation is now focused on the $\mathscr{L}_2$-gain analysis of the supertwisting observer [6]

$$\begin{aligned}\dot{\tilde{x}}_1 &= \tilde{x}_2 + m_1|y - \tilde{x}_1|^{\frac{1}{2}}\,\mathrm{sign}(y - \tilde{x}_1),\\ \dot{\tilde{x}}_2 &= u + m_2\,\mathrm{sign}(y - \tilde{x}_1),\end{aligned} \tag{2.3.64}$$

whose states $\tilde{x}_1$ and $\tilde{x}_2$ estimate the position x_1 of system (2.3.52) and its velocity x_2 based on the available measurement $y = x_1 + \omega_o$, corrupted by the measurement error $\omega_o \in \mathbb{R}$ whose second order time derivative $\ddot{\omega}_o$ is assumed to be piecewise continuous. The observer gains $m_1, m_2 \in \mathbb{R}^+$ are to be selected to ensure a certain disturbance attenuation level.

Setting the observation error $e = \begin{bmatrix} e_1 & e_2 \end{bmatrix}^T$ as

$$e_1 = y - \tilde{x}_1, \quad e_2 = \dot{y} - \tilde{x}_2, \tag{2.3.65}$$

the following second order observation error system

$$\dot{e} = \begin{bmatrix} e_2 - m_1|e_1|^{\frac{1}{2}} \operatorname{sign}(e_1) \\ -m_2 \operatorname{sign}(e_1) + \omega_u + \ddot{\omega}_o \end{bmatrix} \tag{2.3.66}$$

is obtained with the generic matched disturbance

$$\omega = \omega_u + \ddot{\omega}_o. \tag{2.3.67}$$

The above system (2.3.66) proves to be globally asymptotically stable [16] provided that

$$\sup_{t \geq 0} |[\omega_u(t) + \ddot{\omega}_o(t)]| < \min \left\{ \frac{1}{2} m_1, \ \frac{m_1 m_2}{1 + m_1} \right\}.$$

The subsequent $\mathscr{L}_2$-gain analysis of the observation error system is coupled to the preselected output

$$z = e, \tag{2.3.68}$$

and system (2.3.66) is thus represented at the generic form (2.3.1)–(2.3.2), specified with $x = e$ and

$$\varphi(e) = \begin{bmatrix} e_2 - m_1|e_1|^{\frac{1}{2}} \operatorname{sign}(e_1) \\ -m_2 \operatorname{sign}(e_1) \end{bmatrix}, \tag{2.3.69}$$

$$\psi(e) = [\, 0 \quad 1 \,]^T, \tag{2.3.70}$$

$$h(e) = [\, e_1 \quad e_2 \,]^T. \tag{2.3.71}$$

The Hamilton–Jacobi inequality (2.3.17), specified with (2.3.69)–(2.3.71), is now verified for positive definite function

$$V = 2m_2|e_1| + \frac{1}{2} e_2^2 + \frac{1}{2} r^2(e_1, e_2), \tag{2.3.72}$$

where

$$r(e_1, e_2) = e_2 - m_1|e_1|^{\frac{1}{2}} \operatorname{sign}(e_1). \tag{2.3.73}$$

Remark 2.3.2. *Remark 2.3.1 applies here as well.*

Substituting (2.3.69)–(2.3.71) and (2.3.72) into the left-hand side of the Hamilton–Jacobi inequality (2.3.17) yields

$$\mathscr{H} = -\frac{1}{2}m_1 r^2 |e_1|^{-\frac{1}{2}} + m_2 r\,\mathrm{sign}(e_1) - m_2 e_2\,\mathrm{sign}(e_1) + \frac{1}{4\gamma^2}(e_2 + r)^2 + e_1^2 + e_2^2. \tag{2.3.74}$$

Then setting

$$\lambda_3 = 4\left(1 + \frac{1}{\gamma^2}\right) \tag{2.3.75}$$

and using (2.3.73), it follows that

$$\mathscr{H} \le -\left(m_1 m_2 - \frac{1}{4}\lambda_3 m_1^2 |e_1|^{\frac{1}{2}} - |e_1|^{\frac{3}{2}}\right)|e_1|^{\frac{1}{2}} - \frac{1}{2}\left(m_1 - \lambda_3 |e_1|^{\frac{1}{2}}\right) r^2(e_1, e_2)|e_1|^{-\frac{1}{2}}. \tag{2.3.76}$$

The latter inequality validates the negative definiteness of the Hamiltonian $\mathscr{H}$ within the strip

$$\mathscr{S}_{R_o} = \left\{ e \in \mathbb{R}^2 : |e_1| < R_o^2 \right\} \tag{2.3.77}$$

of some width $R_o > 0$ provided that

$$m_1 > \lambda_3 R_o, \tag{2.3.78}$$

$$m_2 > \frac{1}{4}\lambda_3 m_1 R_o + \frac{R_o^3}{m_1}. \tag{2.3.79}$$

The hypothesis **H** is thus locally verified for all $e \in \mathscr{S}_{R_o} \subset \mathbb{R}^2$. By applying Theorem 2.3.1, the following result is established.

Theorem 2.3.4. *Given arbitrary $\gamma > 0$ and radius $R_o > 0$, let the observer gains m_1 and m_2 be chosen to ensure that inequalities* (2.3.78), (2.3.79), *coupled to* (2.3.75), *hold true. Then the observation error system* (2.3.66) *is internally asymptotically stable, and it possesses $\mathscr{L}_2$-gain less than γ with respect to the error output* (2.3.68), *locally within the strip $\mathscr{S}_{R_o}$ given by* (2.3.77).

2.3.4.3 Output feedback SOSM control

Implementing the twisting controller (2.3.51), fed by the supertwisting velocity observer (2.3.64), running in parallel, results in the closed-loop position feedback system (2.3.50), governed by

$$\dot{x}_1 = x_2, \quad \dot{x}_2 = -k_1\,\mathrm{sign}(y) - k_2\,\mathrm{sign}(\tilde{x}_2) + \omega_u. \tag{2.3.80}$$

The overall closed-loop system (2.3.66) and (2.3.80) to be analyzed next admits the following state-space representation

$$\begin{aligned} \dot{y}_1 &= y_2, \\ \dot{y}_2 &= -k_1 \operatorname{sign}(y_1) - k_2 \operatorname{sign}(y_2 - e_2) + \omega, \\ \dot{e}_1 &= e_2 - m_1 |e_1|^{\frac{1}{2}} \operatorname{sign}(e_1), \\ \dot{e}_2 &= -m_2 \operatorname{sign}(e_1) + \omega, \end{aligned} \tag{2.3.81}$$

$$z = \begin{bmatrix} y_1 & y_2 & e_1 & e_2 \end{bmatrix}^T \tag{2.3.82}$$

in terms of $y_1 = y$, $y_2 = \dot{y}$, the observation error (2.3.65), the performance output z, and the generic disturbance ω, given by (2.3.67). The generic representation (2.3.1)–(2.3.2) is thus obtained with the state variable $x = z$ and

$$\varphi = \begin{bmatrix} y_2 \\ -k_1 \operatorname{sign}(y_1) - k_2 \operatorname{sign}(y_2 - e_2) \\ e_2 - m_1 |e_1|^{\frac{1}{2}} \operatorname{sign}(e_1) \\ -m_2 \operatorname{sign}(e_1) \end{bmatrix}, \tag{2.3.83}$$

$$\psi = \begin{bmatrix} 0 & 1 & 0 & 1 \end{bmatrix}^T, \tag{2.3.84}$$

$$h = \begin{bmatrix} y_1 & y_2 & e_1 & e_2 \end{bmatrix}^T. \tag{2.3.85}$$

The $\mathscr{L}_2$-gain analysis is successively applied beyond and on the discontinuity manifolds

$$S_c = \left\{ (y_1, y_2, e_1, e_2) \in \mathbb{R}^4 : y_1 = y_2 = 0 \right\}, \tag{2.3.86}$$

$$S_o = \left\{ (y_1, y_2, e_1, e_2) \in \mathbb{R}^4 : e_1 = e_2 = 0 \right\}. \tag{2.3.87}$$

SMs occur on the former manifold when the available measurement y and its time derivative reach the origin and remain there so that the $\mathscr{L}_2$-gain analysis is confined to that of the observer, whereas on the latter manifold, SMs occur when the observation errors are nullified and the output feedback formally coincides with the state feedback.

2.3.4.3.1 $\mathscr{L}_2$-Gain analysis beyond the discontinuity manifolds

Conditions of Theorem 2.3.1 are first verified beyond the discontinuity manifolds (2.3.86) and (2.3.87). For this purpose, the positive definite function

$$V = \frac{1}{2}(y_1 + y_2)^2 + k_1 |y_1| + 2m_2 |e_1| + \frac{1}{2} e_2^2 + \frac{1}{2} r^2(e_1, e_2) \tag{2.3.88}$$

is introduced with $r(e_1, e_2)$ given by (2.3.73).

By substituting (2.3.83)–(2.3.85) and (2.3.88) into the Hamilton–Jacobi inequality (2.3.17), its left-hand side takes the form

$$\begin{aligned}\mathscr{H} &= y_1y_2 + y_2^2 + k_1y_2\,\mathrm{sign}(y_1) - k_1|y_1| - k_2y_1\,\mathrm{sign}(y_2 - e_2)\\&\quad - k_1y_2\,\mathrm{sign}(y_1) - k_2y_2\,\mathrm{sign}(y_2 - e_2) + 2m_2r\,\mathrm{sign}(e_1)\\&\quad - \frac{1}{2}m_1r^2|e_1|^{-\frac{1}{2}} - m_2e_2\,\mathrm{sign}(e_1) - m_2r\,\mathrm{sign}(e_1)\\&\quad + \frac{1}{2\gamma^2}(y_1 + y_2 + e_2 + r)^2 + y_1^2 + y_2^2 + e_1^2 + e_2^2. \qquad (2.3.89)\end{aligned}$$

Employing the well-known inequalities $\left(\sum_{i=1}^n a_i\right)^2 \le 2\sum_{i=1}^n a_i^2$, $|a_1 - a_2| \ge |a_1| - |a_2|$, $a_1a_2 \le \frac{1}{2}a_1^2 + \frac{1}{2}a_2^2$ and substituting $e_2 = r(e_1, e_2) + m_1|e_1|^{\frac{1}{2}}\,\mathrm{sign}(e_1)$, resulting from (2.3.73), it follows that

$$\begin{aligned}\mathscr{H} &\le \left(\frac{3}{2} + \frac{1}{\gamma^2}\right)y_1^2 - (k_1 - k_2)|y_1| + \left(\frac{5}{2} + \frac{1}{\gamma^2}\right)y_2^2 - k_2|y_2|\\&\quad - \frac{1}{2}(m_1 - 2k_2)\,r^2(e_1, e_2)|e_1|^{-\frac{1}{2}} + \left(2 + \frac{4}{\gamma^2}\right)r^2(e_1, e_2)\\&\quad + \left(2m_1^2 + \frac{1}{\gamma^2}m_1^2\right)|e_1| + e_1^2 - (m_1m_2 - 2k_2m_1 - k_2)\,|e_1|^{\frac{1}{2}}. \qquad (2.3.90)\end{aligned}$$

Taking into account relations (2.3.60) and (2.3.75), determining the parameters λ_1, λ_2, and λ_3, one then derives the inequality

$$\begin{aligned}\mathscr{H} \le &-\left(\lambda_1 - \lambda_2\sqrt{y_1^2 + y_2^2}\right)\sqrt{y_1^2 + y_2^2} - \left(\frac{1}{2}m_1 - 2k_2 - \frac{1}{2}\lambda_3|e_1|^{\frac{1}{2}}\right)\\&\times r^2(e_1, e_2)|e_1|^{-\frac{1}{2}} - \left(m_1m_2 - 2k_2m_1 - k_2 - \frac{1}{2}\lambda_3m_1^2|e_1|^{\frac{1}{2}} - |e_1|^{\frac{3}{2}}\right)|e_1|^{\frac{1}{2}},\end{aligned} \qquad (2.3.91)$$

whose right-hand side is locally negative definite in the region

$$\mathscr{R} = \left\{(y_1, y_2, e_1, e_2) \in \mathbb{R}^4 : y_1^2 + y_2^2 < R_c^2 \text{ and } |e_1| < R_o^2\right\}, \qquad (2.3.92)$$

provided that relations (2.3.59) and (2.3.63) hold, and

$$\begin{aligned}m_1 &> 4k_2 + \lambda_3R_o,\\m_2 &> 2k_2 - \frac{k_2}{m_1} - \frac{1}{2}\lambda_3m_1R_o - \frac{R_o^3}{m_1}.\end{aligned} \qquad (2.3.93)$$

The Hamilton–Jacobi inequality (2.3.17) is thus locally verified and hypothesis **H** is locally validated beyond the discontinuity manifolds (2.3.86) and (2.3.87) within the region $\mathscr{R} \subset \mathbb{R}^4$.

2.3.4.3.2 $\mathscr{L}_2$-Gain analysis on the discontinuity manifolds S_c and S_o

Once the trajectories of (2.3.81) evolve in the SM along the discontinuity manifold (2.3.86), the underlying system (2.3.81) is reduced to the observation error system of the second order which is given by (2.3.66), and the $\mathscr{L}_2$-gain analysis of this resulting system has been summarized in Theorem 2.3.4.

Once the trajectories of (2.3.81) evolve in the SM along the discontinuity manifold (2.3.87), the underlying system (2.3.81) is reduced to the one of the second order which is given by (2.3.52), driven by the twisting state feedback (2.3.51), and Theorem 2.3.3 has summarized the $\mathscr{L}_2$-gain analysis of the resulting SM system.

2.3.4.3.3 $\mathscr{L}_2$-Gain analysis of the overall system

The following result is straightforwardly concluded from derivations of Sections 2.3.4.3.1 and 2.3.4.3.2, coupled to Theorems 2.3.3 and 2.3.4.

Theorem 2.3.5. *Given arbitrary* $\gamma > 0$, $R_c > 0$, *and* $R_o > 0$, *let the controller and observer gains* k_1, k_2 *and* m_1, m_2 *be chosen to ensure that inequalities* (2.3.59), (2.3.63), (2.3.93), *coupled to* (2.3.60) *and* (2.3.75), *hold true. Then, the overall system* (2.3.81) *is internally asymptotically stable, and it possesses* $\mathscr{L}_2$*-gain less than* γ *with respect to the performance output* (2.3.82) *locally within the region* $\mathscr{R}$.

2.3.4.4 Experimental verification on a DC motor

Experimental setup, designed in the research laboratory of CITEDI-IPN, involves an industrial DC motor of inertia $J = 3.11 \times 10^{-5}$ Kg·m^2 [11]. The dynamic model of the motor is as follows

$$J\ddot{q} = u + w, \tag{2.3.94}$$

where $t \in \mathbb{R}$ is the time instant, $q(t) \in \mathbb{R}$ is the angular position of the rotor, $\dot{q}(t) \in \mathbb{R}$ is its angular velocity, $u(t) \in \mathbb{R}$ is the control input, $w(t) \in \mathbb{R}$ is the disturbance involved to count for motor dynamic uncertainties (e.g., friction forces and unmodeled dynamics) and manipulatable input signals.

The performance of the twisting output feedback controller (2.3.51), fed by the supertwisting velocity observer (2.3.64), was tested. In the experimental verification, the desired attenuation level was prespecified with $\gamma = 1$, whereas the operating region of interest was confined to $R_c = 400$ and $R_o = 300$. The controller parameters were then set to

$$k_1 = 0.08, \quad k_2 = 0.05, \quad m_1 = 40, \quad m_2 = 440$$

to ensure that the tuning conditions (2.3.59), (2.3.63), and (2.3.93) of Theorem 2.3.5 were satisfied.

In the first experimental run, the input signal

$$w_1 = 0.5e^{-2t}\cos(20t)N \cdot m \tag{2.3.95}$$

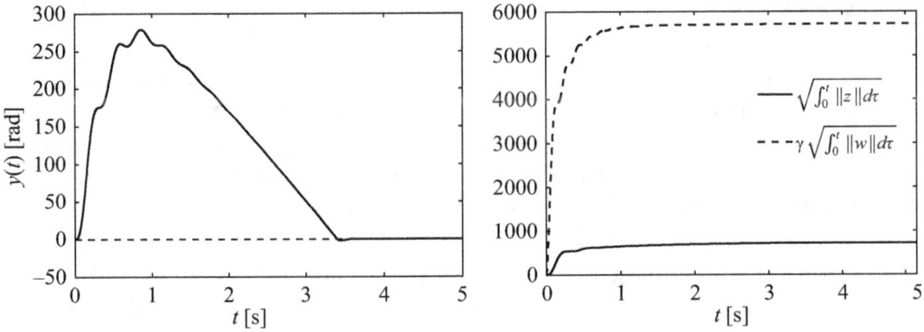

(a) Position evolution (b) $\mathscr{L}_2$-gain performance

Figure 2.3.5 Performance under exponentially decaying harmonic disturbance (2.3.95)

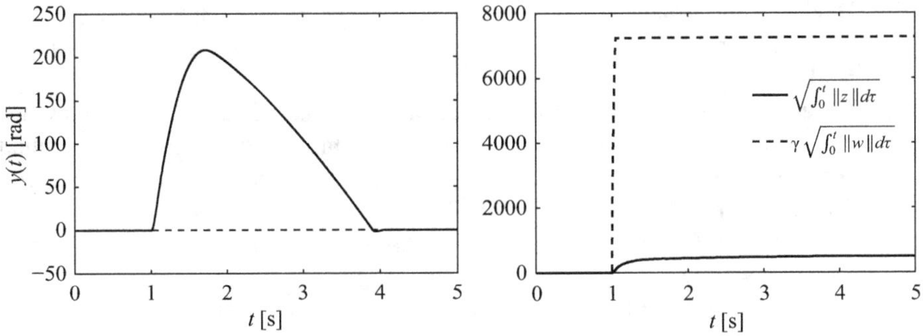

(a) Position evolution (b) $\mathscr{L}_2$-gain performance

Figure 2.3.6 Performance under impulsive-wise disturbance (2.3.96)

additionally contributed to the motor disturbance w. Figure 2.3.5(a) shows the position dynamics, whereas Figure 2.3.5(b) depicts the $\mathscr{L}_2$-gain performance respecting disturbance attenuation level $\gamma = 1$. Good performance is concluded from these figures, the disturbance is not rejected but attenuated only, and the system keeps $\mathscr{L}_2$-gain less than the prespecified attenuation level $\gamma = 1$.

Figures 2.3.6(a,b) reflect the second experimental run when the input signal was changed to the impulsive one

$$w_1 = \begin{cases} 1 & \text{if } 1 \leq t < 0.05, \\ 0 & \text{otherwise.} \end{cases} \tag{2.3.96}$$

Good performance, concluded from these figures, confirms the effectiveness of the developed method of tuning the parameter values in order to keep the resulting $\mathscr{L}_2$-gain less than a prespecified $\gamma = 1$. As in the previous experimental run, the disturbance is not rejected but only attenuated with a reasonable attenuation level, $\gamma = 1$.

References

[1] Basar, T. and Bernhard, P. (1995). *$\mathscr{H}_\infty$-Optimal Control and Related Minimax Design Problems: A Dynamic Game Approach*. Birkhauser. Boston.

[2] Benderradji, H., Benamor, A., Chrifi-Alaoui, L., Bussy, P., and Makouf, A. (2012). Second order sliding mode induction motor control with a new Lyapunov approach. *2012—9th International Multi-Conference on Systems, Signal and Devices*, pp. 1–6.

[3] Bernuaa, E., Efimov, D., Perruquetti, W., and Polyakov, A. (2014). On homogeneity and its application to sliding mode control. *Journal of Franklin Institute*, **351**(4), pp. 1866–1901.

[4] Castaños, F. and Fridman, L. (2011). Dynamic switching surfaces for output sliding mode control: An $\mathscr{H}_\infty$ approach. *Automatica*, **47**(7), pp. 1957–1961.

[5] Clarke, F.H. (1988). *Optimization and Non-Smooth Analysis*, Wiley Interscience, New York.

[6] Davila, J., Fridman, L. and Levant, A. (2005). Second order sliding mode observer for mechanical systems. *IEEE Transactions on Automatic Control*, **50**(11), pp. 1785–1789.

[7] Estrada, A., Loria, A., Santiesteban, R., and Fridman, L. (2011). Lyapunov stability analysis of a twisting based control algorithm for systems with unmatched perturbations. *2011 50th IEEE Conference on Decision and Control and European Control Conference (CDC-ECC)*, pp. 4586–4591.

[8] Filippov, A.F. (1988). *Differential Equations with Discontinuous Right-Hand Side*, Kluwer, Dordrecht, The Netherlands.

[9] Levant, A. (1993). Sliding order and sliding accuracy in sliding mode control. *International Journal of Control*, **58**(6), pp. 1247–1263.

[10] Isidori, A. and Astolfi, A. (1992). Disturbance attenuation and $\mathscr{H}_\infty$-control via measurement feedback in nonlinear systems. *IEEE Transactions on Automatic Control*, **37**(9), pp. 1283–1293.

[11] Leadshine Technology Co., Ltd (n.d.). Datasheet of DCM series brushed servo motors. Retrieved from http://www.leadshine.com/UploadFile/Down/DCM5xxxxd.pdf.

[12] Moreno, J.A. and Osorio, M. (2012). Strict Lyapunov functions for the super-twisting algorithm. *IEEE Transactions on Automatic Control*, **57**(4), pp. 1035–1040.

[13] Orlov, Y. (2005). Finite time stability and robust control synthesis of uncertain switched systems. *SIAM Journal on Control and Optimization*, **43**(4), 1253–1271.

[14] Orlov, Y. (2009). Discontinuous Systems—Lyapunov Analysis and Robust Synthesis Under Uncertainty Conditions. Springer-Verlag, London, UK.

[15] Orlov Y. and Aguilar, L. (2014). *Advanced $\mathscr{H}_\infty$ Control—Towards Nonsmooth Theory and Applications*. Birkhäuser. Springer-Verlag, New York, USA. 2014.

[16] Orlov Y., Aoustin, Y., and Chevallereau, C. (2011). Finite time stabilization of a perturbed double integrator—Part I: Continuous sliding mode-based

output feedback synthesis. *IEEE Transactions on Automatic Control*, **56**(3), pp. 614–618.

[17] Osuna-Altamirano, T. and Orlov, Y. (2014a). $\mathscr{L}_2$-Gain analysis of sliding mode controller. *13th International Workshop on Variable Structure Systems 2014*. Nantes, France. June 29–July 2, 2014.

[18] Osuna-Altamirano, T. and Orlov, Y. (2014b). Análisis de ganancia $\mathscr{L}_2$ de un sistema de orden *n* con controladores de modos deslizantes. *XVI Congreso Latinoamericano de Control Automático (CLCA 2014)*. Cancún, Quintana Roo, México. October 2014, pp. 528–533.

[19] Ponce I.U., Orlov, Y., and Aguilar L. (2013a). Nonsmooth $\mathscr{L}_2$-gain analysis of twisting controller. *2013 6th International Scientific Conference on Physics and Control*. San Luis Potosí, México.

[20] Ponce I.U., Orlov, Y., and Aguilar L. (2013b). Nonsmooth $\mathscr{L}_2$-gain analysis of super-twisting algorithm. *2013 6th Congreso Nacional de Control Automático*. Ensenada, México, pp. 628–633.

[21] Santiesteban, R., Fridman, L., and Moreno J.A. (2010). Finite-time convergence analysis for "twisting" controller via a strict Lyapunov function. *2010 11th International Workshop on Variable Structure Systems*, pp. 1–6.

[22] Van Der Shaft, A.J. (1992). $\mathscr{L}_2$-Gain analysis of nonlinear systems and nonlinear state feedback control. *IEEE Transactions on Automatic Control*, **37**(6), pp. 770–784.

[23] Zhang, J., Shen, T., and Jiao, X. (2009). $\mathscr{L}_2$-Gain analysis and feedback design for discontinuous time-delay systems based on functional differential inclusion. *Joint 48th IEEE Conference on Decision and Control and 28th Chinese Control Conference*, pp. 5114–5119.

[24] Zhao, G. and Wang J. (2013). Finite time stability and $\mathscr{L}_2$-gain analysis for switched linear systems with state-dependent switching. *Journal of the Franklin Institute*, **350**, pp. 1075–1092.

[25] Utkin, V.I. (1992). *Sliding Modes in Control and Optimization*. Springer-Verlag Berlin, Heidelberg.

Chapter 2.4

Analysis of transient motions in variable-structure systems through the dynamic harmonic balance principle

Igor Boiko

Abstract

The conventional harmonic balance principle is a convenient tool for finding periodic solutions in variable-structure systems, which may occur as chattering in sliding mode control or as a normal operating mode in relay systems, limit cycling tests aimed at controller tuning through variable-structure algorithms, etc. In the present book chapter, the conventional harmonic balance principle is extended to transient oscillatory processes in systems. This principle is termed the dynamic harmonic balance principle. It is formulated for the system having one single-valued odd-symmetric nonlinearity and linear plant without zeros in the transfer function. Based on the dynamic harmonic balance, the equations for the amplitude, frequency rate of change, and amplitude rate of change are derived. This principle is then illustrated by analysis of transient motions in a variable-structure system and the decaying motions of a rocking block.

2.4.1 Introduction

Sliding-mode control systems often exhibit oscillatory motion. The process of convergence may be viewed as transient oscillations. If the steady mode is chattering then the process of establishing chattering can be viewed as transient oscillations. However, methods that are available normally allow for analysis of only already established oscillations but not transient motions. The harmonic balance (HB) principle is a convenient tool for finding parameters of self-excited periodic motions. Due to its convenience and simplicity, it is widely used in many areas of science and engineering. This principle is used for deriving the model of the oscillations. We can see that for a system with one nonlinearity and linear dynamics (a Lure system), the HB principle can be illustrated by the Nyquist plot of the linear dynamics and the plot of the negative reciprocal of the describing function (DF) [1] of the nonlinearity in the complex plane and finding the point of intersection of the two plots, which would correspond to the self-excited periodic motion in the system. Therefore, the

HB principle treats the system as a loop connection of the linear dynamics and of the nonlinearity. It is also possible to reformulate the HB so that the format of the system analysed is not a loop connection but the denominator of the closed-loop system. This would imply a different interpretation of the HB, which would allow one to extend the HB principle to analysis of not only self-excited periodic motions but also other types of oscillatory motions.

Examples of vanishing oscillatory motion of variable frequency can be found in the conventional [2] and second-order sliding-mode (SM) control system models [3, 4]. Another example of a transient oscillatory motion is the transient processes in oscillatory mechanical systems. A number of examples are given in References 5 and 6. Examples from different areas of engineering and physics can be found where transient oscillatory motions of variable frequency occur. The problem of the convergence rate assessment, including qualitative (finite time or asymptotic) and quantitative assessment, is of high importance for these systems.

Transient oscillations were studied in References 7 (see also Reference 8), 9 and some recent publications [10–13].

The availability of the model of the system revealing transient oscillations might seem to be sufficient for analysis and not require development of methods other than simulations. However, the frequency-domain approach to assessment of convergence rate would provide a number of advantages over the direct solution of the system differential equations. The most important one is the possibility of explanation of the mechanism of the frequency and amplitude change during the transient. The model of transient oscillations would allow one to estimate the convergence time for the periodic motion to be established in the systems. This model would allow one to compare different methods and correctly select the running time for the test.

In this chapter, a frequency-domain approach to analysis of transient oscillations is presented, which is suitable for analysis of high-order nonlinear systems. The HB principle is extended to the case of transient oscillations and named the *dynamic harmonic balance* (DHB) principle [14].

At first the conventional HB principle is considered. Then a Lure system with a high-order plant is analysed with the use of the DHB involving a quasi-static approach to the frequency of the oscillations. Such characteristics as frequency and amplitude of oscillations as functions of time are derived. After that the condition of the full DHB is derived. Finally, the approach is applied to analysis of the transient motions in a system with the sub-optimal algorithm.

2.4.2 Transient oscillations in Lur'es systems

Let us introduce some ideas relevant to analysis of transient oscillations, and to determine variables that would be necessary to describe a transient process. Consider the system that includes linear dynamics given by

$$\begin{aligned} \dot{\mathbf{x}} &= \mathbf{A}\mathbf{x} + \mathbf{B}u, \\ y &= \mathbf{C}\mathbf{x}, \end{aligned} \qquad (2.4.1)$$

where $\mathbf{x} \in R^n$, $y \in R^1$, $u \in R^1$, $\mathbf{A} \in R^{n\times n}$, $\mathbf{B} \in R^{n\times 1}$ and $\mathbf{C} \in R^{1\times n}$, and a single-valued odd-symmetric nonlinearity $f(y)$:

$$u = -f(y)\,. \tag{2.4.2}$$

We shall refer to (2.4.1) as the linear part of the system. The system (2.4.1), (2.4.2) is a Lure system because it includes linear dynamics and a nonlinearity interconnected in a loop. The transfer function of the linear part is $W_l(s) = \mathbf{C}\,(\mathbf{I}s - \mathbf{A})^{-1}\,\mathbf{B}$, which can be presented as a ratio of two polynomials $W_l(s) = P(s)/Q(s)$. Let us assume that an autonomous mode occurs (no external signals applied), so that the input to the nonlinearity is the output of the linear dynamics, and the output of the nonlinearity is the input to the linear dynamics. Self-excited periodic motions that occur in the system can be found through the use of the HB principle. The conventional HB condition (for periodic motion) is formulated as

$$W_l(j\Omega)N(a) = -1, \tag{2.4.3}$$

where Ω is the frequency and a is the amplitude of the self-excited periodic motion (oscillation) at the input to the nonlinearity, and $N(a)$ is the DF of the nonlinearity. If the linear part's relative degree is higher than *two*, then the Nyquist plot of system (2.4.1) has a point of intersection with the real axis at some finite frequency and, therefore, the HB equation (2.4.3) has a solution. Let us now find the closed-loop transfer function $W_{\mathrm{cl}}(s)$ of system (2.4.1), (2.4.2) using the replacement of the nonlinearity with the DF, $u = -N(a)\cdot y$:

$$W_{\mathrm{cl}}(s) = \frac{W_l(s)N(a)}{1 + W_l(s)N(a)} = \frac{P(s)N(a)}{Q(s) + P(s)N(a)}. \tag{2.4.4}$$

One can see that (2.4.3) is equivalent to

$$R(a, j\Omega) = Q(j\Omega) + P(j\Omega)N(a) = 0, \tag{2.4.5}$$

which means that the denominator of the closed-loop transfer function becomes zero when the frequency and the amplitude become equal to the frequency and the amplitude of the self-excited periodic motion. Equation (2.4.5) is sometimes used for finding a periodic solution via algebraic methods. However, (2.4.5) is not usually attributed to the denominator of the closed-loop transfer function but considered a direct result of (2.4.3). Assuming that $R(a, s)$ can be represented as $R(a, s) = (s - s_1)(s - s_2)\cdots(s - s_n)$, where s_i are roots of the characteristic polynomial, we conclude that there must be at least one pair of complex conjugate roots with zero real parts. It would imply the existence of the conservative component in $W_{\mathrm{cl}}(s)$. Indeed, we can consider the existence of self-excited oscillations as a result of the existence of the component $\left(s^2 + \rho^2\right)$ in the denominator of $W_{\mathrm{cl}}(s)$, where ρ is a parameter that depends on the amplitude a. However, one notices that even if a damped oscillation occurs, so that there exists a pair of complex conjugate roots s_i, s_{i+1}, then $(s - s_i)(s - s_{i+1}) = 0$, and the characteristic polynomial becomes zero, with $s = \sigma \pm j\Omega$, where σ is the decay (strictly speaking, we have a decaying oscillation only if $\sigma < 0$; yet we will refer to this variable as the decay even if $\sigma \geq 0$).

The HB condition (2.4.3) can be viewed as the realisation of the *regeneration principle*, according to which $y(t) = L^{-1}[L[u(t)]\,W_l(s)]$ and $u(t) = f(y(t))$. We can now use the regeneration principle to approach a transient oscillation, in which both $u(t)$ and $y(t)$ are sinusoidal signals with exponentially decaying (or growing) amplitude, and formulate and prove the following property.

Theorem 2.4.1. *If the input signal to the linear dynamics given by the transfer function $W_l(s)$ is the harmonic signal with decaying amplitude of decay σ: $u(t) = e^{\sigma t}\sin(\Omega t)$, the output of the linear dynamics is also a harmonic signal of the same frequency and decay: $y(t) = ae^{\sigma t}\sin(\Omega t + \varphi)$.*

Proof. It follows from the property of the Laplace transform that $L[e^{-at}f(t)] = F(s+a)$. Therefore, for the system input $u(t)$, the Laplace transform will be $L[u(t)] = \Omega/[(s-\sigma)^2 + \Omega^2]$, which will result in the system output (in the Laplace domain) $Y(s) = \Omega\, W_l(s)/[(s-\sigma)^2 + \Omega^2]$. The substitution $s' = s - \sigma$ yields $Y(s') = \Omega\, W_l(s' + \sigma)/[(s')^2 + \Omega^2]$, which means that $y'(t) = L^{-1}[Y(s')]$ is a sinusoid of frequency Ω, amplitude $|W_l(\sigma + j\Omega)|$, and having phase shift $\arg W_l(\sigma + j\Omega)$. In turn, the output signal is $y(t) = e^{\sigma t}y'(t)$, i.e., a decaying sinusoid. □

Therefore, for the analysis of propagation of the decaying sinusoids through linear dynamics, we can use the same transfer functions, in which the Laplace variable should be replaced with $(\sigma + j\Omega)$.

The DF N in the case of a transient oscillation is a function of not only amplitude but of its derivatives too [9] (we disregard possible dependence of the DF on the frequency). Considering the fact that conditions (2.4.3) and (2.4.5) are equivalent and the equality of the denominator of the closed-loop transfer function to zero (for some s) implies the fulfilment of (2.4.3), we rewrite (2.4.3) extending it to the transient oscillation as

$$N(a, \dot{a}, \ldots)W_l(\sigma + j\Omega) = -1. \tag{2.4.6}$$

The use of the derivatives of the amplitude as arguments of the DF is inconvenient. It is more convenient to consider σ and its derivatives than the derivatives of the amplitude because the former is also present as the argument of the frequency response. Therefore, the consideration of σ instead of $\dot{a}$ would allow us to avoid the above-mentioned duplication. Also, we limit our consideration of the DF arguments to the first derivative of the amplitude (or, equivalently, to σ) only, which will correspond to the use of the regeneration principle for decaying sinusoids. We show below that for many nonlinearities the DF is a function of the amplitude only – as it is in the conventional DF analysis. Therefore, we can write the condition of the existence of a transient or steady oscillation as follows:

$$N(a, \sigma)W_l(\sigma + j\Omega) = -1. \tag{2.4.7}$$

Equation (2.4.7) is referred in Reference 15 as the *DHB quasi-static with respect to frequency* condition (equation). It does not account for the derivative of the frequency and for that reason (2.4.7) is referred to as quasi-static.

Assume that the characteristic polynomial of the closed-loop system (which has parametric dependence on the amplitude of the oscillations) has a pair of complex conjugate roots with negative real parts. Then a vanishing oscillation of certain frequency and amplitude occurs. The idea of considering equations of vanishing oscillations is similar to the one of the Krylov–Bogolyubov method [16]. However, the latter can only deal with small 'deviations' from the harmonic oscillator and is limited to second-order systems. In the present approach, the 'equivalent damping' is not limited to small values. Let us consider instantaneous values of the frequency, amplitude and decay and formulate the *DHB quasi-static with respect to frequency* as follows.

At every time, a single-frequency mode transient oscillation can be described as a process of variable (instantaneous) frequency, amplitude and decay that must satisfy (2.4.7).

In (2.4.7) and the formulation given above, we consider only transient oscillations with zero mean and single-frequency mode when the characteristic polynomial (2.4.5) has only one pair of complex conjugate roots.

The DF of an arbitrary nonlinearity for nonharmonic input signal should be computed as suggested in Reference 7 (see also Reference 8).

$$N(a,\sigma,\omega) = q(a,\sigma,\omega) + q'(a,\sigma,\omega)\frac{s-\sigma}{\omega}, \tag{2.4.8}$$

where $s = \frac{d}{dt}$, $q = \frac{1}{2\pi}\int_0^{2\pi} f\,[a\sin\Psi, a\,(\sigma\sin\Psi + \omega\cos\Psi)]\sin\Psi d\Psi$ and $q' = \frac{1}{2\pi}\int_0^{2\pi} f\,[a\sin\Psi, a\,(\sigma\sin\Psi + \omega\cos\Psi)]\cos\Psi d\Psi$, which accounts for the dependence on the frequency, also. It should be noted that for single-valued or hysteretic symmetric nonlinearities, formula (2.4.8) produces conventional DF expressions that are commonly used for harmonic inputs.

The overall motion can now be obtained from the DHB as follows:

$$y(t) = a(t)\,e^{\sigma(t)t}\sin\Psi(t), \tag{2.4.9}$$

where $a(t)$ and $\sigma(t)$ are obtained from the following differential equation:

$$\dot{a}(t) = a(t)\sigma(t), \qquad a(0) = a_0, \tag{2.4.10}$$

and $\Psi(t)$ is the phase computed as $\Psi(t) = \int_0^t \Omega(\tau)d\tau + \phi$, where $\Omega(t)$ is obtained from (2.4.7) and ϕ is selected to satisfy initial conditions.

2.4.3 Motions in the vicinity of a periodic solution

Let us carry out frequency-domain analysis of the transient process of the convergence to the periodic motion in the vicinity of a periodic solution in system (2.4.1) and (2.4.2) using the DHB condition (quasi-static) (2.4.7). The conventional HB condition, which can also be obtained from (2.4.7) when $\sigma = 0$, is written as follows:

$$N(a_0)W_l(j\Omega_0) = -1, \tag{2.4.11}$$

where Ω_0 and a_0 are the frequency and the amplitude of the periodic solution. The DHB (quasi-static) condition for the increments from the periodic solution is given by

$$N(a_0 + \Delta a, \sigma) W_l(\sigma + j(\Omega_0 + \Delta\Omega)) = -1. \tag{2.4.12}$$

We take the derivative from both sides of (2.4.12) with respect to Δa (or a) in the point $a = a_0$, which yields

$$\begin{aligned} &\left(\left. \frac{\partial N(a,\sigma)}{\partial a} \right|_{a=a_0} + \left. \frac{\partial N(a,\sigma)}{\partial \sigma} \right|_{\sigma=0} \left. \frac{\mathrm{d}\sigma}{\mathrm{d}a} \right|_{a=a_0} \right) W_l(j\Omega_0) \\ &\qquad + N(a_0) \left. \frac{\mathrm{d}W(s)}{\mathrm{d}s} \right|_{s=j\Omega_0} \left. \frac{\mathrm{d}s}{\mathrm{d}a} \right|_{a=a_0} = 0. \end{aligned} \tag{2.4.13}$$

Let us limit our analysis only to nonlinearities having DFs that do not depend on σ (e.g., the ideal relay nonlinearity). Express the derivative $\left.\frac{\mathrm{d}s}{\mathrm{d}a}\right|_{a=a_0}$ from (2.4.13):

$$\left. \frac{\mathrm{d}s}{\mathrm{d}a} \right|_{a=a_0} = - \frac{\left. \frac{\mathrm{d}N(a)}{\mathrm{d}a} \right|_{a=a_0} W_l(j\Omega_0)}{N(a_0) \left. \frac{\mathrm{d}W(s)}{\mathrm{d}s} \right|_{s=j\Omega_0}}. \tag{2.4.14}$$

Considering that $s = \sigma + j\Omega$, we can rewrite (2.4.14) as follows:

$$\left. \frac{\mathrm{d}\sigma}{\mathrm{d}a} \right|_{a=a_0} + j \left. \frac{\mathrm{d}\Omega}{\mathrm{d}a} \right|_{a=a_0} = - \frac{\left. \frac{\mathrm{d}N(a)}{\mathrm{d}a} \right|_{a=a_0} W_l(j\Omega_0)}{N(a_0) \left. \frac{\mathrm{d}W(s)}{\mathrm{d}s} \right|_{s=j\Omega_0}}. \tag{2.4.15}$$

Equation (2.4.15) involves complex variables. It can be split into two equations for the real and imaginary parts. However, for finding $\frac{\mathrm{d}\sigma}{\mathrm{d}a}$ one can consider only the real parts of (2.4.15). Once the equation for the real parts is solved and $a(t)$ is found, $\Omega(t)$ can be found, too. Considering that

$$\begin{aligned} \frac{1}{W(s)} \left. \frac{\mathrm{d}W(s)}{\mathrm{d}s} \right|_{s=j\Omega_0} &= \left. \frac{\mathrm{d} \ln W(s)}{\mathrm{d}s} \right|_{s=j\Omega_0} \\ &= \left. \frac{\mathrm{d} \arg W(j\omega)}{\mathrm{d}\omega} \right|_{\omega=j\Omega_0} - j \left. \frac{\mathrm{d} \ln |W(j\omega)|}{\mathrm{d}\omega} \right|_{\omega=j\Omega_0} \end{aligned}$$

and

$$\begin{aligned} -\frac{1}{N(a)} \left. \frac{\mathrm{d}N(a)}{\mathrm{d}a} \right|_{a=a_0} &= \left. \frac{\mathrm{d} \ln \tilde{N}(a)}{\mathrm{d}a} \right|_{a=a_0} \\ &= \left. \frac{\mathrm{d} \ln \left| \tilde{N}(a) \right|}{\mathrm{d}a} \right|_{a=a_0} + j \left. \frac{\mathrm{d} \arg \tilde{N}(a)}{\mathrm{d}a} \right|_{a=a_0}, \end{aligned}$$

where $\tilde{N}(a) = -N^{-1}(a)$, we can write for the real part of (2.4.15):

$$\frac{d\sigma}{da} = \mathrm{Re}\left\{\frac{\dfrac{d\ln\left|\tilde{N}(a)\right|}{da} + j\dfrac{d\arg\tilde{N}(a)}{da}}{\dfrac{d\arg W(j\omega)}{d\omega} - j\dfrac{d\ln|W(j\omega)|}{d\omega}}\right\},$$

which can be rewritten as follows (we skip for brevity the notation of the point at which the derivative is taken):

$$\frac{d\sigma}{da} = \frac{\dfrac{d\ln\left|\tilde{N}(a)\right|}{da}\dfrac{d\arg W(j\omega)}{d\omega} - \dfrac{d\arg\tilde{N}(a)}{da}\dfrac{d\ln|W(j\omega)|}{d\omega}}{\left(\dfrac{d\arg W(j\omega)}{d\omega}\right)^2 + \left(\dfrac{d\ln|W(j\omega)|}{d\omega}\right)^2}. \tag{2.4.16}$$

As a 'side' product of our analysis, stability of a periodic solution can be assessed as satisfaction to the condition $\left.\frac{d\sigma}{da}\right|_{a=a_0} < 0$. This condition means that at amplitudes larger than the amplitude of the periodic motion decay σ must be negative (converging oscillation), and at amplitudes smaller than the amplitude of the periodic motion σ must be positive (diverging oscillation).

2.4.4 DHB accounting for frequency rate of change (full DHB)

In many situations, such as analysis of motions in the vicinity of a periodic solution, the DHB quasi-static with respect to the frequency is quite capable of providing a precise result because the frequency changes insignificantly and the derivative of the frequency can be neglected. In other cases, such as analysis of convergence of systems with second-order sliding modes [12] and finite-time convergence, the system experiences significant changes of the frequency of the oscillations. In this situation, the quasi-static approach to the account of the oscillation frequency may result in the loss of accuracy of analysis. Therefore, inclusion of the frequency derivative would be necessary.

Consider the following illustrative example. Let the linear part be the second-order dynamics and the controller be the ideal relay with amplitude h. The DF of the ideal relay is $N(a) = \frac{4h}{\pi a}$. If we assume that the input to the nonlinearity is a decaying sinusoid of constant frequency $y(t) = a_0 e^{\sigma t}\sin(\omega t)$, then the control amplitude should be $a_u = aN(a)$, the derivative of the control amplitude $\dot{a}_u = \frac{\partial N}{\partial a}\dot{a}a + N\dot{a}$ and the decay

$$\sigma_u = \frac{\dot{a}_u}{a_u} = \frac{-a\dfrac{4h}{\pi a^2}\dot{a} + \dfrac{4h}{\pi a}\dot{a}}{\dfrac{4h}{\pi a}a} = 0.$$

The result shows that $\sigma_u = 0$, which is quite predictable, because the output of the relay controller is an oscillation of constant amplitude, which features zero decay.

However, it is well known (see, e.g., References 1 and 9) that the output of the linear part represents oscillations of decaying amplitude and growing frequency. Now let us analyse the output of the linear dynamics for the input signal having zero decay of the amplitude. Because the input to the linear dynamics has zero decay, we can conclude that the decay in the signal $y(t)$ is a result of not the decay in the control $u(t)$ but of the variable frequency of $u(t)$. Moreover, the nonlinearity may change the decay according to

$$\sigma_u = \frac{\dot{a}_u}{a_u} = \frac{\frac{\partial |N|}{\partial a}\dot{a}a + |N|\dot{a}}{|N|a} = \frac{\partial \ln |N|}{\partial a}\dot{a} + \sigma = \left(\frac{\partial \ln |N|}{\partial \ln a} + 1\right)\sigma, \tag{2.4.17}$$

where the absolute value of N is used (unlike the previous relationship) due to the possibility of N being a complex quantity, and this change is offset by the decay change due to the frequency variation at the signal propagation through the linear part. The described situation is not covered by (2.4.7), which assumes constant frequency. It requires the development of a different model.

Let us consider another example, which is the propagation of a sinusoid of nearly constant amplitude and increasing frequency through the integrator. We assume that the signal of ideally constant amplitude is $u^*(t) = \cos\left(\omega_0 t + \frac{1}{2}\dot{\omega}_0 t^2\right)$, propagates through the integrator producing signal $y^*(t) \approx y(t) = \frac{1}{\omega_0 + \dot{\omega}_0 t}\sin\left(\omega_0 t + \frac{1}{2}\dot{\omega}_0 t^2\right) = \frac{1}{\omega}\sin\Psi$ (the approximate equality is valid if $\dot{\omega}_0$ is small enough). We note that the instantaneous phase of the signal is $\Psi(t) = \omega_0 t + \frac{1}{2}\dot{\omega}_0 t^2$. Therefore, the instantaneous frequency is $\omega(t) = \dot{\Psi}(t) = \omega_0 + \dot{\omega}_0 t$, and the instantaneous derivative of the frequency is $\ddot{\Psi}(t) = \dot{\omega}_0 = \text{const}$. Find the derivative of $y(t)$ as follows:

$$\begin{aligned}\dot{y}(t) &= \cos\left(\omega_0 t + \frac{1}{2}\dot{\omega}_0 t^2\right) + \frac{\dot{\omega}_0}{(\omega_0 + \dot{\omega}_0 t)^2}\sin\left(\omega_0 t + \frac{1}{2}\dot{\omega}_0 t^2\right) \\ &= \cos\Psi + \frac{\dot{\omega}_0}{\omega^2}\sin\Psi.\end{aligned}$$

The integrator introduces the phase lag less than 90° and the equations of the integrator can be represented by the conventional integrator having an additional feedback, which accounts for the effect of the variable frequency. Therefore, the frequency response of the integrator to a decaying sinusoid of variable frequency is $W(\sigma + j\omega, \dot{\omega}_0) = \frac{1}{\sigma + \frac{\dot{\omega}_0}{\omega^2} + j\omega}$. From this formula, one can see that if $\sigma = -\frac{\dot{\omega}_0}{\omega^2}$ then the phase lag introduced by the integrator $\varphi = \arg W(\sigma + j\omega, \dot{\omega}_0)$ is 90°. The amplitude characteristic of the integrator is given by $M = |W(\sigma + j\omega, \dot{\omega}_0)|$, and the decay introduced by the integrator due to variable frequency is $\sigma(t) = \frac{\dot{a}}{a} = -\frac{\dot{\omega}_0}{\omega}$ (given the amplitude of $y(t)$ being $a = \frac{1}{\omega_0 + \dot{\omega}_0 t}$).

Let us use the notion of the *rotating phasor* to analyse propagation of a decaying sinusoid of variable frequency through higher-order linear dynamics. Let the output of the linear dynamics $y(t)$ be given by

$$\bar{y}(t) = a(t)e^{j\Psi(t)} = e^{\ln a(t)}e^{j\Psi(t)} = e^{\ln a(t) + j\Psi(t)}, \tag{2.4.18}$$

where a represents the length of the phasor and Ψ represents the angle between the phasor and the real axis. We can associate either the real or the imaginary part of $\bar{y}(t)$ with the real signal $y(t)$. We assume that the transfer function of the linear part does not have any zeros: $W_l(s) = 1/Q(s) = 1/\left(a_0 + a_1 s + a_2 s^2 + \cdots + a_n s^n\right)$. For the presented transfer function, we can write for $y(t)$:

$$u = a_0 y + a_1 \dot{y} + a_2 \ddot{y} + \cdots + a_n y^{(n)}. \tag{2.4.19}$$

Let us find the derivatives of $y(t)$ as follows:

$$\begin{aligned} \dot{\bar{y}}(t) &= e^{\ln a(t) + j\Psi(t)} \left(\frac{\mathrm{d}\ln a}{\mathrm{d}t} + j\dot{\Psi} \right) = e^{\ln a(t) + j\Psi(t)} \left(\frac{1}{a}\dot{a} + j\dot{\Psi} \right) \\ &= e^{\ln a(t) + j\Psi(t)} (\sigma + j\omega) = (\sigma + j\omega)\, \bar{y}, \end{aligned} \tag{2.4.20}$$

$$\ddot{\bar{y}}(t) = (\dot{\sigma} + j\dot{\omega})\, \bar{y} + (\sigma + j\omega)\, \dot{\bar{y}} = \left[(\dot{\sigma} + j\dot{\omega}) + (\sigma + j\omega)^2 \right] \bar{y}.$$

Considering that we disregard all derivatives of the decay and higher than first derivatives of the frequency in our model, we can rewrite the last formula as

$$\ddot{\bar{y}}(t) \approx \left[(\sigma + j\omega)^2 + j\dot{\omega} \right] \bar{y} \tag{2.4.21}$$

and the formula for the third derivative as

$$\begin{aligned} \dddot{\bar{y}}(t) &\approx \left[2(\sigma + j\omega)(\dot{\sigma} + j\dot{\omega}) + j\ddot{\omega} \right] \bar{y} + \left[(\sigma + j\omega)^2 + j\dot{\omega} \right] (\sigma + j\omega)\, \bar{y} \\ &\approx \left[j3\dot{\omega} + (\sigma + j\omega)^2 \right] (\sigma + j\omega)\, \bar{y} = \left[(\sigma + j\omega)^3 + j3\dot{\omega}\,(\sigma + j\omega) \right] \bar{y}, \end{aligned} \tag{2.4.22}$$

$$\bar{y}^{(4)}(t) \approx \left[(\sigma + j\omega)^4 + j6\dot{\omega}\,(\sigma + j\omega) - 3\dot{\omega}^2 \right] \bar{y}.$$

Higher derivatives can be obtained in a similar manner. It is worth noting that the formulas above are organised to have a term $(\sigma + j\omega)$ to the respective power and the term which is the product of $\dot{\omega}$ and another multiplier. Therefore, we can write for $y(t)$:

$$\bar{u} = \left[a_0 + a_1 (\sigma + j\omega) + a_2 (\sigma + j\omega)^2 + \cdots + a_n (\sigma + j\omega)^n \right] \bar{y} + S(\sigma, \omega, \dot{\omega}) \bar{y},$$

where $S(\sigma, \omega, \dot{\omega})$ includes all terms containing $\dot{\omega}$. This component can be accounted for as an additional feedback – the same way it was done for the integrator.

Let us introduce a certain modified frequency response as

$$W_l^*(\sigma, \omega, \dot{\omega}) = \frac{\bar{y}}{\bar{u}} = \frac{1}{Q(\sigma + j\omega) + S(\sigma, \omega, \dot{\omega})} = \frac{W_l(\sigma + j\omega)}{1 + W_l(\sigma + j\omega) S(\sigma, \omega, \dot{\omega})}$$

and write the DHB equation as

$$N(a) W_l^*(\sigma, \omega, \dot{\omega}) = -1. \tag{2.4.23}$$

Equation (2.4.23) can be split into two equations: for real and imaginary parts, or for equations of the magnitude balance and phase balance. Equation (2.4.23) must be complemented with an equation that relates the difference of the decays at the input and the output of the linear part and the frequency rate of change – the same way it was done in the example of the integrator analysis. We note that the amplitude of the control (the first harmonic) is the $a_u = a|Q(\sigma + j\omega) + S(\sigma, \omega, \dot{\omega})|$, therefore, its time derivative is

$$\dot{a}_u = \dot{a}\,|Q(\sigma + j\omega) + S(\sigma, \omega, \dot{\omega})| + a\frac{\mathrm{d}\,|Q(\sigma + j\omega) + S(\sigma, \omega, \dot{\omega})|}{\mathrm{d}t}.$$

The decay of signal $u(t)$ is computed as

$$\sigma_u = \frac{\dot{a}_u}{a_u} = \frac{\dot{a}\,|Q(\sigma + j\omega) + S(\sigma, \omega, \dot{\omega})| + a\dfrac{\mathrm{d}\,|Q(\sigma + j\omega) + S(\sigma, \omega, \dot{\omega})|}{\mathrm{d}t}}{a\,|Q(\sigma + j\omega) + S(\sigma, \omega, \dot{\omega})|}$$

$$= \sigma + \frac{\mathrm{d}\ln|Q(\sigma + j\omega) + S(\sigma, \omega, \dot{\omega})|}{\mathrm{d}t}.$$

Because we disregard $\dot{\sigma}$ and $\ddot{\omega}$, we can rewrite the last formula as follows:

$$\sigma_u = \sigma + \frac{\partial \ln|Q(\sigma + j\omega) + S(\sigma, \omega, \dot{\omega})|}{\partial\omega}\dot{\omega} = \sigma - \frac{\partial \ln\left|W_l^*(\sigma, \omega, \dot{\omega})\right|}{\partial\omega}\dot{\omega}.$$

The derivative in the last formula defines the slope of the magnitude–frequency characteristic of $W_l^*(\sigma, \omega, \dot{\omega})$. Considering also the formula for the decay of $u(t)$ derived above through the DF, we can now write the condition of the balance of the decays in the closed-loop system as

$$\frac{\partial \ln\left|W_l^*(\sigma, \omega, \dot{\omega})\right|}{\partial \ln\omega}\frac{\dot{\omega}}{\omega} = -\frac{\partial \ln|N(a)|}{\partial \ln a}\sigma. \tag{2.4.24}$$

We can now formulate the DHB principle as the following theorem, the proof of which is given above.

Theorem 2.4.2. *A single-frequency transient oscillation can be described by four variables: instantaneous frequency, instantaneous amplitude, instantaneous decay and instantaneous rate of change of the frequency (time derivative), which at every time must satisfy equations (2.4.23) and (2.4.24).*

It can be noted that $\frac{\partial \ln\left|W_l^*(\sigma,\omega,\dot{\omega})\right|}{\partial \ln\omega} = 0.05\frac{\mathrm{d}M(\omega)}{\mathrm{d}\log\omega}$, where $M(\omega)$ is the magnitude frequency response (dB) (the Bode plot) for the transfer function $W_l^*(\sigma, \omega, \dot{\omega})$. Therefore, this term gives the slope of the Bode plot. Both $\dot{\omega}/\omega$ and $\sigma = \dot{a}/a$ are relative rates of change of the frequency and the amplitude, respectively. Equation (2.4.24), therefore, is establishing the balance between these two rates of change, and will be referred to as the *rates balance*.

The overall solution for the time evolution of the system output is obtained via integration of the first-order differential equation (2.4.10) with σ expressed through a using the three algebraic equations presented above.

2.4.5 Analysis of transient motions of rocking block through DHB

Consider a rather challenging example of application of the DHB principle to analysis of the rocking motions of a block on a flat surface (Figure 2.4.1). The block rocks alternately about pivot points A and B revealing vanishing oscillations of increasing frequency.

Mathematical model of the dynamics of the rocking block was presented in Reference 14. After certain transformations, the model can be given by the block diagram in Figure 2.4.2, with parameter values: $k_1 = 9.76$, $k_2 = 0.981$, $\rho = 1$, and initial value of the state vector $x = 0.05, \dot{x} = 0$. The complexity of the dynamics of the system being considered is related to the fact that oscillations generated in the system have zero or very low frequency at the beginning and approach infinite frequency at the end. For the amplitude, vice versa: it starts from some finite value and approaches

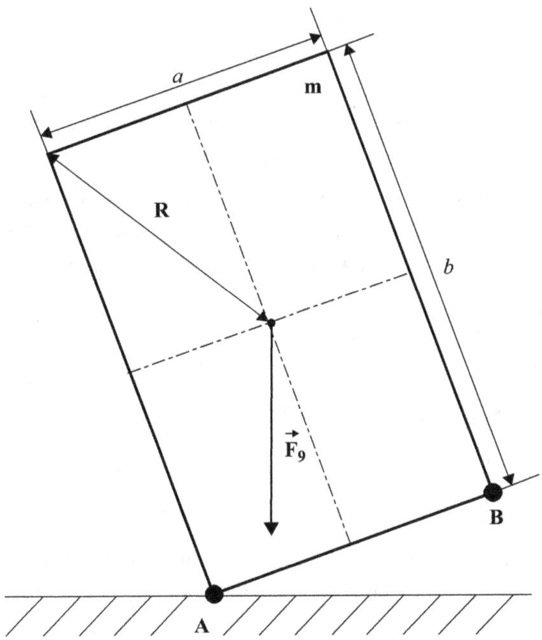

Figure 2.4.1 Rocking block

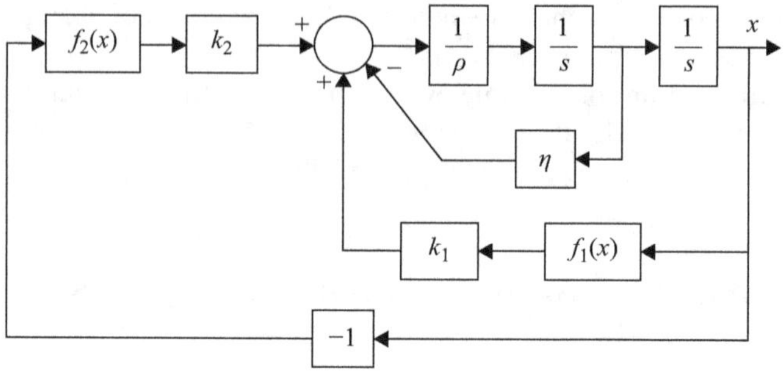

Figure 2.4.2 Dynamic model of system

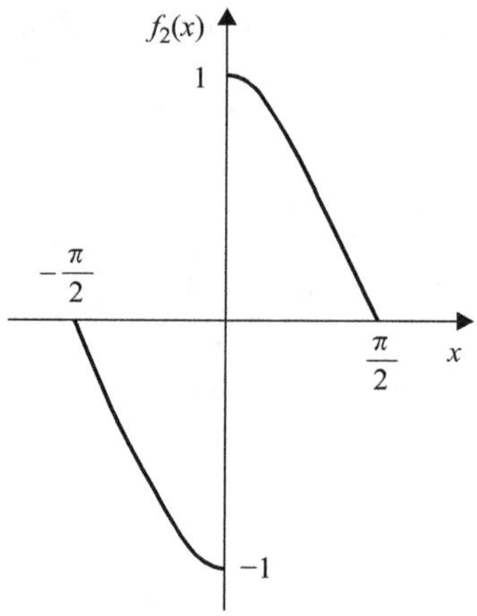

Figure 2.4.3 Second nonlinearity

zero with time approaching infinity. The nonlinearities of the system are $f_1(x) = \sin x$ and $f_2(x) = \mathrm{sign}(x)\cos(x)$. The second nonlinearity is shown in Figure 2.4.3.

The DFs for these two nonlinear functions are obtained as follows [1, 14]:

$$N_1(a) = \frac{2J_1(a)}{a},$$

where $J_1(a)$ is the Bessel function of the first order, and

$$N_2(a) \approx \frac{4}{\pi a} - \frac{8a}{3\pi}.$$

We can now carry out analysis of the motions in the system using the DHB principle. The transfer function of the linear part is given as follows:

$$W_l(s) = \frac{1}{(\rho s + \eta)s}.$$

The modified frequency response can be written for $W_l(s)$ as

$$W_l^*(\sigma, \omega, \dot{\omega}) = \frac{1}{\rho(\sigma + j\omega)^2 + \eta(\sigma + j\omega) + j\dot{\omega}}.$$

The nonlinearity in the Lure system representation can be given as the sum of the two nonlinearities. Therefore, the DF of the combined nonlinearity is

$$N(a) = k_2\left(\frac{4}{\pi a} - \frac{8a}{3\pi}\right) - k_1\frac{2J_1(a)}{a}.$$

Complex equation (2.4.23) can be rewritten as equations for the real and imaginary parts:

$$2\rho\sigma\omega + \eta\omega + \dot{\omega} = 0, \tag{2.4.25}$$

$$\rho(\sigma^2 - \omega^2) + \eta\sigma = -N(a). \tag{2.4.26}$$

We obtain the third algebraic equation per (2.4.24), considering that

$$\frac{\partial \ln\left|W_l^*(\sigma, \omega, \dot{\omega})\right|}{\partial\omega} = \frac{2\omega\rho}{\rho(\sigma^2 - \omega^2) + \eta\sigma},$$

$$\frac{\mathrm{d}J_1(a)}{\mathrm{d}a} = \frac{J_0(a) - J_2(a)}{2},$$

with $J_0(a)$ and $J_2(a)$ being the Bessel functions of zero and second order, respectively, and

$$\frac{\partial \ln N}{\partial \ln a} = \frac{1}{N(a)}\left[-k_2\left(\frac{4}{\pi a} + \frac{8a}{3\pi}\right) + k_1\left(\frac{2J_1(a)}{a} - J_0(a) + J_2(a)\right)\right],$$

as follows:

$$\begin{aligned}\sigma - \frac{2\omega\rho}{\rho(\sigma^2 - \omega^2) + \eta\sigma}\dot{\omega} = \Bigg[\frac{1}{N(a)}\Bigg(&-k_2\left(\frac{4}{\pi a} + \frac{8a}{3\pi}\right)\\ &+ k_1\left(\frac{2J_1(a)}{a} - J_0(a) + J_2(a)\right)\Bigg) + 1\Bigg]\sigma.\end{aligned} \tag{2.4.27}$$

Expressing $\dot{\omega}$ from (2.4.25) and substituting into (2.4.27), we obtain two equations with three unknown variables. We also denote

$$g_1(a) = -N(a) = -k_2\left(\frac{4}{\pi a} - \frac{8a}{3\pi}\right) + k_1\frac{2J_1(a)}{a}$$

and

$$g_2(a) = \frac{1}{N(a)}\left[-k_2\left(\frac{4}{\pi a} + \frac{8a}{3\pi}\right) + k_1\left(\frac{2J_1(a)}{a} - J_0(a) + J_2(a)\right)\right] + 1$$
$$= \frac{1}{N(a)}k_1\left(J_2(a) - J_0(a)\right).$$

With this notation, the set of the two equations can be written as

$$\rho(\sigma^2 - \omega^2) + \eta\sigma = g_1(a),$$
$$\sigma + \frac{2\omega^2\rho(2\rho\sigma + \eta)}{\rho(\sigma^2 - \omega^2) + \eta\sigma} = g_2(a)\sigma. \tag{2.4.28}$$

We solve (2.4.28) as a set of equations for σ and ω, considering amplitude a as a given parameter. With this approach, expressing ω from the first equation of (2.4.28) and substitution in the second equation yields one algebraic equation with one unknown variable σ:

$$\frac{2\rho(2\rho\sigma + \eta)\left(\sigma^2 - \frac{g_1(a)-\eta\sigma}{\rho}\right)}{\sigma g_1(a)} = g_2(a) - 1,$$

which can be transformed into a third-order algebraic equation for σ:

$$c_0 + c_1\sigma + c_2\sigma^2 + c_3\sigma^3 = 0, \tag{2.4.29}$$

where $c_0 = -2g_1(a)\eta$, $c_1 = -4\rho g_1(a) + 2\eta^2 - g_1(a)(g_2(a) - 1)$, $c_2 = 6\rho\eta$ and $c_3 = 4\rho^2$.

Solution of the cubic equation (2.4.29) is well known. In the considered example, the Cardano's formula can be used, which allows one to analytically find the real root of Equation (2.4.29). This becomes possible due to the fact that σ is a real quantity. With a and σ available, the differential equation (2.4.10) can be numerically integrated. After that ω is computed through the formula that was used for the substitution (based on (2.4.26)). Equations (2.4.25)–(2.4.27) are thus solved at every step of the integration of (2.4.10), and the time response of the system is obtained.

The initial amplitude $a(0)$ can be accepted equal to $x(0)$ if $\dot{x}(0) = 0$ and the former is, therefore, considered the point of maximum of $x(t)$. The transient process in the system is shown in Figure 2.4.4 for the presented DHB-based solution and simulations based on the solution of the differential equations.

One can see that the presented approach provides a good estimate of the transient motions even for this challenging dynamic system. In fact, there is only some phase mismatch accumulated at the beginning of the transient. This phase shift does not change afterwards.

The DHB-based approach distinguishes from the solution of the original differential equations by the possibility of explanation of the mechanism of the evolution of

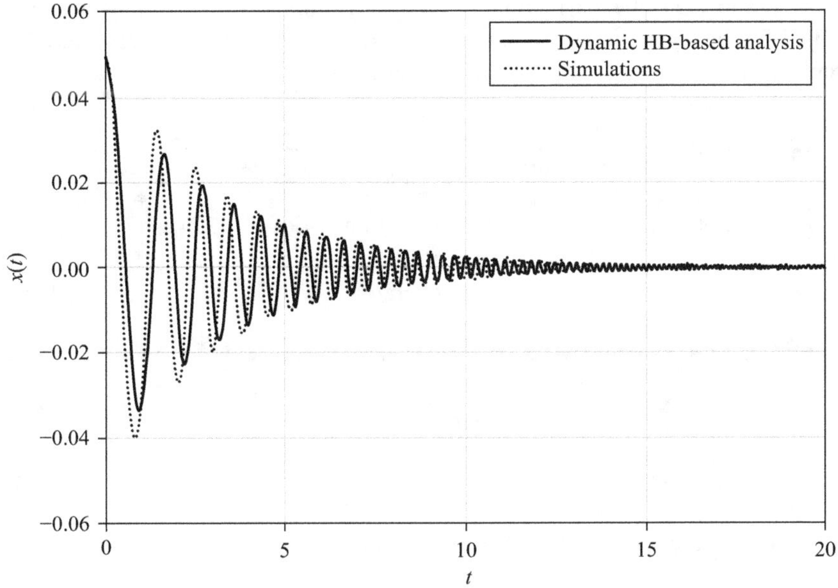

Figure 2.4.4 Transient motions in the system

the oscillations parameters and the possibility of the analysis of asymptotic behaviour of the system, which is hard to do by numeric integration due to extremely large and extremely low numbers involved. It can be found in particular that, in accordance with the criterion proposed in Reference 12, which involves the notion of the phase deficit, oscillation amplitude has asymptotic convergence to zero in the analysed system. One can also see that if the amplitude of the oscillations decreases due to the dissipation of energy then the frequency, in accordance with the equation of the balance of the rates of change (2.4.27), must increase. Overall the presented model conveniently and precisely explains transient processes in the rocking block dynamics.

2.4.6 Analysis of asymptotic second-order SM system using DHB principle

Let us consider motions in the second-order system given by the following linear part:

$$a_2\ddot{x} + a_1\dot{x} + a_0x = u, \tag{2.4.30}$$

or equivalently by the transfer function $W(s) = 1/\left(a_2s^2 + a_1s + a_0\right)$. And the nonlinear control is the ideal relay:

$$u = -h \cdot \text{sign}(x). \tag{2.4.31}$$

Let us apply the DHB principle for analysis of motions in this relay system. The modified frequency response for the linear part can be written, as presented above, by replacing s with $\sigma + j\omega$ and adding the term containing $\dot{\omega}$:

$$W^*(\sigma, \omega, \dot{\omega}) = \frac{1}{a_2(\sigma + j\omega)^2 + a_1(\sigma + j\omega) + a_0 + j\dot{\omega}}. \tag{2.4.32}$$

Let us denote the polynomial of the denominator of (2.4.32) as Q:

$$Q = a_2(\sigma + j\omega)^2 + a_1(\sigma + j\omega) + a_0 + j\dot{\omega}. \tag{2.4.33}$$

Splitting (2.4.33) into the real and imaginary parts yields:

$$\operatorname{Re} Q = a_2\left(\sigma^2 - \omega^2\right) + a_1\sigma + a_0, \tag{2.4.34}$$

$$\operatorname{Im} Q = 2a_2\sigma\omega + a_1\omega + \dot{\omega}. \tag{2.4.35}$$

As a result, the magnitude of the modified frequency response can be written as

$$|W^*(\sigma, \omega, \dot{\omega})| = \frac{1}{\sqrt{\operatorname{Re}^2 Q + \operatorname{Im}^2 Q}}, \tag{2.4.36}$$

where the real and the imaginary parts are given by (2.4.34) and (2.4.35), respectively. The derivative of the magnitude with respect to the frequency is given by

$$\frac{\partial\, |W^*(\sigma, \omega, \dot{\omega})|}{\partial \omega} = \left(\operatorname{Re}^2 Q + \operatorname{Im}^2 Q\right)^{-\frac{3}{2}} \left[2a_2\omega \operatorname{Re} Q - (2a_2\sigma + a_1)\operatorname{Im} Q\right]. \tag{2.4.37}$$

The DF of the ideal relay is given as follows:

$$N(a) = \frac{4h}{\pi a}, \tag{2.4.38}$$

which also results in

$$\frac{\partial \ln N(a)}{\partial \ln a} = -1. \tag{2.4.39}$$

Let us now write the three equations of the DHB. The **phase balance** condition of $\arg N(a) + \arg W^*(\sigma, \omega, \dot{\omega}) = -\pi$, due to the fact that $\arg N(a) = 0$, can be written for the considered system simply as

$$2a_2\sigma\omega + a_1\omega + \dot{\omega} = 0, \tag{2.4.40}$$

which in fact is the equality of $\operatorname{Im} Q$ to zero.

The **magnitude balance** condition of $\frac{4h}{\pi a}\,|W^*(\sigma,\omega,\dot{\omega})| = 1$ is transformed into the following equation due to the fact that $\operatorname{Im} Q = 0$:

$$-\frac{4h}{\pi a} = a_2\left(\sigma^2 - \omega^2\right) + a_1\sigma + a_0. \tag{2.4.41}$$

And the **rates balance** condition is obtained from (2.4.24) as follows:

$$-\frac{2a_2\omega^2(2a_2\sigma + a_1)}{a_2\left(\sigma^2 - \omega^2\right) + a_1\sigma + a_0} = \sigma. \tag{2.4.42}$$

In deriving (2.4.42), it was assumed that $\operatorname{Re} Q < 0$, which led to the following reduction of (2.4.37):

$$\frac{\partial \ln|W^*(\sigma,\omega,\dot{\omega})|}{\partial \ln \omega} = \frac{2a_2\omega^2}{a_2\left(\sigma^2 - \omega^2\right) + a_1\sigma + a_0}.$$

Now let us analyse the convergence of the oscillations in the considered second-order system, using the DHB equations (2.4.40)–(2.4.42). It is known from the previous research [9, 17] that the frequency of the oscillations tends to infinity and the amplitude tends to zero when time tends to infinity. However, the convergence character can be analysed with the presented model of the DHB.

Let us proceed from the condition of $\omega \to \infty$. Divide the phase balance condition (2.4.40) by ω and express from it $\dot{\omega}$ to ω ratio:

$$\frac{\dot{\omega}}{\omega} = -(2a_2\sigma + a_1). \tag{2.4.43}$$

This is a differential equation for the frequency ω, which has an exponential solution. Assuming that the decay σ is negative and finite, we conclude that $\dot{\omega} \to \infty$ (if $\dot{\omega}$ were finite then the right-hand side of (2.4.43) would be zero). The magnitude balance condition allows us to find a:

$$a_\infty = \lim_{\omega\to\infty} a = \lim_{\omega\to\infty} \frac{-4h}{\pi\left[a_2\left(\sigma^2 - \omega^2\right) + a_1\sigma + a_0\right]} = 0,$$

subject to $\lim_{\omega\to\infty}(\sigma/\omega) = 0$, which agrees with the above-noted assumptions.

Now from the rates balance condition (2.4.42), the decay can be found as follows. Find the limit of the left-hand side of (2.4.42):

$$\sigma_\infty = \lim_{\omega\to\infty} \frac{-2a_2\omega^2(2a_2\sigma_\infty + a_1)}{a_2\left(\sigma_\infty^2 - \omega^2\right) + a_1\sigma_\infty + a_0} = 2(2a_2\sigma_\infty + a_1). \tag{2.4.44}$$

Considering (2.4.44) as an equation, we can find σ as follows:

$$\sigma_\infty = \frac{2a_1}{1 - 4a_2} < 0. \tag{2.4.45}$$

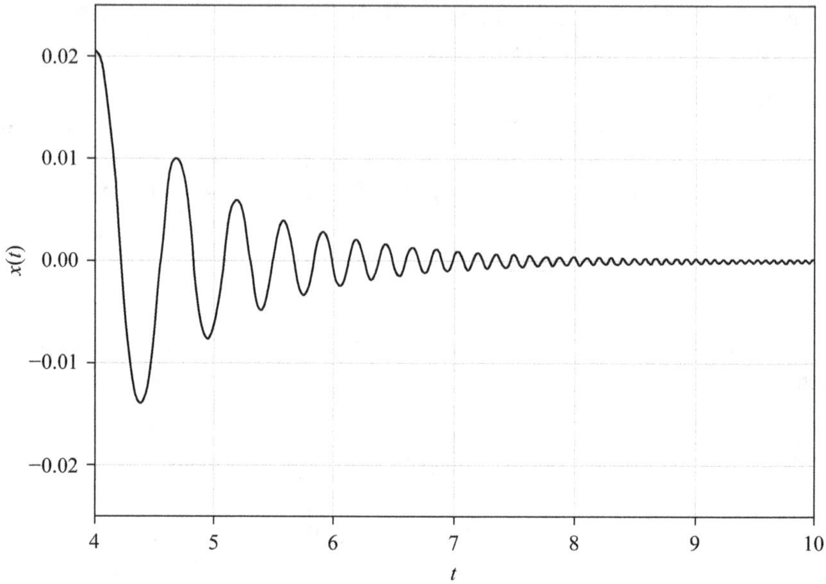

Figure 2.4.5 Second-order asymptotic SM transient motions

We can now rewrite (2.4.43) using the formula for σ_∞:

$$\frac{\dot{\omega}}{\omega} = -\sigma_\infty/2.$$

Therefore, the system reveals converging oscillations with frequency approaching infinity as $\omega(t') = \omega_{\text{in}} e^{-0.5\sigma_\infty t'}$ (where t' is time variable shifted with respect to time t by a sufficiently large value, and 'in' indicates a value at $t' = 0$), amplitude approaching zero as $a(t') = a_{\text{in}} e^{\sigma_\infty t'}$, frequency rate of change approaching infinity and the decay approaching the value of $2a_1/(1-4a_2) < 0$, as time tends to infinity. The process of convergence is asymptotic because the decay σ_∞ is finite.

Simulations for $a_0 = a_2 = h = 1$, $a_1 = 1.6$ (Figure 2.4.5) show that the analytical solution is precise. The decay of the amplitude evaluated as $\sigma = \dot{a}/a \approx \frac{a_{i+1}-a_i}{T_i}\frac{2}{a_{i+1}+a_i}$, where a_i is the instantaneous amplitude at time corresponding to ith maximum of $x(t)$, and T_i is the interval between the two maxima, provides the values of $\sigma \in [-0.97, -1.05]$ for various points of maximum on the time interval $t \in [5.5\ \text{s}, 9.5\ \text{s}]$ given in Figure 2.4.5, whereas the theoretical value computed per (2.4.45) is $\sigma = -1.07$.

As in the rocking block dynamics, the *phase deficit* (the angle between the high-frequency asymptote of the Nyquist plot and the plot of negative reciprocal of the DF) is zero. Therefore, according to the criterion proposed in Reference 12, the convergence must be asymptotic. This fully agrees with the presented analysis.

2.4.7 Conclusions

In this chapter, transient oscillations are considered. The analysis of the transient oscillations is carried out through application of the DHB principle, which is also presented in this chapter. The DHB is treated as an extension of the conventional HB applicable to a periodic motion, to the analysis of an oscillatory motion of variable frequency and amplitude. The equations of the magnitude balance and phase balance are modified and completed with an equation of balance of rates of change of amplitude and frequency. Application of the DHB principle is illustrated by examples of analysis of transient motions in a rocking block and in a second-order nonlinear dynamic system.

The presented theory, supported by the examples, provides an approach to analysis of such complex motions as transient oscillations in nonlinear systems. It is shown that, despite the approximate character of analysis, the problem can be solved very efficiently. The conditions are formulated as a principle that involves satisfaction to three equations. It is believed that the proposed approach provides many new opportunities in analysis of transient motions in nonlinear systems.

References

[1] Atherton, D.P. (1975). *Nonlinear Control Engineering – Describing Function Analysis and Design*. Wokingham, Berkshire: van Nostrand Company Limited.

[2] Utkin, V. (1992). *Sliding Modes in Control and Optimization*. Berlin: Springer-Verlag.

[3] Levant, A. (Levantovsky, L.V.) (1993). Sliding order and sliding accuracy in sliding mode control. *International Journal of Control*, 58 (6), 1247–1263.

[4] Bartolini, G., Ferrara, A., and Usai, E. (1998). Chattering avoidance by second-order sliding mode control. *IEEE Transactions on Automatic Control*, 43 (2), 241–246.

[5] Ogata, K. (2004). *System Dynamics* (4th ed.). Englewood Cliffs, NJ: Prentice-Hall.

[6] Meirovitch, L. (2001). *Fundamentals of Vibrations*. Boston, MA: McGraw-Hill.

[7] Popov, E.P. and Paltov, I.P. (1960). *Approximate Methods of Analysis of Nonlinear Control Systems*. Moscow: Fizmatgiz (in Russian).

[8] Vukic, Z., Kuljaca, L., Donlagic, D., and Tesnjak, S. (2003). *Nonlinear Control Systems*. New York, NY: Marcel Dekker.

[9] Gelb, A. and Vander Velde, W.E. (1968). *Multiple-Input Describing Functions and Nonlinear System Design*. New York, NY: McGraw-Hill.

[10] Boiko, I. (2008). Extension of harmonic balance principle and its application to analysis of convergence rate of second-order sliding-mode control algorithms. *2008 American Control Conference*, Seattle, WA, USA, pp. 4691–4696.

[11] O'Toole, M., Bouazza-Marouf, K., and Kerr, D. (2010). Chatter suppression in sliding mode control: strategies and tuning methods. In *ROMANSY 18 – Robot Design, Dynamics and Control*. Chapter 1, Springer, Vienna, pp. 109–116.

[12] Boiko, I. (2011). On frequency-domain criterion of finite-time convergence of second-order sliding mode control algorithms. *Automatica*, 47 (9), 1969–1973.

[13] Rothig, A. (2013). Extension of the Krylov–Bogolyubov method and its application to the decay rate analysis of nonlinear control algorithms. *2013 European Control Conference*, Zurich, Switzerland, pp. 1693–1698.

[14] Boiko, I. (2012). Dynamic harmonic balance principle and analysis of rocking block motions. *Journal of the Franklin Institute*, 349 (3), 1198–1212.

[15] Boiko, I. (2011). Dynamic harmonic balance and its application to analysis of convergence of second-order sliding mode control algorithms. *Proceedings of the 2011 American Control Conference*, San Francisco, CA, USA, pp. 208–213.

[16] Krylov, N.M. and Bogolyubov, N.N. (1937). *Introduction to Nonlinear Dynamics*. Kiev: Academy of Science of the Ukrainian SSR.

[17] Boiko, I. (2009). *Discontinuous Control Systems: Frequency-Domain Analysis and Design*. Boston, MA: Birkhäuser.

Section 3

Discretization of sliding-mode controllers

Chapter 3.1

On discretization of high-order sliding modes

Arie Levant, Miki Livne, and Davin Lunz

Abstract

Output-feedback high-order sliding-mode (HOSM) controls include HOSM-based differentiators and, therefore, possess complicated discontinuous dynamics. Their practical application naturally involves discrete noisy output sampling and numeric integration of the internal variables. Resulting hybrid systems are shown to be stable, and the corresponding asymptotic sliding-mode accuracies are calculated in the presence of Euler integration and discrete sampling, whereas both might feature variable or constant time steps. Discrete differentiators are developed which restore the optimal accuracy of their continuous-time counterparts. Numeric criteria detect the end of the differentiator transient. Simulation confirms the presented results.

3.1.1 Introduction

According to the sliding-mode control (SMC) approach, the dynamics uncertainty is to be removed keeping an appropriate constraint $\sigma = 0$. Due to the system uncertainty, the control exactly keeping the constraint is unknown, and the control is switched providing for returning to the constraint each time the equality $\sigma = 0$ turns out to be violated. It results in high-frequency switching of the control, and the corresponding motion mode $\sigma \equiv 0$ features theoretically infinite control switching frequency and is called sliding mode (SM), whereas σ is called the sliding variable [16, 49, 52]. Further for simplicity σ is assumed to be a scalar function.

The advantages and disadvantages of the approach are obvious. On the one hand, the control is simple and effective. It is especially simple, if $\dot{\sigma}$ contains control (the relative-degree-1 case), and the control takes the form of a simple relay $u = -\alpha \operatorname{sign} \sigma$. The constraint $\sigma = 0$ is established in finite time (FT), and the uncertainty is effectively diminished [16, 49, 52]. On the other hand, the control is discontinuous on the constraint manifold, and the high-frequency switching can cause dangerous system vibrations (the chattering effect [5, 20, 52]).

High-order sliding modes (HOSMs) [8, 27, 29, 41, 45, 50] were historically proposed to overcome the above chattering-effect problem and to solve the problem of establishing the constraint in FT also in the case when the relative degree differs from 1.

Suppose that the equality $\sigma = 0$ is kept on the solutions of a closed-loop system. The sliding order r is the lowest integer r, such that the rth-order total time derivative $\sigma^{(k)}$ is not a continuous function of the state variables and time [27, 29]. The corresponding motion $\sigma \equiv 0$ is called rth-order SM, and for brevity is called r-sliding mode (r-SM).

The chattering attenuation is obtained by artificially increasing the number k of the derivative $\sigma^{(k)}$ which contains the discontinuity. Consider $u^{(l)}$ as the virtual control, and suppose that $\sigma^{(r)}$ is the first total time derivative of σ to contain the control u, then $\sigma^{(r+l)}$ is to be the first to contain the virtual control $u^{(l)}$. Thus, one has to establish the $(r+l)$-SM $\sigma = 0$ in FT by means of discontinuous $u^{(l)}$ [7, 11, 27]. It has been shown [33] that in this case the high-energy chattering is removed.

Note that the chattering reduction is not due to the continuity of the corresponding control $u(t)$, but due to simultaneously keeping continuous functions of system variables $\sigma, \dot{\sigma}, \ldots, \sigma^{(k+l-1)}$ at zero [33]. Homogeneous HOSMs also feature high accuracy in the presence of small switching imperfections and noises [27, 30].

Standard SMs [16, 52] are of the first order, i.e., already $\dot{\sigma}$ contains discontinuous control u, $r = 1$. Thus, the chattering attenuation is obtained using 2-SMs and discontinuous $\dot{u}$ [7, 8, 27].

Relative degree is defined as the lowest total derivative order of the output σ which explicitly contains control [24]. Families of universal controls are recursively constructed and solve the problem for any relative degree r [30, 49] of the output σ by means of discontinuous r-SM control. In particular, the FT stabilization of σ is possible by means of control, continuous everywhere except the manifold $\sigma = \dot{\sigma} = \ldots = \sigma^{(r-1)} = 0$ [31]. The controllers are complemented by the robust exact SM-based differentiators in FT providing for the unavailable derivatives $\sigma, \ldots, \sigma^{(r-1)}$ [29].

SM control is proved to be insensitive to disturbances in the control channel (matched disturbances), robust with respect to sampling noises and small delays. Homogeneous SMs [30] are proved to be robust to small disturbances, including those which change the relative degree [39], and to the presence of fast stable sensors and actuators [21, 33].

It is natural to introduce appropriate nonlinear constraints, describing FT stable differential equations $\Sigma\left(t, \sigma, \ldots, \sigma^{(r-1)}\right) = 0$. Then keeping $\Sigma \equiv 0$ in 1-SM would lead to the establishment of the r-SM $\sigma \equiv 0$ in FT. Unfortunately, this idea does only easily work with $r = 2$, when $\sigma = \dot{\sigma} + |\sigma|^{1/2} \operatorname{sign} \sigma$ can be taken. For higher r one typically encounters the problem of unbounded gradients of the respective functions Σ, which results in singular unbounded 1-SM controls [41]. The problem is usually solved in the framework of the homogeneity theory [9, 23, 30, 46].

The results described above constitute a solid foundation for extensive applications of SMC for solution of various control and observation problems under uncertainty conditions [2, 6, 8, 11, 14, 15, 17, 19, 22, 41, 44, 45, 50]. Lyapunov functions are found and used for HOSM controllers [13, 42, 47, 48].

The asymptotic accuracy of the r-SM was studied in Reference 27. It has been found there that the best possible accuracy of a SM in the presence of discrete switching is directly defined by its sliding order r. In particular, the best possible asymptotic accuracy with the sampling time interval $\tau > 0$ is $\sigma^{(j)} = O\left(\tau^{r-j}\right), j = 0, 1, \ldots, r-1$.

And indeed the homogeneity technique [30] provides for that accuracy. Moreover, the accuracy is preserved, if the derivatives are estimated by homogeneous differentiators [29]. More exactly, the asymptotic accuracy $\sigma^{(j)} = O\left(\max\left(\tau^{r-j}, \varepsilon^{(r-j)/r}\right)\right)$ is obtained for the sampling accuracy $\varepsilon > 0$ of the σ-measurements. Unfortunately, this result is restricted to the ideal case, when the system is described by the Filippov differential equations with zero-hold measurements.

The output-feedback HOSM control contains differentiators (i.e., observers) in the feedback. Also, the chattering attenuation procedure described above inserts integrators in the feedback. Thus, the produced feedback often (actually almost always) has its own dynamics. In reality, the control input is produced by a modern computer unit, integrators are replaced by some discrete integral approximation, most usually by one-step Euler approximation. The resulting closed-loop system features complex interaction of the discrete controller with the continuous-time dynamic system, i.e., becomes a heterogeneous, hybrid system. The area was practically *terra incognita* 2–3 years ago. Not only the accuracy, even the stability of the hybrid system was not proved. The authors have got considerable advances in this field. This chapter presents an intermediate summary of the current state of the research.

3.1.2 Preliminaries: sliding order and SM accuracy

Realization of SM control inevitably includes discrete switching and noisy measurements. Here, we estimate the worst and the best possible realization accuracy of HOSMs in the presence of noises and discrete switching, and show that possible practical SM accuracies are strictly determined by the numbers of the output derivatives in which the discontinuity appears for the first time.

3.1.2.1 Accuracy of SMs in the absence of noises

The following lemma actually describes SM accuracy in the presence of discrete control switching. It extends a similar result of Reference 27 and is similarly proved.

Lemma 3.1.1. *Let $\omega(t)$ be a scalar function having continuous derivative $\omega^{(l)}$ on the segment $[0, \tau]$, $\tau > 0$. Then for each natural number l, there exist such $c_0, c_1, \ldots, c_{l-1} > 0$ and $d_1, d_2, \ldots, d_{l-1} > 0$ that for any $\delta > 0$.*

1. *If $\left|\omega^{(l)}\right| \geq \delta$ holds on the segment, then*

$$\max|\omega| \geq c_0\delta\tau^l, \max|\dot{\omega}| \geq c_1\delta\tau^{l-1}, \ldots, \max\left|\omega^{(l-1)}\right| \geq c_{l-1}\delta\tau. \tag{3.1.1}$$

2. *If $\left|\omega^{(l)}\right| \leq \delta$ and $|\omega| \leq d_0\delta\tau^l$ hold over the segment $[0, \tau]$ for some d_0, then*

$$\max|\dot{\omega}| \leq (d_0d_1 + 1)\delta\tau^{l-1}, \ldots, \max\left|\omega^{(l-1)}\right| \leq (d_0d_{l-1} + 1)\delta\tau. \tag{3.1.2}$$

Obviously, the second statement of the lemma provides for a rather crude estimation, since with $d_0 = 0$ all derivatives of ω should vanish. Lemma 3.1.1 shows that the accuracy of keeping an output $\sigma = 0$ is directly connected with the number of its continuous total time derivatives. This naturally implies the following definition [27], directly formulated for the vector output σ.

Definition 3.1.1. *Suppose the constraint $\sigma(x) = 0$, $\sigma : \mathbb{R}^k \to \mathbb{R}^m$, is identically kept on some solutions of a dynamic system $\dot{x} = v(x)$, $x \in \mathbb{R}^k$, understood in the Filippov sense, $v(x)$ is any Lebesgue-measurable locally bounded vector function. Then the solutions keeping $\sigma(x) = 0$ are said to be in the $(r_1, r_2, \ldots, r_m)$th-order sliding mode, if*

1. *the total time derivatives $\sigma_i^{(j)}(x)$ are continuous functions of x, $j \le r_i - 1$;*
2. *the r-sliding set $L_r = \left\{x \mid \sigma_i^{(j)}(x) = 0, j \le r_i - 1,\ i = 1, \ldots, m\right\}$ is not empty and locally consists of Filippov solutions;*
3. *$\sigma_1^{(r_1)}, \sigma_2^{(r_2)}, \ldots, \sigma_m^{(r_m)}$ are discontinuous functions of x or do not exist.*

In the non-autonomous case, the time t is considered as an additional coordinate, and the equation $\dot{t} = 1$ is formally added.

Recall that a scalar output $\sigma(t,x)$ of a smooth single-input single-output (SISO) system

$$\dot{x} = a(t,x) + b(t,x)u \tag{3.1.3}$$

has the relative degree r, if the rth total time derivative of σ is the first to explicitly contain the control, and the corresponding control coefficient does not vanish. In the multi-input multi-output (MIMO) case, the dimensions of u and σ are to be equal, for each component σ_i of the output, a partial relative degree r_i is to exist with respect to some control component, and the matrix $\left(\frac{\partial \sigma_i^{(r_i)}}{\partial u_j}\right)$ is to be non-singular [24].

It follows from the sliding-order definition that SM order is component-wise larger or equal than vector relative degree, if the latter exists. It can be higher, if, for example, the control itself features some discontinuous dynamics. An r-SM is called unstable, asymptotically stable, FT stable, etc., if the r-sliding manifold L_r features the same property.

Consider a MIMO dynamic system (3.1.3). Let the output $\sigma(t,x)$ and the input u be vectors, $\sigma : \mathbb{R}^{n+1} \to \mathbb{R}^m$, $u \in \mathbb{R}^l$, a, b be smooth. The system is assumed to have the partial relative degrees $r = (r_1, \ldots, r_m)$, $r_i > 0$, which means that the successive total time derivatives $\sigma_i^{(j)}, j = 0, 1, \ldots, r_i - 1, i = 1, \ldots, m$, do not contain controls, but controls appear in $\sigma_i^{(r_i)}$. Respectively, get a vector equation

$$\sigma^{(r)} = h(t,x) + g(t,x)u, \tag{3.1.4}$$

where $\sigma^{(r)}$ denotes $\left(\sigma_1^{(r_1)}, \ldots, \sigma_m^{(r_m)}\right)^T$. As a direct consequence of Lemma 3.1.1 obtain the following theorem.

Theorem 3.1.1. *Let system (3.1.3) be smooth with partial relative degrees* $r = (r_1, \ldots, r_m)$*. Then for some constants* $c_{r_i,0}, \ldots, c_{r_i,r_i-1}$*,* $d_{r_i,1}, \ldots, d_{r_i,r_i-1} > 0$*,* $i = 1, \ldots, m$*, the following is true for each component* σ_i *over any time interval of the length* τ *with continuous control* $u(t) \in \mathbb{R}^l$*.*

1. *If* $\left|\sigma_i^{(r_i)}\right| \geq \delta_i$ *holds on the segment for some* $\delta_i > 0$*, then*

$$\max|\sigma_i| \geq c_{r_i,0}\delta_i\tau^{r_i}, \max|\dot{\sigma}_i| \geq c_{r_i,1}\delta_i\tau^{r_i-1}, \ldots, \max\left|\sigma^{(r_i-1)}\right| \geq c_{r_i,r_i-1}\delta_i\tau. \tag{3.1.5}$$

2. *If* $\left|\sigma_i^{(r_i)}\right| \leq \delta_i$ *and* $|\sigma_i| \leq d_{r_i,0}\delta_i\tau^r$ *hold over the segment for some* $d_{r_i,0}$ *and* $\delta_i > 0$*, then*

$$\max|\dot{\sigma}| \leq \left(d_{r_i,0}d_{r_i,1} + 1\right)\delta_i\tau^{r_i-1}, \ldots, \max\left|\sigma^{(r_i-1)}\right| \leq \left(d_{r_i,0}d_{r_i,r_i-1} + 1\right)\delta_i\tau. \tag{3.1.6}$$

In particular, in the case of the SISO SMC problem, it follows from the theorem that no one can expect an accuracy better than $\sigma = O(\tau^r), \dot{\sigma} = O(\tau^{r-1}), \ldots, \sigma^{(r-1)} = O(\tau)$ in the SM $\sigma \approx 0$, if $\sigma^{(r)}$ is separated from zero between the switchings. On the other hand, if $\sigma^{(r)}$ exists and is bounded, then keeping $\sigma \approx 0$ implies that also $\sigma^{(j)} \approx 0, j = 1, \ldots, r-1$.

As follows from the second statement of the theorem, the r-SM accuracy can be higher than $\sigma^{(j)} = O\left(\tau^{r-j}\right)$, if $\sigma^{(r)}$ is kept close to zero. For example, the implicit Euler method [1] actually increases the order of the real (i.e., approximate) SM due to the online estimation of the equivalent control, which allows to decrease the discontinuous component of the control. Unfortunately, such estimation requires some additional system knowledge.

3.1.2.2 Accuracy of SMs in the presence of noises

Once more consider the uncertain SMC problem (3.1.3), (3.1.4). Recall that it is possible to provide for the exact FT establishment of the r-SM $\sigma \equiv 0$ using only output measurements [29, 30].

Theorem 3.1.2. *Suppose that the control, based on the input measurements only, provides for the exact finite-time establishment of the r-SM* $\sigma \equiv 0$ *for any function h, satisfying* $\|h\| \leq C$*. Let* σ_i *be measured with a Lebesgue-measurable noise* $\eta_i(t)$ *of the maximal magnitude* $\varepsilon_i \geq 0$*,* $\eta_i(t) \leq \varepsilon_i$*, with unknown features,* $i = 1, \ldots, m$*. Then the worst-case SM accuracy cannot be better than*

$$\begin{gathered}|\sigma_i| \leq \varepsilon_i, |\dot{\sigma}| \leq \tilde{c}_{i,1}\varepsilon_i^{\frac{r_i-1}{r_i}}, \ldots, \left|\sigma_i^{(r_i-1)}\right| \leq \tilde{c}_{i,r_i-1}\varepsilon_i^{\frac{1}{r_i}}, \\ \tilde{c}_{i,j} = \left(\frac{C}{m^{\frac{1}{2}}}\right)^{\frac{j}{r_i}}, \quad j = 1, \ldots, r_i, \; i = 1, \ldots, m.\end{gathered} \tag{3.1.7}$$

Proof. Let the output satisfy the equation $\sigma^{(r)} = g(t,x)u$, i.e., (3.1.4) with $\|h\| \leq C$, $h \equiv 0$. Let now the measured signal $\hat{\sigma}_i$ be of the form $\hat{\sigma}_i(t,x) = \sigma_i(t,x) + \varepsilon_i \cos\left(\left(m^{-1/2}C/\varepsilon\right)^{1/r_i} t\right)$, i.e., the noise be equal $\varepsilon \cos\left(\left(m^{-1/2}C/\varepsilon\right)^{1/r_i} t\right)$. Then the noisy signal $\hat{\sigma}_i$ satisfies

$$\sigma_i^{(ri)} = \left(\cos\left(\left(\frac{C}{\varepsilon m^{\frac{1}{2}}}\right)^{\frac{1}{r}} t\right)\right)^{(r)} + g(t,x)u, \quad \left|\left(\cos\left(\left(\frac{C}{\varepsilon m^{\frac{1}{2}}}\right)^{\frac{1}{r}} t\right)\right)^{(r)}\right| \leq \frac{C}{m^{\frac{1}{2}}}.$$

Respectively, according to the assumptions, the control will successfully establish and keep $\hat{\sigma}_i \equiv 0$, which corresponds to $\left|\sigma_i^{(j)}\right| \leq \left(\frac{C}{\sqrt{m}}\right)^{\frac{j}{r}} \varepsilon^{\frac{r-j}{r}}, j = 0, \ldots, r-1$. □

Note that Theorems 3.1.1 and 3.1.2 are true for any τ and ε_i, neither τ nor ε_i needs to be small.

Under the conditions of Theorem 3.1.2 let the output σ be measured with noises of the magnitudes $\varepsilon_i > 0$ at some discrete time instants, and let the control be updated at each sampling instant and remain constant between the sampling moments. Then the inequalities (3.1.5) hold independently of the noise presence over each sampling time interval of the length τ on which the inequality $\left|\sigma_i^{(ri)}\right| > \delta_i > 0$ is held.

The situation is more complicated, if the maximal sampling step tends to zero. Some additional assumptions are needed to ensure that the corresponding solutions uniformly converge to solutions with continuous sampling. If such a convergence takes place, then, according to (3.1.7), the worst-case SM accuracy is not better than $\left|\sigma_i^{(j)}\right| = O\left(\varepsilon_i^{\frac{r-j}{r}}\right)$.

Example. The output σ of the SISO system (3.1.3), (3.1.19) of the relative degree r is traditionally nullified by keeping the constraint $\Sigma = \left(\frac{d}{dt} + \lambda\right)^{r-1} \sigma = 0$ in 1-SM $\Sigma \equiv 0$ [16, 52]. Let τ be the sampling step, then $\Sigma = O(\tau)$ is the only possible accuracy according to Theorem 3.1.1. The respective overall r-SM accuracy is $\sigma = O(\tau), \dot{\sigma} = O(\tau), \ldots, \sigma^{(r-1)} = O(\tau)$ [51]. It definitely satisfies (3.1.5), but is much worse than the best possible accuracy (3.1.6).

3.1.3 Accuracy of homogeneous differential inclusions

Recall that a solution of a differential inclusion (DI) $\dot{x} \in F(x)$, $F(x) \subset \mathbb{R}^n$, is defined as any absolutely continuous function $x(t)$, satisfying the DI for almost all t. We call a DI $\dot{x} \in F(x)$ *Filippov DI*, if $F(x) \subset \mathbb{R}^n$ is non-empty, compact and convex for any x, and F is an upper-semicontinuous set function. The latter means that the maximal distance of the points of $F(x)$ from the set $F(y)$ tends to zero, as $x \to y$.

It is well known that such DIs feature most standard features, i.e., existence and extendability of solutions, except the uniqueness of solutions [18]. Asymptotically stable Filippov DIs have smooth Lyapunov functions [12].

3.1.3.1 Weighted homogeneity of DIs

Introduce the weights $m_1, m_2, \ldots, m_n > 0$ of the coordinates $x_1, x_2, \ldots, x_n$ in $\mathbb{R}^n$. Define the dilation

$$d_\kappa : (x_1, x_2, \ldots, x_n) \mapsto (\kappa^{m_1} x_1, \kappa^{m_2} x_2, \ldots, \kappa^{m_n} x_n),$$

where $\kappa > 0$. Recall References 4, 25 that a function $f : \mathbb{R}^n \to \mathbb{R}$ is said to have the homogeneity degree (weight) $q \in \mathbb{R}$, $\deg f = q$, if the identity $f(x) = \kappa^{-q} f(d_\kappa x)$ holds for any x and $\kappa > 0$.

Definition 3.1.2 ([30]). *A vector-set field $F(x) \subset \mathbb{R}^n$ (DI $\dot{x} \in F(x)$), $x \in \mathbb{R}^n$, is called homogeneous of the degree $q \in \mathbb{R}$, if the identity $F(x) = \kappa^{-q} d_\kappa^{-1} F(d_\kappa x)$ holds for any x and $\kappa > 0$.*

Consider a differential equation $\dot{x} = f(x)$, $\dot{x}_i = f_i(x)$, as a particular case of DI, when the set $F(x)$ contains only one vector $f(x)$. Then the above definition is reduced to the standard definition $\deg \dot{x}_i = \deg f_i = m_i + q$ [4, 25]. Note that the non-zero homogeneity degree q of a vector-set field can always be scaled to ± 1 by an appropriate proportional change of the weights $m_1, \ldots, m_n$.

Also note that the homogeneity of a vector-set field $F(x)$ can equivalently be defined as the invariance of the DI $\dot{x} \in F(x)$ with respect to the combined time-coordinate transformation

$$G_\kappa : (t, x) \mapsto (\kappa^p t, d_\kappa x), \qquad \kappa > 0,$$

where p, $p = -q$, might naturally be considered as the weight of t. Indeed, the homogeneity condition can be rewritten as

$$\dot{x} \in F(x) \Leftrightarrow \frac{d(d_\kappa x)}{d(\kappa^p t)} \in F(d_\kappa x).$$

Theorem 3.1.3 ([30, 34, 43]). *Let a Filippov DI be homogeneous of a negative homogeneity degree. Then FT stability, asymptotic stability and contractivity features are equivalent. The maximal (minimal) stabilization time is a well-defined upper (lower) semi-continuous function of the initial conditions.*

Here, the upper (lower) semi-continuity of a scalar function φ means that $\limsup_{x \to y} \varphi(x) \leq \varphi(y)$ $\left(\liminf_{x \to y} \varphi(x) \geq \varphi(y)\right)$. The contractivity [30] is equivalent to the existence of $T > 0$, $R > r > 0$, such that all solutions starting in the ball $\|x\| \leq R$ at the time 0 are in the smaller ball $\|x\| \leq r$ at the time T. It can be also proved that FT stability of $\dot{x} \in F(x)$ implies the inequalities $\deg F = q < 0$, $\deg \dot{x}_i = \deg x_i + \deg F = m_i + q \geq 0$, $i = 1, \ldots, n$.

3.1.3.2 Accuracy of disturbed homogeneous DIs

It is well known that FT-stable homogeneous DIs feature robustness with respect to various disturbances, delays, and sampling errors [10, 30, 32, 39, 40, 43]. Estimate the steady-state accuracy of a disturbed DI

$$\dot{x} \in F(x,\gamma), \qquad x \in \mathbb{R}^n, \;\; \gamma \in \mathbb{R}^\nu, \tag{3.1.8}$$

where γ is the vector disturbance parameter. The set field $F(x,\gamma) \subset \mathbb{R}^n$ is a non-empty compact convex set-valued function, upper-semicontinuous at all points $(x,0)$, $x \in \mathbb{R}^n$.

Introduce the dilations

$$\begin{aligned} d_\kappa &: (x_1,\ldots,x_n) \mapsto (\kappa^{m_1}s_1,\ldots,\kappa^{m_n}s_n), \quad m_1,\ldots,m_n > 0, \\ \Delta_\kappa &: (\gamma_1,\ldots,\gamma_\nu) \mapsto \left(\kappa^{\omega_1}\gamma_1,\ldots,\kappa^{\omega_\mu}\gamma_\mu\right), \quad \omega_1,\ldots,\omega_\nu > 0. \end{aligned}$$

Inclusion (3.1.8) is assumed homogeneous in both x and γ, while the undisturbed inclusion $\dot{x} \in F(x,0)$ is assumed FT stable with the homogeneity degree $q = -p$, $p > 0$. Hence, $m_i \geq p$. The homogeneity of (3.1.8) means that the transformation

$$(t,x,\gamma) \mapsto (\kappa^p t, d_\kappa x, \Delta_\kappa \gamma), \qquad \kappa > 0, \tag{3.1.9}$$

establishes a one-to-one correspondence between the solutions of the inclusion (3.1.8) with different parameters γ. In other words, $F(x,\gamma) = \kappa^p\, d_\kappa^{-1} F(d_\kappa x, \Delta_\kappa \gamma)$. In particular, the standard homogeneity $F(x,0) = \kappa^p\, d_\kappa^{-1} F(d_\kappa x, 0)$ is obtained for $\gamma = 0$.

In its turn $\gamma \in \Gamma(\rho,x) \subset \mathbb{R}^\nu$, where Γ is a homogeneous compact non-empty set-valued function with the magnitude parameter $\rho \geq 0$, i.e., $\forall \kappa, \rho > 0\, \forall x \in \mathbb{R}^n$: $\Gamma(\kappa^{m_\rho}\rho, d_\kappa x) = \Delta_\kappa\, \Gamma(\rho,x)$, $m_\rho > 0$. The function Γ monotonously increases with respect to the parameter ρ, i.e., $0 \leq \rho \leq \hat{\rho}$ implies $\Gamma(\rho,x) \subset \Gamma(\hat{\rho},x)$. It is also assumed that $\Gamma(0,x) = \{0\} \subset \mathbb{R}^n$ and $\Gamma(\rho,x)$ is Hausdorff-continuous in ρ, x at the points $(0,x)$.

It is easy to see that the time-coordinate-parameter transformation

$$\tilde{G}_\kappa : (t,\rho,x) \mapsto (\kappa^p t, \kappa^{m_\rho}\rho, d_\kappa x) \tag{3.1.10}$$

establishes a one-to-one correspondence between the solutions of $\dot{x} \in F(x,\Gamma(\rho,x))$ with different values of ρ.

Obviously, due to the homogeneity of Γ and the compactness of the disc $\|x\| \leq R$, for any $R > 0$ and any $\varepsilon > 0$ there exists $\rho > 0$, such that $\|x\| \leq R$ implies that $\forall z \in \Gamma(\rho,x)$: $\|z\| < \varepsilon$. Also, with any fixed $\rho \geq 0$ the function Γ maps bounded sets to bounded sets.

Now, consider the general retarded DI

$$\dot{x} \in F\left(x(t-\tau[0,1]), \Gamma(\rho, x(t-\tau[0,1]))\right), \tag{3.1.11}$$

where $\tau \geq 0$ is the maximal possible time delay.

The presence of the delays in (3.1.11) requires some initial conditions

$$x(t) = \xi(t), \quad t \in [-\tau, 0], \quad \xi \in \Xi(\tau, \rho, x_0). \tag{3.1.12}$$

The sets $\Xi(\tau, \rho, x_0)$ should possess some natural homogeneity properties, which are automatically satisfied, provided $\Xi = \tilde{\Xi}_{\varpi}(\tau, \rho, x_0)$, where $\tilde{\Xi}_{\varpi}(\tau, \rho, x_0)$ is comprised of the solutions of the simple Filippov DI

$$\begin{aligned} &\dot{\xi}_i \in \varpi \left(\|\xi\|_h + \rho^{1/m_\rho} \right)^{m_i - p} [-1, 1], \\ &i = 1, \ldots, n, \ \xi(0) = x_0, \ -\tau \le t \le 0. \end{aligned} \tag{3.1.13}$$

Recall that $m_i \ge p$. It is also formally assumed here that $\forall c \ge 0 : c^0 \equiv 1$. Inclusion (3.1.13) is homogeneous (i.e., invariant) with respect to the transformation $(t, \tau, \rho, \xi) \mapsto (\kappa t, \kappa^p \tau, \kappa^{m_\rho} \rho, d_\kappa \xi)$. The parameter ϖ is chosen sufficiently large to include the initial conditions of a considered concrete system.

Obviously, regular solutions of $\dot{x} \in F(x, 0)$ always satisfy (3.1.11), i.e., solutions of (3.1.11) always exist. Also, solutions of the inclusion with "discrete measurements" and uniformly bounded "noises" always exist. They correspond to the solutions with the right-hand side of the inclusion frozen between the "sampling instants", $\dot{x}(t) = \dot{x}(t_k) \in F(x(t_k), \Gamma(\rho, x(t_k)))$, $t \in [t_k, t_{k+1}]$, with the time periods $t_{k+1} - t_k \le \tau$. Both types of solutions are compatible with the above construction (3.1.13) of initial conditions.

Theorem 3.1.4 ([30, 39]). *After a finite-time transient all solutions of the disturbed DI (3.1.11) enter the region* $|x_i(t)| \le \mu_i \delta^{m_i}$, $\delta = \max\left\{\rho^{1/m_\rho}, \tau^{1/p}\right\}$, *to stay there forever. The constants* $\mu_i > 0$ *do not depend on* $\rho \ge 0$.

3.1.3.3 Accuracy of FT-stable homogeneous systems

This subsection demonstrates application of Theorem 3.1.4.

3.1.3.3.1 Accuracy of homogeneous FT-stable systems

Consider an FT-stable Filippov homogeneous DI $\dot{x} \in F(x)$, $\deg t = 1$, $\deg x_i = m_i$, $i = 1, \ldots, n_x$. Let x be measured at discrete time instants t_k, $t_{k+1} - t_k \le \tau$ with the measurement errors $|\eta_i(t_k)| \le \varepsilon_i$. Then the solutions satisfy the DI $\dot{x} \in F(x(t - \tau[0, 1]) + \varepsilon[-1, 1])$. Respectively, the accuracy $x_i = O(\rho^{m_j}), j = 1, \ldots, n_x$, is established in FT, where $\rho = \max\left\{\tau, \max_i \varepsilon_i^{\frac{1}{m_i}}\right\}$. Note that that estimation corresponds to the simplest case, when the process is described by a DI between the sampling instants. No discrete dynamics is involved.

3.1.3.3.2 Accuracy of a hybrid system

Consider an FT-stable homogeneous DI

$$\dot{x} \in F(x,z), \tag{3.1.14}$$

$$\dot{z} \in \Psi(h(x),z). \tag{3.1.15}$$

Here, (3.1.15) represents a possibly discontinuous "feedback" understood in the Filippov sense. The system is assumed to describe error dynamics of some controlled process. The uncertainties and time dependence are taken into account by some appropriate expansion of the sets on the right-hand side. Let the system homogeneity degree be -1, $\deg t = 1$, and $\deg x_i = m_i$, $i = 1,\dots,n_x$, $\deg z_i = m_{zi}$, $i = 1,\dots,n_z$. The output h is a homogeneous vector function, $\deg h_i = m_{hi} > 0$, $i = 1,\dots,n_h$. The vector variable $z \in \mathbb{R}^{n_z}$ is the internal controller state.

In "practice" the output $h(x)$ is sampled at the time instants t_k, $0 \le t_{k+1} - t_k \le \tau$, with the error $\eta \in \mathbb{R}^{n_h}$, $\eta_i \in \varepsilon_i[-1,1]$, $i = 1,\dots,n_h$, and some "actuator" delay $\tilde{\tau}_k$. The dynamic feedback (3.1.15) becomes a discrete system defined at the discrete time instants $t_{k,j}$, $j = 0,\dots,l_k$, $t_{k,0} = t_k + \tilde{\tau}_k$, $t_{k,l_k} = t_{k+1,0} = t_{k+1} + \tilde{\tau}_{k+1}$. Naturally, $t_{k+1} + \tilde{\tau}_{k+1} > t_k + \tilde{\tau}_k$ is assumed to hold. It is also assumed that $0 < t_{k,j+1} - t_{k,j} = \underline{\tau}_{k,j} < \underline{\tau}(\tau)$, i.e., the maximal integration step $\underline{\tau}$ can be chosen in dependence on an upper bound τ of the sampling periods τ_k. Also $\tilde{\tau}_k \le \tilde{\tau}$ holds.

The resulting feedback is approximated by the Euler integration, producing the hybrid system

$$\begin{aligned} \dot{x} &\in F\left(x, z\left(t_{k,j}\right)\right), \\ z\left(t_{k,j+1}\right) &\in z\left(t_{k,j}\right) + \left[\Psi\left(h\left(x\left(t_k + \tilde{\tau}_k\right)\right)\right) + \eta_i\left((t_k + \tilde{\tau}_k), z\left(t_{k,j}\right)\right)\right] \underline{\tau}_{k,j}, \\ j &= 0,\dots,l_k - 1. \end{aligned} \tag{3.1.16}$$

Note that a random element is taken from the set Ψ at each time step k,j. The solution component $z\left(t_{k,j}\right)$ of (3.1.16) can be equivalently described by solutions of equations with piece-wise-constant right-hand sides, which take the same values at the time instants $t_{k,j}$. Indeed, the intermediate values of z do not affect the process component x changing continuously in time. Thus, the components $x(t)$ of solutions of (3.1.16) coincide with the x components of some solutions of the DI

$$\begin{aligned} \dot{x} &\in F\left(x, z\left(t + (\tilde{\tau} + \tau)[-1,0]\right)\right), \\ \dot{z}(t) &\in \Psi\left(h\left(x\left(t + (\tilde{\tau} + \tau)[-1,0]\right)\right) + \varepsilon[-1,1], z\left(t + \underline{\tau}[-1,0]\right)\right). \end{aligned} \tag{3.1.17}$$

In its turn these solutions satisfy

$$\begin{aligned} \dot{x} &\in F(x, z(t + \rho[-1,0]))), \\ \dot{z}(t) &\in \Psi(h(x(t + \rho[-1,0])) + \vec{\rho}^{\,m_h}[-1,1], z(t + \rho[-1,0])), \end{aligned} \tag{3.1.18}$$

where $\rho = \max\left\{\tilde{\tau} + \tau, \max_i \varepsilon_i^{\frac{1}{m_{hi}}}\right\}$ and $\vec{\rho}^{\,m_h} = \left(\rho^{m_{h1}},\dots,\rho^{m_{hn_h}}\right)^T$. Introduce also some appropriate initial values as it is done in Section 3.1.3.2.

It follows now from Theorem 3.1.4 that the accuracy $x_j = O(\rho^{m_j}), j = 1, \ldots, n_x$, is established in FT.

3.1.4 Homogeneous continuous-time SM control

In this section, we develop SMC which realizes the best possible asymptotic SM accuracy (3.1.6), (3.1.7) calculated in the previous section. All over this section, the accuracies are calculated under the assumption that, whereas the sampling is discrete, the differential equations/inclusions still take place between the samplings.

3.1.4.1 Homogeneous SISO SM control

Consider the SISO SMC problem (3.1.3) with smooth functions a and b. The system is understood in the Filippov sense [18]. Let the smooth scalar output $\sigma(t,x)$ have the relative degree r, which means that

$$\sigma^{(r)} = h(t,x) + g(t,x)u, \tag{3.1.19}$$

where h, g are uncertain smooth functions, $g(t,x) \neq 0$. As usual [27, 29, 30] assume that h, g are bounded,

$$|h(t,x)| \leq C, \quad 0 < K_m \leq g(t,x) \leq K_M. \tag{3.1.20}$$

Such bounds exist at least for any compact operational region. Any solution of (3.1.3) is assumed infinitely extendable in time, provided σ, its derivatives and u remain bounded along the solution.

The above are the only system conditions needed in Section 3.1.4.1. The system uncertainty is defined by the numbers r, C, K_m, K_M which are supposed to be known.

The uncertain dynamics (3.1.19) can be replaced by the concrete DI

$$\sigma^{(r)} \in [-C, C] + [K_m, K_M]u. \tag{3.1.21}$$

Most r-SM controllers are built as controllers for (3.1.21) making $\vec{\sigma} = (\sigma, \dot{\sigma}, \ldots, \sigma^{(r-1)})$ vanish in FT.

In order to use the results from Section 3.1.3.1, the closed-loop inclusion is to be homogeneous with negative homogeneity degree. Scaling the system homogeneity degree to -1, $\deg t = 1$, obtain that with $C > 0$ inevitably $\deg \sigma^{(r)} = 0$, on the other hand, $\deg \sigma^{(r)} = \deg \sigma^{(r-1)} - \deg t$. Thus, the only possible homogeneity weights are $\deg \sigma = r, \ldots, \deg \sigma^{(r-1)} = 1$. This homogeneity is called r-sliding homogeneity [30]. Respectively, the control

$$u = U_r(\vec{\sigma}) \tag{3.1.22}$$

is called r-sliding homogeneous, if $\deg u = 0$, i.e.,

$$U_r\big(\sigma, \dot{\sigma}, \ldots, \sigma^{(r-1)}\big) \equiv U_r\big(\kappa^r \sigma, \kappa^{r-1} \dot{\sigma}, \ldots, \kappa \sigma^{(r-1)}\big) \tag{3.1.23}$$

holds for any $\kappa > 0$, $\vec{\sigma} \in \mathbb{R}^r$. Since the control is required to be locally bounded [18], due to (3.1.23) it is also **globally bounded**. The right-hand side of the inclusion (3.1.21), (3.1.22) is assumed minimally enlarged at the points of the discontinuity of (3.1.22) to satisfy the Filippov conditions [18].

Replace $\sigma \in \mathbb{R}$ with $\omega \in \mathbb{R}$ in the following formulas, enabling the further usage of the controllers for different components of the vector output σ in the MIMO case. Let $\beta_{1,r}, \ldots, \beta_{r-1,r}$ be some predefined positive coefficients, and α be the chosen control magnitude. Then a simple family of r-sliding homogeneous controllers of the form

$$u = -\alpha \Psi_{r-1,r}\left(\omega, \dot{\omega}, \ldots, \omega^{(r-1)}\right), \tag{3.1.24}$$

called embedded SM controllers [29], are provided by the following recursion. Let $d \geq r$, define

$$\begin{aligned} \varphi_{0,r} &= \omega,\ N_{0,r} = |\omega|^{1/r},\ \Psi_{0,r} = \operatorname{sign}\omega; \\ \varphi_{j,r} &= \omega^{(j)} + \beta_{j,r} N_{j-1,r}^{r-j} \Psi_{j-1,r},\ \Psi_{j,r} = \operatorname{sign}\varphi_{j,r}; \\ N_{j,r} &= \left(|\omega|^{d/r} + |\dot{\omega}|^{d/(r-1)} + \cdots + \left|\omega^{(j-1)}\right|^{d/(r+1-j)}\right)^{1/d}. \end{aligned} \tag{3.1.25}$$

The following are valid parametric sets $\{\beta_{1,r}, \ldots, \beta_{r-1,r}\}, d_r$ for $r = 2, \ldots, 4$: $r = 2$, $\{1\}$, $d_1 = 1$; $r = 3$, $\{1, 2\}$, $d_2 = 6$; $r = 4$, $\{0.5, 1, 3\}$, $d_4 = 12$. It is further assumed that $\beta_{1,r}, \ldots, \beta_{r-1,r}$ are always properly chosen, which means that the differential equations $\varphi_{r-1,r} = 0$ are FT stable [38].

Provided the parameters $\beta_{j,r}$ are properly chosen [38] and α is sufficiently large, under the above assumptions (3.1.19), (3.1.20) the listed r-SM controllers solve the stated problem of FT establishing and keeping $\sigma \equiv 0$ [29] by means of uniformly bounded control and for any initial conditions.

Another well-known family of SM controllers, called quasi-continuous SM controllers [31], also features control continuous everywhere except the r-sliding set $\omega = \dot{\omega} = \cdots = \omega^{(r-1)} = 0$. Such controllers feature considerably less chattering. Other constructions of similar homogeneous HOSM controllers and the choice of parameters are considered in References 30, 38.

3.1.4.1.1 Differentiator

Any r-sliding homogeneous controller can be combined with an $(r-1)$th-order differentiator [29] producing an output feedback controller. Its applicability in this case is possible, since $\sigma^{(r)}$ is bounded due to the boundedness of the feedback function $u = -\alpha \Psi_{r-1,r}(\vec{\sigma})$ in (3.1.21).

Let the input signal $f(t)$ consists of a bounded Lebesgue-measurable noise with unknown features, and an unknown basic signal $f_0(t)$, whose n_dth derivative has a known Lipschitz constant $L > 0$. These are the only restrictions on the input of the differentiator. While the number L is to be known, one does not need to know the noise magnitude.

The outputs z_j of the following differentiator estimate of the derivatives $f_0^{(j)}$, $j = 0, \ldots, n_d$. The recursive form of the differentiator is

$$\begin{aligned}
\dot{z}_0 &= -\lambda_{n_d} L^{\frac{1}{n_d+1}} |z_0 - f(t)|^{\frac{n_d}{n_d+1}} \operatorname{sign}(z_0 - f(t)) + z_1, \\
\dot{z}_1 &= -\lambda_{n_d-1} L^{\frac{1}{n_d}} |z_1 - \dot{z}_0|^{\frac{n_d-1}{n_d}} \operatorname{sign}(z_1 - \dot{z}_0) + z_2, \\
&\vdots \\
\dot{z}_{n_d-1} &= -\lambda_1 L^{\frac{1}{2}} \left|z_{n_d-1} - \dot{z}_{n_d-2}\right|^{\frac{1}{2}} \operatorname{sign}\left(z_{n_d-1} - \dot{z}_{n_d-2}\right) + z_{n_d}, \\
\dot{z}_{n_d} &= -\lambda_0 L \operatorname{sign}\left(z_{n_d} - \dot{z}_{n_d-1}\right).
\end{aligned} \tag{3.1.26}$$

Parameters λ_i of differentiator (3.1.26) are chosen in advance for each n_d. An infinite sequence of parameters λ_i can be built, valid for all natural n_d [29]. In particular, one can choose $\lambda_0 = 1.1$, $\lambda_1 = 1.5$, $\lambda_2 = 2$, $\lambda_3 = 3$, $\lambda_4 = 5$, $\lambda_5 = 8$ [31] or $\lambda_0 = 1.1$, $\lambda_1 = 1.5$, $\lambda_2 = 3$, $\lambda_3 = 5$, $\lambda_4 = 8$, $\lambda_5 = 12$, which is enough for $n_d \leq 5$. In the absence of noises, the differentiator provides for the exact estimations in FT. Equation (3.1.26) can be rewritten in the usual non-recursive form

$$\begin{aligned}
\dot{z}_0 &= -\tilde{\lambda}_{n_d} L^{\frac{1}{n_d+1}} |z_0 - f(t)|^{\frac{n_d}{n_d+1}} \operatorname{sign}(z_0 - f(t)) + z_1, \\
\dot{z}_1 &= -\tilde{\lambda}_{n_d-1} L^{\frac{2}{n_d+1}} |z_0 - f(t)|^{\frac{n_d-1}{n_d+1}} \operatorname{sign}(z_0 - f(t)) + z_2, \\
&\vdots \\
\dot{z}_{n_d-1} &= -\tilde{\lambda}_1 L^{\frac{n_d}{n_d+1}} |z_0 - f(t)|^{\frac{1}{n_d+1}} \operatorname{sign}(z_0 - f(t)) + z_{n_d}, \\
\dot{z}_{n_d} &= -\tilde{\lambda}_0 L \operatorname{sign}(z_0 - f(t)).
\end{aligned} \tag{3.1.27}$$

It is easy to see that $\tilde{\lambda}_0 = \lambda_0$, $\tilde{\lambda}_{n_d} = \lambda_{n_d}$, and $\tilde{\lambda}_j = \lambda_j \tilde{\lambda}_{j+1}^{j/(j+1)}$, $j = n_d - 1, n_d - 2, \ldots, 1$.

Notation. Assuming that the sequence λ_j, $j = 0, 1, \ldots$, is the same over the whole chapter, denote (3.1.27) by the equality $\dot{z} = D_{n_d}(z, f, L)$. Also for any $w \neq 0$ and $\gamma > 0$ denote $\lfloor w \rceil^\gamma = |w|^\gamma \operatorname{sign} w$; $\lfloor 0 \rceil^\gamma = 0$; $\lfloor w \rceil^0 = \operatorname{sign} w$.

Let the noise be absent. Subtracting $f^{(i+1)}(t)$ from the both sides of the equation for $\dot{z}_i$ of (3.1.27), denoting $\sigma_{d,i} = z_i - f^{(i)}$, $i = 0, \ldots, n_d$, and using $f^{(n_d+1)}(t) \in [-L, L]$ obtain the differentiator error dynamics

$$\begin{aligned}
\dot{\sigma}_{d,0} &= -\tilde{\lambda}_{n_d} L^{\frac{1}{n_d+1}} \lfloor \sigma_{d,0} \rceil^{\frac{n_d}{n_d+1}} + \sigma_{d,1}, \\
\dot{\sigma}_{d,1} &= -\tilde{\lambda}_{n_d-1} L^{\frac{2}{n_d+1}} \lfloor \sigma_{d,1} \rceil^{\frac{n_d-1}{n_d+1}} + \sigma_{d,2}, \\
&\vdots \\
\dot{\sigma}_{d,n_d-1} &= -\tilde{\lambda}_1 L^{\frac{n_d}{n_d+1}} \lfloor \sigma_{d,n_d-1} \rceil^{\frac{1}{n_d+1}} + \sigma_{d,n_d}, \\
\dot{\sigma}_{d,n_d} &\in -\tilde{\lambda}_0 L \operatorname{sign} \sigma_{d,n_d} + L[-1, 1].
\end{aligned} \tag{3.1.28}$$

It is homogeneous with $\deg t = -1$, $\deg \sigma_{d,i} = n_d + 1 - i$ [29]. Thus, according to Section 3.1.3.3.1 with sampling time periods not exceeding $\tau > 0$ and the maximal possible sampling error $\varepsilon \geq 0$, the differentiation accuracy $\left|z_j - f_0^{(j)}\right| \leq \mu_j L \rho^{n_d+1-j}$, $\rho = \max\left(\tau^{n_d+1-j}, (\varepsilon/L)^{(n_d+1-j)/(n_d+1)}\right)$, is ensured, where the constant numbers $\mu_j > 0$ only depend on the parameters $\lambda_0, \ldots, \lambda_{n_d}$ of the differentiator. This accuracy is known to be asymptotically optimal in the presence of noises [26, 28], which means that only the coefficients μ_j can be improved.

3.1.4.1.2 Differentiator initialization

Although one can take arbitrary initial values of differentiator for its feedback application, it may considerably destroy the initial system transient, since at the beginning the differentiator outputs will have no resemblance to the right derivatives. The overall performance can be drastically improved if the initial values of the differentiator are chosen right.

The most simple method is to take $z_0(t_0) = f(t_0)$ and $z_i(t_0) = 0$, $i = 1, \ldots, n_d$, where t_0 is the first sampling time. Then one just provides some reasonable time for the differentiator convergence prior to the control application.

Another method, *which we consider preferable if the noise magnitude is available*, is to choose some initial time increments of the length Δt, consisting of a number of real sampling intervals. The $n_d + 1$ sampling values of the input f are stored for n_d such successive time increments, and then the initial values of the differentiator are calculated by divided differences. During all this period the control is not applied, i.e., kept at zero. Then the differentiator is practically already in the steady state from the very beginning. This initialization process is robust with respect to noises of the magnitude of the order Δt^{n_d+1}, i.e., Δt is to be chosen with respect to the maximal possible noise. A small additional time for the initial error elimination can still be considered.

One can also apply non-homogeneous differentiator modifications [3, 13, 35] with faster convergence. In that case, the global system homogeneity is lost.

3.1.4.1.3 Differentiator with variable Lipschitz parameter *L*

It is proved that if L continuously changes in time, then if at some moment differentiation errors are zero, they will stay at zero forever. Obviously, such statement cannot withstand a practice exam.

Practically important result is that if L is differentiable, and $|\dot{L}|/L \leq M$ for some M, then for some $\delta > 0$ the differentiator converges provided the initial errors satisfy $\left|z_j - f_0^{(j)}\right| \leq \delta L$ [36]. The accuracy $\left|z_j - f_0^{(j)}\right| \leq \mu_j L \max\left(\tau^{n_d+1-j}, \varepsilon^{(n_d+1-j)/(n_d+1)}\right)$ is kept provided the noise satisfies $|\eta(t)| \leq L(t)\varepsilon$ [36].

Globally convergent differentiator with fast convergence and $|\dot{L}|/L \leq M$ has been recently presented [35]. Note that its parameters depend on M.

3.1.4.1.4 Output feedback control

Consider the system (3.1.3) of the relative degree r under the conditions (3.1.19), (3.1.20) with a bounded control (3.1.22). Incorporating the $(r-1)$th-order differentiator into the feedback equations, obtain the SISO output-feedback r-sliding controller

$$u = U_r(z), \qquad \dot{z} = D_{r-1}(z, \sigma, L), \tag{3.1.29}$$

where $L \geq C + K_M \sup |U_r|$. Suppose that (3.1.21), (3.1.22) is FT stable. Then the output-feedback controller (3.1.29) ensures the FT establishment of the r-SM $\vec{\sigma} = 0$. Moreover [30], if (3.1.22) is r-sliding homogeneous, the closed-loop inclusion (3.1.21), (3.1.29) is homogeneous with $\deg z_i = \deg \sigma^{(i)} = r - i$ and the system homogeneity degree -1. Respectively, due to Section 3.1.3.3.1, if σ is sampled with the accuracy $\varepsilon \geq 0$ and the sampling intervals not exceeding $\tau > 0$, then the asymptotic SM accuracy $\sigma^{(j)} = O\left(\max\left(\tau^{r-j}, \varepsilon^{(r-j)/r}\right)\right)$ is obtained.

3.1.4.2 Homogeneous MIMO SM control

Once more consider dynamic system (3.1.3),

$$\dot{x} = a(t,x) + b(t,x)u, \qquad \sigma = \sigma(t,x), \tag{3.1.30}$$

but let now σ and u be vectors, $\sigma : \mathbb{R}^{n+1} \to \mathbb{R}^m$, $u \in \mathbb{R}^m$. The system is assumed to have the vector relative degree $r = (r_1, \ldots, r_m)$, $r_i > 0$. It means that the successive total time derivatives $\sigma_i^{(j)}, j = 0, 1, \ldots, r_i - 1, i = 1, \ldots, m$, do not contain controls, and can be used as a part of new coordinates [24]. Respectively, (3.1.19) turns to be a vector equation,

$$\sigma^{(r)} = h(t,x) + g(t,x)u, \tag{3.1.31}$$

where $\sigma^{(r)}$ denotes $\left(\sigma_1^{(r_1)}, \ldots, \sigma_m^{(r_m)}\right)^T$, the functions h, and g are unknown and smooth. The function g is a non-singular matrix. It is often called high-frequency gain matrix.

Let g be represented in the form $g = K\bar{g}$, where $K > 0$ defines the "size" of the matrix g, and $\bar{g}$ corresponds to the matrix "direction". A nominal "direction" matrix $G(t,x)$ is assumed non-singular and available in real time, so that

$$g(t,x) = K(t,x)(G(t,x) + \Delta g(t,x)), \qquad \left\|\Delta g G^{-1}\right\|_1 \leq p < 1. \tag{3.1.32}$$

Here, Δg is the uncertain deviation of $\bar{g}$ from G, and the norm $\|\cdot\|_1$ of the matrix $A = (a_{ij})$ is defined as $\|A\|_1 = \max_i \sum_j |a_{ij}|$. The estimation G can be any Lebesgue-measurable function, p is a known constant. Mark that similar assumptions are adopted in Reference 14.

Similarly to (3.1.20), assume that the uncertain vector function h and the scalar function K are bounded,

$$\|h(t,x)\| \leq C, \quad 0 < K_m \leq K(t,x) \leq K_M, \tag{3.1.33}$$

where C, K_m, K_M are known constants.

It is also assumed that trajectories of (3.1.30) are infinitely extendible in time for any Lebesgue-measurable control with uniformly bounded $\|gu\|/H$.

The above are the only system conditions needed in Section 3.1.4.2. The system uncertainty is defined by r, C, K_m, K_M, p, and $G(t,x)$ which are, respectively, assumed to be known or available.

Note that the availability of $G(t,x)$ in real time does not necessarily mean that $x(t)$ is available, and G is known analytically. For example, the aerodynamic characteristics of an aircraft are usually available as approximate table functions of the observable dynamic pressure and altitude.

Introduce a virtual control v,

$$u = G(t,x)^{-1}v. \tag{3.1.34}$$

Then dynamics (3.1.31) take the form

$$\sigma^{(r)} = h(t,x) + K(t,x)\left(I + \Delta g(t,x)G^{-1}(t,x)\right)v, \tag{3.1.35}$$

where I is the unit matrix.

Introduce the notation $\vec{\sigma}_i = \left(\sigma_i, \ldots, \sigma_i^{(r_i-1)}\right), \vec{\sigma} = (\vec{\sigma}_1, \ldots, \vec{\sigma}_m)$. Choose the components of $v = (v_1, \ldots, v_m)^T$ in the form of the embedded r_i-sliding homogeneous controller (3.1.24), (3.1.25)

$$v_i = -\alpha\Psi_{r_i-1,r_i}(\vec{\sigma}_i), \qquad i = 0, 1, \ldots, m, \quad u = G(t,x)^{-1}v, \tag{3.1.36}$$

where $\alpha > 0$. Now the closed-loop system satisfies the decoupled $(r_1, r_2, \ldots, r_m)$-sliding homogeneous inclusion

$$\sigma_i^{(r_i)} \in [-C, C] - \alpha[K_m(1-p), K_M(1+p)]\,\Psi_{r_i-1,r_i}(\vec{\sigma}_i), \quad i = 1, \ldots, m, \tag{3.1.37}$$

with the weights $\deg \sigma_i^{(j)} = r_i - j$. According to Section 3.1.4.1, (3.1.37) is FT stable with sufficiently large α.

Respectively, the output-feedback control gets the form

$$\begin{aligned} v_i &= -\alpha\Psi_{r_i-1,r_i}(z_i), && i = 0, 1, \ldots, m, \ \ u = G(t,x)^{-1}v, \\ \dot{z}_i &= D_{r_i-1}(L, \sigma_i, z_i), && L \geq C + 2K_M\alpha. \end{aligned} \tag{3.1.38}$$

The closed-loop inclusion is still homogeneous with $\deg \sigma_i^{(j)} = \deg z_{i,j} = r_i - j$. The following theorem follows from Section 3.1.3.3.1.

Theorem 3.1.5. *Let the MIMO system (3.1.30), (3.1.31) satisfy conditions (3.1.32), (3.1.33). Then output-feedback control (3.1.38) provides for the finite-time establishment and keeping of the r-SM $\sigma = 0$. Let σ_i be measured with the sampling accuracy $\varepsilon_i \geq 0$, $i = 1, 2, \ldots, m$, and the sampling intervals not exceeding $\tau > 0$, then the asymptotic SM accuracy $\sigma_i^{(j)} = O\left(\max\left(\tau^{r_i-j}, \varepsilon_i^{(r_i-j)/r_i}\right)\right)$ is obtained.*

As we have seen, the obtained SM asymptotics are the best possible. Note that one can here use quasi-continuous controllers [31, 38], but the corresponding technique is more complicated [37], though provides for superior performance.

3.1.5 Discretization of SM differentiators

In reality, described differentiators are realized by means of computers. This turns a real-time differentiator into a discrete dynamic system. In this section, we present the discretization methods for SM-based differentiators, analyse their accuracy and online detection of their convergence.

3.1.5.1 Discrete differentiators and their accuracy

Consider the differentiator (3.1.26) or (3.1.27), which is represented as $\dot{z} = D_{n_d}(f, z, L)$. Let the basic input $f_0(t)$ be sampled at the time instants t_k, $0 \le t_{k+1} - t_k = \tau_k \le \tau$, with the error $\eta \in \mathbb{R}$, $\eta \in \varepsilon[-1,1]$, $f = f_0 + \eta$, $\left|f_0^{(n_d+1)}\right| \le L$. Differentiator (3.1.27) is a discontinuous dynamic system. Therefore, its only reliable numeric integration is based on the Euler method. Also, the discretization is naturally to be based on the Euler integration.

The simplest way is to perform one Euler integration step between each two successive measurements. The respective discretization is

$$z(t_{k+1}) = z(t_k) + D_{n_d}(f(t_k), z(t_k), L)\tau_k. \tag{3.1.39}$$

The accuracies obtained in Section 3.1.4.1.1 correspond to the case when between the measurements the differentiator is described by differential equations. It corresponds to the infinite number of infinitesimally small Euler-integration steps between the measurements at t_k, t_{k+1}. In practice only finite number l_k of such steps is taken. Let the Euler steps take place at the discrete time instants $t_{k,j}, j = 0, \ldots, l_k$, $t_{k,0} = t_k$, $t_{k,l_k} = t_{k+1} = t_{k+1,0}$. Thus, all sampling instants are also the instants of the integration subdivision. It is also assumed that $0 < t_{k,j+1} - t_{k,j} = \underline{\tau}_{k,j} < \underline{\tau}(\tau)$, i.e., in general $\underline{\tau}$ may depend on τ.

It is natural to take the discretization

$$z(t_{k,j+1}) = z(t_{k,j}) + D_{n_d}(f(t_k), z(t_{k,j}), L)\,\underline{\tau}_{k,j}, \qquad j = 0, \ldots, l_k - 1.$$

It is *not* the discretization we use. The idea is never to use the differences of the input signal f and z_0 taken at *different* time instants. Thus, the proposed discretization is

$$z(t_{k,j+1}) = z(t_{k,j}) + \Delta_{n_d}(f(t_k),\ z(t_{k,j}), L)\,\underline{\tau}_{k,j}, \qquad j = 0, \ldots, l_k - 1, \tag{3.1.40}$$

where the vector function $\Delta_{n_d}\left(f(t_k), z\left(t_{k,j}\right), L\right)$ has the components

$$\begin{aligned}
\Delta_{n_d,0} &= -\tilde{\lambda}_{n_d} L^{\frac{1}{n_d+1}} \lfloor z_0(t_k) - f(t_k) \rceil^{\frac{n_d}{n_d+1}} + z_1\left(t_{k,j}\right), \\
\Delta_{n_d,1} &= -\tilde{\lambda}_{n_d-1} L^{\frac{2}{n_d+1}} \lfloor z_0(t_k) - f(t_k) \rceil^{\frac{n_d-1}{n_d+1}} + z_2\left(t_{k,j}\right), \\
&\vdots \\
\Delta_{n_d,n_d-1} &= -\tilde{\lambda}_1 L^{\frac{n_d}{n_d+1}} \lfloor z_0(t_k) - f(t_k) \rceil^{\frac{1}{n_d+1}} + z_{n_d}\left(t_{k,j}\right), \\
\Delta_{n_d,n_d} &= -\tilde{\lambda}_0 L \ \mathrm{sign}\ (z_0(t_k) - f(t_k)).
\end{aligned} \tag{3.1.41}$$

One can expect that the resulting accuracy is worse than the standard differentiator accuracy from Section 3.1.4.1.1, but it is to be reclaimed for $\underline{\tau} \to 0$, $l_k \to \infty$.

Unfortunately, the results of Section 3.1.3.3.1 are not valid for such systems. One needs to use the technique described in Section 3.1.3.3.2 to estimate the resulting accuracy. Following are the currently known results, some of which are declared here for the first time. In particular, the standard accuracy from Section 3.1.4.1.1 is always preserved for the standard first-order differentiator, $n_d = 1$ [40].

1. Let the integration steps be equal, $t_{k,j+1} - t_{k,j} = \underline{\tau}_{k,j} = \underline{\tau}$. Let $\rho = \max\left[(\varepsilon/L)^{1/(n_d+1)}, \tau\right]$. Also suppose that the derivatives $f_0^{(i)}$ of the orders $2, 3, \ldots, n_d + 1$, are bounded: $\left|f_0^{(i)}\right| \leq D_i$, $D_{n_d+1} = L$. Then there exist such constants $\mu_i > 0$ that independently of the sampling intervals' choice the following inequalities hold after an FT transient:

$$\begin{aligned}
&\left|z_0\left(t_{k,j}\right) - f_0\left(t_{k,j}\right)\right| \leq \mu_0 L \rho^{n_d+1}; \\
&\left|z_i\left(t_{k,j}\right) - f_0^{(i)}\left(t_{k,j}\right)\right| \leq \mu_i L \rho^{n_d+1-i} + j\underline{\tau} D_{i+1}, \qquad i = 1, 2, \ldots, n_d.
\end{aligned} \tag{3.1.42}$$

Note that this result is published in Reference 40 for the case when the integration steps and the sampling intervals coincide, $l_k = 1$, $\underline{\tau} = \tau$. The proof is very similar to one presented in Reference 40.

2. Let the maximal integration and sampling steps, $\underline{\tau}$ and τ, be small enough. Also suppose that the derivatives $f_0^{(i)}$ of the orders $2, 3, \ldots, n_d + 1$, are bounded: $\left|f_0^{(i)}\right| \leq D_i$, $D_{n_d+1} = L$. Then there exist such constants $\mu_i > 0$ that independently of the sampling and integration intervals' choice, the inequalities

$$\left|z_i\left(t_{k,j}\right) - f_0^{(i)}\left(t_{k,j}\right)\right| \leq \mu_i L \rho^{n_d+1-i}, \qquad i = 0, 1, \ldots, n_d, \tag{3.1.43}$$

hold after an FT transient. Here, $\rho = \max\left[(\varepsilon/L)^{1/(n_d+1)}, \underline{\tau}^{1/n_d}, \tau\right]$. Note that contrary to other cases, here μ_i depend on $D_2/L, \ldots, D_{n_d}/L$.

Note that the standard asymptotics of Section 3.1.4.1.1 are restored for $\underline{\tau} \leq \tau^{n_d}$. Also this result is published in Reference 40 for the case when the integration steps and the sampling intervals coincide, $l_k = 1$, $\underline{\tau} = \tau$. The proof follows the general lines of Section 3.1.3.3.2. One has to represent the errors of the discrete system (3.1.40) as solutions of some disturbed FT-stable inclusion sampled at the times $t_{k,j}$.

As we see, in general, the asymptotic accuracy of the continuous-time differentiator with discrete measurements is lost, when the differential equations are replaced with discrete Euler integration and the differentiation order exceeds 1. It is restored if the maximal integration step $\underline{\tau}$ and the maximal sampling interval τ satisfy the inequality $\underline{\tau} \leq \tau^{n_d}$. That choice of the integration step still can be feasible for $n_d = 2$, but usually becomes impractical already for $n_d = 3$. Also the requirement for derivatives $f_0^{(i)}$, $i = 2, \ldots, n_d$, to be bounded is restrictive. The following discrete differentiator resolves all these issues.

3. *Homogeneous discrete differentiator.* Choose

$$\begin{aligned}
\Delta_{n_d,i}\left(t_{k,j}\right) &= -\tilde{\lambda}_{n_d-i} L^{\frac{i+1}{n_d+1}} \lfloor z_0(t_k) - f(t_k) \rceil^{\frac{n_d-i}{n_d+1}} + z_{i+1}\left(t_{k,j}\right) \\
&\quad + \sum_{s=i+2}^{n_d-1} \frac{z_s\left(t_{k,j}\right) \underline{\tau}_{k,j}^{s-1}}{(s-i)!}, \qquad i = 0, 1, \ldots, n_d - 1; \qquad (3.1.44) \\
\Delta_{n_d,n_d}\left(t_{k,j+1}\right) &= -\tilde{\lambda}_0 L \,\mathrm{sign}\,(z_0(t_k) - f(t_k)).
\end{aligned}$$

New terms appear in the second line of (3.1.44) only if $n_d > 1$. Note that (3.1.44) can be also rewritten in the recursive form [35]. Let the maximal integration and sampling steps, $\underline{\tau} \leq \tau$, **be any positive numbers**. Then there exist such constants $\mu_i > 0$ that independently of the function f_0 and the choice of the sampling intervals and integration steps, the inequalities (3.1.43) hold after an FT transient for $\rho = \max\left[(\varepsilon/L)^{1/(n_d+1)}, \tau\right]$.

Obviously, discrete differentiator (3.1.40), (3.1.44) completely reclaims the accuracy of its continuous-time analogue. This result has been published in Reference 40 for the case when the integration steps and the sampling intervals coincide. It also seems that additional integration steps do not cause any noticeable accuracy improvement.

3.1.5.2 Convergence criteria

It is practically important to detect the moment when the differentiator starts to produce reliable derivative estimations. Choose some parameter $\rho > 0$. Taking into account the homogeneity features of the differentiator, we expect that if the noise $\eta(t)$, the maximal sampling interval τ, and the maximal integration step $\underline{\tau}$ satisfy

$$|\eta| \leq k_\eta \rho^{n_d+1}, \qquad \tau \leq k_\tau \rho, \qquad (3.1.45)$$

and, if integration steps are variable and differ from the sampling steps,

$$\underline{\tau} \leq \underline{k}_\tau \rho^{n_d}, \qquad (3.1.46)$$

then, respectively, to the chosen discretization scheme, the differentiation accuracy (3.1.42) or (3.1.43) is to be obtained in the steady state.

The only available real-time information consists here of the observed sampled differences $z_0(t_k) - f(t_k)$, which are corrupted by noise and taken at discrete times t_k. According to the results of the previous subsection, we can expect the steady-state accuracy $\left|z_0\left(t_{k,j}\right) - f\left(t_{k,j}\right)\right| \leq k_f L\rho^{n_d+1}$ to be kept.

The idea is that provided this accuracy is observed for sufficiently long time, one can assume that the transient is over. Due to the homogeneity reasoning that observation time is to be proportional to ρ. Thus one expects

$$|z_0(t_k) - f(t_k)| \leq k_f L\rho^{n_d+1}, \qquad t_k \in [t - k_t\rho, t] \tag{3.1.47}$$

to hold for some $k_t > 0$ starting from some moment t.

The general convergence criterion. For any sufficiently small $\rho > 0$

1. for any choice of positive parameters $\mu_0, \ldots, \mu_{n_d}$, there exists a set of positive parameters k_η, k_τ, k_f, k_t, and, maybe, $\underline{k}_\tau$,
2. for any choice of positive parameters k_η, k_τ, k_f, k_t, and, maybe, $\underline{k}_\tau$, there exists a set of positive parameters $\mu_0, \ldots, \mu_{n_d}$,

such that for any input $f(t)$, $\left|f_0^{(n_d+1)}\right| \leq L$ (plus boundedness of $\left|f_0^{(i)}\right|$, $i \geq 2$ for the discretization scheme (3.1.41)), provided (3.1.47) holds at the time moment $t = t_*$, it also holds for any $t \geq t_*$, and the accuracies (3.1.42) or (3.1.43) (respectively, to the chosen discretization scheme) are kept starting from the time $t_* - k_t\rho$.

Let us concretize the above criterion to remove any doubts.

- Continuous-time criterion [2], differentiator (3.1.26). The numeric integration is excluded, (3.1.45) are the only restrictions. The accuracy (3.1.43) is observed *for any* $\rho > 0$ (not only small).
- Equal integration steps, scheme (3.1.40), (3.1.41). Only (3.1.45) is required. The observed accuracy is (3.1.42).
- Variable integration steps, scheme (3.1.40), (3.1.41). Both (3.1.45) and (3.1.46) are required. The observed accuracy is (3.1.43). Choice of k_η, k_τ, k_f, k_t, and $\underline{k}_\tau$ if sampling and integration steps do not coincide, depends on $D_2/L, \ldots, D_{n_d}/L$.
- Improved scheme (3.1.40), (3.1.44). Only (3.1.45) is required. The accuracy (3.1.43) is observed *for any* $\rho > 0$ (not only small).

All the above cases yield the same asymptotic accuracy in the case $n_d = 1$.

3.1.5.3 Discrete differentiator with variable Lipschitz parameter L

Let L be a variable function of t, and $|\dot{L}|/L \leq M$ hold for some M, then all the above schemes (3.1.40), (3.1.41), or (3.1.44) make sense [36]. The only difference is that each time when the requirement of the boundedness of $f_0^{(j)}$ appears in the constant-L case, it is replaced by the boundedness of $f_0^{(j)}(t)/L(t)$. Let $|\eta(t)| \leq L(t)\varepsilon$, $\rho = \max\,[\varepsilon^{\frac{1}{n_d+1}}, \tau]$, then the same accuracy (3.1.43) or (3.1.42) is obtained and kept for sufficiently small τ, ε. The globally convergent differentiator with variable L [35] features the same accuracy.

The above convergence criteria are literally true for such differentiators, but ρ and τ are required to be sufficiently small in all cases. Naturally, the parameter L appearing in the accuracies (3.1.42) or (3.1.43) depends on time. The corresponding result has meantime been published for the improved discretization scheme (3.1.40), (3.1.44) with coinciding integration and sampling steps [35].

3.1.5.4 Simulation results

Due to the large scope of the presented results, only a few illustrative examples are provided. Consider the input function

$$f_0 = \tfrac{1}{5}\sin(3t + 0.3) - \tfrac{2}{5}\sin(2.2t + 1.5) + \tfrac{1}{10}\sin(0.5t + 4.7), \tag{3.1.48}$$

which obviously has bounded derivatives. Assign $L = 4, 10, 26, 70$, and 200 for $n_d = 1, 2, 3, 4$, and 5, respectively. Choose the parameters $\lambda_0 = 1.1, \lambda_1 = 1.5, \lambda_2 = 3, \lambda_3 = 5, \lambda_4 = 8$, and $\lambda_5 = 12$ of differentiators. Recall that $\underline{\tau}$ and τ are, respectively, the maximal values of the integration steps $\underline{\tau}_{k,j}$ and the sampling steps τ_k; ε is the noise magnitude. Naturally, $\underline{\tau}_{k,j} \le \tau_k$, $\underline{\tau} \le \tau$ hold. The variable values of τ_k and $\underline{\tau}_{k,j}$ are calculated online as random numbers uniformly distributed in the range from 0 to the corresponding upper bound.

One of the main presented results is that the theoretical asymptotically optimal differentiation accuracy $z_i - f_0^{(i)} = O(\rho^{n_d+1-i})$, $\rho = \max\left(\varepsilon^{\frac{1}{n_d+1}}, \tau\right)$ of the continuous-time differentiator is restored by the Euler-integration discrete differentiator with variable integration and sampling steps, provided $\underline{\tau}$ is of the order of τ^{n_d} or higher.

Let $n_d = 5$, $\varepsilon = 0$ for simplicity. For $\tau = 0.01$, the ideal accuracy reclamation would require taking $\underline{\tau}$ proportional to 10^{-12} which is practically impossible. Instead fix a reasonable value $\underline{\tau} = 0.0001$ and gradually increase τ starting from $\tau = \underline{\tau}$ calculating the corresponding accuracies $\sup\left|z_i - f_0^{(i)}\right| = \left\|z_i - f_0^{(i)}\right\|_\infty$ on a sufficiently long steady-state time interval. One can expect that starting from some moment the accuracies obey the above standard asymptotics.

It is seen from Figure 3.1.1 that the standard asymptotics are restored for $\tau > \tau_c$, where the critical value $\tau_c \approx e^{-5} \approx 0.007$. It corresponds to $\underline{\tau} \le 6 \times 10^6\tau^5$. This relation is expected to hold for any τ. In particular, one needs to take $\underline{\tau} \le 0.0006$ for $\tau = 0.01$, which is still feasible, and $\underline{\tau} \le 6 \times 10^{-9}$ for $\tau = 0.001$, which is already practically impossible.

It is also seen from the graphs that the accuracies are solely defined by $\underline{\tau}$ if $\tau < \tau_c$. In other words, if $\underline{\tau} \ge 6 \times 10^6\tau^5$, too large integration steps influence the accuracy stronger than the sampling interval itself. According to this reasoning $\tau_c \approx 0.45\underline{\tau}^{1/5}$. Note that τ_c depends not only on the parameters of the differentiator, but also on the upper bounds of the input derivatives starting from the order 2.

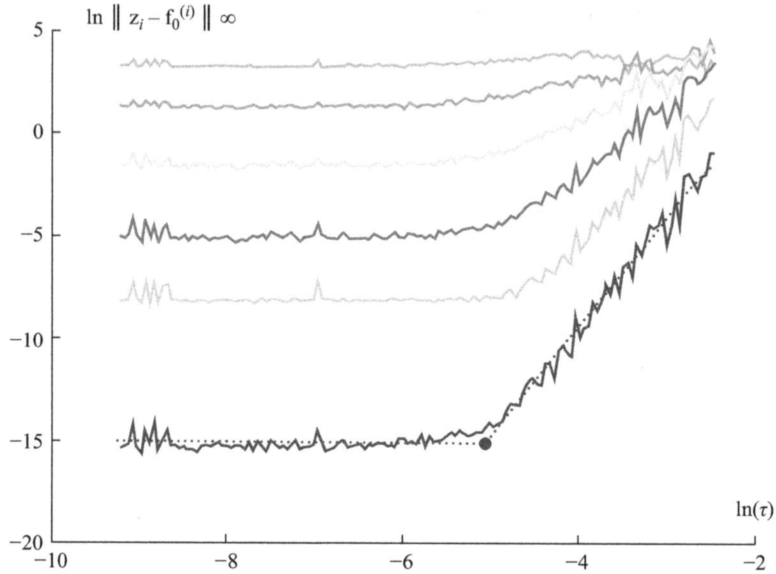

Figure 3.1.1 Logarithmic graphs of the fifth-order differentiator accuracies with $\underline{\tau} = 0.0001$*. The accuracies' lines correspond to the derivative orders 0, 1, 2, 3, 4, and 5 from the bottom to the top. Integration and sampling steps are variable*

In order to check the above reasoning, the simulation was carried out for differentiators of the orders 1, 2, 3, and 4 for the same input (3.1.48). The critical values τ_c are found as functions of $\underline{\tau}$. The hypothetical relation $\tau_c \approx \gamma_c \underline{\tau}^{1/n_d}$ is confirmed by the graphs in Figure 3.1.2. The higher the differentiation order the more difficult and less reliable is the detection of τ_c. Thus, there is no graph corresponding to $n_d = 5$.

3.1.6 Discretization of SMs

Implementation of HOSM controllers usually requires the control values to be calculated and fed to the system at discrete time instants. In the case, when the controller does not involve its own dynamics, in particular, if all needed derivatives are directly sampled, the resulting system is adequately described by variable sampling noises and delays, and by the analysis from Section 3.1.3.3.1.

The situation changes when the output feedback is applied, which incorporates a dynamic observer. Let the sampling take place at the time instants t_k, $0 < t_{k+1} - t_k = \tau_k \leq \tau$. As we have seen (Section 3.1.5), the differentiator accuracy deteriorates, if the differentiator is replaced by its Euler approximation. Therefore, one would expect that the accuracy of the output-feedback r-SM control also deteriorates. In fact, it is not the case due to the overall homogeneity of the closed-loop error dynamics.

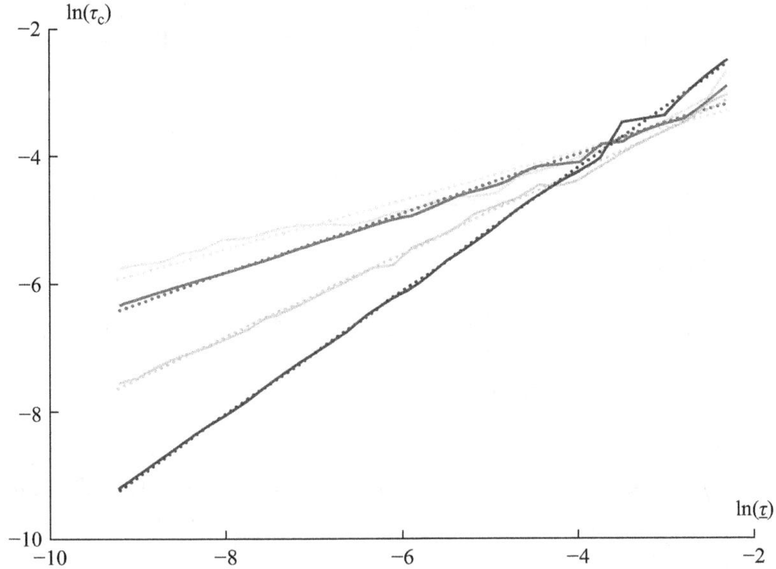

Figure 3.1.2 Logarithmic graphs of the critical sampling period τ_c vs. $\underline{\tau}$ for $n_d = 1, 2, 3, 4$ from the bottom to the top. Integration and sampling steps are variable

3.1.6.1 SISO case

Consider the system (3.1.3) of the relative degree r under the conditions (3.1.19), (3.1.20) with a bounded control (3.1.22). Let the output-feedback control (3.1.29) be applied, and the differentiator be replaced by its simplest Euler discretization. Let the sampling take place at the time instants t_k, $0 < t_{k+1} - t_k = \tau_k \le \tau$. The hybrid error dynamics satisfies the DI

$$\begin{aligned} &\sigma^{(r)} \in [-C, C] + [K_m, K_M]U_r(z(t_k)), \qquad t \in [t_k, t_{k+1}], \\ &z(t_{k+1}) = z(t_k) + \tau_k D_{r-1}\,(z(t_k), \sigma(t_k), L), \end{aligned} \tag{3.1.49}$$

where $L \ge C + K_M \sup |U_r|$. Let σ be sampled with the accuracy $\varepsilon \ge 0$. Applying the reasoning of Section 3.1.3.3.1 obtain that after an FT transient, the system features the asymptotic accuracies $\sigma^{(j)} = O\big(\max\big(\tau^{r-j}, \varepsilon^{(r-j)/r}\big)\big)$.

Consider now the chattering attenuation procedure based on the artificial increase of the relative degree. Differentiating (3.1.19) we obtain

$$\sigma^{(r+1)} = h_e(t, x, u) + g(t, x)\dot{u}. \tag{3.1.50}$$

Assume that the functions h_e and $h'_x b + g'_x bu$ are bounded in a vicinity of the $(r+1)$-SM $\sigma \equiv 0$,

$$|h_e(t, x, u)| \le C_e, \quad \big|h'_x(t,x)b(t,x) + g'_x(t,x)b(t,x)u\big| \le C_{1e}. \tag{3.1.51}$$

This assumption is natural, since u is close to $u_{\text{eq}} = -h/g$ in the vicinity of the $(r+1)$-SM, and is therefore bounded at least locally. The applied feedback control gets the form

$$\dot{u}(t) = U_{r+1}(z(t_k)), \qquad \dot{z} = D_r(z, \sigma, L_e), \tag{3.1.52}$$

where $z \in \mathbb{R}^{r+1}$, $L_e \geq C_e + K_M\alpha$, $\alpha = \sup |U_{r+1}|$. According to Section 3.1.4.1.4, the accuracy $\sigma^{(j)} = O\big(\max\big(\tau^{r+1-j}, \varepsilon^{(r+1-j)/(r+1)}\big)\big)$ is to be obtained.

Its discretization produces

$$u(t_{k+1}) = u(t_k) + \tau_k U_{r+1}(z(t_k)), \qquad z(t_{k+1}) = z(t_k) + \tau_k D_r(z(t_k), \sigma(t_k), L_e).$$

The following theorem summarizes these two cases.

Theorem 3.1.6 ([37]). *Discretization does not destroy the asymptotic closed-system accuracy in the SISO case under the standard conditions (3.1.20). Under additional conditions (3.1.51), the same is true for the chattering attenuation procedure with sufficiently small* τ.

Note that the first part of the theorem has actually been proved above. The second part is restricted to the case when only one integrator is inserted in the feedback and the relative degree increases from r to $r+1$. The result is probably true also for the introduction of any number of integrators, but the proof is still not available.

3.1.6.2 MIMO case

Like in the SISO case, we differentiate (3.1.31) and obtain

$$\sigma^{(r+(1,\ldots,1))} = h_e(t,x,u) + g(t,x)\dot{u},$$

where $h_e(t,x,u) \in \mathbb{R}^m$. Similarly, an additional assumption is needed. The functions h_e and $h'_x b + g'_x bu$ are bounded in a vicinity of the $(r+(1,\ldots,1))$-SM $\vec{\sigma} \equiv 0$,

$$\|h_e(t,x,u)\| \leq C_e, \quad \big\|h'_x(t,x)b(t,x) + g'_x(t,x)b(t,x)u\big\| \leq C_{1e}. \tag{3.1.53}$$

The theorem exactly analogous to Theorem 3.1.6 is also true in the MIMO case due to the effective decoupling (3.1.35) of the system. Mark that the proposed homogeneous SM control (3.1.34), (3.1.36), or (3.1.38) is applied here.

3.1.6.3 Example

Consider a simple second-order system

$$\begin{aligned} \dot{x}_0 &= \sin t + x_1, \\ \dot{x}_1 &= \cos x_0 + u, \qquad \sigma = x_0, \end{aligned}$$

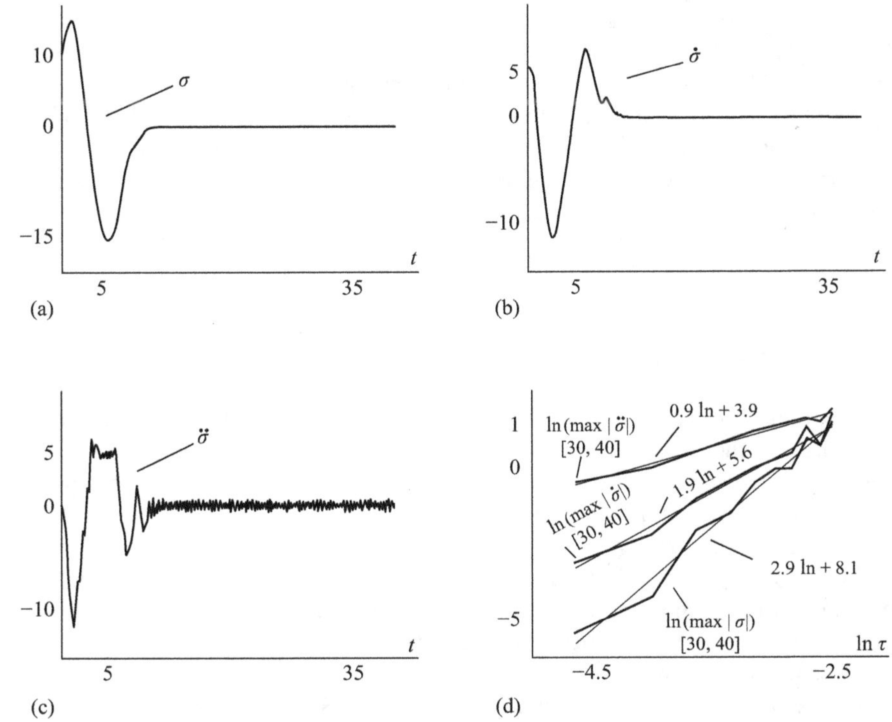

Figure 3.1.3 *(a)–(c) Stabilization of $\sigma, \dot{\sigma}, \ddot{\sigma}$ with $\tau = 0.01$; (d) asymptotics of $\sigma, \dot{\sigma}, \ddot{\sigma}$ for the maximal sampling steps $\tau = 0.01, 0.02, \ldots, 0.1$*

to be stabilized by continuous SM control. The solution is to apply the dynamic control $\dot{u} = -\alpha\Psi_{2,3}(z)$, $\dot{z} = D_2(z, \sigma, L)$. Its discretization yields

$$\begin{aligned}
\dot{x}_0 &= \sin t + x_1, \\
\dot{x}_1 &= \cos x_0 + u(t_k), \qquad \sigma = x_0, \\
u(t_{k+1}) &= \nu u(t_k) - \tau_k \alpha \Psi_{2,3}\left(z_0(t_k), z_1(t_k), z_2(t_k)\right), \\
z(t_{k+1}) &= z(t_k) + \tau_k D_2\left(z(t_k), \sigma(t_k), L\right),
\end{aligned} \tag{3.1.54}$$

where the standard embedded 3-SM controller is taken,

$$\Psi_{2,3}(z) = \operatorname{sign}\Big(z_2 + 2\left(|z_1|^3 + |z_0|^2\right)^{1/6} \operatorname{sign}\big(z_1 + |z_0|^{2/3} \operatorname{sign} z_0\big)\Big).$$

The continuous-time part of system (3.1.54) was integrated by the Euler method with the integration step 10^{-4} and initial values $x_0(0) = 10$, $x_1(0) = 5$. The discrete-time subsystem in (3.1.54) has the parameters $\lambda_0 = 1.1$, $\lambda_1 = 1.5$, $\lambda_2 = 2$, $L = 16$, $\alpha = 8$, and $z_0(0) = z_1(0) = z_2(0) = 0$.

Take random positive sampling steps τ_k, uniformly distributed in the segment $[10^{-4}, \tau]$. The stabilization of σ, $\dot{\sigma}$, $\ddot{\sigma}$ with $0 < \tau_k \leq \tau = 0.01$ is demonstrated in Figure 3.1.3(a)–(c), respectively. Now let the maximal sampling step τ take values $0.01, 0.02, \ldots, 0.1$. Logarithmic plots of $\max_{[30,40]} \left|\sigma^{(i)}\right|$, $i = 0, 1, 2$, together with the corresponding best-fitting lines $2.9 \ln \tau + 8.1$, $1.9 \ln \tau + 5.6$ and $0.9 \ln \tau + 3.9$ are shown in Figure 3.1.3(d). According to Theorem 3.1.6, the worst-case accuracy orders correspond to the slope values 3, 2, and 1, respectively. Thus, the simulation results are in good compliance with the theory.

3.1.7 Conclusions

The accuracy of disturbed homogeneous FT-stable DIs is analysed and the results are applied to the accuracy analysis of homogeneous SMC systems. The current knowledge on the accuracy and performance of dynamic systems closed by discretized dynamic homogeneous SM controllers is presented.

Different discretization schemes of homogeneous SM-based differentiators are considered, and their accuracy is analysed. For the first time, the internal numeric Euler integration is considered between the sampling instants, and the corresponding effect on the accuracy is studied. Differentiator convergence criteria are presented. All the results are extended to the case of the variable Lipschitz parameter.

Discretization of output-feedback homogeneous SM controllers is shown not to destroy the overall system accuracy in the presence of noises and discrete sampling. The same is true if one discrete Euler integrator is introduced in the feedback in order to effectively attenuate system chattering by the artificial increase of the relative degree. The results are true for both SISO and MIMO cases.

References

[1] V. Acary, B. Brogliato, and Y.V. Orlov. Chattering-free digital sliding-mode control with state observer and disturbance rejection. *IEEE Transactions on Automatic Control*, 57(5):1087–1101, 2012.

[2] M. Angulo, L. Fridman, and A. Levant. Robust exact finite-time output based control using high-order sliding modes. *International Journal of Systems Science*, 42(11):1847–1857, 2011.

[3] M.T. Angulo, J.A. Moreno, and L.M. Fridman. Robust exact uniformly convergent arbitrary order differentiator. *Automatica*, 49(8):2489–2495, 2013.

[4] A. Bacciotti and L. Rosier. *Liapunov Functions and Stability in Control Theory*. Springer-Verlag, London, 2005.

[5] G. Bartolini. Chattering phenomena in discontinuous control systems. *International Journal of Systems Science*, 20:2471–2481, 1989.

[6] G. Bartolini, A. Damiano, G. Gatto, I. Marongiu, A. Pisano, and E. Usai. Robust speed and torque estimation in electrical drives by second-order sliding

modes. *IEEE Transactions on Control Systems Technology*, 11(1):84–90, Jan. 2003.

[7] G. Bartolini, A. Ferrara, and E. Usai. Chattering avoidance by second-order sliding mode control. *IEEE Transactions on Automatic Control*, 43(2):241–246, Feb. 1998.

[8] G. Bartolini, A. Pisano, E. Punta, and E. Usai. A survey of applications of second-order sliding mode control to mechanical systems. *International Journal of Control*, 76(9/10):875–892, 2003.

[9] E. Bernuau, D. Efimov, W. Perruquetti, and A. Polyakov. On homogeneity and its application in sliding mode control. *Journal of the Franklin Institute*, 351(4):1866–1901, 2014.

[10] E. Bernuau, A. Polyakov, D. Efimov, and W. Perruquetti. Verification of ISS, iISS and IOSS properties applying weighted homogeneity. *Systems & Control Letters*, 62(12):1159–1167, 2013.

[11] I. Boiko and L. Fridman. Analysis of chattering in continuous sliding-mode controllers. *IEEE Transactions on Automatic Control*, 50(9):1442–1446, Sep. 2005.

[12] F.H. Clarke, Y.S. Ledayev, and R.J. Stern. Asymptotic stability and smooth Lyapunov functions. *Journal of Differential Equations*, 149(1):69–114, 1998.

[13] E. Cruz-Zavala, J.A. Moreno, and L. Fridman. Uniform robust exact differentiator. *IEEE Transactions on Automatic Control*, 56(11):2727–2733, 2012.

[14] M. Defoort, T. Floquet, A. Kokosy, and W. Perruquetti. A novel higher order sliding mode control scheme. *Systems & Control Letters*, 58(2):102–108, 2009.

[15] F. Dinuzzo and A. Ferrara. Higher order sliding mode controllers with optimal reaching. *IEEE Transactions on Automatic Control*, 54(9):2126–2136, 2009.

[16] C. Edwards and S.K. Spurgeon. *Sliding Mode Control: Theory and Applications*. Taylor & Francis, London, 1998.

[17] Y. Feng, X. Yu, and Z. Man. Non-singular terminal sliding mode control of rigid manipulators. *Automatica*, 38(12):2159–2167, 2002.

[18] A.F. Filippov. Differential equations with discontinuous right-hand sides. Kluwer Academic Publishers, Dordrecht, 1988.

[19] T. Floquet, J.P. Barbot, and W. Perruquetti. Higher-order sliding mode stabilization for a class of nonholonomic perturbed systems. *Automatica*, 39(6):1077–1083, 2003.

[20] L. Fridman. Chattering analysis in sliding mode systems with inertial sensors. *International Journal of Control*, 76(9/10):906–912, 2003.

[21] L. Fridman and A. Levant. Accuracy of homogeneous sliding modes in the presence of fast actuators. *IEEE Transactions on Automatic Control*, 55(3):810–814, 2010.

[22] K. Furuta and Y. Pan. Variable structure control with sliding sector. *Automatica*, 36(2):211–228, 2000.

[23] Y.G. Hong and Z.P. Jiang. Finite-time stabilization of nonlinear systems with parametric and dynamic uncertainties. *IEEE Transactions on Automatic Control*, 51(12):1950–1956, 2006.

[24] A. Isidori. *Nonlinear Control Systems*, 2nd ed. Springer-Verlag, New York, NY, 1989.

[25] M. Kawski. Homogeneous stabilizing feedback laws. *Control Theory and Advanced Technology*, 6:497–516, 1990.

[26] A.N. Kolmogoroff. On inequalities between upper bounds of consecutive derivatives of an arbitrary function defined on an infinite interval. *American Mathematical Society Translations, Series 1*, 2:233–242, 1962.

[27] A. Levant. Sliding order and sliding accuracy in sliding mode control. *International Journal of Control*, 58(6):1247–1263, 1993.

[28] A. Levant. Robust exact differentiation via sliding mode technique. *Automatica*, 34(3):379–384, 1998.

[29] A. Levant. Higher order sliding modes, differentiation and output-feedback control. *International Journal of Control*, 76(9/10):924–941, 2003.

[30] A. Levant. Homogeneity approach to high-order sliding mode design. *Automatica*, 41(5):823–830, 2005.

[31] A. Levant. Quasi-continuous high-order sliding-mode controllers. *IEEE Transactions on Automatic Control*, 50(11):1812–1816, 2005.

[32] A. Levant. Robustness of homogeneous sliding modes to relative degree fluctuations. In *Proceedings of the Sixth IFAC Symposium on Robust Control Design*, June 16–18, 2009, Haifa, Israel, Volume 6, Part 1, pages 167–172, 2009.

[33] A. Levant. Chattering analysis. *IEEE Transactions on Automatic Control*, 55(6):1380–1389, 2010.

[34] A. Levant. On fixed and finite time stability in sliding mode control. In *Proceedings of the 52nd IEEE Conference on Decision and Control*, Florence, Italy, December 10–13, 2013, 2013.

[35] A. Levant. Globally convergent fast exact differentiator with variable gains. In *Proceedings of the European Conference on Control*, Strasbourg, France, June 24–27, 2014, 2014.

[36] A. Levant and M. Livne. Exact differentiation of signals with unbounded higher derivatives. *IEEE Transactions on Automatic Control*, 57(4):1076–1080, 2012.

[37] A. Levant and M. Livne. Uncertain disturbances' attenuation by homogeneous MIMO sliding mode control and its discretization. *IET Control Theory & Applications*, 9(4):515–525, 2015.

[38] A. Levant and Y. Pavlov. Generalized homogeneous quasi-continuous controllers. *International Journal of Robust and Nonlinear Control*, 18(4–5):385–398, 2008.

[39] M. Livne and A. Levant. Accuracy of disturbed homogeneous sliding modes. In *Proceedings of the 13th International Workshop on Variable Structure Systems*, Nantes, France, June 29–July 2, 2014.

[40] M. Livne and A. Levant. Proper discretization of homogeneous differentiators. *Automatica*, 50:2007–2014, 2014.

[41] Z. Man, A.P. Paplinski, and H. Wu. A robust MIMO terminal sliding mode control scheme for rigid robotic manipulators. *IEEE Transactions on Automatic Control*, 39(12):2464–2469, 1994.

[42] J.A. Moreno and M. Osorio. Strict Lyapunov functions for the super-twisting algorithm. *IEEE Transactions on Automatic Control*, 57:1035–1040, 2012.

[43] Y. Orlov. Finite time stability of switched systems. *SIAM Journal of Control and Optimization*, 43(4):1253–1271, 2005.

[44] A. Pisano and E. Usai. Sliding mode control: A survey with applications in math. *Mathematics and Computers in Simulation*, 81(5):954–979, 2011.

[45] F. Plestan, A. Glumineau, and S. Laghrouche. A new algorithm for high-order sliding mode control. *International Journal of Robust and Nonlinear Control*, 18(4/5):441–453, 2008.

[46] A. Polyakov, D. Efimov, and W. Perruquetti. Finite-time and fixed-time stabilization: Implicit Lyapunov function approach. *Automatica*, 51:332–340, 2015.

[47] A. Polyakov and A. Poznyak. Lyapunov function design for finite-time convergence analysis: "Twisting" controller for second-order sliding mode realization. *Automatica*, 45(2):444–448, 2009.

[48] A. Polyakov and A. Poznyak. Unified Lyapunov function for a finite-time stability analysis of relay second-order sliding mode control systems. *IMA Journal of Mathematical Control and Information*, 29(4):529–550, 2012.

[49] Y. Shtessel, C. Edwards, L. Fridman, and A. Levant. *Sliding Mode Control and Observation*. Birkhäuser, Boston, MA, 2014.

[50] Y.B. Shtessel and I.A. Shkolnikov. Aeronautical and space vehicle control in dynamic sliding manifolds. *International Journal of Control*, 76(9/10):1000–1017, 2003.

[51] J.-J.E. Slotine and W. Li. *Applied Nonlinear Control*. Prentice-Hall, Englewood Cliffs, NJ, 1991.

[52] V.I. Utkin. *Sliding Modes in Control and Optimization*. Springer-Verlag, Berlin, 1992.

Chapter 3.2

Experimental results on implicit and explicit time-discretization of equivalent control-based sliding mode control

Olivier Huber, Bernard Brogliato, Vincent Acary, Ahcene Boubakir, Franck Plestan, and Bin Wang

Abstract

This chapter presents a set of experimental results concerning the sliding mode control of an electropneumatic system. The controller is implemented via a micro-processor as a discrete-time input. Three discrete-time control strategies are considered for the implementation of the discontinuous part of the sliding mode controller: explicit discretizations with and without saturation, and an implicit discretization (i.e., very easy to implement as a projection on the interval $[-1, 1]$). While the explicit implementation is known to generate numerical chattering, the implicit one is expected to significantly reduce chattering while keeping the accuracy. The experimental results reported in this work remarkably confirm that the implicit discrete-time sliding mode supersedes the explicit ones with several important features: chattering in the control input is almost eliminated (while the explicit and saturated controllers behave like high-frequency bang–bang inputs), the input magnitude depends only on the perturbation size and is "independent" of the controller gain and sampling time. On the contrary the explicit controller shows obvious chattering for all sampling times, its magnitude increases as the controller gain increases, and it does not reduce when the sampling period augments. The tracking errors are comparable for both methods, though the implicit method keeps the precision when the control gain increases, which is not the case for the explicit one. Introducing a saturation in the explicit controller does not allow to significantly improve the explicit controller behavior if one does not take care of the saturation width.

3.2.1 Introduction

Sliding-mode control has very attractive features like robustness and simplicity of implementation, with few gains to tune. Its main drawback is the existence of the so-called chattering phenomenon, which may be due to actuators limitations, unmodeled dynamics, or time-discretization. Several works recently focussed on the time-discretization effects, showing that an *explicit* implementation of either Euler of ZOH discretizations yields limit cycles [9, 10, 30, 32], while the *implicit* form suppresses, in theory, the numerical chattering [1, 2, 31] due to the time-discretization. The goal of this chapter is to show that the superiority of the implicit discretization is verified in practice. Note that even if in the following the sliding variable is always scalar (e.g., the sliding surface co-dimension is one), the implicit method works with an arbitrary number of sliding surfaces. This is in sharp contrast with the previous approaches. Before going further, let us briefly recall what is meant by explicit and implicit discrete-time sliding mode controllers.

3.2.1.1 Explicit versus implicit discrete sliding mode control

To illustrate the difference between the explicit and the implicit discretizations, we consider the scalar system $\dot{x}(t) = u(t) + d(t)$, with $u(t) \in -\text{sgn}(x(t))$, where $\text{sgn}(\cdot)$ is the set-valued signum function: $\text{sgn}(0) = [-1, 1]$, $\text{sgn}(x) = \{1\}$ if $x > 0$, and $\text{sgn}(x) = \{-1\}$ if $x < 0$. Let the disturbance $d(t)$ satisfy $|d(t)| \leq \delta < 1$ for some δ. Recall that using Filippov's mathematical framework of differential inclusions [8], one deduces that for any $x(0)$, the state $x(t)$ reaches the "sliding surface" $x = 0$ in a finite time t^*, and then $x(t) = 0$ for all $t \geq t^*$. Using terminology from the differential inclusions literature, $u(t)$ is defined from $\xi(t)$, a *selection* of $\text{sgn}(0)$ (the interval $[-1, 1]$) for $t \geq t^*$, and it satisfies $\xi(t) = u(t) = -d(t)$ after t^*. In a sense, the set-valued controller acts as a disturbance observer once the sliding mode is attained. It is clear that if one multiplies the signum by a gain $a > 0$, i.e., $u(t) \in -a\ \text{sgn}(x(t))$, then one still has $u(t) = -d(t)$ in the sliding phase after t^*. However, this time the value of the selection $\xi(t)$ inside the set-valued part of $\text{sgn}(x(t))$ is divided by a, i.e., $\xi(t) = -\frac{d(t)}{a}$.

Let us now consider the Euler discretization of this system. It reads $x_{k+1} = x_k + hu_k + hd_k$, where $f_k = f(t_k)$ for a function $f(\cdot)$; and $t_k = t_0 + kh$, $k \in \mathbb{N}$ are the sampling times; and $h > 0$ is the sampling period. In such a simple case, the Euler and ZOH discretizations are the same, except for the disturbance $d_k = \int_{t_k}^{t_{k+1}} d(t)\,\mathrm{d}t$ for the ZOH method. Our focus is on how to choose u_k. The explicit method yields $u_k \in -\text{sgn}(x_k)$, yielding the closed loop $x_{k+1} - x_k - hd_k \in -h\ \text{sgn}(x_k)$. As alluded to above, limit cycles exist and create oscillations around the sliding surface (here the origin), known as the *numerical chattering in the output*. One of the consequences is that the explicit controller keeps switching between the two values 1 and -1, and never attains any point in $(-1, 1)$. In particular the explicit controller cannot approximate the continuous-time selection $\xi(\cdot) = u(\cdot)$ when the system evolves close to the sliding surface. If a gain $a > 0$ premultiplies $u(\cdot)$, then the explicit controller switches between two discrete values a and $-a$, the switching frequency being inversely proportional to the sampling period: this is the *numerical chattering in the input*. It is noteworthy that the mere notion of a sliding surface does not exist in this case, since the discrete trajectories cannot attain the origin, and the controller cannot take values in the set-valued part equal to $(-1, 1)$. One then has to resort to so-called quasi-sliding surfaces [27].

The implicit method is implemented as follows. Since $d(t)$ is unknown, one first constructs a *nominal unperturbed system* with state $\tilde{x}_k$, from which the input is computed: $\tilde{x}_{k+1} = x_k + hu_k,\ u_k \in -\mathrm{sgn}(\tilde{x}_{k+1})$. This is a so-called *generalized equation* with unknown $\tilde{x}_{k+1}$. Its solution yields after few manipulations $u_k = \mathrm{proj}\left([-1,1]; -\frac{x_k}{h}\right)$, i.e., the projection on the interval $[-1,1]$ of $-\frac{x_k}{h}$, and is a causal input (not depending on any future values of the state). In such a simple case, the closed-loop expression of u_k is the same as the saturation function with a width h^{-1}. However, in case of sliding surfaces of higher co-dimension, such an analogy with the well-known saturated sign function, is not trivial. Notice that in the unperturbed case, $\tilde{x}_k$ and x_k are the same. As proved in References 1 and 2, the implicit controller guarantees convergence of $\tilde{x}_k$ to the origin in a finite number of steps (the chattering is suppressed), and a disturbance attenuation by a factor h during the sliding mode (defined here from the fact that $u_k \in (-1,1)$). Most importantly, the control input takes values in $(-1,1)$ once $\tilde{x}_k$ has reached the origin, as may be seen from the generalized equation from which it is calculated, and one has during that phase $u_k = -d_k$: u_k is a selection τ_{k+1} of the discrete-time differential inclusion $\tilde{x}_{k+1} = x_k + hu_k,\ u_k \in -\mathrm{sgn}(\tilde{x}_{k+1})$, and the discrete-time input observes the disturbance when the sliding mode is attained. Similarly to the continuous-time case, if the controller is multiplied by a gain $a > 0$, then the selection $\tau_k = \frac{-d_k}{a}$. To conclude this chapter, let us reproduce here one of the results in Reference 2. Let us start by considering the following scalar sliding mode system:

$$\begin{cases} \dot{x}(t) = -a\tau(t) + d(t), \\ \tau(t) \in \mathrm{sgn}(x(t)), \end{cases} \tag{3.2.1}$$

where $d(\cdot)$ is a Lebesgue measurable perturbation such that $\|d\|_\infty < \rho < a$. The control input is here $u(t) = \tau(t)$. The discrete-time sliding mode system is implemented as follows:

$$\begin{cases} \tilde{x}_{k+1} = x_k - ah\tau_{k+1}, \\ \tau_{k+1} \in \mathrm{sgn}(\tilde{x}_{k+1}), \\ x_{k+1} = x_k - ah\tau_{k+1} + hd_k. \end{cases} \tag{3.2.2}$$

The first two lines of (3.2.9) may be considered as the *nominal unperturbed* plant, from which one computes the input at time t_k. The input is said implicit since it involves $\tilde{x}_{k+1}$ in the sign multifunction. It is however a causal input as shown next, and $\tilde{x}_{k+1}$ is just an intermediate variable which does not explicitly enter into the controller. The third line is the Euler approximation of the plant, on which the disturbance is acting. One has $u(t) = \tau_{k+1}$ on the time-interval $[t_k, t_{k+1}]$.

Proposition 3.2.1. *Let x_0 be the given initial state. Then, after a finite number of steps k_0, one obtains that $\tilde{x}_k = 0$ and $x_k = hd_{k-1}$, for all $k > k_0 > 0$. In other words, the disturbance is attenuated by a factor h. Moreover the approximated derivative of the state satisfies $\frac{x_{k+1}-x_k}{h} = d_k - d_{k-1}$ for all $k > k_0 + 1$, whereas $\frac{\tilde{x}_{k+1}-\tilde{x}_k}{h} = 0$ for all $k > k_0$. The control input takes values inside the sign multifunction multivalued part on the sliding surface for all $k > k_0$.*

Proof: Let us start with the case $|x_0| > ah > 0$. The generalized equation $\tilde{x}_{k+1} = x_k - ah\tau_{k+1}$ and $\tau_{k+1} \in \text{sgn}(\tilde{x}_{k+1})$ is equivalent, using the material in the Appendix 1, to the inclusion $\tau_{k+1} - \frac{x_k}{ah} \in -N_{[-1,1]}(\tau_{k+1})$, and to $\tau_{k+1} = \text{proj}\left([-1,1]; \frac{x_k}{ah}\right)$. Thus one obtains

- If $x_k > ah$ then $\tilde{x}_{k+1} = x_k - ah$ and $\text{sgn}(\tilde{x}_{k+1}) = 1$,
- If $x_k < -ah$ then $\tilde{x}_{k+1} = x_k + ah$ and $\text{sgn}(\tilde{x}_{k+1}) = -1$,
- If $x_k \in [-ah, ah]$ then $\tilde{x}_{k+1} = 0$ and $\text{sgn}(\tilde{x}_{k+1}) \in [-1, 1]$.[1]

From the above, we infer that

- If $x_k > ah$ then $x_{k+1} = x_k + hd_k - ah = x_k + h(d_k - a) < x_k + h(\rho - a)$. Since $\rho - a < 0$ the state is strictly decreased from step k to step $k+1$.
- If $x_k < -ah$ then $x_{k+1} = x_k + hd_k + ah = x_k + h(d_k + a) > x_k + h(a - \rho)$. Since $a - \rho > 0$ the state is strictly increased from step k to step $k+1$.

One deduces that if the initial data satisfy $|x_0| > ah$, then after $k_0 = \left\lceil \frac{x_0}{h|a-\rho|} \right\rceil$ Step 1 gets $\tilde{x}_{k_0} = 0$, where $\lceil \cdot \rceil$ is the ceiling function. Indeed at k_0 the state x_k reaches the interval $[-ah, ah]$ and then the unique solution for $\tilde{x}_k$ is zero. From $\tilde{x}_{k_0} = 0$ one deduces that $|x_{k_0}| < ah$. In the case that $|x_0| \leq ah$, it is easy to check that $k_0 = 1$. Indeed, one has to solve the generalized equation

$$\begin{cases} \tilde{x}_{k_0+1} = x_{k_0} - ah\tau_{k_0+1}, \\ \tau_{k_0+1} \in \text{sgn}(\tilde{x}_{k_0+1}), \end{cases} \tag{3.2.3}$$

whose unique solution is found by inspection to be $\tilde{x}_{k_0+1} = 0$.[2] The reasoning can be repeated to conclude that $\tilde{x}_k = 0$, for all $k \geq k_0$. Therefore, $\frac{\tilde{x}_{k+1} - \tilde{x}_k}{h} = 0$ for all $k > k_0$. Now let us assume that for $k \geq k_0$, we have

$$\tilde{x}_{k+1} = x_k - ah\tau_{k+1} = 0, \quad k \geq k_0, \tag{3.2.4}$$

i.e., $\tau_{k+1} = \frac{x_k}{ha}$. In this case, the state x_{k+1} is given by $x_{k+1} = hd_k$, and therefore $x_k = hd_k$, $\tau_{k+1} = \frac{d_k}{a}$ for all $k \geq k_0 + 1$, so that $\frac{x_{k+1} - x_k}{h} = d_k - d_{k-1}$, for all $k > k_0 + 1$.

Remark 3.2.1. *The implicit discretization of set-valued sign controllers has been independently introduced in Reference 18 with the so-called proxy-based sliding mode control. The authors noted that the implicit Euler discretization allows one to obtain a perfect (at the machine precision) vanishing of the sliding variable. Related work is also done in Reference 12, which however only applies to a simple*

[1] This replaces the third and fourth items in the proof of Proposition 1 in Reference 2, which contains an unfortunate error.

[2] The underlying crucial property that makes this hold is the maximal monotonicity of the sign multifunction.

scalar case. These methods have their roots in the numerical analysis and simulation of mechanical systems with unilateral constraints, impacts, and Coulomb friction [21, 22] and of linear complementarity systems for switched circuits with ideal diodes [4, 5, 20].

Therefore, the implicit controller has the same features as its continuous-time counterpart. We may summarize them as follows:

1. When there is no perturbation, the sliding surface is reached after a finite number of steps and there is no chattering.
2. When a perturbation acts on the system, the state of the nominal system reaches the sliding surface after a finite number of steps, while the perturbation effect is attenuated by a factor h on the state of the system.
3. Despite the system's state, x_k never attains its sliding surface due to the disturbance, the notion of discrete-time sliding mode *does* exist, and corresponds to the nominal system's state $\tilde{x}_k$ vanishing or equivalently to the set-valued controller evolving strictly inside the interval $[-1, 1]$. In this mode the controller compensates for the disturbance, with a delay due to the discretization. Therefore, its magnitude is independent, in the sliding phase, of the controller gain, and there is no need to adapt the gain (denoted as a above, and as G in the sequel) online.
4. Theoretically there is no numerical chattering during the sliding mode, neither in the sliding variable nor in the input.
5. The discrete-time controller keeps the simplicity of its continuous-time counterpart, with no added gain to tune.
6. Computing the input at each step boils down to solving a simple generalized equation, equivalently a projection on $[-1, 1]$ in the codimension 1 sliding surface case. With more sliding surfaces, we can either compute the control input by enumeration (for a co-dimension up to 3), or solve an optimization problem like a quadratic program. This is quite easy to implement in a code.
7. The implicit discretization allows to prove Lyapunov stability of the sliding variable dynamics, mimicking the continuous-time Lyapunov functions [15, 16].

The main objective of this work is to confirm these features experimentally, while they have been analyzed and numerically illustrated in References 1, 2 and 13–17. Preliminary experimental results were presented in Reference 31.

The chapter is organized as follows: in Section 3.2.2, the dynamics and the various controllers of the electropneumatic actuators are detailed. Section 3.2.3 is dedicated to the experimental results: the explicit and the implicit discrete-time algorithms are applied to the system and compared in terms of their overall performance, comprising the tracking accuracy, the input chattering, the input magnitude, and the disturbance rejection, when the controller gain and the sampling period are varied. In addition the saturated explicit controller is also tested.

3.2.2 Dynamics of the plant and controllers

3.2.2.1 Implicit controller design

To start, let us explain in detail how the so-called implicit controller (which might be also named the *projected sliding-mode controller*) is calculated in case of tracking of a reference output $y_d(t)$ (only the regulation case has been described above). We consider the same scalar system as in the Introduction, i.e., $\dot{x}(t) = u$, $y = x$, and first we disregard the disturbance for the sake of simplicity. In this case, the Euler and ZOH discretizations are the same. Let the sliding variable be $\sigma = x - x_d$. The controller is set to $u(x, t) \in -\text{sgn}(\sigma) + \dot{x}_d(t)$, so that the closed-loop system is $\dot{\sigma}(t) \in -\text{sgn}(\sigma(t))$. The plant discretization is

$$\frac{x_{k+1} - x_k}{h} = u_k \tag{3.2.5}$$

and the implicit input is set equal to

$$u_k \in -\text{sgn}(x_{k+1} - x_{d,k+1}) + \frac{x_{d,k+1} - x_{d,k}}{h} \tag{3.2.6}$$

where the last term accounts for the Euler approximation of $\dot{x}_d(t)$. The discrete-time sliding variable is given by $\sigma_k = x_k - x_{d,k}$. We therefore obtain

$$x_{k+1} - x_k \in -h \ \text{sgn}(\sigma_{k+1}) + x_{d,k+1} - x_{d,k} \Leftrightarrow \sigma_{k+1} - \sigma_k \in -h \ \text{sgn}(\sigma_{k+1}). \tag{3.2.7}$$

Let $w_{k+1} \stackrel{\Delta}{=} \sigma_{k+1} - \sigma_k$. Using convex analysis, we may write $w_{k+1} \in -h \ \text{sgn}(\sigma_{k+1}) \Leftrightarrow \sigma_{k+1} \in -N_{[-h,h]}(w_{k+1})$, where $N_{[-h,h]}(w_{k+1})$ is the normal cone to $[-h, h]$ calculated at $w_{k+1} \in [-h, h]$, given in this case by

$$N_{[-h,h]}(w_{k+1}) = \begin{cases} 0 & \text{if} \quad |w_{k+1}| < h \\ [0, +\infty) & \text{if} \quad w_{k+1} = h \\ (-\infty, 0] & \text{if} \quad w_{k+1} = -h. \end{cases} \tag{3.2.8}$$

Inserting this in (3.2.7) yields

$$w_{k+1} + \sigma_k \in -N_{[-h,h]}(w_{k+1}) \Leftrightarrow -N_{[-h,h]}(w_{k+1}) - \sigma_k \ni w_{k+1}. \tag{3.2.9}$$

By using basic convex analysis, one finds equivalently

$$w_{k+1} = \text{proj}([-h, h]; -\sigma_k) = h \ \text{proj}\left([-1, 1]; -\frac{\sigma_k}{h}\right) \tag{3.2.10}$$

where proj denotes the orthogonal projection (details on the above manipulations can be found in Appendix A of Reference 3, and in Appendix 1). From (3.2.5), we have

$$u_k = \frac{1}{h} w_{k+1} + \frac{x_{d,k+1} - x_{d,k}}{h}.$$

Thus we obtain

$$u_k = \frac{1}{h}\text{proj}([-h,h];-\sigma_k) + \frac{x_{d,k+1}-x_{d,k}}{h} = \text{proj}\left([-1,1];-\frac{\sigma_k}{h}\right) + \frac{x_{d,k+1}-x_{d,k}}{h}$$

$$= \begin{cases} -\frac{\sigma_k}{h} + \frac{x_{d,k+1}-x_{d,k}}{h} & \text{if} \quad |\sigma_k| \le h \\ -1 + \frac{x_{d,k+1}-x_{d,k}}{h} & \text{if} \quad \sigma_k > h \\ 1 + \frac{x_{d,k+1}-x_{d,k}}{h} & \text{if} \quad \sigma_k < -h. \end{cases} \tag{3.2.11}$$

The implicit controller is thus bounded whatever the value of the sampling period $h > 0$. It is obviously quite easy to implement in a code. When a disturbance acts on the system $\dot{x}(t) = u + d(t)$, $d(t) \le \delta < 1$ for some known δ, then a similar procedure applied on a nominal unperturbed system yields the generalized equation:

$$\begin{cases} \tilde{\sigma}_{k+1} = \sigma_k + hu_k - x_{d,k+1} + x_{d,k} \\ u_k \in -\text{sgn}(\tilde{\sigma}_{k+1}) + \frac{x_{d,k+1}-x_{d,k}}{h} \end{cases} \tag{3.2.12}$$

from which the controller is calculated again to be $u_k = \text{proj}\left([-1,1];-\frac{\sigma_k}{h}\right) + \frac{x_{d,k+1}-x_{d,k}}{h}$. Following the developments in Reference 2 briefly recalled in the Introduction, such a controller guarantees interesting properties of the closed-loop system like disturbance attenuation by a factor h, and *convergence to the nominal state sliding surface* $\tilde{\sigma}_k = 0$ *in a finite number of steps*. Moreover, it also has the features listed in 1 to 7 above, and this will be confirmed by the experimental results given in Section 3.2.3. It is noteworthy that an explicit implementation of the input yields

$$u_k = -\text{sgn}(\sigma_k) + \frac{x_{d,k+1} - x_{d,k}}{h}. \tag{3.2.13}$$

It is not necessary to write an inclusion $u_k \in -\text{sgn}(\sigma_k) + \frac{x_{d,k+1}-x_{d,k}}{h}$ in (3.2.13), because the *multivalued part of the sign function cannot be realized with an explicit controller*. Indeed the fact is not only that the zero value does not exist numerically but even if it did, one would also not be able to choose in a unique way the controller value inside $[-1, 1]$ (in numerical analysis of differential inclusions, this is known as the *selection procedure* [3, §9.2, 9.4]). On the contrary the implicit implementation *does* realize the set-valuedness of the input. Moreover the computed controller value is the unique selection of the discrete-time inclusion, as a result of solving the above generalized equation.

Remark 3.2.2 (Chattering). *The oscillations around the sliding surface, which are solely due to the time discretization, are known as the* output numerical chattering. *They have been analysed with explicit discretizations in References 9, 10, 30 and 32. The input numerical chattering is focussed on in this work.*

Remark 3.2.3 (Controller commutations). *It is widely accepted in the control literature that sliding mode controllers have to be implemented through infinite-frequency* *commutations of some actuator and that the infinitely fast switching strategy is necessary to approximate the continuous-time solution obtained from Filippov's mathematical framework. This is false when one considers the implicit implementation which is briefly summarized above. The implicit implementation takes the form of a projection onto a finite interval (a hypercube if there is more than one switching surface) as shown in (3.2.11), hence it is a Lipschitz continuous function of the state. As it will be shown in Section 3.2.3, the* input numerical chattering *is drastically reduced when the implicit controller is used. It is quite possible that the input shape may be in practice as important as the output shape, because input chattering is known to demand a lot of the actuators.*

Remark 3.2.4 (Sliding surface with codimension $m \geq 2$). *The above implicit discretization method extends to the case when more than one attractive sliding surface is designed: then one computes the controller by solving a generalized equation that extends (3.2.12) (see equation 15 in Reference 2). Its well-posedness relies on CB (or a discretized version of it, if the zero-order-hold method is used [2, 15, 17]) being a P-matrix. However, CB need not be diagonal:* couplings between the sliding surfaces are allowed. *It also extends to the twisting controller [14, 15, 17], and a preliminary analysis for the super-twisting algorithm can be found in Lemmas 5 and 6 of Reference 2. As shown above, in case $m = 1$, the implicit discretization gives a saturated discrete-time input, with a specific saturation width. In higher codimensions, such an interpretation is not trivial; however, one may say that the implicit discretization provides a suitable regularization of the controller which allows to almost suppress chattering, to guarantee some Lyapunov stability as well as finite-time convergence properties [16]: this proves that it is a correct approximation of the set-valued continuous-time controller.*

3.2.2.2 Plant dynamics and controllers

The electropneumatic system used for the controllers evaluation consists of two actuators which are controlled by two servodistributors (Figure 3.2.1). Each actuator is composed by two chambers denoted by P (positive) and N (negative). The controllers proposed in the sequel are designed in order to control the position of one of these two actuators, named "Main actuator," whereas the second actuator, named "Perturbation actuator" and mechanically connected to the Main one, is used in order to produce an external perturbation force. With a nominal 7-bar source pressure, the maximum produced force is 2720 N; furthermore, both actuators have the same physical features: piston diameter is 80 mm and rod diameter is 25 mm. The external perturbation force controller is not under interest in this chapter and has been designed and tuned by Sitia Co. (http://www.sitia.fr/) which has built the experimental set-up. The air mass flow rates q_m entering in the chambers are modulated by two three-way servodistributors. The pneumatic jack horizontally moves a load carriage of mass M. Under some assumptions detailed in Reference 26, the dynamic model of the pneumatic actuator can be written as a nonlinear system which is affine in the control input $[u_P\ u_N]^T$, u_P (respectively u_N) being the control input of the servodistributor connected to the

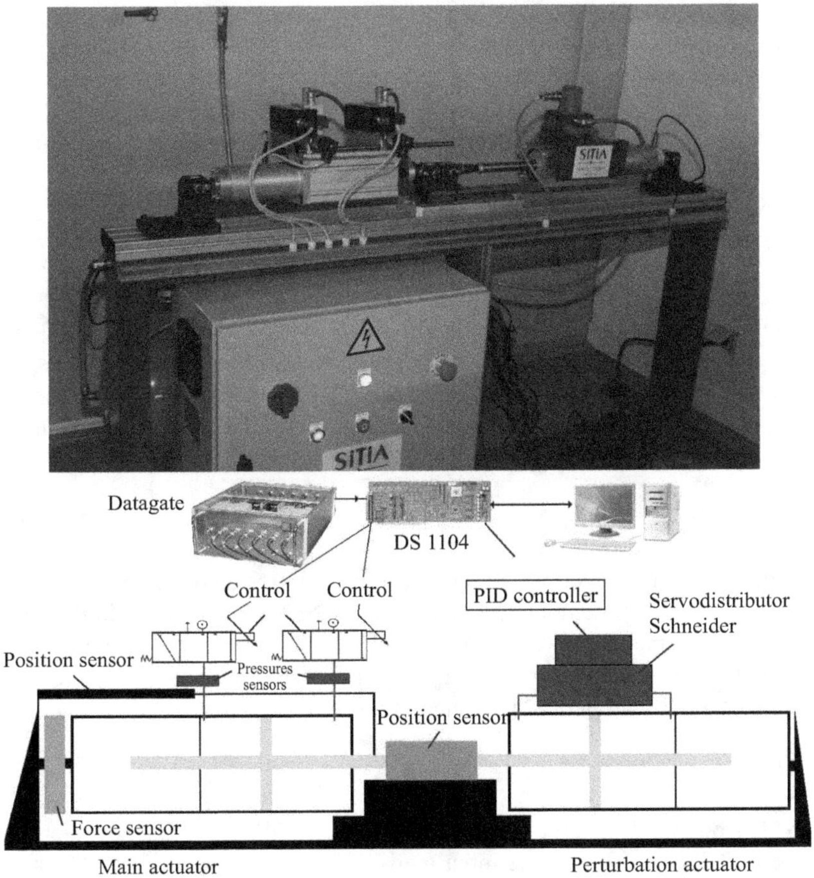

Figure 3.2.1 Photography and scheme of the electropneumatic system [26]

P (respectively N) chamber. The model is divided in two parts: two first equations concern the pressure dynamics in each chamber, whereas the motion of the actuator is described by the two last equations. Then the model of the electropneumatic experimental set-up reads as

$$\begin{cases} \dot{p}_P &= \dfrac{krT}{V_P(y)}\left[\varphi_P + \psi_P \cdot u_P - \dfrac{S}{rT}p_P v\right], \\ \dot{p}_N &= \dfrac{krT}{V_N(y)}\left[\varphi_N + \psi_N \cdot u_N + \dfrac{S}{rT}p_N v\right], \\ \dot{v} &= \dfrac{1}{M}[S\,(p_P - p_N) - b_v v - F], \\ \dot{y} &= v \end{cases} \tag{3.2.14}$$

with p_P (respectively p_N) the pressure in the P (respectively N) chamber, y and v being the position and velocity of the actuator. The force F is a disturbance that takes into account dry friction and unknown external forces. Note that the previous system appears to have two control inputs given that there is one servo distributor connected to each chamber. In the sequel, only the main actuator position is controlled: given that there is a single control objective, one states[3]:

$$u = u_P = -u_N.$$

The constant k is the polytropic constant, r is the ideal gas constant, T is the temperature which is supposed the same inside or outside the chambers and b_v is the viscous friction constant. V_P and V_N are the volumes in both chambers. These volumes depend on the actuator position y. S is the piston section and is constant. Finally, φ_X and ψ_X (X being P or N) are both fifth order polynomial functions versus p_X [25] and allow to model the mass flow rate q_X in the chamber X such that

$$q_X = \varphi_X(p_X) + \psi_X(p_X)u_X. \tag{3.2.15}$$

This kind of system is uncertain: in fact, the uncertainties on the polytropic constant, on the mass flow, on the temperature, on the mass, on the viscous friction coefficient, and on the disturbance force can be modeled by additive bounded functions added to the nominal part of each parameter [11]. As an example, the mass M can be viewed as the sum of a nominal part and an uncertain one

$$M = M_n + \Delta M,$$

with ΔM a bounded uncertainty and M_n the nominal value. By considering that the system (3.2.14) with a single input reads as $\dot{x} = f(x) + g(x)u$ with $f(\cdot)$ and $g(\cdot)$ uncertain vector fields defined as

$$f(x) = \begin{bmatrix} \dfrac{krT}{V_P(y)}\left[\varphi_P - \dfrac{S}{rT}p_P v\right] \\ \dfrac{krT}{V_N(y)}\left[\varphi_N + \dfrac{S}{rT}p_N v\right] \\ \dfrac{1}{M}[S\,(p_P - p_N) - b_v v - F] \\ v \end{bmatrix}, \quad g(x) = \begin{bmatrix} \dfrac{krT}{V_P(y)}\psi_P \\ -\dfrac{krT}{V_N(y)}\psi_N \\ 0 \\ 0 \end{bmatrix}, \tag{3.2.16}$$

it may be rewritten as

$$\dot{x} = (f_n + \Delta f)(x) + (g_n + \Delta g)(x)\, u \tag{3.2.17}$$

[3] Multivariable control can be designed [11], in case of position and pressure (in a chamber) control; an advantage of control pressure is that the rigidity of the actuator is improved.

with $f_n(\cdot)$, $g_n(\cdot)$ being the *nominal* dynamics parts of $f(\cdot)$ and $g(\cdot)$, and Δf, Δg the *uncertainties* and *perturbations*. Due to the presence of uncertainties and perturbations, a robust controller is required in order to get high performances (e.g., in terms of accuracy). The choice of sliding mode controller [28, 29] has been made because of its intrinsic features of robustness. Let us define the so-called *sliding variable* as

$$\sigma(x,t) = \ddot{e} + \lambda_1 \dot{e} + \lambda_0 e \tag{3.2.18}$$

with $e \triangleq y - y_d(t)$, $y_d(t)$ being the desired trajectory, supposed to be sufficiently differentiable. The coefficients λ_1, λ_0 are defined such that, given z a complex variable, the polynomial $Q(z) = z^2 + \lambda_1 z + \lambda_0$ is Hurwitz. The first and second derivatives of e are computed by direct numerical differentiation with appropriate first-order filters (see **Remark 3.2.6** on the influence of those filters on the closed-loop behavior). The idea of the continuous-time sliding mode controller is the following: the control ensures, in spite of uncertainties and perturbations and thanks to a discontinuous term, the finite time convergence to the so-called *sliding surface* $\sigma = 0$ (if the controller is well-tuned). Once the system trajectories have reached this domain, they are evolving on it and the closed-loop system dynamics is governed by the definition of σ, i.e., when $\sigma = 0$, one has $\ddot{e} = -\lambda_1 \dot{e} - \lambda_0 e$, which ensures exponential convergence to $(e, \dot{e}) = (0, 0)$. Note that once $\sigma = 0$, this convergence is not influenced by the uncertainties or perturbations. One gets

$$\begin{aligned}
\dot{\sigma} &= e^{(3)} + \lambda_1 \ddot{e} + \lambda_0 \dot{e} \\
&= \frac{1}{M}\left[S(\dot{p}_P - \dot{p}_N) - b_v \dot{v} - \dot{F}\right] - y_d^{(3)}(t) + \frac{\lambda_1}{M}\left[S(p_P - p_N) - b_v v - F\right] \\
&\quad - \lambda_1 \ddot{y}_d(t) + \lambda_0\left(\dot{y} - \dot{y}_d(t)\right),
\end{aligned} \tag{3.2.19}$$

where we assumed that the disturbance $F(\cdot)$ is differentiable, for simplicity (rigorously, dry friction may introduce some nondifferentiability at zero relative tangential velocity, depending on the used model). As shown in References 11 and 19 and given the system (3.2.17), the first time derivative of σ in (3.2.19) can be written as

$$\begin{aligned}
\dot{\sigma} &= \Psi(x,t) + \Phi(x) u \\
&= \Psi_n(x,t) + \Delta\Psi(t) + \left[\Phi_n(x) + \Delta\Phi(t)\right] u
\end{aligned} \tag{3.2.20}$$

such that Ψ_n, Φ_n are the nominal functions and $\Delta\Psi$, $\Delta\Phi$ are the uncertain terms. From References 11 and 19, the functions Ψ and Φ are bounded in the physical working domain (which gives that the uncertain terms are also bounded). Furthermore, one supposes that $\Delta\Phi$ is sufficiently small with respect to Φ_n to ensure that $1 + \frac{\Delta\Phi}{\Phi_n} > 0$. From a practical point of view, this assumption is not too strong: it simply means that the uncertainties are small compared to the nominal values. Let us consider the control law[4]:

$$u = \frac{1}{\Phi_n}[-\Psi_n + v]. \tag{3.2.21}$$

By applying (3.2.21) in (3.2.20), one gets

$$\dot{\sigma} = \frac{\Delta\Phi}{\Phi_n}\Psi_n + \Delta\Psi + \left[1 + \frac{\Delta\Phi}{\Phi_n}\right]v. \tag{3.2.22}$$

The controller v is a set-valued input defined as

$$v \in -G\text{sgn}(\sigma) \tag{3.2.23}$$

with G tuned sufficiently large[5] to ensure the sliding condition [28, 29] $\sigma\dot{\sigma} \leq -\eta|\sigma|$ ($\eta > 0$). Each controller has been implemented under its discrete forms as follows (with $k \geq 0$, $\sigma_k = \sigma(kh)$, h being the sampling period)

- Explicit sliding mode control (with sgn$(\cdot)$ function)

$$v_k = -G\text{sgn}(\sigma_k), \tag{3.2.24}$$

- Explicit saturated sliding mode control (with sat$(\cdot)$ function)

$$v_k = -G\text{sat}(\sigma_k, \varepsilon), \tag{3.2.25}$$

with

$$\text{sat}(\sigma_k, \varepsilon) = \begin{cases} \text{sgn}(\sigma_k) \text{ if } |\sigma_k| \geq \varepsilon \\ \sigma_k \text{ if } |\sigma_k| < \varepsilon. \end{cases} \tag{3.2.26}$$

- Implicit sliding mode control (with sgn$(\cdot)$ multifunction)

$$v_k \in -G\text{sgn}(\sigma_{k+1}) \tag{3.2.27}$$

(implemented with a projection as indicated in Section 3.2.2.1).

[4] As shown in Reference 6, such a control law allows to reduce the magnitude of the sliding mode controller by using the nominal information in the controller.

[5] Following the sliding condition, the gain has to be tuned as $G > \dfrac{\max\left|\frac{\Delta\Phi}{\Phi_n}\Psi_n + \Delta\Psi\right| + \eta}{\min\left[1 + \frac{\Delta\Phi}{\Phi_n}\right]}$. By a similar way than Reference 23, it can be shown that over the trajectories and in the working domain, the term $\frac{\Delta\Phi}{\Phi_n}\Psi_n + \Delta\Psi$ is bounded, whenever $1 + \frac{\Delta\Phi}{\Phi_n} > 0$.

Table 3.2.1 Comparisons of position error e, when G = 10⁴

h	**2 ms**	**5 ms**	**10 ms**	**15 ms**
(a) Explicit control				
Avg(abs(e))	0.32601	0.4791	0.97802	3.8759
Range of e	−2.0701 to 2.0189	−3.0067 to 2.1572	−3.2599 to 4.4023	−10.3767 to 12.1426
Standard deviation of e	0.4274	0.6327	1.0366	4.2553
(b) Saturation control				
Avg(abs(e))	0.28053	0.51399	0.99017	3.6119
Range of e	−1.7255 to 1.1288	−1.7006 to 2.4793	−4.5846 to 2.6004	−14.4069 to 12.1394
Standard deviation of e	0.3319	0.6132	1.1394	4.4131
(c) Implicit control				
Avg(abs(e))	0.71254	1.7138	3.3861	5.1387
Range of e	−1.6760 to 1.2200	−3.4182 to 3.2213	−7.9230 to 6.4083	−9.4997 to 6.5194
Standard deviation of e	0.7731	1.8780	3.7182	5.4749

3.2.3 Experimental results

This section is devoted to analyze the experimental data. The controllers have been implemented with three feedback gains $G = 10^4$, $G = 10^5$, $G = 10^6$ and five sampling times 1, 2, 5, 10, and 15 ms. The length of the interval of study is 20 seconds. The saturation input has been tested for six different values of the saturation width, with the sampling time $h = 1$ ms. In the data reported below, the unitless width of the saturation is $\varepsilon = 0.1$ (the other widths which have been tested yielded similar results and the results obtained with them are therefore omitted). The comparisons are mainly made with respect to the magnitude and chattering of the inputs u and v, and the tracking error e.

3.2.3.1 Comparison of the tracking errors e

Data in Tables 3.2.1–3.2.3 characterize the position tracking error e obtained by the three different implementation methods, from the aspects of average, range, standard deviation, and variation with four different sampling periods. The symbol Avg denotes the average of the tracking error over the duration of the test, abs is the absolute value of tracking error. The total variation of a real-valued function $f(\cdot)$ defined on an interval $[a, b] \subset R$ is approximated by the quantity

$$Var_{[a,b]}(f) = \sum_{i=0}^{N-1} |f(t_{i+1}) - f(t_i)|, \tag{3.2.28}$$

Table 3.2.2 Comparisons of position error e when G = 10⁵

h	2 ms	5 ms	10 ms	15 ms
(a) Explicit control				
Avg(abs(e))	0.98336	1.1363	2.4372	5.5254
Range of e	−4.3911 to 3.9936	−4.7722 to 3.9665	−11.3641 to 6.6129	−17.7670 to 19.0185
Standard deviation of e	1.2430	1.3412	2.8063	6.4330
(b) Saturation control				
Avg(abs(e))	1.2502	1.7987	4.4362	5.4374
Range of e	−4.2085 to 4.9032	−2.3505 to 7.6094	−4.7248 to 14.8659	−11.9105 to 19.3981
Standard deviation of e	1.5220	1.4996	2.8328	6.6223
(c) Implicit control				
Avg(abs(e))	0.72598	1.7017	3.2844	5.0835
Range of e	−1.8663 to 2.3094	−5.8677 to 4.6001	−8.1843 to 6.3261	−9.2313 to 8.1833
Standard deviation of e	0.7941	1.9237	3.5816	5.4152

Table 3.2.3 Variation of position error e

h	2 ms	5 ms	10 ms	15 ms
(a) $G = 10^4$				
Explicit	1.7838e+03	904.1336	844.2871	1.4462e+03
Saturation	1.6527e+03	914.4627	838.3387	1.6821e+03
Implicit	1.6452e+03	657.6504	428.0244	196.0669
(b) $G = 10^5$				
Explicit	2.5724e+03	1.7742e+03	1.6081e+03	2.5070e+03
Saturation	2.5691e+03	2.0749e+03	2.1638e+03	2.5756e+03
Implicit	1.6360e+03	650.2710	480.1660	228.8022

where the set of time instants $\{t_0, t_1, \ldots, t_N\}$ is a partition of $[a, b]$. In the following, the variations of the position error e for the three different implementation methods with the different gains G, have been calculated by choosing the partition times t_i in (3.2.28), as the sampling times.

Remark 3.2.5. *The variation in (3.2.28) as a quantity to characterize the analyzed signals is not common in Control Engineering. It is thought here in the context of sliding mode control that such a quantity is useful to measure the chattering level of a signal, since it does represent how much the signal varies. However, due to the partition that has been chosen (the sampling times) the results are not comparable*

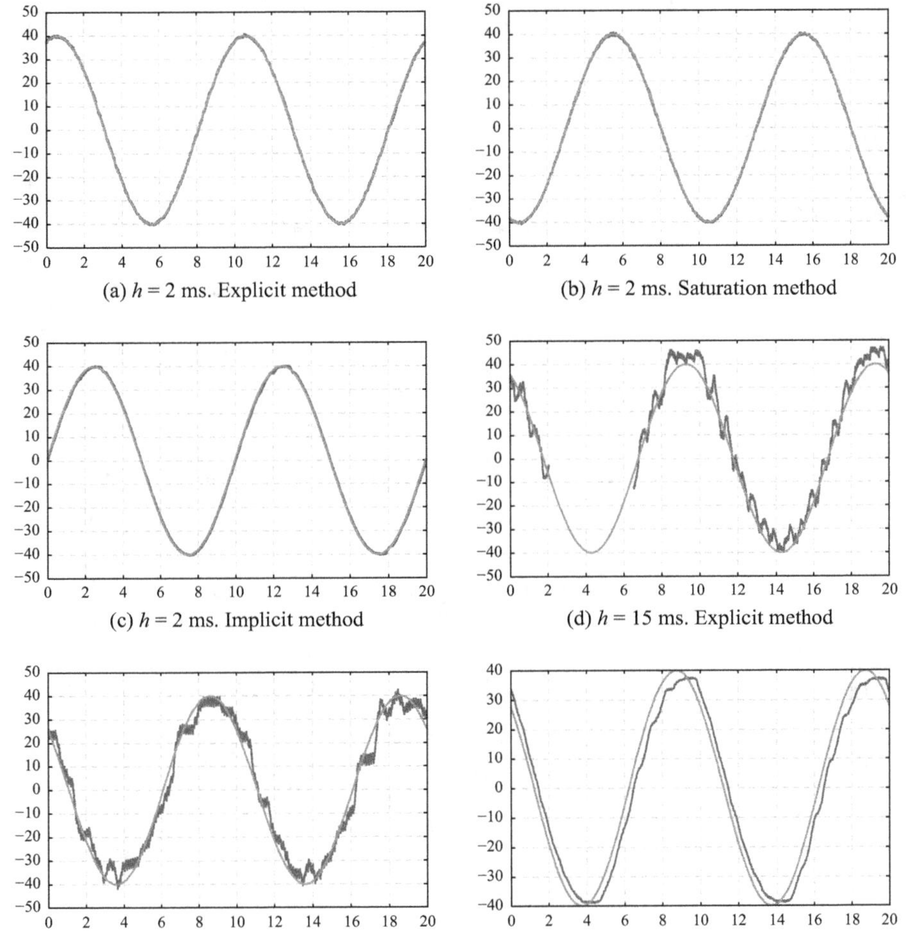

(a) $h = 2$ ms. Explicit method (b) $h = 2$ ms. Saturation method

(c) $h = 2$ ms. Implicit method (d) $h = 15$ ms. Explicit method

(e) $h = 15$ ms. Saturation method (f) $h = 15$ ms. Implicit method

Figure 3.2.2 Real position y (in millimeters) in bold and y_d (in millimeters), under $h = 2$ and $h = 15$ ms for $G = 10^4$. Real position y in bold and y_d

from one sampling period to the next, but only between the three controllers for a fixed h. In other words, in Table 3.2.3, data have to be compared inside a single column but not from one column to another one.

We first compare the controllers performance in terms of the tracking error, for two gains $G = 10^4$ and $G = 10^5$. All the data concerning e are reported in Tables 3.2.1–3.2.3 and on Figures 3.2.2 and 3.2.3. Table 3.2.1 data and Figures 3.2.2(a–c) show that when $G = 10^4$ and $h = 2$ or 5 ms, the implicit method does not bring any improvement over the explicit ones, but has lower precision capabilities for small time steps. It is only for the larger time step $h = 15$ ms that the results for the implicit controller

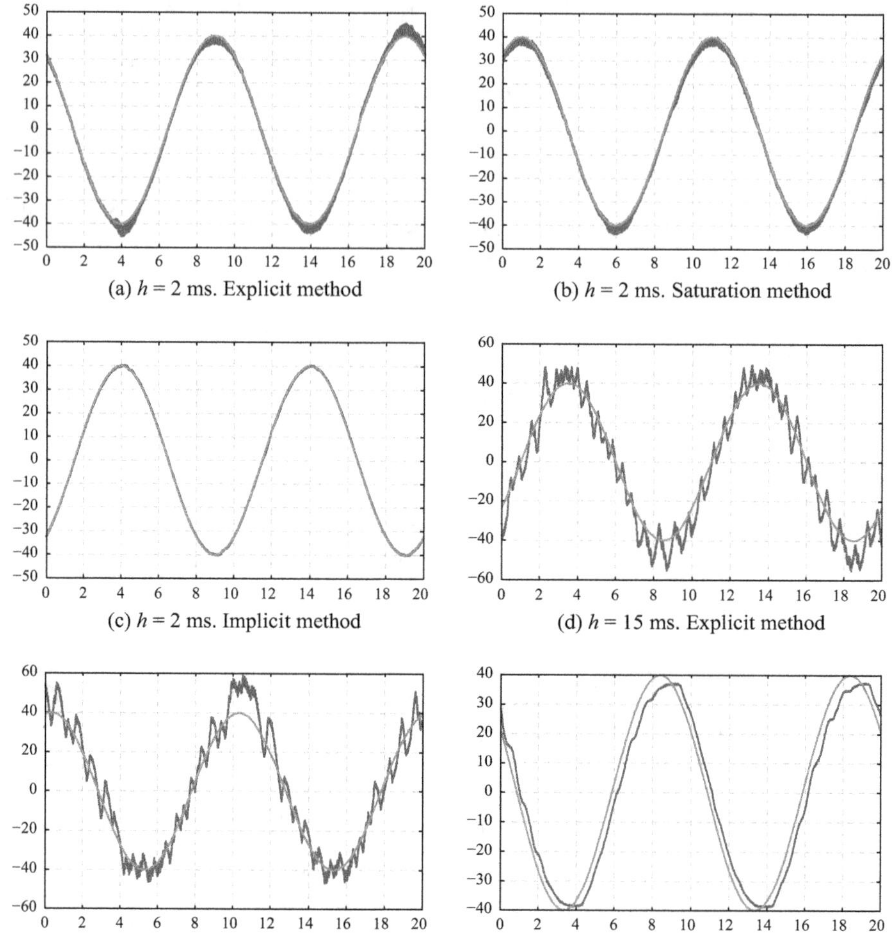

(a) $h = 2$ ms. Explicit method
(b) $h = 2$ ms. Saturation method
(c) $h = 2$ ms. Implicit method
(d) $h = 15$ ms. Explicit method
(e) $h = 15$ ms. Saturation method
(f) $h = 15$ ms. Implicit method

Figure 3.2.3 Real position y (in millimeters) in bold and y_d (in millimeters), under $h = 2$ ms and $h = 15$ ms for $G = 10^5$

(Table 3.2.1, last column) become the same as those of the other two controllers. However, it is confirmed in Table 3.2.3(a) that the variation of the implicit input starts to be significantly smaller than that of the other two, for $h \geq 5$ ms, the improvement being huge for $h = 15$ ms. These first data tend to indicate that in the case of the implicit input, its variation is drastically smaller for larger sampling periods (for $h = 15$ ms: 1.4462×10^3 for the explicit method, 196 for the implicit one with $G = 10^4$), confirming that chattering on e is reduced when the implicit controller (3.2.27) is used. The fact that the output signal is smooth for the implicit method,

Table 3.2.4 Switching function, gain $G = 10^4$

h	**2 ms**	**5 ms**	**10 ms**	**15 ms**
(a) Range of the switching function				
Explicit (3.2.24)	(−1.000, 1.000)	(−1.000, 1.000)	(−1.000, 1.000)	(−1.000, 1.000)
Saturation (3.2.25)	(−1.000, 1.000)	(−1.000, 1.000)	(−1.000, 1.000)	(−1.000, 1.000)
Implicit (3.2.27)	(−0.4635, 0.5385)	(−0.3247, 0.3338)	(−0.2969, 0.3117)	(−0.1935, 0.2194)
(b) Variation of the switching function				
Explicit (3.2.24)	6926	2822	2258	1936
Saturation (3.2.25)	6.6197e+03	2.7224e+03	2.2199e+03	2008
Implicit (3.2.27)	1.8416e+03	357.9547	211.4038	79.1096

while it chatters for the other two controllers for large sampling time, is obvious in Figures 3.2.2(d–f).

Table 3.2.2 concerns $G = 10^5$, i.e., the gain is now multiplied by 10. All three methods show similar results in terms of average, range, and standard deviation of e, the implicit one providing slightly better results. One infers that augmenting the gain G from 10^4 to 10^5 allows to significantly improve the tracking performance of the implicit control (3.2.21) (3.2.27) compared to that of the explicit inputs, especially in terms of the variation which is a good quantification of the chattering. In fact, comparing Table 3.2.1(c) and Table 3.2.2(c), one sees that the performance of the implicit input is almost unchanged when the gain is multiplied by 10, which is not the case of the other two methods: for these both latter, e is clearly increased. We shall observe this insensitivity property of the implicit method, again in Section 3.2.3.2. In addition the output produced by the implicit method is smoothed, even for small $h = 1$ ms (Figure 3.2.3(a–c), or 3.2.3(d–f)).

The variation values are given in Table 3.2.3(b) with $G = 10^5$, and is quite visible in Figures 3.2.3(d–f): the variation of e with the implicit input is much smaller than with the other two controllers, except for $h = 1$ ms where the obtained values are of same order. This indicates that the chattering on e is drastically reduced with the implicit input (3.2.21) (3.2.27).

⇝ *A first conclusion, that will be strengthened in the next paragraph, is that the implicit control method allows to take larger gains without decreasing the performance. This means that it is possible to reject/counteract larger perturbations/uncertainties without more chattering, and makes the process of gain G tuning much easier. The performance of implicit control is better when G is larger, while it is less good with the explicit and saturation controllers.*

Table 3.2.5 Switching function, gain $G = 10^5$

h	2 ms	5 ms	10 ms	15 ms
(a) Range of the switching function				
Explicit (3.2.24)	−1.000 to 1.000	−1.000 to 1.000	−1.000 to 1.000	−1.000 to 1.000
Saturation (3.2.25)	−1.000 to 1.000	−1.000 to 1.000	−1.000 to 1.000	−1.000 to 1.000
Implicit (3.2.27)	−0.0606 to 0.0545	−0.0360 to 0.0417	−0.0289 to 0.0349	−0.0173 to 0.0247
(b) Variation of the switching function				
Explicit (3.2.24)	2980	2050	1932	1836
Saturation (3.2.25)	2.3486e+03	1.9858e+03	1902	1860
Implicit (3.2.27)	183.1965	34.7510	25.2005	8.1039

Table 3.2.6 Comparisons of u when $G = 10^4$

h	2 ms	5 ms	10 ms	15 ms
(a) Range of u				
Explicit	−7.8876 to 8.4594	−8.1550 to 8.6118	−8.7349 to 8.3970	−10 to 10
Saturation	−8.0737 to 8.1963	−7.9095 to 8.0899	−8.5541 to 8.7543	−10 to 10
Implicit	−3.2500 to 3.5871	−1.9990 to 2.6204	−1.9399 to 2.1267	−1.8990 to 1.9484
(b) Variation of u				
Explicit	4.1102e+04	1.7731e+04	1.3816e+04	1.2759e+04
Saturation	4.0209e+04	1.6864e+04	1.3838e+04	1.3671e+04
Implicit	9.5190e+03	1.5731e+03	963.2736	609.5058
(c) Standard deviation of u				
Explicit	5.7570	5.8144	5.8437	6.3526
Saturation	5.7270	5.6808	5.8811	6.5001
Implicit	1.1183	0.8915	0.8519	0.8650

3.2.3.2 *Comparison of control inputs u (3.2.21) and v (3.2.24), (3.2.25), and (3.2.27)*

The features of the control inputs is a key point in this work, given that one of the objectives is to show the influence of implicit control to the chattering effect. Let us now pass to the control inputs comparisons, with data reported in Tables 3.2.4–3.2.7 and on Figures 3.2.4 and 3.2.5. Data given in Tables 3.2.4 and 3.2.5 characterize the "switching functions" for these three methods by considering the range and variation

Table 3.2.7 Comparisons of u when G = 10^5

h	**2 ms**	**5 ms**	**10 ms**	**15 ms**
(a) Range of u				
Explicit	−10 to 10	−10 to 10	−10 to 10	−10 to 10
Saturation	−10 to 10	−10 to 10	−10 to 10	−10 to 10
Implicit	−3.2541 to 3.8092	−2.0772 to 2.6066	−2.0325 to 2.3656	−1.9642 to 1.9461
(b) Variation of u				
Explicit	29800	20500	19320	18360
Saturation	2.3516e+04	1.9846e+04	19020	18600
Implicit	9.3245e+03	1.5389e+03	1.1560e+03	629.0904
(c) Standard deviation of u				
Explicit	10.0004	9.9729	9.9903	9.9547
Saturation	9.9746	9.9080	9.9262	9.9561
Implicit	1.0974	0.9113	0.9037	0.8630

for each of them. Remark 3.2.5 applies also for the variation of the control, so that in Tables 3.2.4 (b), 3.2.5 (b), 3.2.6 (b), and 3.2.7 (b), data have to be compared inside a single column but not from one column to another one.

⇝ *What we call the* switching functions *are* $sgn(\sigma_k)$ *in (3.2.24),* $sat(\sigma_k, \varepsilon)$ *in (3.2.25), and the selection (remember* ξ *in the introduction) of* $sgn(\sigma_{k+1})$ *in (3.2.27). For the implicit controller, this is what we called the selection* ξ_k *in Introduction. This is not to be confused with the discontinuous control v in (3.2.23).*

Comparisons of the inputs u in three methods are given in Tables 3.2.6 and 3.2.7 from three aspects, i.e., range, variation, and standard deviation. In addition, the three controllers are depicted on Figures 3.2.4 and 3.2.5, for various time steps and gains.

Globally, the experimental results show that the implicit method drastically reduces the input chattering and magnitude compared with the other two methods. The explicit and saturation switching inputs keep oscillating between the maximum and minimum values like a bang–bang controller (see data in Tables 3.2.4(a) and 3.2.5(a), and Figures 3.2.4(a,b)). This results in a large amplitude of inputs u as well (Tables 3.2.6(a) and 3.2.7(a), as well as Figures 3.2.5(a–c,f–h)). Notice that the explicit and saturation inputs behave slightly better when the time step increases. This is visible by comparing Figures 3.2.5(a) and 3.2.5(f), 3.2.5(b) and 3.2.5(g), 3.2.5(c) and 3.2.5(h) which all concern $h = 2$ ms and $h = 15$ ms, respectively, for both gains G. However, the magnitude of the implicit input is far much better in all cases (Figures 3.2.5(d) 3.2.5(e) 3.2.5(i) 3.2.5(j)).

The magnitude of the ranges of the switching function and control u in the implicit method is much less than the other two methods, see Tables 3.2.6(a) and 3.2.7(a). These facts are well supported by Figures 3.2.4 and 3.2.5. Consider Figure 3.2.5: when $h = 15$ ms, while the ranges of the control law u in explicit method and explicit saturation method are both between −10 and 10 (Figures 3.2.5(h) and 3.2.5(g)), the

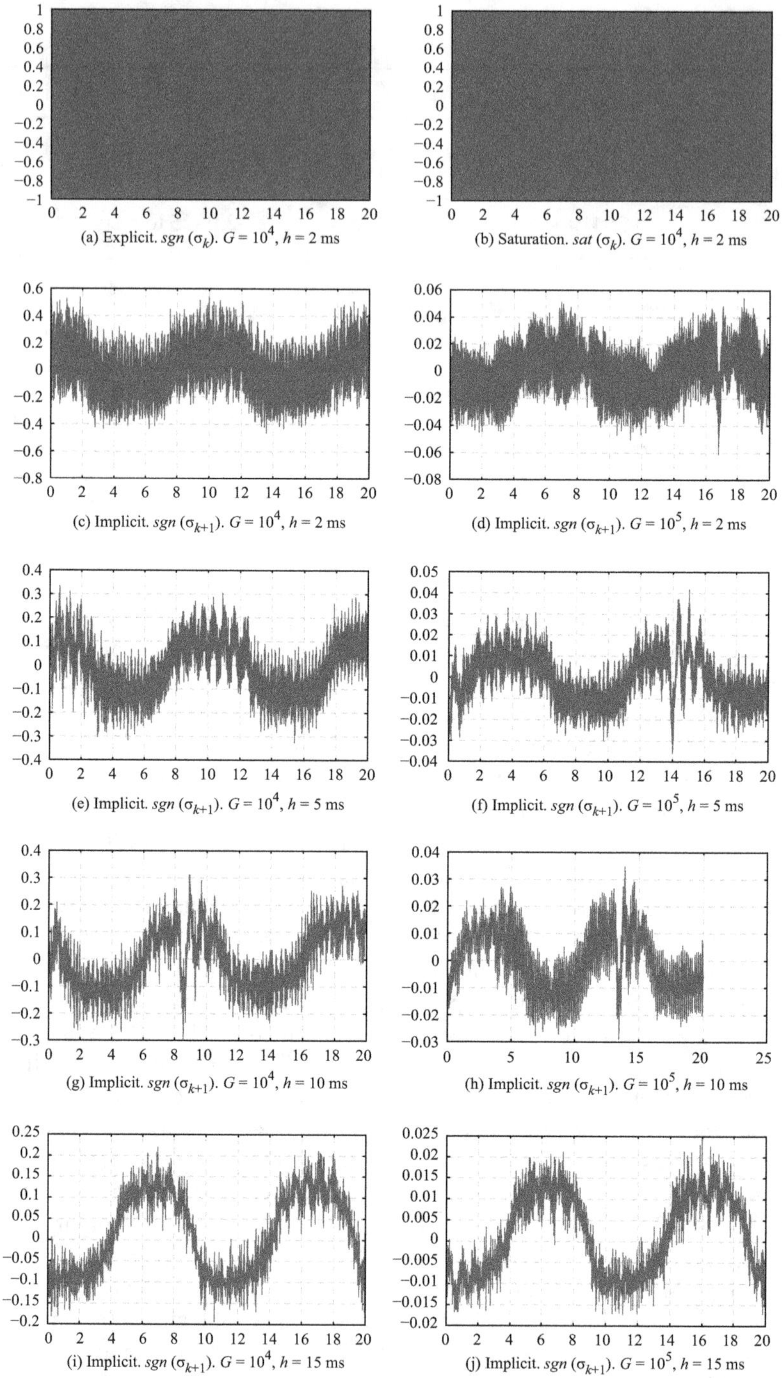

Figure 3.2.4 Switching function: comparison between explicit ($sgn(s_k)$), saturation ($sat(s_k)$), and implicit ($sgn(s_{k+1})$) algorithms

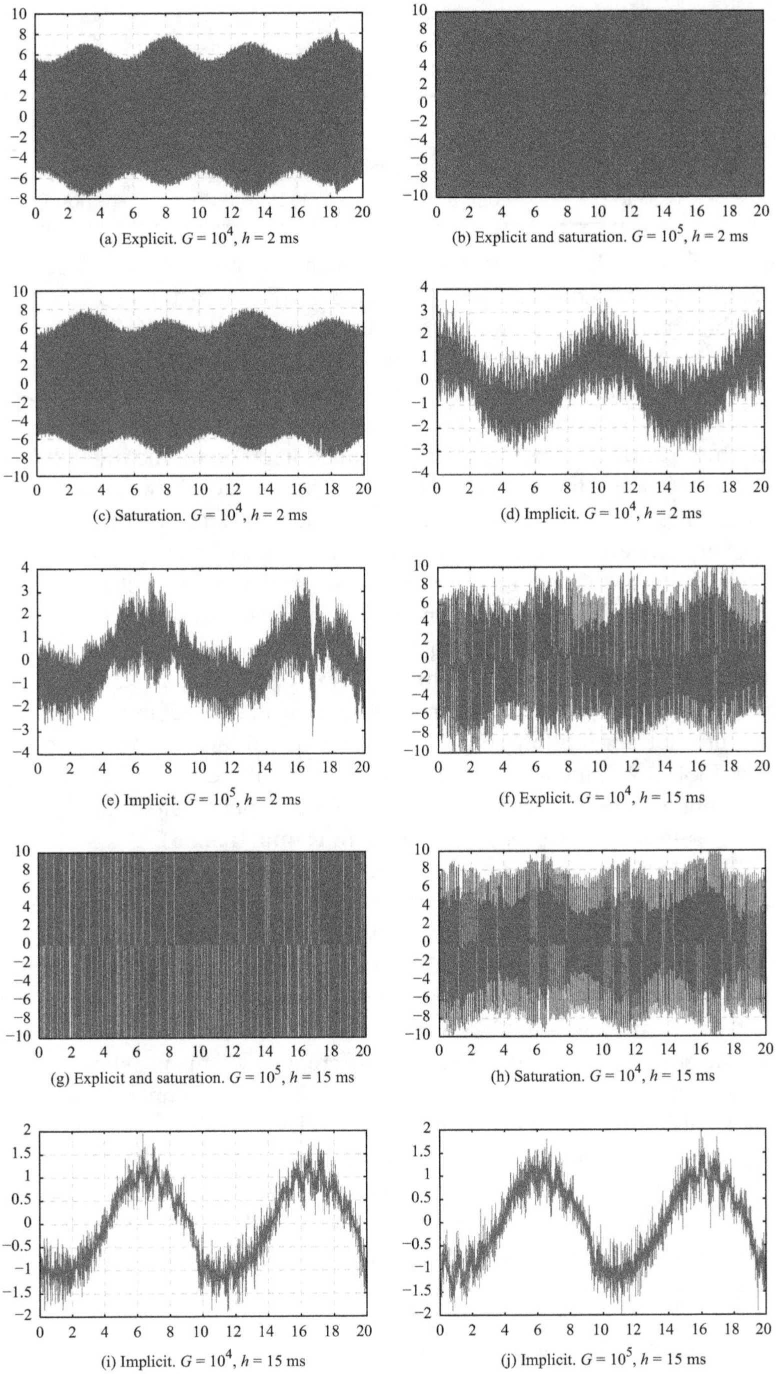

(a) Explicit. $G = 10^4$, $h = 2$ ms
(b) Explicit and saturation. $G = 10^5$, $h = 2$ ms
(c) Saturation. $G = 10^4$, $h = 2$ ms
(d) Implicit. $G = 10^4$, $h = 2$ ms
(e) Implicit. $G = 10^5$, $h = 2$ ms
(f) Explicit. $G = 10^4$, $h = 15$ ms
(g) Explicit and saturation. $G = 10^5$, $h = 15$ ms
(h) Saturation. $G = 10^4$, $h = 15$ ms
(i) Implicit. $G = 10^4$, $h = 15$ ms
(j) Implicit. $G = 10^5$, $h = 15$ ms

Figure 3.2.5 Comparison of control u between explicit, saturation, and implicit methods

Table 3.2.8 Magnitude of implicit switching function $sgn(x_{k+1})$ *for varying gains G and sampling period h*

G	**10^4**	**10^5**	**10^6**	**5.10^6**
$h = 5$ ms	−0.3 to 0.35	−0.03 to 0.035	−0.003 to 0.003	−0.0006 to 0.00065
$h = 10$ ms	−0.25 to 0.3	−0.025 to 0.03	−0.0025 to 0.0025	−0.0005 to 0.0005

range of u for the implicit case is strictly between −2 and 2 (Figures 3.2.5(i,j)). The comparison between Figures 3.2.4(c) through 3.2.4(j), 3.2.5(d) and 3.2.5(e), 3.2.5(i) 3.2.5(j), which concern the implicit controller switching function for various gains and sampling times show that for $h \geq 2$ ms, the implicit input v in (3.2.27) is largely independent of the gain and sampling time. From Tables 3.2.4(a) and 3.2.5(a), the data in the rows corresponding to the implicit controller allow to obtain a confirmation of this fact. Furthermore, the switching function range for the implicit controller is divided by 10 when the gain G passes from 10^4 to 10^5, which implies that the sliding mode input v_k in (3.2.27) has a magnitude that does not vary with the gain (recall that what we call the switching function has to be multiplied by the gain G to obtain the input v). This is in very good agreement with theoretical predictions (item 3) in the Introduction. One can also have a look at Tables 3.2.6(a–c), and 3.2.7(a–c), to obtain the same conclusions, that the range (magnitude), the variation, and the standard deviation (chattering) of u for (3.2.27) are drastically smaller than for (3.2.24) and (3.2.25). The magnitudes of the switching function for the implicit controller, for 6 different gains G and two different sampling periods h, are reported in Table 3.2.8. It confirms that the magnitude of the input v in (3.2.23), which is the switching function times the gain G, does not depend neither on G nor on h in this range of sampling times (see a comment in **Remark 3.2.6**).

⇝ *This insensitivity property is believed to be a fundamental property of the implicit method introduced in References 1 and 2, compared to explicit implementations which drastically differ when h and/or G are varied.*

The results depicted on Figures 3.2.4 and 3.2.5 clearly demonstrate that while the explicit and saturation controllers tend to approximate a signal that switches infinitely fast between two extreme values like bang–bang inputs, this is not at all the case for the implicit controller that behaves in a totally different way. This is a nice confirmation of both theoretical and numerical predictions [1, 2] that the implicit controller does represent the discrete-time approximation of the selection of the differential inclusion according to Filippov's mathematical framework.

Input chattering is also visible in Tables 3.2.4(b), 3.2.5(b), 3.2.6(b,c), and 3.2.7(b,c). Variation of the implicit switching function is much smaller than the other two, and standard deviation of u as well. These results demonstrate that the switching function chattering and magnitude strongly influence the input u in (3.2.21).

Remark 3.2.6. *All the results tend to show that when the sampling period is too small (typically in our experiments $h = 2$ ms), then the implicit controller performance (output precision and chattering, input magnitude and chattering) is* *decreasing. This is visible on Figure 3.2.4 with the evolution of the implicit signum function from Figure 3.2.4 (c–g) for gain* $G = 10^4$*, and from Figure 3.2.4 (h)–(l) when the gain* $G = 10^5$*. In theory the implicit switching function should not vary by changing the sampling period. This phenomenon is due to bandwidth limitations in the first order filters used to estimate velocities and accelerations from position measurement, in order to calculate the sliding variable in (3.2.18). It results in a deterioration of the closed-loop performance and controller chattering. Further work on proper tuning of these filters to accommodate for smaller sampling periods h proves that the performance of the implicit discretization can be further improved for small sampling times.*

3.2.3.3 Summary

These extensive experimental tests prove that items 3–7 in the Introduction are not only theoretical and numerical predictions obtained in References 1 and 2 but also significantly influence the discrete-time implemented sliding-mode controller. The implicit method (3.2.27) allows to drastically reduce the input chattering and magnitude, while enhancing the tracking capabilities (output chattering is almost entirely eliminated) and simplifying the tuning of the gain G. It also allows the designer to choose larger sampling periods, which may be of strong interest in practice, where one wants to make less calculations to save time and energy. Perhaps counter-intuitively for control engineers, the performance and robustness increase when the gain G increases, which is thought to considerably simplify the controller gain tuning process. A video of the experimental tests is available at http://nullptr.fr/pages/videos.html.

3.2.4 Numerical analysis of the saturation controller

Saturating the set-valued signum function is often presented in the literature as the absolute remedy to the chattering effects. However, no analysis seems to be available to confirm this assertion. In order to corroborate our above conclusions on the saturation input (which is shown in general not to decrease the input and output numerical chattering compared to the explicit discretization without saturation), we have led a preliminary simulation study on a perturbed system to analyze the effect of adding a saturation, on both the tracking error and the input chattering. The dynamics is given by

$$\begin{cases} \dot{x}(t) = A(t) + Bu(t) + \sin(4\pi t), & A = \begin{pmatrix} 0 & 1 \\ 19 & -2 \end{pmatrix}, \\ \sigma = Cx & \\ u(t) = u^{eq}(t) + u^s(t), & B = \begin{pmatrix} 0 \\ 1 \end{pmatrix}, \; C^T = \begin{pmatrix} 1 \\ 1 \end{pmatrix}. \end{cases} \tag{3.2.29}$$

where $u^{eq}(x) = -(CB)^{-1}CAx$, $u^s(x) \in -\text{sgn}(Cx)$. On Figures 3.2.6 and 3.2.7 the performance index is the sum of the $|\sigma_k|$ for the last 20 seconds. On Figures 3.2.8 and 3.2.9 the performance index is the sum of the $|u^s_{k+1} - u^s_k|$ for the last 20 seconds. The

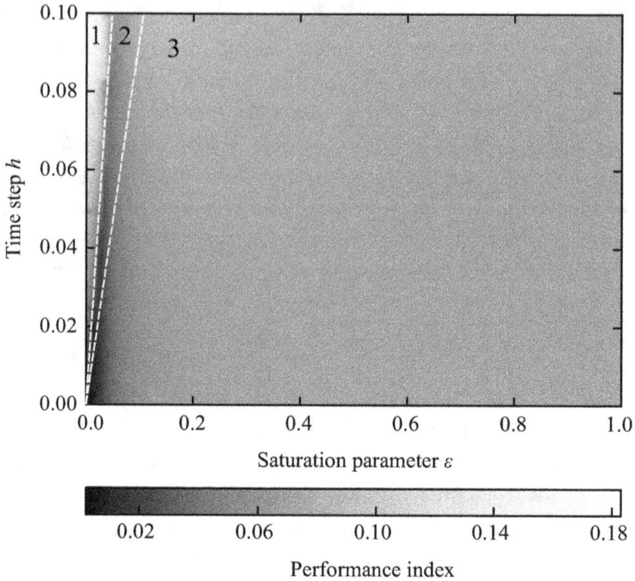

Figure 3.2.6 Simulation results with 100 regularly spaced values for the sampling period h and 100 logarithmically spaced values for the saturation parameter ε

conclusions should not be considered as generic because the study has been made for only one disturbance: results might change if another disturbance is considered. Nevertheless Figures 3.2.6–3.2.9 allow us to draw some conclusions. With both indexes, we can divide the space into 3 cones, numbered 1, 2, and 3 on Figures 3.2.6–3.2.9. This separation helps us to compare both controllers. In Figure 3.2.6 the performance in terms of chattering is presented. For large values of ε, the chattering does not change when the sampling period varies: the control action does not attenuate the effect of the perturbation. With a small ε, the behavior is more complex as depicted in Figure 3.2.7: the overall best performance is obtained with small values for both ε and h. But for small values of ε, the performance can rapidly degrade if the sampling period h is not small enough as seen in Region 1. The dark points indicate that for each value of h, the pair (ε, h) of parameters yielding the best performance. It seems that there is a linear relationship between those values. However, it is unclear if this observation on one particular system remains valid with a different perturbation. The level sets in Figure 3.2.7 are used to compare the performance of the implicit and the saturated explicit controllers. On Figures 3.2.8 and 3.2.9, the performance in terms of control cost is presented. The best performance is achieved for large ε since the slope of the saturated function is gentle. On the other hand, in Figure 3.2.9, with a small ε, the cost increases and explodes with ε close to 0, as in Region 1. The level sets indicate that the difference between the costs of the two different controllers. It is worth noting that in Region 2 where the saturated controller is better in Figure 3.2.7,

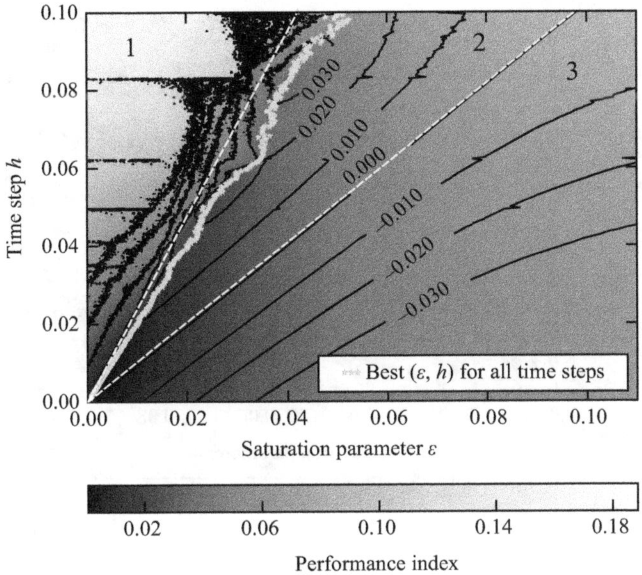

Figure 3.2.7 Detail of Figure 3.2.6, 300 values for h and 1000 values for ε, forming a regular grid. Level sets were also added to show the difference in performance between the implicit discretization and the explicit one with saturation. If the difference is positive, the explicit saturated control is performing better than the implicit one

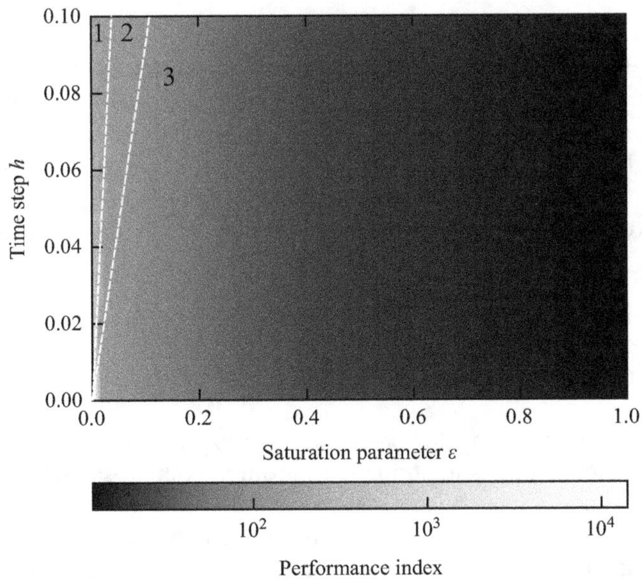

Figure 3.2.8 Simulation results with 100 regularly spaced values for the sampling period h and 100 logarithmically spaced values for the saturation parameter ε

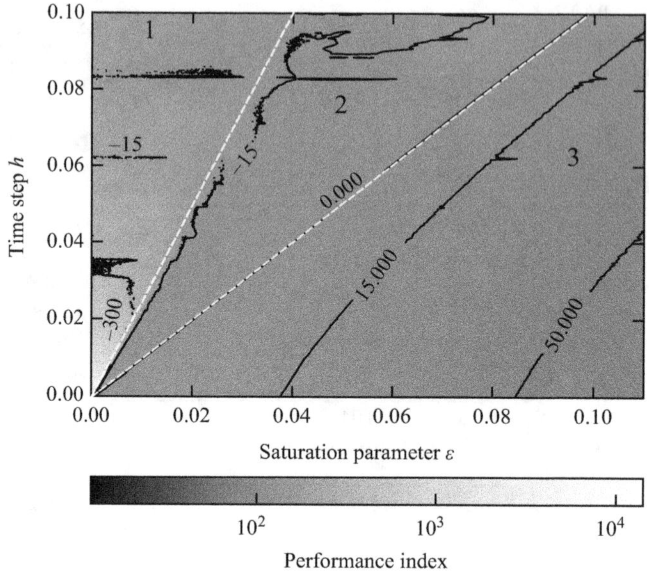

Figure 3.2.9 Detail of Figure 3.2.8, 300 values for h and 1000 values for ε, forming a regular grid. Level sets were also added to show the difference in performance between the implicit discretization and the explicit one with saturation. If the difference is positive, the explicit saturated control is performing better than the implicit one

it has a higher cost in term of control (Figure 3.2.9). In Region 3, where the saturated controller performs less in terms of chattering (Figure 3.2.6), it has a smaller cost in terms of control (Figure 3.2.8). Indeed with a large ε, the control input is small when the closed-loop system is close to the sliding manifold. The cost is then very small, but the disturbance is not attenuated at all. The implicit controller appeals to us as the best compromise between the input and output chattering. It is also very easy to use, since it requires no particular tuning with respect to the sampling period or the perturbation.

3.2.5 Conclusion

Experiments have been conducted on an electropneumatic system, with three different implementations of the sliding mode controller: explicit, saturated explicit, and implicit discretizations. The results demonstrate that the theoretical and numerical predictions of References 1 and 2 are true: the implicit implementation, which is very easy to implement in a code, drastically supersedes the other two. The output and input chattering are reduced in a significant way without changing the controller's basic structure (i.e., no additional filter, observer, or dynamic controller is added compared to the original basic sliding mode controller) and keeping its simplicity (in particular the gain tuning is easy, which is a strong feature of the ECB–SMC method). The main feature of the implicit discretization is that it keeps, in discrete-time, the multivalued feature of the theoretical continuous-time sliding-mode controller, as it is mathematically imposed in Filippov's framework. The proposed implicit discretization method is generic in the sense that it could apply to any kind of sliding mode set-valued control. These conclusions have been confirmed elsewhere on another experimental setup for both the ECB–SMC and the twisting controllers [14, 15, 17].

Appendix 1 Some basic convex analysis tools

In this section we provide few results which are useful to calculate the controller in Section 3.2.2.1. From page 115 of Reference 24 it follows that the conjugate function of the indicator function of the set $[-1, 1]$, denoted as $\psi_{[-1,1]}(\cdot)$, is the support function $\psi^\star_{[-1,1]}(\cdot)$ of the set $[-1, 1]$, given by the absolute value function $x \mapsto |x|$. From Theorem 23.5 in Reference 24, one has that the subdifferentials (in the sense of convex analysis) of these two conjugate functions satisfy: $x \in \partial\psi_{[-1,1]}(z) \Leftrightarrow z \in \partial\psi^\star_{[-1,1]}(x)$. From the definition of the subgradient, one has $\partial\psi^\star_{[-1,1]}(x) = \mathrm{sgn}(x)$ where sgn is the multivalued signum function as defined in the introduction. By definition of the subdifferential of a convex set, $\partial\psi_{[-1,1]}(z)$ is the normal cone to the set $[-1, 1]$ at z. These results allow one to derive (3.2.9) from (3.2.7).

Consider now the inclusion $x - y \in -N_C(x)$ for some convex, nonempty closed set C of $\mathbf{R}^n$, and two vectors x and y of $\mathbf{R}^n$. Using Theorem 1.5.5 of Reference 7, one finds that x is the Euclidean projection of y onto C. This allows us to deduce (3.2.10) from (3.2.9).

Acknowledgments

This work has been performed with the support of the French National Research Agency (ANR) project ChaSliM (ANR 2011 BS03 007 01).

References

[1] V. Acary and B. Brogliato, *Implicit Euler numerical scheme and chattering-free implementation of sliding mode systems*, Systems and Control Letters, vol. 59, pp. 284–293, 2010.

[2] V. Acary, B. Brogliato, and Y. Orlov, *Chattering-free digital sliding-mode control with state observer and disturbance rejection*, IEEE Transactions on Automatic Control, vol. 57, no 5, pp.1087–1101, 2012.

[3] V. Acary and B. Brogliato, *Numerical methods for nonsmooth dynamical systems. Applications in mechanics and electronics,* Lecture Notes in Applied and Computational Mechanics. vol. 35, Springer, Berlin, xxi, p. 525, 2008.

[4] V. Acary, O. Bonnefon, and B. Brogliato, *Nonsmooth modeling and simulations for switched circuits*, Lecture Notes in Electrical Engineering, vol. 69, Springer Verlag, Heidelberg, 2011.

[5] M.K. Camlibel, W.P.M.H. Heemels, and J.M. Schumacher, *Consistency of a time-stepping method for a class of piecewise linear networks*, IEEE Transactions on Circuits and Systems I, vol. 49, no 3, pp. 349–357, 2002.

[6] R. Castro-Linarès, S. Laghrouche, A. Glumineau, and F. Plestan, *Higher order sliding mode observer-based control*, IFAC Symposium on System, Structure and Control, Oaxaca, Mexico, 2004.

[7] F. Facchinei and J.S. Pang, *Finite-dimensional variational inequalities and complementarity problems*. Springer Series in Operations Research, vol. 1, Springer-Verlag, New-York, 2003.

[8] A.F. Filippov, *Differential equations with discontinuous righthand sides*, Mathematics and its Applications, Kluwer Academic Publishers, Dordsecht, 1988.

[9] Z. Galias and X. Yu, *Complex discretization behaviours of a simple sliding-mode control system*, IEEE Transactions on Circuits and Systems-II: Express Briefs, vol. 53, no 8, pp. 652–656, 2006.

[10] Z. Galias and X. Yu, *Analysis of zero-order holder discretization of two-dimensional sliding-mode control systems*, IEEE Transactions on Circuits and Systems-II: Express Briefs, vol. 55, no 12, pp. 1269–1273, 2008.

[11] A. Girin, F. Plestan, X. Brun, and A. Glumineau, *Robust control of an electropneumatic actuator: application to an aeronautical benchmark*, IEEE Transactions on Control Systems Technology, vol. 17, no 3, pp. 633–645, 2009.

[12] G. Golo, A.J. van der Schaft, and Č. Milosavljević, *Discretization of control law for a class of variable structure control systems*, In: X. Yu and J.-X. Xu (editors). Advances in Variable Structure Systems: Analysis, Integration and Applications, pp. 45–54, World Scientific, Singapore, 2000.

[13] O. Huber, V. Acary, and B. Brogliato, *Comparison between explicit and implicit discrete-time implementations of sliding-mode controllers*, Proceedings of the 52nd IEEE Conference on Decision and Control, pp. 2870–2875, 2013, Florence, Italy.

[14] O. Huber, V. Acary, B. Brogliato, and F. Plestan, *Discrete-time twisting controller without numerical chattering: analysis and experimental results with an implicit method*, Proceedings of the 53rd IEEE Conference on Decision and Control, pp.4373–4378, December 15–17, 2014, Los Angeles, USA.

[15] O. Huber, *Analysis and implementation of discrete-time sliding mode control*, PhD Thesis, University Grenoble-Alpes, 2015 (https://hal.inria.fr/tel-01194430/document).

[16] O. Huber, V. Acary, and B. Brogliato, *Analysis of explicit and implicit discrete-time equivalent-control based sliding mode controllers*, INRIA Research Report 8383, 2013 (http://arxiv.org/pdf/1310.6004v1.pdf).

[17] O. Huber, V. Acary, B. Brogliato, and F. Plestan, *Implicit discrete-time twisting controller without numerical chattering: analysis and experimental results*, Control Engineering Practice, Vol. 46, no 1, pp. 129–141, DOI: 10.1016/j.conengprac.2015.10.013.

[18] R. Kikuuwe, S. Yasukouchi, H. Fujimoto, and M. Yamamoto, *Proxy-based sliding mode control: a safer extension of PID position control*, IEEE Transactions on Robotics, vol. 26, no 4, pp. 670–683, 2010.

[19] S. Laghrouche, M. Smaoui, F. Plestan, and X. Brun, *Higher order sliding mode control based on optimal approach of an electropneumatic actuator*, International Journal of Control, vol. 79, no 2, pp. 119–131, 2006.

[20] D. Leenaerts, *On linear dynamic complementary systems*, IEEE Transactions on Circuits and Systems-I: Fundamental Theory and Applications, vol. 46, no 8, pp. 1022–1026, 1999.

[21] J.J. Moreau, *Standard inelastic shocks and the dynamics of unilateral constraints*, In: CISM Courses and Lectures, no 288, Springer-Verlag, Wien, New-York, pp. 173–221, 1985.

[22] J.J. Moreau, *Unilateral contact and dry friction in finite freedom dynamics*, In: J.J. Moreau, P.D. Panagiotopoulos (editors), Nonsmooth Mechanics and Applications, CISM Courses and Lectures no 302, International Centre for Mechanical Sciences, Springer-Verlag, Wien, New-York, pp. 1–82, 1988.

[23] F. Plestan, Y. Shtessel, V. Brégeault, and A. Poznyak, *Sliding mode control with gain adaptation—application to an electropneumatic actuator*, Control Engineering Practice, vol. 21, no 5, pp. 679–688, 2013.

[24] R.T. Rockafellar, *Convex Analysis*, Princeton University Press, Princeton Landmarks in Mathematics, Princeton, 1970.

[25] S. Sesmat and S. Scavarda, *Static characteristics of a three way servovalve*, Proc. Conf. Fluid Power Tech., Aachen, Germany, 1996.

[26] Y. Shtessel, M. Taleb, and F. Plestan, *A novel adaptive-gain supertwisting sliding mode controller: methodology and application*, Automatica, vol. 48, no 5, pp. 759–769, 2012.

[27] H. Sira–Ramirez, *Non-linear discrete variable structure systems in quasi–sliding mode*, International Journal of Control, vol. 54, no 5, pp. 1171–1187, 1991.

[28] V. Utkin, *Variable structure systems with sliding mode*, IEEE Transactions on Automatic Control, vol. 22, no 2, pp. 212–222, 1977.

[29] V. Utkin, *Sliding modes in control and optimization*, Springer-Verlag, Berlin, Communications and Control Engineering, 1992.

[30] B. Wang, X. Yu, and G. Chen, *ZOH discretization effect on single-input sliding mode control systems with matched uncertainties*, Automatica, vol. 45, pp. 118–125, 2009.

[31] B. Wang, B. Brogliato, V. Acary, A. Boubakir, and F. Plestan, *Experimental comparisons between implicit and explicit implementations of discrete-time sliding mode controllers: towards input and output chattering suppression*, IEEE Transactions on Control Systems Technology, vol. 23, no 5, pp. 2071–2075, 2015, DOI: 10.1109/TCST.2015.2396473.

[32] X. Yu, B. Wang, Z. Galias, and G. Chen, *Discretization effect on equivalent control-based multi-input sliding-mode control systems*, IEEE Transactions on Automatic Control, vol. 53, no 6, pp. 1563–1569, 2008.

Chapter 3.3

A generalized reaching law for discrete-time sliding mode

Bijnan Bandyopadhyay and Sohom Chakrabarty

Abstract

The discrete application of a continuous control designed using continuous time model of a plant is common practice. The work in this chapter evaluates new conditions required to be satisfied by the parameters of a sliding mode control so designed, such that the plant response remains bounded under such discrete application of the control. This work is taken forward to propose a generalized reaching law for discrete time sliding mode designs. This generalized reaching law is more flexible in its choice of functions and parameters, enabling it to deal with potentially new problem scenarios. A new problem in which the disturbance affecting the system is bounded by known functions, instead of a constant bound, is solved utilizing this generalized reaching law.

3.3.1 Introduction

There has been much focus toward developing control for continuous models of plants. However, with the advent in computation technology, control computation and application are done only after a finite interval of time. This time period depends on the computational speed of the computer as well as the sensor and actuator speed. With a relatively small time period, the plant response roughly matches the expected response of the continuous-time model. However, this time period is not small in all cases, and then the plant response deviates from the expected response of the continuous model and can even become unstable.

This has led to the interest in the theory of discrete time sliding mode control, which was developed in References 1–9. Among recent works, much focus is toward discretization of higher order sliding mode controllers and differentiators [10–13], as well as implicit discretization and application of sliding mode control [14, 15]. All the above are important theoretical advancements in the direction of discrete application of sliding mode control. However they have not answered the specific question of analyzing a zero order hold (ZOH)-discretized continuous plant which is fed with a sliding mode control designed for the continuous plant model and bring out the complex dynamics of the sliding variable arising due to such discretization.

There are several works like References 16–19, to mention a few, which implement such a continuous-time sliding mode control in a digital environment, thereby functioning as a ZOH-discretized system. However, no rigorous analysis is reported about the behavior and stability conditions of the implemented system. It has been inherently assumed that the discretization step in such cases is relatively small and would not lead to a major deviation in the plant response. But there is always a deviation, as discretization never gives rise to robustness on the sliding surface. One can only expect robustness inside an ultimate band upon discretization, and hence it is important to know the width of this ultimate band. There may arise additional design criteria on the controller parameters which one needs to keep in mind in order to achieve this ultimate boundedness of the sliding variable. In this chapter, the band approach method of stability analysis [20, 21] has been utilized to obtain a relationship between the ultimate band and the parameters of the continuous-time sliding mode control, so that the sliding motion remains inside a chosen ultimate band. Also, it is shown that the sliding motion thus achieved would be stable with the same surface parameters selected for continuous-time sliding mode control.

It is found that upon discretization of the closed-loop dynamics, the sliding variable dynamics becomes different to the proportional plus switching reaching law used to derive the control. Along with the sliding variable, it also had the states of the system mixed in it [22]. The band approach method has been applied to analyze this dynamics. The range of the ultimate band was found out, and once selected from this range, the controller parameters were calculated in relation with the ultimate band. This work puts forward a more complicated dynamics of the sliding variable including the states of the system, which is subjected to the band approach method of analysis. This inspired the author to work further and propose a more generalized dynamics that the band approach method can handle. Thereby, a generalized discrete reaching law is laid down which not only comprises of functions in the sliding variable but also includes functions in other known variables.

With the help of these different functions in the generalized reaching law, one may carefully design the dynamics of the sliding variable. These designs can vary from one problem to another. In this chapter, one such problem is discussed. It studies the presence of a matched uncertainty bounded by known functions in the reaching algorithm [23]. With appropriate choices of the functions and the parameters in the generalized reaching law, one can obtain a bound about the sliding surface into which the sliding variable converges for the problem. Such a problem had not been discussed in earlier literature of discrete sliding mode control systems.

One needs to mention at this point that there had been several works in the literature which deal with modifying the already popular reaching laws [24–26]. Recently also works on reaching laws have been published [27–29]. However, there has not been any proposal for any generalized reaching algorithm, which can be chosen for the purpose of studying various discrete time sliding mode control problems.

The chapter is organized in the following way: In Section 3.3.2, preliminaries to help understand the band approach method are laid down. This consists of definitions of the bands used and a pictorial representation of them in the space of the sliding variable. In Section 3.3.3, a continuous-time sliding mode control for a linear time invariant (LTI) system is derived using the proportional plus switching reaching law, and it is applied in a digitized environment, which gives rise to a discretized model of the system and a closed-loop sliding variable dynamics which is analyzed via band approach method. In Section 3.3.4, simulation results are shown to compare the performance due to a continuous-time sliding mode control derived from the continuous plus switching reaching law with controller parameters selected as per analysis for continuous system stability to a continuous-time sliding mode control where controller parameters of the continuous plus switching reaching law are selected via the band approach method. In Section 3.3.5, the generalized reaching law is laid down. In Section 3.3.6, the problem with an uncertainty bounded by known functions is discussed, and simulation results supporting the theory are shown. In Section 3.3.7, the work presented in this chapter is summarized.

3.3.2 Definition of the bands

When we are dealing with sliding mode control for discrete time systems, the main idea is that the reaching law should bring each element of the sliding variable vector $s(k)$ toward zero, i.e., each $s_i(k)$ toward zero, and finally settling inside an ultimate band. To achieve this, $|s(k+1)| < |s(k)|$ must happen for certain positions of $s(k)$ such that $s_i(k)$ is beyond some pre-specified band. If $s_i(k)$ crosses zero at any instance, we may wish to avoid divergence during crossing by limiting the crossing within some pre-specified band on the other side of zero. Also, there must be an ultimate band inside which the trajectory of $s_i(k)$ will be confined after it enters this band for the first time. Here and in the sequel, $|v|$ denotes a vector with each element as the absolute value of the respective element of vector v.

For discussing the definitions and the analysis with the band approach method, we would consider for simplicity that $s(k) \in \mathbb{R}$. When $s(k) \in \mathbb{R}^m$, then the definitions and analysis would be applicable element-wise to the sliding variable vector.

Definition 3.3.1. *The **decrement band** B_d is such that whenever $|s(k)| > B_d$, $|s(k+1)| < |s(k)|$ is ensured.*

Note that there is no guarantee for monotonic decrement of $|s(k)|$ everywhere in the space of the sliding variable but limiting it to everywhere outside $[-B_d, B_d]$.

Definition 3.3.2. *The **limit to cross band** B_c is such that whenever $|s(k)| > B_d$ and $sign(s(k+1)) = -sign(s(k))$, $|s(k+1)| < B_c$ is ensured.*

This means that $|s(k+1)|$ cannot cross B_c on the other side of $s(k) = 0$ when $|s(k)| > B_d$.

Definition 3.3.3. *The **maximum crossing band** B_s is such that whenever $0 \leq |s(k)| \leq B_d$ and $sign(s(k+1)) = -sign(s(k))$, $|s(k+1)| < B_s$ is ensured.*

This means that $|s(k+1)|$ cannot cross B_s on the other side of $s(k) = 0$ when $0 \leq |s(k)| \leq B_d$. Note the difference in the definitions of B_c and B_s. Both are zero-crossing limits but when $s(k)$ is outside and inside $[-B_d, B_d]$, respectively.

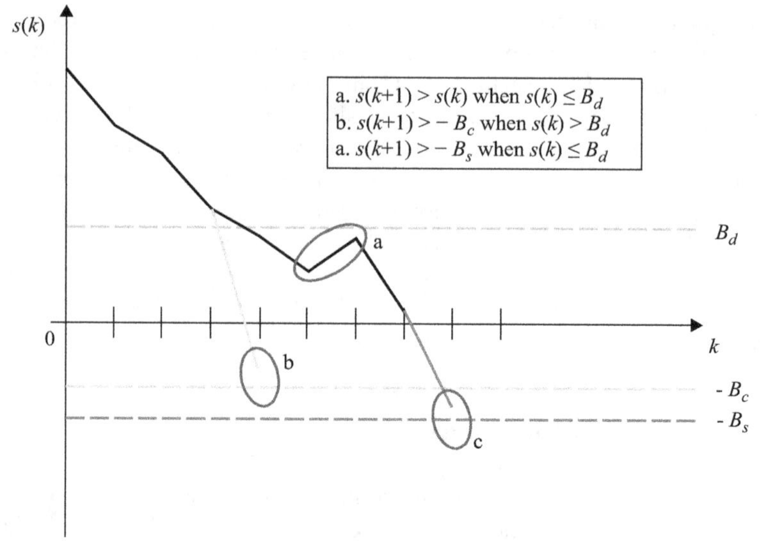

Figure 3.3.1 The different bands

Definition 3.3.4. *The* ***ultimate band*** δ *is such that* $|s(k)| \leq \delta \Rightarrow |s(k+1)| \leq \delta$.

Note the difference in the definitions of B_c, B_s, and δ. This means that ultimate band may exist even if the sliding variable does not cross zero at any instant. This is a relaxation from Gao's approach where crossing–recrossing must happen at every instant.

Such relaxation had earlier been attempted in Reference 5. However, in Reference 5, the definitions, which determine crossing–recrossing, are all centered around the ultimate band δ, which need not be the case. In the present work, there is no requirement for crossing–recrossing, and the dynamics of the sliding variable is centered around the four bands as defined earlier. All these may or may not be equal to δ. A pictorial representation of these bands (except the ultimate band) in the space of the sliding variable has been shown in Figure 3.3.1.

3.3.3 Digital application of continuous sliding mode control

Let a continuous-time LTI model of a plant be

$$\dot{x} = Ax + B(u + f), \tag{3.3.1}$$

where $x \in \mathbb{R}^n$ are the states, $u \in \mathbb{R}^m$ is the control, and $f \in \mathbb{R}^m$ is the disturbance matched with the input and bounded as $|f(t)| \leq f_m$. A and B are constant matrices of appropriate dimensions.

For continuous-time systems, a proportional plus switching reaching law is stated in Reference 30. In presence of an uncertainty in the input channel (matched uncertainty), this reaching law is modified as

$$\dot{s} = -\mu s - \varepsilon \operatorname{sign}(s) + d(t), \tag{3.3.2}$$

where $-d_m \leq d(t) \leq d_m$ is a bounded uncertainty derived from the uncertainty in the input channel of the system concerned, and $\mu > 0$ and $\varepsilon > d_m$ are parameters that would appear in the control. It is easy to show that under such conditions, the reaching law as (3.3.2) would satisfy the η-reachability condition [31], i.e., the sliding variable will reach the surface $s(t) = 0$ in finite time. Once $s(t) = 0$, the system (3.3.1) attains sliding mode, and one has complete disturbance rejection.

However, when the control u is applied through a ZOH, and assuming the disturbance f acting in the input channel before the ZOH, one arrives at the discretized system as

$$x(k+1) = Gx(k) + H(u(k) + f(k)), \tag{3.3.3}$$

where

$$G = e^{AT}, \quad H = \int_0^T e^{A\tau} B d\tau, \tag{3.3.4}$$

T being the sampling time or the time through which the ZOH holds the control u and the input disturbance f to a constant value $u(k)$ and $f(k)$, at the kth time step, i.e., during $kT \leq t < (k+1)T$.

It may seem that the assumption of a constant disturbance during the sampling interval is restrictive and of little importance, but it does not affect the analysis in the sequel. This is because of the following. Let us assume that the disturbance is not constant but varying during the interval $kT \leq t < (k+1)T$. However, by the nature of the disturbance, it is bounded as $|f(t)| \leq f_m$. Hence, while taking the integration during ZOH discretization, if one substitutes $-f_m$ or f_m for every $f(t)$ inside the interval, then

$$Gx(k) + H(u(k) - f_m) \leq x(k+1) \leq Gx(k) + H(u(k) + f_m), \tag{3.3.5}$$

allowing one to analyze for the worst-case conditions by considering the equalities. Any $|f(t)| < f_m$ will give rise to a solution point lying inside these worst-case scenarios. Considering either of the equalities, we get the same bound of the disturbance appearing in the sliding variable dynamics (3.3.10) which considers $f(t) = f(kT)$ to be constant throughout the interval. Since one is concerned about the bound of the disturbance and not the actual values of it in order to fix the controller parameters and determine bounded stability, the analysis for both cases ($f(t)$ held constant and $f(t)$ varying in the interval) can proceed in the same way.

The analysis centers around the application of a continuous-time sliding mode control developed using the continuous reaching law (3.3.2) to the continuous-time system (3.3.1) through a ZOH with holding time T. This is equivalent to a system where either the computation speed is $1/T$ or the sensor rate is $1/T$ readings per second or the actuator frequency is limited by $1/T$. In any case, the control applied is held constant throughout the interval T till the next value of the control signal is generated. This is equivalent to dealing with the discretized system (3.3.3) where the continuous control value calculated at $t = kT$, i.e., $u(kT) = u(k)$ is applied and held constant throughout the interval $kT \leq t < (k+1)T$.

Without loss of generality, one can assume the continuous-time system model (3.3.1) is in the regular form [31], since there always exists a transformation which would bring the model in a regular form. Then, the system model can be re-written as

$$\begin{aligned} \dot{x_1} &= A_{11}x_1 + A_{12}x_2, \\ \dot{x_2} &= A_{12}x_1 + A_{22}x_2 + B_2(u+f). \end{aligned} \tag{3.3.6}$$

One can design a linear sliding surface for the above continuous-time system as

$$s(t) = c^T x(t) = c_1 x_1 + c_2 x_2 \tag{3.3.7}$$

such that $c^T B \neq 0$.

Then the control calculated using the reaching law (3.3.2) is

$$u(t) = -(c^T B)^{-1}(c^T Ax(t) + \mu c^T x(t) + \varepsilon \text{sign}(c^T x(t))) \tag{3.3.8}$$

which would bring the sliding variable $s(t)$ to the surface $s(t) = 0$ in finite time and thereby maintain it on this manifold. The resulting system would then be asymptotically stable once the condition

$$\sigma(A_{11} - A_{12}c_2^{-1}c_1) < 0 \tag{3.3.9}$$

is satisfied by the surface matrix c^T [31], where $\sigma(X)$ is the spectrum of the matrix X.

However, the control value is changed only at defined intervals of time kT, $k = 0, 1, 2, \ldots$, and it is held constant for the next time interval T. Hence, at $t = kT$, one has $u(t) = u(kT) = u(k)$ and $x(t) = x(kT) = x(k)$. The system dynamics then is given by the discretized model (3.3.3) of the plant. Putting $u(t) = u(kT) = u(k)$ in it, we get in any interval $kT \leq t < (k+1)T$,

$$\begin{aligned} & x(k+1) = Gx(k) - H(c^T B)^{-1}(c^T Ax(k) + \mu c^T x(k) + \varepsilon \text{sign}(c^T x(k))) + Hf(k) \\ \text{or} \quad & s(k+1) = \Theta x(k) + (1 - \xi\mu)s(k) - \xi\varepsilon \text{sign}(s(k)) + c^T Hf(k), \end{aligned} \tag{3.3.10}$$

where $\Theta = (c^T G - \xi c^T A - c^T)$ and $\xi = (c^T H)(c^T B)^{-1}$. The disturbance term is $d(k) = c^T Hf(k)$, and hence $d_m = |c^T H| f_m$ is the absolute maximum of $d(k)$.

The sliding variable dynamics (3.3.10) is similar in structure to the Gao's reaching law

$$s(k+1) = (1 - \mu\tau)s(k) - (d_m + \varepsilon\tau)\text{sign}(s(k)) + d(k) \tag{3.3.11}$$

given in Reference 1, but only with the additional term $\Theta x(k)$ in the RHS. In the sequel, lemmas and the main theorem will be proposed which will guarantee stability of the above dynamics (3.3.10) inside an ultimate band. Also, stability of the sliding mode thus achieved will be proved, as it is not apparent that the sliding surface (3.3.7) designed for the continuous-time system (3.3.6) would still be able to guarantee stability of the sliding mode as $s(t) = 0$ is never attained due to the application of the control input through the ZOH.

3.3.3.1 Reaching phase and ultimate band

In this subsection, we shall derive the bounded stability of the sliding variable, thus finding out the ultimate band about the sliding surface. For this, we shall take help of the following lemmas and theorem.

Lemma 3.3.1. *Suppose the following relations hold:*

$$\text{(i)}\ \ B_d \in [d_m, 2d_m), \tag{3.3.12a}$$

$$\text{(ii)}\ \ \xi\mu \in (0, 1], \tag{3.3.12b}$$

$$\text{(iii)}\ \ \xi\varepsilon = \Theta x(k) sign(s(k)) - \xi\mu B_d + d_m. \tag{3.3.12c}$$

Then the reaching law (3.3.10) *satisfies* $|s(k+1)| < |s(k)|$ *whenever* $|s(k)| > B_d$.

Proof. Using (3.3.12c) and $\xi\mu > 0$ in the region $s(k) > B_d$ and considering the maximum value of uncertainty $d(k) = d_m$ in (3.3.10), we get

$$\begin{aligned} s(k+1) &= s(k) - \xi\mu(s(k) - B_d) + d(k) - d_m \\ \Rightarrow s(k+1) &< s(k). \end{aligned} \tag{3.3.13}$$

Similarly, in the region $s(k) < -B_d$ and considering the minimum value of uncertainty $d(k) = -d_m$, we would have arrived at $s(k+1) > s(k)$, i.e., $|s(k+1)| < |s(k)|$.

Let us define

$$B_c := 2d_m - B_d. \tag{3.3.14}$$

Using (3.3.12c) and $\xi\mu \leq 1$ in the region $s(k) > B_d$ and considering the minimum value of uncertainty $d(k) = -d_m$ in (3.3.10), we get

$$\begin{aligned} s(k+1) &= (1 - \xi\mu)s(k) + \xi\mu B_d + d(k) - d_m \\ \Rightarrow s(k+1) &> (1 - \xi\mu)B_d + \xi\mu B_d - 2d_m \\ &= -B_c. \end{aligned} \tag{3.3.15}$$

Similarly, in the region $s(k) < -B_d$ and considering the maximum value of uncertainty $d(k) = d_m$, we would have arrived at $s(k+1) < B_c$.

Now, from (3.3.12a) and (3.3.14), we get $B_c \in (0, d_m]$. This implies $B_c \leq B_d$. Hence, $|s(k+1)| < |s(k)|$ is ensured even when $s(k)$ crosses zero. □

Lemma 3.3.2. *Let the relations (3.3.12b) and (3.3.12c) hold for the reaching law (3.3.10). Additionally, suppose*

$$\xi\mu = \frac{2d_m}{B_d} - 1. \tag{3.3.16}$$

Then for $s(k) \in [-B_d, B_d]$, $|s(k+1)| \leq B_d$ *is satisfied.*

Proof. In the region $0 < s(k) \leq B_d$, using (3.3.12c) and (3.3.12b) in (3.3.10), we get

$$\begin{aligned} s(k+1) &= (1 - \xi\mu)s(k) + \xi\mu B_d - d_m + d(k) \\ &\leq (1 - \xi\mu)B_d + \xi\mu B_d \\ &= B_d \end{aligned} \tag{3.3.17}$$

by taking the maximum value of the disturbance $d(k) = d_m$. Similarly, in the region $0 > s(k) \geq -B_d$, we would get $s(k+1) \geq -B_d$ taking the minimum value of the disturbance $d(k) = -d_m$.

Let us define

$$B_s := 2d_m - \mu\tau B_d. \tag{3.3.18}$$

In the region $0 < s(k) \leq B_d$, using (3.3.12c) and taking $s(k) \to 0$ in (3.3.10), we get

$$\begin{aligned} s(k+1) &= \xi\mu B_d - d_m + d(k) \\ &\geq \xi\mu B_d - 2d_m \\ &= -B_s \end{aligned} \tag{3.3.19}$$

by taking the minimum value of the disturbance $d(k) = -d_m$. Similarly, in the region $0 > s(k) \geq -B_d$, we would get $s(k+1) \leq B_s$ taking the maximum value of the disturbance $d(k) = d_m$.

However, since $\xi\mu$ is given by (3.3.16), we get $B_s = B_d$. Hence $|s(k+1)| \leq B_d$ even when $\{s(k)\}$ crosses zero while $s(k) \in [-B_d, B_d]$. □

Theorem 3.3.1. *The sequence* $\{s(k)\}$ *following the reaching law (3.3.10) will be ultimately bounded by* $\delta = B_d \; \forall \, k \geq K \in \mathbb{N}$, *with*

1. $B_d \in [d_m, 2d_m)$,
2. $\xi\mu = \frac{2d_m}{B_d} - 1$ *and*
3. $\xi\varepsilon = \Theta x(k) sign(s(k)) - \xi\mu B_d + d_m$,

excepting the cases where $|s(0)| > B_d$ *and the disturbance sequence* $\{d(k)\} = d_m sign(s(0)) \, \forall \, k \geq 0$ *with* $B_d > d_m$ *chosen. In those cases,* $\delta = (1+\rho)B_d$, *with* $\rho > 0$, *however small.*

Proof. From Lemma 3.3.2, the ultimate band δ must be B_d if $|s_0| \leq B_d$, as in this case $|s(k)| \leq B_d \; \forall \, k \geq 0$ shall hold. But for $|s_0| > B_d$, Lemma 3.3.1 only guarantees

$|s(k+1)| < |s(k)|$ and not of $\{s(k)\}$ entering $[-B_d, B_d]$ at some finite $k = K \in \mathbb{N}$. However, for $B_d = d_m$ chosen, we have the values of the controller parameters as $\xi\mu = 1$ and $\xi\varepsilon = \Theta x(k) sign(s(k))$ from (3.3.16) and (3.3.12c), respectively, we have the reaching law (3.3.10) becoming same as Utkin's reaching law

$$s(k+1) = d(k) \tag{3.3.20}$$

in the presence of an uncertainty. Then it is easy to see that $s(k) \leq B_d$ will happen at the first instant since $|d(k)| \leq d_m = B_d$. Then Lemma 3.3.2 will follow $\forall\ k \geq 1$ and hence $\delta = B_d$ can be chosen.

But for other choices $B_d > d_m$, $\{s(k)\}$ entering $[-B_d, B_d]$ whenever $|s_0| > B_d$ cannot be guaranteed. In fact, mathematically, we can see from (3.3.10) and (3.3.12c) that in the region $s(k) > B_d$, for any uncertainty $d(k)$,

$$\begin{aligned} & s(k+1) \leq (1-\xi\mu)s(k) + (\xi\mu B_d - d_m) + d(k) \\ & \Rightarrow s(k) \leq (1-\xi\mu)^k s_0 + \sum_{i=0}^{k-1}(1-\xi\mu)^i(\xi\mu B_d - d_m) + \sum_{i=0}^{k-1}(1-\mu\tau)^i d_{k-1-i}. \end{aligned} \tag{3.3.21}$$

Now, for $d(k) = d_m\ \forall\ k \geq 0$, the above expression gives

$$\begin{aligned} & s(k) \leq (1-\xi\mu)^k s_0 + \sum_{i=0}^{k-1}(1-\xi\mu)^i(\xi\mu B_d - d_m) + \sum_{i=0}^{k-1}(1-\mu\tau)^i d_m \\ & \Rightarrow \lim_{k\to\infty} s(k) = \frac{1}{\xi\mu}\xi\mu B_d = B_d \end{aligned} \tag{3.3.22}$$

meaning that $\{s(k)\}$ never crosses B_d in finite time starting from $s(0) > B_d$. For all other $\{d(k)\}$, it is obvious from (3.3.21) that $\limsup_{k\to\infty} s(k) < B_d$, ensuring finite time entry of $\{s(k)\}$ in $[-B_d, B_d]$.

For $s_0 < -B_d$, the same analysis would have given us

$$\liminf_{k\to\infty} s(k) \geq -B_d \tag{3.3.23}$$

with $\lim_{k\to\infty} s(k) = -B_d$ happening for $\{d(k)\} = -d_m\ \forall\ k \geq 0$. Hence, we cannot guarantee $\{s(k)\}$ entering $[-B_d, B_d]$ in finite time for all disturbances $\{d(k)\}$. So for completeness, we need to make the ultimate band $\delta = (1+\rho)B_d$ where $\rho > 0$ (however small) when $B_d > d_m$ is chosen. Then we can be sure that $s(k) \leq \delta$ in some finite step $K \in \mathbb{N}$ and $s(k+1) \leq \delta$ will be ensured thereafter by Lemma 3.3.2, thus satisfying the definition of the ultimate band. □

Remark 3.3.1. *For almost all practical cases, we will have* $\{d(k)\} \neq d_m sign(s_0)$ *at least for one instant k. Then,* $\limsup_{k\to\infty} s(k) < B_d$ *and* $\liminf_{k\to\infty} s(k) > -B_d$*, guaranteeing* $\{s(k)\}$ *to enter* $[-B_d, B_d]$ *for some initial* $|s(0)| > B_d$*. Hence the ultimate band can be chosen* $\delta = B_d$ *in almost all practical cases.*

Remark 3.3.2. *The switching gain ε is adaptive in nature as per the analysis presented, because of the $\Theta x(k)$ term. Thus designing ε as a constant, as done for continuous time systems using proportional plus switching reaching law (Gao's design), one may not achieve convergence of the sliding variable to the specified band.*

3.3.3.2 Stability of sliding mode

Once sliding mode is established, as shown in the previous subsection, one needs to ensure that it is stable, in the sense that the states become bounded as time goes to infinity. Since the sliding surface is not reached exactly, and the system remains only inside an ultimate band about the surface, one cannot expect the states to reach zero asymptotically but to remain inside some bound of their own.

Considering the $x_1(t)$ dynamics from (3.3.6) and the sliding surface (3.3.7), one can write the following dynamics for the reduced-order system in the sliding mode:

$$\dot{x_1} = (A_{11} - A_{12}c_2^{-1}c_1)x_1 + A_{12}c_2^{-1}s, \tag{3.3.24}$$

Let us define $\widetilde{A} = A_{11} - A_{12}c_2^{-1}c_1$ and suppose $k = K$ be the time instant (step number) at which the sliding variable enters the ultimate band. Hence, we have $|s| \le \delta\, \forall\, k \ge K$. One can then write from (3.3.24) the following:

$$\begin{aligned} x_1(K+1) &= e^{\widetilde{A}T}x_1(K) + \int_0^T e^{\widetilde{A}t}A_{12}c_2^{-1}s(t)dt \\ \text{or}\quad |x_1(K+1)| &\le e^{\widetilde{A}T}|x_1(K)| + \widetilde{H}\delta, \end{aligned} \tag{3.3.25}$$

where $\widetilde{H} = \int_0^T e^{\widetilde{A}t}dt|A_{12}c_2^{-1}|$. Proceeding successively for future time steps, one can show that

$$x_1(K+n) \le e^{\widetilde{A}nT}|x_1(K)| + \sum_{i=0}^{n-1} e^{\widetilde{A}iT}\widetilde{H}\delta, \tag{3.3.26}$$

Thus, as $n \to \infty$, one gets $x_1(k)$ bounded by $(I - e^{\widetilde{A}T})^{-1}\widetilde{H}\delta$ since $\sigma(\widetilde{A}) < 0$. As x_2 is algebraically related to x_1 and s in the sliding mode, hence $x_2(k)$ is also bounded. This proves that if the sliding mode can be reached by discretized application of the control input, the obtained sliding mode would be stable for the same design of the sliding surface as for the continuous-time system for which the control was designed.

3.3.4 Simulation example

In this section, two examples are provided, one according to the proposed design and one according to Gao's design. We shall see that the design proposed in this chapter leads the sliding variable inside the ultimate band as chosen, whereas in Gao's design, the sliding variable dynamics is inferior in the sense that they are not bounded inside the band as chosen by our proposed method. To note, there is no proper analysis according to Gao's design on what would be the ultimate band in such a case.

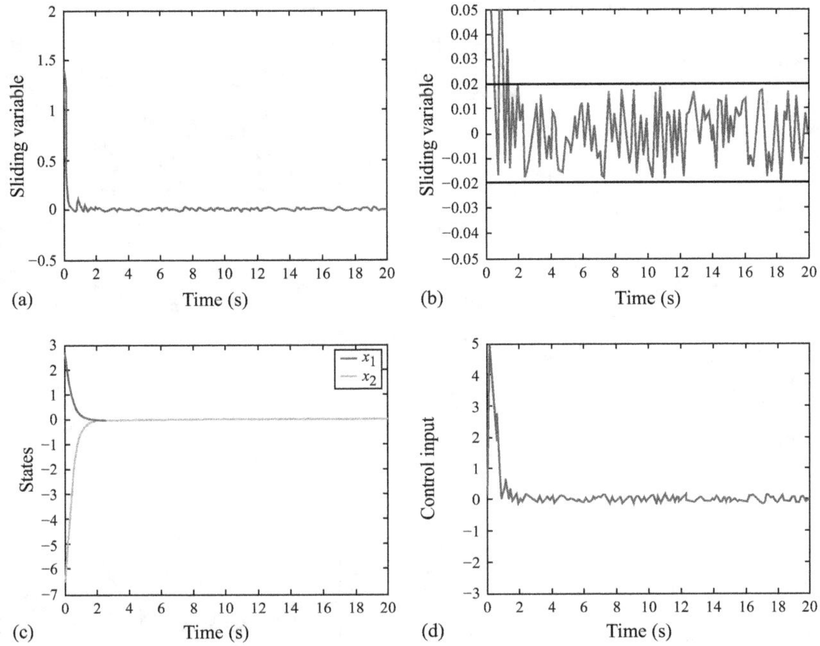

Figure 3.3.2 Simulations with band approach method. (a) Sliding variable dynamics, (b) zoomed view with ultimate band, (c) internal state dynamics, and (d) control input

3.3.4.1 Proposed design

An inherently unstable continuous-time system is considered with a matched uncertainty $f(t)$, random in nature but bounded as $-0.1 \leq f(t) \leq 0.1$ and initial state $[3 \quad -7]^T$.

$$\dot{x} = \begin{bmatrix} 0 & 1 \\ 3 & -0.5 \end{bmatrix} x + \begin{bmatrix} 0 \\ 1 \end{bmatrix} (u + f). \tag{3.3.27}$$

A time period $T = 0.15$ is chosen, which is on the higher side than normal, but keeping in mind the thumb rule to not exceed one-tenth of the system time constant. The surface matrix c^T is designed for closed-loop reduced-order system pole to be at -3.

$$s(k) = c^T x(k) = \begin{bmatrix} 3 & 1 \end{bmatrix} x(k). \tag{3.3.28}$$

With these, we get $d_m = 0.0179$. Hence, $\delta = 1.1 d_m = 0.0197$ is chosen. This yields the value of the controller parameter $\mu = 4.5645$.

Simulation results are shown in Figure 3.3.2 for the sliding surface, with zooming in to show that the sliding variable remains inside the chosen ultimate band, the state dynamics and the control.

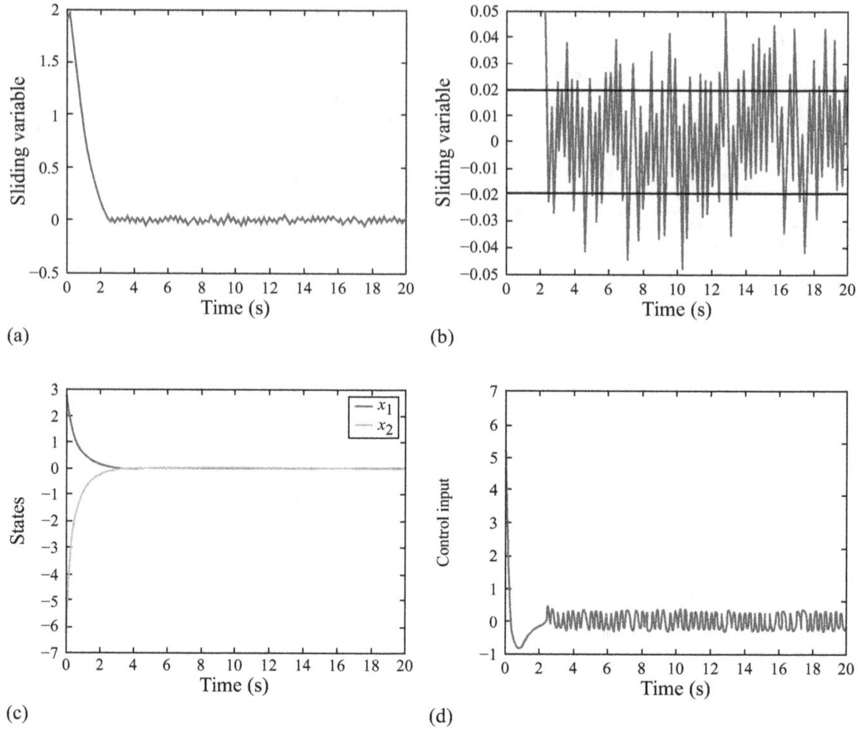

Figure 3.3.3 Simulations with Gao's method. (a) Sliding variable dynamics, (b) zoomed view with ultimate band, (c) internal state dynamics, and (d) control input

3.3.4.2 Gao's design

For the same system in the above subsection, the DSMC is designed as per Gao's method. The same sliding surface is chosen for the same time period and initial condition. However, we choose the controller parameter value $\mu = 1$ arbitrarily. One needs $\varepsilon > c^T B f_m = 0.1$ as per Gao's design. Hence $\varepsilon = 0.2$ is chosen. Note that there is no specific method of selection of these parameters following Gao's design. For this, simulation results are shown in Figure 3.3.3 for the sliding surface, the state dynamics, and the control. Also, a zoomed view of the sliding surface dynamics is shown, and for the purpose of reference, the ultimate band as obtained in the previous subsection (proposed design) is superimposed on it. This clearly shows that the system behavior using Gao's design method is inferior to the one proposed in this chapter.

3.3.4.3 Instability with classical design

With the same system, it is also found out that the classical design with proportional plus switching reaching law (3.3.2) as per Gao's method leads to instability with

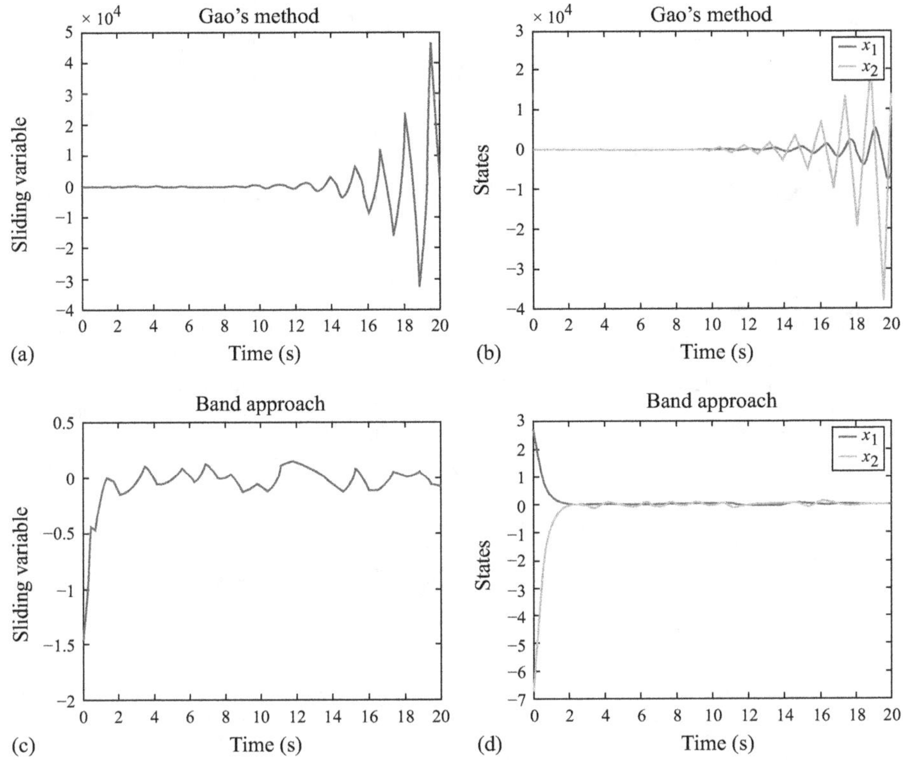

Figure 3.3.4 Simulations Gao's method and band approach method for T = 0.7 s and reduced-order pole at −3

increase in sampling time and/or moving the reduced-order pole farther left. Instability due to the former is apparent as the discrete time system deviates much from the continuous-time system with increase in time period. However moving the reduced-order pole farther left should give faster convergence of the states in the sliding mode, which can be a desirable choice for the control engineer. However, this leads to instability, even for a low sampling time. But when we apply our design methods on the sliding variable dynamics (3.3.10) obtained due to ZOH discretization of the plant, instability does not appear with increase of sampling time and/or moving the reduced-order pole farther left. This is shown in Figures 3.3.4 and 3.3.5, respectively.

3.3.5 The generalized reaching algorithm

From the above work, it was found that the band approach method is able to deal with a sliding variable dynamics containing even the states of the system. This naturally led to the thought of a generalized reaching law which not only contains functions of

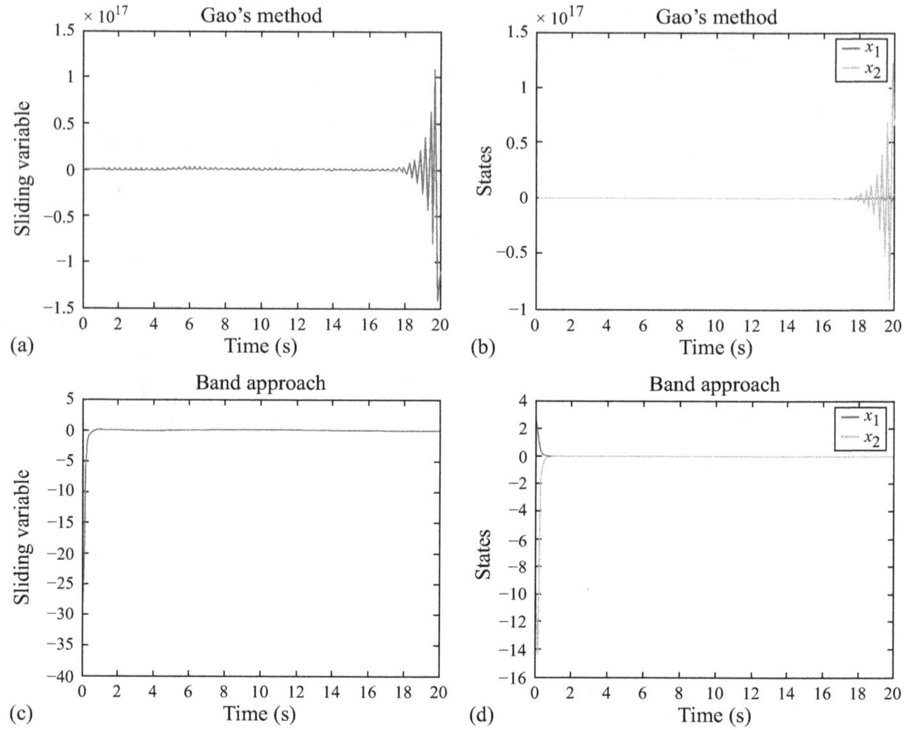

Figure 3.3.5 Simulations Gao's method and band approach method for T = 0.15 s and reduced-order pole at −15

the sliding variable but also contains functions of other known variables. This puts forward a potential to analyze complex dynamics of a system previously not dealt in sliding mode control literature. One such problem is analyzed in the sequel.

Let us propose a discrete time reaching law as

$$s(k+1) = f_1(s(k)) + f_2(\xi(k), k) + f_3 sign(s(k)) + d(k), \tag{3.3.29}$$

where f_1 and f_2 are functions in the variables mentioned, f_3 can either be a constant or a varying gain depending on a particular problem. Here, $s(k) \in \mathbb{R}$ and $\xi(k) \in \mathbb{R}^q$ are taken with k denoting the sample count, where $\xi(k)$ can be any variable other than $s(k)$, which is known at all k. The uncertainty $d(k)$ is assumed to vary only at the sampling instants and bounded. Also, $d(k)$ is assumed to be bounded by known functions and not by constant values, unlike in most cases where we deal matched disturbances using sliding mode control. Hence, $d(k) \in [d_-(k), d_+(k)]$ is assumed, $d_-(k)$ and $d_+(k)$ are known and bounded functions such that $d_-(k) < d_+(k)\ \forall\ k$. This means that $d(k)$ is any value with mean $d_0(k) = (d_+(k) + d_-(k))/2$ and spread $d_1(k) = (d_+(k) - d_-(k))/2$ at every instant k. Obviously, $d_0(k)$ and $d_1(k)$ are also known but varying functions.

Now, let us choose f_1 and f_2 in the generalized reaching algorithm (3.3.29) in such a way that it suits the designer to guarantee ultimately bounded stability of the sliding variable. In that direction, let $f_1(s(k)) = f_0 s(k)$ and $f_2(\xi(k), k) = -d_0(k)$ be chosen in the reaching law (3.3.29). Rewriting the same below, we get

$$s(k+1) = f_0 s(k) - d_0(k) + f_3 sign(s(k)) + d(k). \tag{3.3.30}$$

3.3.5.1 Main analysis with band approach method

With the help of two lemmas and a theorem, it will be shown that for the given problem, the sliding variable $s(k)$ following the reaching law (3.3.30) is ultimately bounded for some selection of f_0 and f_3. The analysis and proofs will be done for cases with $s(k) > 0$. The results obtained will be same if one had analyzed taking $s(k) < 0$ because of the symmetric nature of the reaching law.

Lemma 3.3.3. *Suppose the following relations hold:*

$$\textit{(i)}\ B_d(k) \in [d_1(k), 2d_1(k)), \tag{3.3.31a}$$

$$\textit{(ii)}\ f_0 \in [0, 1), \tag{3.3.31b}$$

$$\textit{(iii)}\ f_3 = (1 - f_0)B_d(k) - d_1(k). \tag{3.3.31c}$$

Then the reaching law (3.3.30) *satisfies* $|s(k+1)| < |s(k)|$ *whenever* $|s(k)| > B_d(k)$.

Proof. It is clear from the choice of $B_d(k)$ as per (3.3.31a) that it is always a non-negative function. Hence in the region $s(k) > B_d(k)$, we shall have $sign(s(k)) = 1$. Using (3.3.31c) and $f_0 < 1$ in the region $s(k) > B_d(k)$ and considering the maximum bound of the uncertainty $d(k) = d_+(k)$ in (3.3.30), we get

$$\begin{aligned} s(k+1) &= f_0 s(k) - d_0(k) + (1 - f_0)B_d(k) - d_1(k) + d_+(k) \\ &= s(k) - (1 - f_0)(s(k) - B_d(k)) \\ \Rightarrow s(k+1) &< s(k). \end{aligned} \tag{3.3.32}$$

Let us define

$$B_c(k) := 2d_1(k) - B_d(k). \tag{3.3.33}$$

Using (3.3.31c) and $f_0 \geq 0$ in the region $s(k) > B_d(k)$ and considering the minimum bound of uncertainty $d(k) = d_-(k)$ in (3.3.30), we get

$$\begin{aligned} s(k+1) &= f_0 s(k) - d_0(k) + (1 - f_0)B_d(k) - d_1(k) + d_-(k) \\ \Rightarrow s(k+1) &> f_0 B_d(k) + (1 - f_0)B_d(k) - 2d_1(k) \\ &= -B_c(k). \end{aligned} \tag{3.3.34}$$

Now, from (3.3.31a) and (3.3.33), we get $B_c(k) \in (0, d_1(k)]$. This implies $B_c(k) \leq B_d(k)$. Hence, (3.3.7) ensures that $|s(k+1)| < |s(k)|$ even when $s(k)$ crosses zero. □

Lemma 3.3.4. *Let the relations (3.3.31b) and (3.3.31c) hold for the reaching law (3.3.30). Additionally, suppose*

$$f_0 = 2 - \frac{2d_1(k)}{B_d(k)}. \tag{3.3.35}$$

Then $|s(k+1)| \leq B_d(k)$ *is satisfied whenever* $|s(k)| \leq B_d(k)$. *Further,* $|s(k+1)| \leq B_d(k+1)$ *is ensured if we choose*

$$B_d(k+1) \geq |s(k)| - (1 - f_0)(|s(k)| - B_d(k)). \tag{3.3.36}$$

Proof. In the region $0 < s(k) \leq B_d(k)$, using (3.3.31c) and (3.3.31b) in (3.3.30), we get

$$\begin{aligned} s(k+1) &= f_0 s(k) - d_0(k) + (1 - f_0)B_d(k) - d_1(k) + d(k) \\ &\leq f_0 B_d(k) + (1 - f_0)B_d(k) \\ &= B_d(k) \end{aligned} \tag{3.3.37}$$

by taking the maximum bound of the disturbance $d(k) = d_+(k)$.

Let us define

$$B_s(k) := 2d_1(k) - (1 - f_0)B_d(k). \tag{3.3.38}$$

In the region $0 < s(k) \leq B_d(k)$, using (3.3.31c) and taking $s(k) \to 0$ in (3.3.30), we get

$$\begin{aligned} s(k+1) &= -d_0(k) + (1 - f_0)B_d(k) - d_1(k) + d(k) \\ &\geq (1 - f_0)B_d(k) - 2d_1(k) \\ &= -B_s(k) \end{aligned} \tag{3.3.39}$$

by taking the minimum bound of the disturbance $d(k) = d_-(k)$. But since f_0 is given by (3.3.35), we get $B_s(k) = B_d(k)$. Hence $|s(k+1)| \leq B_d(k)$ even when $\{s(k)\}$ crosses zero.

Further, in the region $0 < s(k) \leq B_d(k)$, using (3.3.31c) and (3.3.31b) in (3.3.30), we get

$$\begin{aligned} s(k+1) &\leq f_0 s(k) - d_0(k) + (1 - f_0)B_d(k) - d_1(k) + d_+(k) \\ &= f_0 s(k) + (1 - f_0)B_d(k) \\ &= s(k) - (1 - f_0)(s(k) - B_d(k)) \\ &\leq B_d(k+1) \end{aligned} \tag{3.3.40}$$

as per assumption (3.3.36). This ensures that $|s(k)|$ continues to be inside $[-B_d(k), B_d(k)]$ once it is inside this set. □

Theorem 3.3.2. *Suppose* $B_d(k) \in [d_1(k), 2d_1(k))$ *is chosen following the recursion (3.3.36)* $\forall\, k \geq K \in \mathbb{N}$. *Then the sequence* $\{s(k)\}$ *following the reaching law (3.3.30) will be ultimately bounded by* $\delta(k) = B_d(k)\ \forall\, k > K \in \mathbb{N}$ *with*

1. $f_0 = 2 - \frac{2d_1(k)}{B_d(k)}$ *and*
2. $f_3 = (1 - f_0)B_d(k) - d_1(k)$.

Proof. Lemma 3.3.3 guarantees $|s(k+1)| < |s(k)|$ whenever $|s(k)| > B_d(k)$. Now, for $s(k) > B_d$, from the reaching law (3.3.30) and using (3.3.31c), we have

$$\begin{aligned} s(k+1) &= f_0 s(k) - d_0(k) + (1 - f_0)B_d(k) - d_1(k) + d(k) \\ &\leq f_0 s(k) - d_0(k) + (1 - f_0)B_d(k) - d_1(k) + d(k) \\ &= s(k) - (1 - f_0)(s(k) - B_d(k)) \\ &\leq B_d(k+1), \end{aligned} \tag{3.3.41}$$

where $\forall\, k \geq K \in \mathbb{N}$ as per (3.3.36). Hence, $s(k) \leq B_d(k)$ happens for the first time at $k = K + 1$ and then onward, Lemma 3.3.4 will follow, thus guaranteeing $|s(k)| \leq B_d(k)\ \forall\, k > K$. Hence the ultimate bound is $\delta(k) = B_d(k)$, where $B_d(k)$ is evaluated as per the recursion (3.3.36) $\forall\, k > K \in \mathbb{N}$. □

Remark 3.3.3. *The bound $B_d(k)$ is dynamically evaluated from $k = K + 1$ onward as per (3.3.36), depending on the values of $B_d(k)$ and $s(k)$ of the previous instant. But whatever the value of $B_d(k)$ is evaluated, we shall always have $|s(k+1)| \leq B_d(k)$ from Lemma 3.3.4 and the evaluation of $B_d(k+1)$ as per (3.3.36) would ensure $|s(k+1)| \leq B_d(k+1)$ as well, thereby keeping $|s(k)|$ within the envelope of $\{B_d(k)\}$.*

Remark 3.3.4. *We find $\{s(k)\}$ to be theoretically bounded inside a known varying envelope $[-B_d(k), B_d(k)]$ rather than a constant ultimate band. The known function $B_d(k)$ is obviously bounded since $d_1(k)$ is a bounded function and $B_d(k) \in [d_1(k), 2d_1(k)]$. This means that $0 \leq |s(k)| \leq B_d(k)$ at each instant $k \geq K$. The concept of an ultimate band is hereby relaxed by the concept of ultimate boundedness by a known function. However, if one needs to put an ultimate band around $s(k) = 0$, then it would be the global maximum of $B_d(k)$ over the entire time interval. This is not necessary, but only to have an understanding of some ultimate band, as in traditional literature of discrete sliding mode control.*

Remark 3.3.5. *If the disturbance is assumed to be bounded by a constant value, as is traditionally done in literature, then in the presented problem this value will be the absolute maximum of the two given functions that bound the disturbance. Hence, a control calculated considering that maximum bound of the disturbance would require more energy to drive the sliding variable inside an ultimate band than a control calculated as per the theory presented in this chapter. This is because in the presented theory, the disturbance bound considered at most instants will be less than the absolute maximum of the two given functions.*

3.3.5.2 Simulation example

Convergence of the reaching law (3.3.30) is shown, with $d_+(k)$ as a repeating saw-tooth wave with range from 0.1 to 0.2 covered in every 2.5 s and $d_-(k)$ as a sine wave with zero bias, amplitude 0.1, and frequency 2 rad/s. The global absolute maximum value of the disturbance is hence 0.2.

Simulation is performed as per the theory in the above subsection (Figure 3.3.6) and also assuming the disturbance to have bounds ± 0.2 (Figure 3.3.7). The theory for the latter using band approach method is discussed in Reference 21. The initial value of the sliding variable is taken as 5, and the time period as 0.1 s.

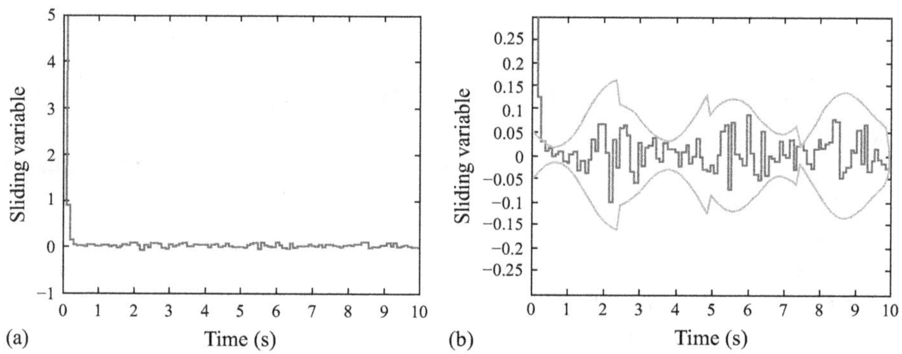

Figure 3.3.6 Simulations showing the sliding variable dynamics with uncertainty bounds $d_+(k)$ and $d_-(k)$. (a) Sliding variable dynamics and (b) zoomed view with ultimate bounding function

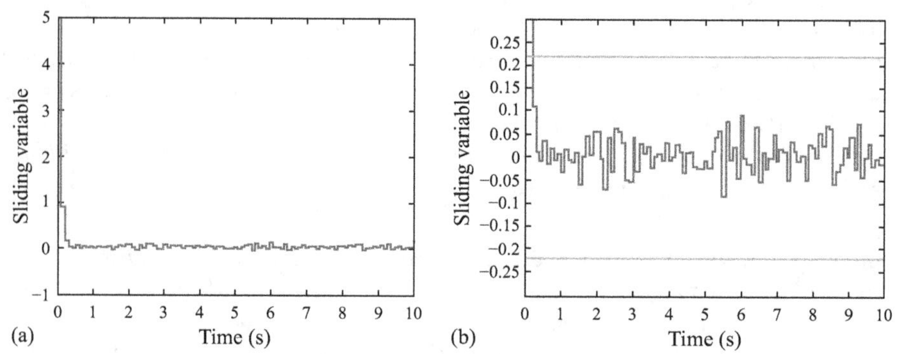

Figure 3.3.7 Simulations showing the sliding variable dynamics with uncertainty bounds ± 0.2. (a) Sliding variable dynamics and (b) zoomed view with ultimate band

3.3.6 Summary

In this chapter, the authors investigate the dynamics of a continuous-time system, when a sliding mode control designed for the continuous system is applied to the system in a discretized manner, viz., through a ZOH. Convergence of the sliding variable to an ultimate band is proved, when the controller parameters are chosen as per the band approach method. Such investigation into the ultimate band when continuous sliding mode control is applied in a discretized fashion is the novelty of this work. Also, the author proposes the calculation of the controller parameters as per a chosen ultimate band. It is found that the switching gain needs to be adaptive in nature to give rise to the ultimate band as desired, unlike the constant switching gain design of traditional sliding mode control as per Gao's method. Also, a generalized discrete time sliding mode control algorithm is proposed, which contains functions not only in the sliding variable but also in the other known variables. With the help of the proposed reaching law, a sliding variable dynamics with an uncertainty bounded by known functions is studied. By appropriate choices of the functions in the generalized reaching law and using the band approach method, the boundedness of the sliding variable inside a chosen function is ensured, which is bounded in nature. The algorithm parameters are calculated as per the chosen function, and all of them are varying functions. Boundedness of the sliding variable inside such a chosen function has not been encountered in previous literature of discrete sliding mode control. The generalized reaching law as proposed in this chapter gives us the power and flexibility to design the reaching law parameters and functions in such a way as to find out a solution to such novel problems.

References

[1] W. Gao, Y. Wang, and A. Homaifa, "Discrete-time variable structure control systems," *IEEE Transactions on Industrial Electronics*, vol. 42, no. 2, pp. 117–122, 1995.

[2] K. Furuta, "Sliding mode control of a discrete system," *Systems and Control Letters*, vol. 14, no. 2, pp. 145–152, 1990.

[3] S. Janardhanan and B. Bandyopadhyay, "Multirate output feedback based robust quasi-sliding mode control of discrete-time systems," *IEEE Transactions on Automatic Control*, vol. 52, no. 3, pp. 499–503, 2007.

[4] G. Bartolini, A. Ferrara, and V. Utkin, "Adaptive sliding mode control in discrete-time systems," *Automatica*, vol. 31, no. 5, pp. 769–773, 1995.

[5] A. Bartoszewicz, "Discrete-time quasi-sliding-mode control strategies," *IEEE Transactions on Industrial Electronics*, vol. 45, no. 4, pp. 633–637, 1998.

[6] V.I. Utkin and S.V. Drakunov, "On discrete-time sliding mode control," *Proceedings of IFAC Symposium on Nonlinear Control Systems (NOLCOS)*, Jul. 1989, pp. 484–489, 1989.

[7] B. Bandyopadhyay and S. Janardhanan, *Discrete-Time Sliding Mode Control: A Multirate Output Feedback Approach*, vol. 323, LNCIS, Berlin Heidelberg: Springer-Verlag, 2006.

[8] C. Miloslavjevic, "General conditions for the existence of a quasi-sliding mode on the switching hyperplane in discrete variable structure systems," *Automation and Remote Control*, vol. 46, pp. 679–684, 1985.

[9] S.Z. Sarpturk, Y. Istefanopulos, and O. Kaynak, "On the stability of discrete-time sliding mode control systems," *IEEE Transactions on Automatic Control*, vol. 32, no. 10, pp. 930–932, 1987.

[10] M. Livne and A. Levant, "Proper discretization of homogeneous differentiators," *Automatica*, vol. 50, pp. 2007–2014, 2014.

[11] Y. Yan, X. Yu, and C. Sun, "Discretization behaviors of a super-twisting algorithm based sliding mode control system," *2015 International Workshop on Recent Advances in Sliding Modes (RASM)*, Apr. 2015, pp. 1–5, 2015.

[12] O. Huber, V. Acary, B. Brogliato, and F. Plestan, "Discrete-time twisting controller without numerical chattering: Analysis and experimental results with an implicit method," *2014 IEEE 53rd Annual Conference on Decision and Control (CDC)*, Dec. 2014, pp. 4373–4378, 2014.

[13] A. Levant, "Sliding order and sliding accuracy in sliding mode control," *International Journal of Control*, vol. 58, no. 6, pp. 1247–1263, 1993.

[14] V. Acary and B. Brogliato, "Implicit Euler numerical simulations of sliding mode systems," *Systems and Control Letters*, vol. 59, pp. 284–293, 2010.

[15] B. Wang, B. Brogliato, V. Acary, A. Boubakir, and F. Plestan, "Experimental comparisons between implicit and explicit implementations of discrete-time sliding mode controllers: Towards chattering suppression in output and input signals," *2014 13th International Workshop on Variable Structure Systems (VSS)*, Jun. 2014, pp. 1–6, 2014.

[16] V.I. Utkin, "Sliding mode control design principles and applications to electric drives," *IEEE Transactions on Industrial Electronics*, vol. 40, no. 1, pp. 23–36, Feb. 1993.

[17] R. Venkataramanan, A. Sabanovic, and S. Cuk, "Sliding mode control of dc-to-dc converter," *Proceedings IEEE IECON*, pp. 251–258, 1985.

[18] M. Carpita, M. Marchesoni, M. Oberti, and L. Puglisi, "Power conditioning system using sliding mode control," *IEEE Power Electronics Specialists Conference*, pp. 626–633, 1986.

[19] A. Sabanovic, N. Sabanovic, and O. Music, "Sliding mode control of dc–ac converter," *IEEE Power Electronics Specialists Conference*, pp. 560–566, 1986.

[20] S. Chakrabarty and B. Bandyopadhyay, "Discrete time sliding mode control with minimal quasi sliding mode band," *12th International Workshop on Variable Structure Systems*, 12–14 Jan. 2012, pp. 404–409, 2012.

[21] S. Chakrabarty and B. Bandyopadhyay, "Minimum ultimate band design of discrete sliding mode control," *Asian Journal of Control*, vol. 17, no. 5, pp. 1889–1897, 2015.

[22] S. Chakrabarty and B. Bandyopadhyay, "On digital implementation of continuous sliding mode control," *39th Annual Conference on IEEE Industrial Electronics Society*, pp. 3640–3644, 2011.

[23] S. Chakrabarty and B. Bandyopadhyay, "A generalized reaching law for discrete time sliding mode control," *Automatica*, vol. 52, pp. 83–86, 2015.

[24] Q. Zhu, T. Wang, M. Jiang, and Y. Wang, "A new design scheme for discrete-time variable structure control systems," *International Conference on Mechatronics and Automation*, 9–12 Aug. 2009, pp. 3475–3479, 2009.

[25] Y. Niu, D.W.C. Ho, and Z. Wang, "Improved sliding mode control for discrete-time systems via reaching law," *IET Control Theory and Applications*, vol. 4, no. 11, pp. 2245–2251, 2010.

[26] L. Yuan, J. Shen, F. Xiao, and H. Wang, "A novel reaching law approach of quasi-sliding-mode control for uncertain discrete-time systems," *Journal of Central South University*, vol. 19, no. 9, pp. 2514–2519, 2012.

[27] A. Bartoszewicz and P. Lesniewski, "Reaching law-based sliding mode congestion control for communication networks," *IET Control Theory & Applications*, vol. 8, no. 17, pp. 1914–1920, 2014.

[28] A. Bartoszewicz and P. Latosinski, "Discrete time sliding mode control with reduced switching—A new reaching law approach," *International Journal of Robust and Nonlinear Control*, vol. 26, no. 1, pp. 47–68, 2016.

[29] A. Bartoszewicz and P. Lesniewski, "A new reaching law based sliding mode flow controller for connection-oriented data transmission networks," *13th IEEE Workshop on Variable Structure Systems*, Jun.–Jul., pp. 1–6, 2014.

[30] J.Y. Hung, W. Gao, and J.C. Hung, "Variable structure control: A survey," *IEEE Transactions on Industrial Electronics*, vol. 40, no. 1, pp. 2–22, 1993.

[31] C. Edwards and S. Spurgeon, *Sliding Mode Control: Theory and Applications*, London: Taylor and Francis, 1998.

Section 4

Applications

Chapter 4.1

Conventional and adaptive second-order sliding mode control of a wind energy conversion system

Carolina Evangelista, Alessandro Pisano, Paul Puleston, and Elio Usai

Abstract

In this chapter, an adaptive second-order sliding mode technique is exploited to optimize the efficiency of a variable-speed wind turbine. A revisited form of a recently proposed adaptation algorithm is proposed to deal with the characteristics and control requirements of the wind energy conversion system (WECS), particularly model uncertainties and fastly varying disturbances due to gusty wind effects. The revisited algorithm enhances the reactivity of the adaptation strategy against fast varying uncertainties. The proposed approach is successfully used to control a doubly-fed induction generator (DFIG)-based wind turbine topology proving its suitability for this application area. The design and convergence analysis of the adaptive controller are developed considering a reduced-order model of the DFIG. However, the performance of the closed-loop system is extensively assessed through computer simulations made also over a full-order realistic model of the WECS under study.

4.1.1 Introduction

Motivated by environmental concerns and strategic reasons, increasing attention is being paid to wind energy, which is one of the most promising renewable clean energy sources [1, 2].

From a control point of view, a wind turbine is a complex system involving nonlinear dynamics with strongly coupled internal variables, external disturbances, and parameter uncertainties, and it is driven by a stochastic input, such as the wind. Furthermore, modern wind turbines are controlled not only to maximize power production but also, e.g., to meet power quality grid specifications and reducing structural loading [3], which give rise to conflicting requirements. Despite the growth in the installed capacity of wind turbines, many unsolved aspects remain and advanced nonlinear control techniques are required to meet the challenge [4, 5].

Sliding mode (SM) control proves to be capable to cope with complex characteristics such as nonlinear uncertain dynamics, model uncertainties, and unmodeled perturbations. Since its origins [6, 7], SM control evolved into a robust and powerful design control technique for a wide range of applications [8–12] The peculiar aspect of conventional SM control is the discontinuous nature of the underlying control action, providing excellent system performance, which includes insensitivity to given uncertainty factors and finite-time convergence.

In spite of the excellent theoretical properties, direct practical application of such discontinuous effort is not adequate for some actual plants. Additionally, it can generate undesirable output chattering, which deteriorates the robustness of conventional SM. To attenuate this problem, the concept of higher-order SMs was introduced and, specifically, several second-order sliding mode (SOSM) algorithms were presented [13–15]. Since then, the number of publications on SOSM theory and applications has grown exponentially (see, e.g., [16, 17] and references therein).

Among the spectrum of SOSM schemes, a number of interesting SOSM solutions relying on Lyapunov-based adaptive mechanisms were presented in the last decade (see, for instance, [18–23]). A different adaptation philosophy was recently undertaken, and as a result, a readily implementable and straightforward adaptive SOSM control (SOSMC) strategy was introduced in Reference 24. Instead of a Lyapunov-based gain adaptation (as in the aforementioned works), this adaptation mechanism takes advantage of the inherent nature of real (nonideal) SM, existing in actual systems operating at finite switching frequency. Its adaptation policy depends on counting the zero-crossings of the sliding variable during appropriate adaptation time windows. Then, the occurrence of real SM is verified by checking whether such count is large enough in accordance to an SM existence criterion. The resulting adaptive controller is endowed with the capability of online adjusting the discontinuous control gain, maintaining the control magnitude at the minimum admissible level (rather than the worst-case over-conservative level, as in fixed-gain SM). Moreover, without adding a great degree of extra complexity to the real-time computations, the proposed adaptation preserves the robustness of its parental fixed-gain SOSM algorithm, while allowing an effective response to unexpected changes in the working conditions and performance enhancement of the controlled system (e.g., actuators' stress alleviation and chattering reduction).

This adaptive SOSMC algorithm has been applied in References 25 and 26 to adjust the parameters of the twisting SOSM algorithm [13] in robotic and automotive applications, respectively. It proved to be specially suitable for implementation in standard microprocessor devices. In fact, it was experimentally verified in Reference 25 by means of an industrial manipulator. More recently in Reference 27, the same logic, combined with a switched adaptation algorithm (varying gains depending on the operating region in the state space), was applied to a suboptimal SOSM algorithm [15], which requires the recording of a limited number of past measurements of the available sliding variable for its implementation, thereby needing no differentiation.

The objective of this chapter is to design, analyze, and assess the feasibility of novel adaptive SOSM techniques to control variable-speed wind turbines. To face the stringent specifications of the control problem in the WECS scenario, where the uncertainties are subject to fast variations due to the wind effect, a revisited form of the adaptation logic is sought. It is based on appropriate receding horizon adaptation time windows rather than on fixed, adjacent, and nonoverlapping ones, as were used in the previous proposals. This modification, which enhances the reactivity of the adaptation strategy against fast varying uncertainties, represents the main theoretical novelty of this work.

The chapter is structured as follows. Section 4.1.2 describes the WECS under study and states the associate control problem. Section 4.1.2.1 presents the reduced-order mathematical model that will be used for controller design. Section 4.1.3 outlines the suggested sliding manifold design and presents the underlying sliding variable dynamics. Section 4.1.4 describes the proposed adaptive controller and provides the relevant performance analysis of the closed-loop system. Section 4.1.5 presents some simulation results, whereas Section 4.1.6 discusses concluding remarks and outlines possible future perspectives of the present line of research. Finally, two appendices (A.1 and A.2) present, respectively, the full-order dynamical model of the WECS that is considered in the simulative analysis and the nominal values of the underlying electromechanical parameters.

4.1.2 Wind energy conversion system

The mechanical power a real turbine can extract, P_t, is a fraction of the available power in the wind, P_ν. It is determined by the power coefficient of the turbine, $C_p(\lambda, \beta)$, [28]:

$$P_t = C_p(\lambda, \beta)P_\nu = C_p(\lambda, \beta)0.5\pi\rho R^2\nu^3, \tag{4.1.1}$$

where ν is the wind speed, ρ is the air density, and R is the blades length. The power coefficient $C_p(\cdot)$, depicted in Figure 4.1.1, depends on the shape and geometrical dimensions of the turbine. Particularly, it is a nonlinear function of the pitch angle β of the blades and of the tip speed ratio $\lambda = \frac{\Omega_g R}{k_{gb}\nu}$, where Ω_g is the angular speed of the generator rotor. A rigid drive train is assumed, and all the turbine variables are referred to the fast shaft of generator side through the transmission relation k_{gb} of the speed multiplier (see Section A.2). The curve $C_p(\lambda, \beta)$ presents a unique maximum, $C_{p\max}$, for $\lambda = \lambda_{\text{opt}}$ and $\beta = \beta_{\text{opt}}$, corresponding to the condition of maximum power extraction (Figure 4.1.2).

The operation of a wind turbine can be divided into four zones depending on the wind speed. The objective of this work is to attain maximum energy conversion efficiency when operating below the rated speed, within the so-called zone II, as it is shown in Figure 4.1.3 [28]. The control in this region is usually performed electronically, maintaining the pitch angle fixed and aiming to attain $\lambda = \lambda_{\text{opt}}$ by adjusting the generator speed accordingly. Above the rated speed, the extracted power should be limited, and the existence of a power limiting mechanism for operation in this zone is assumed in the present work.

The WECS topology considered in this study is based on a DFIG with slip power recovery known as *static Kramer drive*, which is schematically depicted in Figure 4.1.4. In this configuration, the stator power is directly fed to the grid, whereas

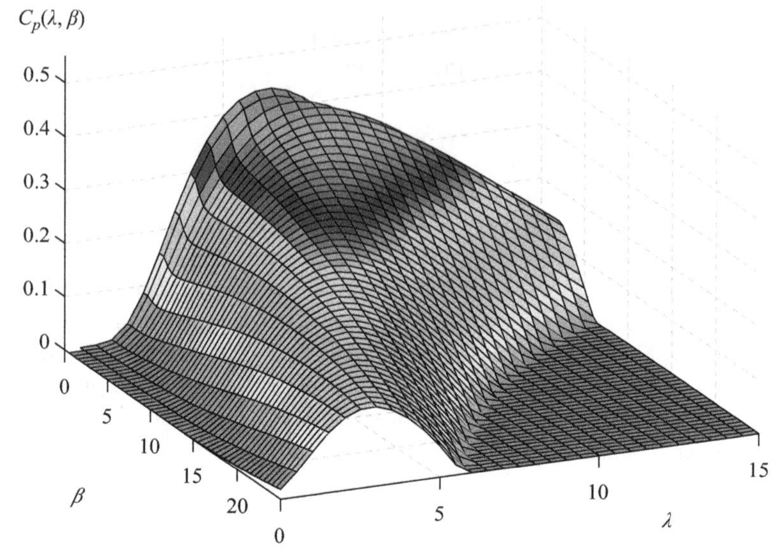

Figure 4.1.1 Power coefficient of the turbine, $C_p(\lambda, \beta)$

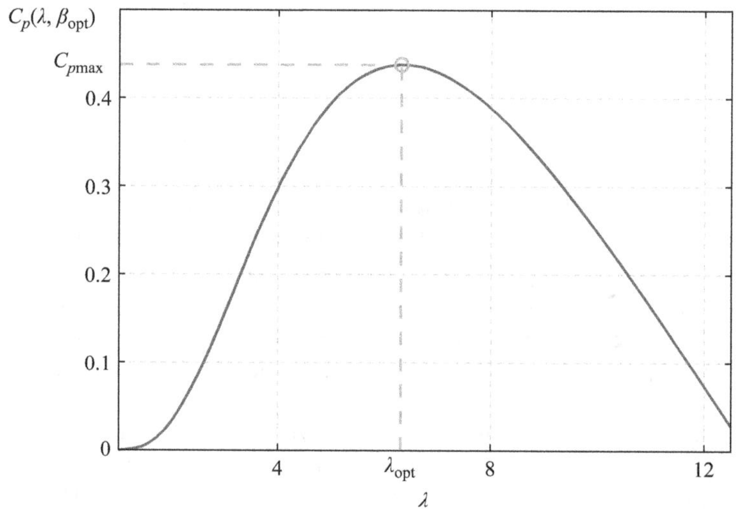

Figure 4.1.2 Unique maximum power extraction condition when $\beta = \beta_{opt}$: C_{pmax}

the rotor power is partially recovered through the electronic drive. The latter comprises an uncontrolled bridge rectifier, a line-commutated inverter, and a smoothing reactor. The generator torque, and hence the system operation speed, can be controlled by modifying the inverter firing angle α_{fa}.

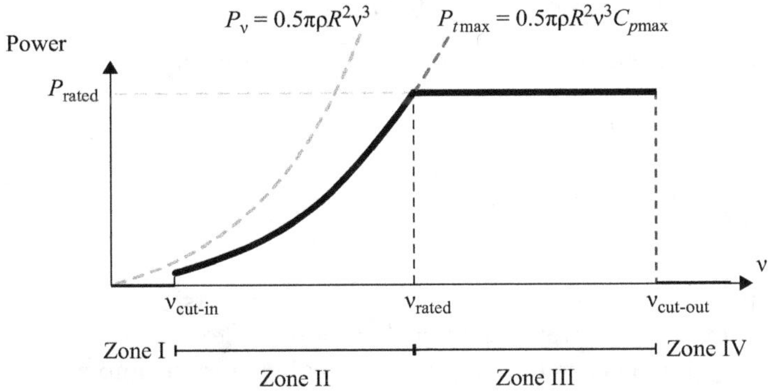

Figure 4.1.3 *Zones of operation for a wind turbine. The objective in zone II is to extract maximum power $P_{tmax} = P_v C_{pmax}$*

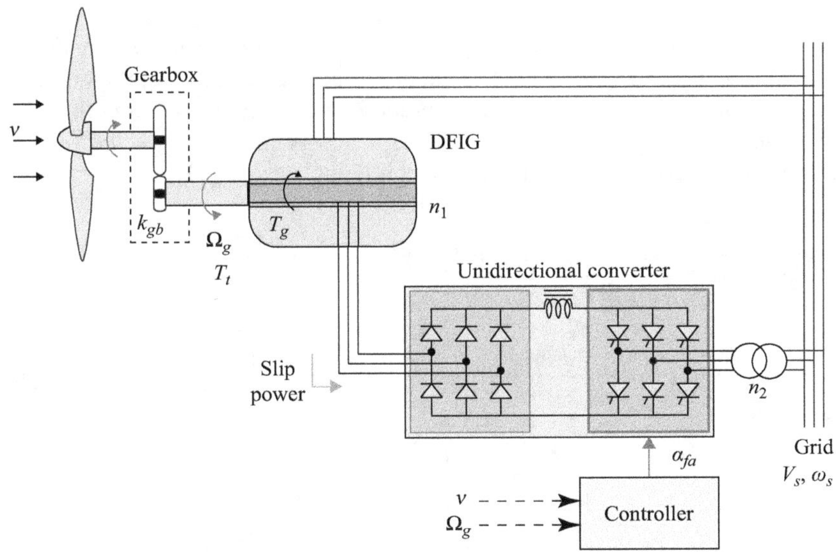

Figure 4.1.4 *DFIG with slip power recovery (configuration type: static Kramer drive)*

A complete dynamical description of the system comprises five state variables, including the mechanical rotational speed (Ω_g) and four electric currents Appendix A.1 and A.2. Given that the electrical dynamics are considerably faster than the mechanical one, a reduced-order model including only the mechanical variable dynamics is used for the control design process, whereas the simulation testing and validation of the method will be carried out using both the simplified and complete models.

4.1.2.1 Reduced-order model for control design

The nonlinear differential equation that describes the mechanical dynamics of the WECS is

$$\dot{\Omega}_g = \frac{1}{J}(T_t(\nu, \Omega_g) + T_g(\Omega_g, u)), \tag{4.1.2}$$

where J is the inertia of the whole combined rotating parts, T_t is the torque exerted by the wind on the blades (referred to the high speed side), and $T_g < 0$, the electrical resistant torque of the generator. The expression of the turbine torque is

$$T_t(\nu, \Omega_g) = \frac{P_t}{\Omega_g} = \frac{\pi \rho R^3}{2k_{gb}} C_t(\lambda)\nu^2, \tag{4.1.3}$$

where $C_t(\lambda) = C_p(\lambda)/\lambda$ is the torque coefficient which can be modeled as $C_t(\lambda) = c_3\lambda^3 + c_2\lambda^2 + c_1\lambda + c_0$ for some appropriate choice of the coefficients c_i [28].

As previously stated, the generator torque is modified by adjusting the inverter firing angle α_{fa}, and the manipulable control action is chosen as $u = |cos(\alpha_{fa})|$ as it simplifies the underlying mathematical design. Thus, we get [29]

$$T_g(\Omega_g, u) = \frac{3V_s^2}{\Omega_s} \frac{R_{eq}/s}{[(R_s + R_{eq}/s)^2 + x_s^2]}, \tag{4.1.4}$$

$$R_{eq}/s = \frac{R_b n_2^2 s + R_s n_1^2 u^2 - n_1 u \sqrt{|\psi|}}{n_2^2 s^2 - n_1^2 u^2}, \tag{4.1.5}$$

$$\psi = n_2^2(sR_s + R_b)^2 + x_s^2(n_2^2 s^2 - n_1^2 u^2), \tag{4.1.6}$$

where $\Omega_s = \omega_s/p_p$, V_s and ω_s are the grid line voltage and frequency, p_p is the number of pole pairs of the generator, $s = (\Omega_s - \Omega_g)/\Omega_s$ is the generator slip, R_s, R_r, and R_f are the resistances of the stator, rotor, and dc-link, respectively. L_s and L_r are the leakage inductances of the stator and rotor windings, n_1, n_2 are the turns ratios of the generator and of the step-down transformer, respectively, and it has used $R_b = R_r + R_f\pi^2/18$ and $x_s = \omega_s(L_s + L_r)$.

Given that power efficiency maximization is obtained when $\lambda = \lambda_{\text{opt}}$, the control strategy consists in tracking a time-varying optimum speed reference $\Omega_{\text{ref}}(t)$, depending on the actual wind speed, having the following expression:

$$\Omega_{\text{ref}}(t) = \frac{\lambda_{\text{opt}} k_{gb}}{R} \nu(t). \tag{4.1.7}$$

4.1.3 Sliding manifold design

Since the control task is to force the angular speed of the generator to track the optimal reference (4.1.7), the following sliding variable is considered

$$\sigma(t) = \Omega_g(t) - \Omega_{\text{ref}}(t) = \Omega_g(t) - \frac{\lambda_{\text{opt}} k_{gb}}{R} \nu(t). \tag{4.1.8}$$

By virtue of (4.1.2), the time derivative of the sliding variable takes the form

$$\dot{\sigma}(t) = \dot{\Omega}_g(t) - \frac{\lambda_{\text{opt}} k_{gb}}{R} \dot{\nu}(t) = \frac{1}{J} T_t(t) + \frac{1}{J} T_g(t) - \frac{\lambda_{\text{opt}} k_{gb}}{R} \dot{\nu}(t). \tag{4.1.9}$$

Note that the control input $u(t)$ enters nonlinearly the right-hand side of (4.1.9) through the electrical resistant torque term $T_g(t)$ which is given by (4.1.4)–(4.1.6). Differentiating further, (4.1.9) yields

$$\ddot{\sigma}(t) = \frac{1}{J} \dot{T}_t(t) + \frac{1}{J} \dot{T}_g(t) - \frac{\lambda_{\text{opt}} k_{gb}}{R} \ddot{\nu}(t), \tag{4.1.10}$$

which can be rewritten in the more convenient form

$$\ddot{\sigma}(t) = f(\Omega_g, \nu, \dot{\nu}, \ddot{\nu}, u) + g(\Omega_g, u)\dot{u} \tag{4.1.11}$$

highlighting the affine dependence of $\ddot{\sigma}$ on the control derivative $\dot{u}$, and where the drift term $f(\cdot)$ and the control gain $g(\cdot)$ are expressed as functions of the respective affecting signals. We can write down in different form the second-order nonlinear uncertain system (which is often referred to as the "auxiliary" system) governing the dynamics of the sliding variable, which takes the form

$$\begin{aligned} z_1(t) &= \sigma(t), \\ \dot{z}_1(t) &= z_2(t), \\ \dot{z}_2(t) &= f(\cdot) + g(\cdot) w(t), \end{aligned} \tag{4.1.12}$$

where $w(t) = \dot{u}(t)$ is the control signal, and $f(\cdot)$ and $g(\cdot)$ are uncertain functions satisfying the following inequalities:

$$0 < G_m \leq g(\cdot) \leq G_M, \qquad |f(\cdot)| \leq F. \tag{4.1.13}$$

The control problem is formulated as follows: given system (4.1.12), where $f(\cdot)$ and $g(\cdot)$ satisfy (4.1.13) for some apriori known constants F, G_m, G_M, design a control signal $w(t)$ capable of steering both $z_1(t)$ and $z_2(t)$ to zero in finite time.

In order to evaluate appropriate uncertainty bounds, the drift term and the control gain have to be derived explicitly. By differentiating (4.1.3) and (4.1.4), and considering the relations (4.1.2) and (4.1.6), it is not difficult to derive the next explicit

Table 4.1.1 Ranges for the uncertain time varying signals

Ω_g:	$\in [\Omega_s\,;\, 2\Omega_s]$ (rad/s)
$-T_g$:	$\in [35\,;\, 450]$ (Nm)
ν:	$\in [6\,;\, 13]$ (m/s)
$\dot{\nu}$:	$\in [-0.5\,;\, 0.5]$ (m/s^2)
$\ddot{\nu}$:	$\in [-0.1\,;\, 0.1]$ (m/s^3)

expressions for the drift term and control gain functions (arguments of time-varying signals are omitted for brevity)

$$
\begin{aligned}
f(\cdot) = {} & \frac{\pi\rho R^3}{2Jk_{gb}}\left(2c_0\nu + \frac{c_1R\Omega_g}{k_{gb}} - \frac{c_3R^3\Omega_g^3}{k_{gb}^3\nu^2}\right)\dot{\nu} - \frac{\lambda_{\text{opt}}k_{gb}\nu}{R}\ddot{\nu} \\
& + \frac{T_t + T_g}{J}\left[\frac{\pi\rho R^3}{2k_{gb}^2}\left(c_1R\nu + \frac{2c_2R^2\Omega_g}{k_{gb}} + \frac{3c_3R^3\Omega_g^2}{k_{gb}^2\nu}\right)\right. \\
& + \frac{3V_s^2}{J\Omega_s^2}\frac{\left[R_s^2 - (R_{eq}/s)^2 + x_s^2\right]}{\left[(R_s + R_{eq}/s)^2 + x_s^2\right]^2}\cdot\left(2n_2^2s\frac{R_bn_2^2s + R_sn_1^2u^2 - n_1u\sqrt{|\psi|}}{(n_2^2s^2 - n_1^2u^2)^2}\right. \\
& \left.\left. + \frac{R_bn_2^2\sqrt{|\psi|}\text{sign}(\psi) - n_1n_2^2u[(R_s^2 + x_s^2)s + R_bR_s]}{\sqrt{|\psi|}\text{sign}(\psi)(n_2^2s^2 - n_1^2u^2)}\right)\right],
\end{aligned}
\tag{4.1.14}
$$

$$
\begin{aligned}
g(\cdot) = {} & \frac{3V_s^2}{J\Omega_s}\frac{\left[R_s^2 - (R_{eq}/s)^2 + x_s^2\right]}{\left[(R_s + R_{eq}/s)^2 + x_s^2\right]^2}\cdot\left(\frac{R_bn_2^2\sqrt{|\psi|}\text{sign}(\psi)u - n_1\psi + n_1^3x_s^2u^2}{\sqrt{|\psi|}\text{sign}(\psi)(n_2^2s^2 - n_1^2u^2)}\right. \\
& \left. + 2n_1^2u\frac{R_bn_2^2s + R_sn_1^2u^2 - n_1u\sqrt{|\psi|}}{(n_2^2s^2 - n_1^2u^2)^2}\right).
\end{aligned}
\tag{4.1.15}
$$

To compute the bounding constants in (4.1.13), first, the nominal values of the constant parameters (reported in Appendices A.1 and A.2) were considered along with the realistic intervals reported in Table 4.1.1 for the time-varying signals entering (4.1.14) and (4.1.15).

Numerical calculations of (4.1.14) and (4.1.15) were then made sweeping through all the possible combinations of the system variables within their admissible ranges of variation (Table 4.1.1). Finally, the obtained uncertainty bounds were further enhanced by 20% to account for unconsidered uncertainties in the parameters and other possible external disturbance effects as well. The following values were obtained:

$$
F = 478, \quad G_m = 368, \quad G_M = 1414. \tag{4.1.16}
$$

4.1.4 Adaptive SOSM design

To stabilize infinite time the uncertain "auxiliary system" (4.1.12)–(4.1.13), where only the sliding variable $z_1 = \sigma$ is supposed to be available for measurements, the next discontinuous algorithm, called "suboptimal algorithm," was proposed in the literature [15]:

$$w(t) = \dot{u}(t) = -\alpha(t)V\text{sign}\,(\sigma - \sigma_M/2), \tag{4.1.17}$$

$$\alpha(t) = \begin{cases} \alpha^* & \text{if } (\sigma - \sigma_M/2)(\sigma_M - \sigma) > 0, \\ 1 & \text{otherwise}, \end{cases} \tag{4.1.18}$$

where σ_M is last extremal value of the sliding variable σ (more precisely, the most recent local maximum, minimum, or horizontal flex point of the sliding variable, which can be evaluated with sufficient precision using measurements of σ only [30]), and α^* and V are tuning constants to be selected in accordance with the next inequalities

$$\alpha^* \in (0, 1] \cap \left(0, \frac{3G_m}{G_M}\right), \tag{4.1.19}$$

$$V > \frac{F}{G_m}\Gamma, \qquad \Gamma = \max\left\{\frac{1}{\alpha^*}, \frac{4G_m}{3G_m - \alpha^* G_M}\right\}. \tag{4.1.20}$$

Such constant-gain controller is modified by implementing an adaptation mechanism which properly adjusts the gain parameter V, now denoted by $V(t)$, thereby changing the first equation of (4.1.17) into

$$w(t) = \dot{u}(t) = -\alpha(t)V(t)\text{sign}\,(\sigma - \sigma_M/2). \tag{4.1.21}$$

The adaptation method that is going to be developed is inspired to that introduced in Reference 24 for the twisting SOSMC algorithm, and it was recently tailored to the suboptimal algorithm in Reference 27. It is worth noting that in this chapter we develop a different form of the adaptation mechanism as compared to that suggested in Reference 27.

The basis of the adaptation logic is a suitable "*sliding mode existence criterion.*" Receding horizon time intervals of fixed length T are considered of the form

$$\mathcal{T}_k \equiv [kT_a - T, kT_a] \qquad k = k^* + 1, k^* + 2, \ldots \tag{4.1.22}$$

where $k^* = int\left(\frac{T}{T_a}\right)$, $T_a << T$ is the measurement sampling period of the gain-adjustment mechanism, and int(x) denotes the integer part of the real number x. At the end of every interval $\mathcal{T}_k$, the number of zero crossings of the quantity

$$\mu(t) = \sigma - \sigma_M/2 \tag{4.1.23}$$

during such an interval is computed. Note that function $\mu(t)$ appears as argument of the sign function in the definition (4.1.21) of the suboptimal algorithm control law, hence when the corresponding number of zero crossings is sufficiently large, it means that a real (or "practical") SM behavior is occurring in a small vicinity of the sliding manifold, which, in turns, implies that the control authority is large enough to counteract the actual uncertainties. Thus, if the practical SM existence criterion is fulfilled, it is sensible to diminish the control gain at the end of the time interval $\mathcal{T}_k$. On the other hand, when the practical SM existence criterion is violated then the control gain is increased.

The proposed adaptation works as follows. The piecewise constant gain $V(t)$ is defined according to

$$V(t) = V_M^k, \qquad t \in [kT_a, (k+1)T_a). \tag{4.1.24}$$

We adjust the amplitude parameter V_M^k according to

$$\begin{aligned} &V_M^i = V, \quad i = 0, 1, 2, ..., k^* - 1; \\ &V_M^{k+1} = \begin{cases} \max(V_M^k - \Lambda_1 T_a, 0) \ \ if \ N_{sw}^k(\mu) \geq N^*, \\ \min(V_M^k + \Lambda_2 T_a, V) \ \ if \ N_{sw}^k(\mu) < N^*, \end{cases} \qquad k \geq k^*; \end{aligned} \tag{4.1.25}$$

where $N_{sw}^k(\mu)$ is the number of sign commutations of $\mu(t)$ in the interval $\mathcal{T}_k$, and N^* is an appropriate integer.

Roughly speaking, at the end of each time interval $\mathcal{T}_k$ we decrement the control magnitude $V(t)$ stepwise by $\Lambda_1 T_a$ if the SM existence criterion is fulfilled, otherwise we increment it stepwise by $\Lambda_2 T_a$. The adaptation logic also includes lower and upper bounds for the control magnitude (0 and V, respectively, the latter being the gain amplitude value fulfilling (4.1.20)).

We preliminarily recall a useful Lemma which will be invoked within the convergence proof of the algorithm.

Lemma 1 *Consider the second-order auxiliary system (4.1.12) with the suboptimal algorithm (4.1.17)–(4.1.18) as control law, and let $N_{sw}^k(\mu)$ be the number of zero crossings of $\mu(t)$ during the time interval $\mathcal{T}_k$ of length T. If condition*

$$N_{sw}^k(\mu) \geq 3 \tag{4.1.26}$$

is satisfied, and, additionally, $|\ddot{z}_1(t)| \leq a_1$, for some $a_1 > 0 \ \forall t \in \mathcal{T}_k$, then there exist constants ρ_1 and ρ_1, independent of T, such that the next inequalities hold

$$|z_1(t)| \leq \rho_1 T^2 \qquad |z_2(t)| \leq \rho_2 T \quad \forall t \in \mathcal{T}_k. \tag{4.1.27}$$

Proof of Lemma 1. If signal $\mu(t) = z_1(t) - z_{1M}/2$ exhibits more than two zero crossings in the interval $\mathcal{T}_k$, then, by the definition of z_{1M}, which stores the most recent extremal value (local maxima or minima, or horizontal flexes) of $z_1(t)$, the corresponding time derivative $z_2(t)$ features at least two zero crossing in the same interval.

In addition, the time between two adjacent extremal values, say $z_{1M,h} = z_1(t_{M_h})$ and $z_{1M,h+1} = z_1(t_{M_{h+1}})$, is bounded ([30]) as

$$c_1\sqrt{|z_{1M,h}|} \le t_{M_{h+1}} - t_{M_h} \le c_2\sqrt{|z_{1M,h}|} \tag{4.1.28}$$

for some constants c_1 and c_2. Furthermore, by construction, there is $t_z \in [t_{M_h}, t_{M_{h+1}}]$, such that $\mu(t_z) = 0$. Thus, the time interval occurring between three consecutive zero crossings of $\mu(t)$, denoted as $T_{3\mu}$, will be such that

$$c_1\left(\sqrt{|z_{M,h}|} + \sqrt{|z_{M,h+1}|}\right) \le T_{3\mu} \le 3c_2\sqrt{|z_{M,h}|}. \tag{4.1.29}$$

This means that

$$c_1\left(\sqrt{|z_{M,h}|} + \sqrt{|z_{M,h+1}|}\right) \le T, \tag{4.1.30}$$

which in turn implies that $\sup_{t\in\mathcal{T}_k} |z_{1M}| \le \frac{1}{4c_1^2}T^2$. Since z_{1M} stores the local minima and maxima of $z_1(t)$, one has that $|z_1(t)| \le \sup_{t\in\mathcal{T}_k} |z_{1M}| \le \frac{1}{4c_1^2}T^2 \ \forall t \in \mathcal{T}_k$, which proves the left inequality in (4.1.27). By virtue of the fact that $z_2(t)$ features at least two zero crossings, and taking into account that $|\ddot{z}_1(t)| < a_1$ by assumption, simple time-integration yields the right inequality in (4.1.27). Lemma 1 is proved. $\triangle$

The last equation of (4.1.12) can be rewritten as follows

$$\dot{z}_2(t) = g(t)\,[\eta(t) + w(t)] \tag{4.1.31}$$

with

$$\eta(t) = \frac{f(t)}{g(t)}. \tag{4.1.32}$$

We assume that a positive constant P exists such that

$$\left|\frac{d}{dt}\eta(t)\right| \le P, \qquad t \ge 0. \tag{4.1.33}$$

We are now in position to state the next theorem.

Theorem 4.1.1. Consider system (4.1.12). Assume that the uncertain functions $f(t)$ and $g(t)$ satisfy the inequalities constraints (4.1.13) and (4.1.32)–(4.1.33). Apply the adaptive suboptimal control law (4.1.21), (4.1.18) (4.1.24), and (4.1.25) with

$$\Lambda_1 > 0, \qquad \Lambda_2 > \Lambda_1 + 3P, \qquad N^* \ge 3. \tag{4.1.34}$$

Then the next relations are achieved after a finite-time transient process for some positive constants b_1, b_2 independent on T

$$|z_1(t)| \le b_1 T^2, \qquad |\dot{z}_2(t)| \le b_2 T. \tag{4.1.35}$$

Proof of Theorem 4.1.1. Unlike the initial conditions already belong to a close vicinity of the origin of the z_1Oz_2 plane, during the initial transient there are no frequent sign commutations of $\mu(t)$ and, hence, the magnitude $V(t)$ keeps constant to its initial value V. According to the convergence properties of the fixed-gain version of the suboptimal algorithm, the trajectory will converge toward the origin. While approaching the origin the frequency of the sign commutations of $\mu(t)$ will progressively increase, according to (4.1.29) and considering the contraction condition $|z_{1M,h+1}| \le \gamma |z_{1M,h}|$, $\gamma < 1$, enforced by the suboptimal algorithm. Theoretically, the frequency of sign commutations of $\mu(t)$ tends to infinity while approaching the origin. Hence, condition (4.1.26) is fulfilled at some $k = M_1 \ge k^*$, and the stepwise reduction of V_M^k will be then activated starting from the end of the time interval $\mathcal{T}_{M_1}$. The gain dominance condition (4.1.20) can be specialized to a restricted time interval $\mathcal{T}_k$ as follows:

$$V > \Gamma N_k, \quad N_k = \sup_{t \ge \mathcal{T}_k} |\eta(t)| \le F/G_m. \tag{4.1.36}$$

By relying on the fact that $N_{sw}^{M_1}(\mu) \ge N^*$, there is $\tau_1 \in \mathcal{T}_{M_1}$ such that the adaptive control gain $V(M_1 T_a) = V_M^{M_1}$ will be dominating the actual upper bound of $|\eta(\tau)|$, in accordance with (4.1.36), as

$$V_M^{M_1} \ge \Gamma |\eta(\tau_1)|, \quad \tau_1 \in \mathcal{T}_{M_1}. \tag{4.1.37}$$

Starting from $k = M_1$, the process of reducing the control gain is activated, i.e., $V_M^{M_1+\ell} = V_M^{M_1+\ell-1} - \Lambda_1 T_a$, $\ell = 1, 2, \ldots$.. Thus, the dominance over the uncertainties (formalized by condition (4.1.37)) will be lost after a finite number of intervals, and at some $k = M_2 > M_1$ the 2-sliding criterion (4.1.26) will be violated. It implies that along the preceding time interval $\mathcal{T}_{M_2-1}$, a dominance inequality analogous to (4.1.37) holds

$$V_M^{M_1} \ge \Gamma |\eta(\tau_2)| \quad \tau_2 \in \mathcal{T}_{M_2-1}. \tag{4.1.38}$$

By Lemma 1, along the time interval $\mathcal{T}_{M_2-1}$ the variables z_1 and z_2 are bounded as in (4.1.27) with

$$\rho_1 = \sup_{t \in \mathcal{T}_{M_2-1}} |\dot{z}_2(t)| = G_M [N_{M_2-1} + V_M^{M_2-1}]. \tag{4.1.39}$$

Along the interval $\mathcal{T}_{M_2+1}$, i.e., one interval after the violation of the 2-sliding criterion (4.1.26), the magnitude of the uncertainty $\eta(t)$ will be such that

$$|\eta(\tau_3)| \le |\eta(\tau_2)| + 3PT, \qquad \forall \tau_3 \in \mathcal{T}_{M_2+1}, \tag{4.1.40}$$

which is derived by taking into account (4.1.33). On the other hand, the adaptive magnitude will be increased at the end of the interval $\mathcal{T}_{M_2} - 1$ and decreased at the end of the successive interval $\mathcal{T}_{M_2+1}$, which means that

$$V_M^{M_2+1} > V_M^{M_2-1} + T(\Lambda_2 - \Lambda_1). \tag{4.1.41}$$

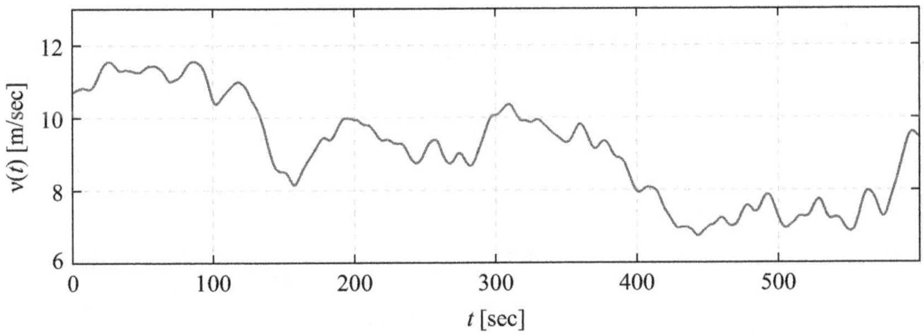

Figure 4.1.5 Considered wind speed profile

Therefore, considering (4.1.38) and (4.1.40)–(4.1.41), if the Λ_2 parameter is such that

$$\Lambda_2 > \Lambda_1 + 3P, \tag{4.1.42}$$

it follows that the dominance condition (4.1.36) will be already restored along the interval $\mathscr{T}_{M_2+1}$, i.e., one interval after the violation of the SM existence criterion (4.1.26). While V_M^k continues to grow, which will happen for finite number of adaptation intervals until the SM existence criterion (4.1.26) will be restored, contractive rotations of the system trajectories in the $z_1 - z_2$ plane will take place, which can be evaluated by studying the piecewise parabolic limit trajectories of the suboptimal algorithm ([30]) starting from the initial condition (4.1.27). Lengthy, but straightforward, computations show that the transient deviations of z_1 and z_2 fulfill inequalities analogous to (4.1.35), with the constants b_1 and b_2 do not depend on T. The process of loosing, and successively restoring, the dominance over the uncertainties will iteratively continue, thereby preserving inequalities (4.1.35). Theorem 1 is proved. $\triangle$

4.1.5 Simulation results

The performance of the proposed adaptive SOSMC system has been assessed via computer simulations. The system under study was a three-bladed horizontal-axis WECS having rated power of 60 kW. Its full-order model and the nominal values of all the parameters can be found in the Appendices A.1 and A.2. The simulation runs correspond to 10 minutes of system operation, using the wind profile of Figure 4.1.5 and considering uncertainty in the electromagnetic parameters (variations up to 20% with respect to their nominal values). A friction torque including a random-varying band-limited noise component was applied at time $t = 300$ seconds, and a time-varying disturbance modifying the line voltage from time $t = 450$ seconds was considered as well (Figure 4.1.6).

An overview of the performed simulations is given. Initially, the reduced-order model (4.1.2)–(4.1.6) of the WECS was considered, and for the sake of comparison,

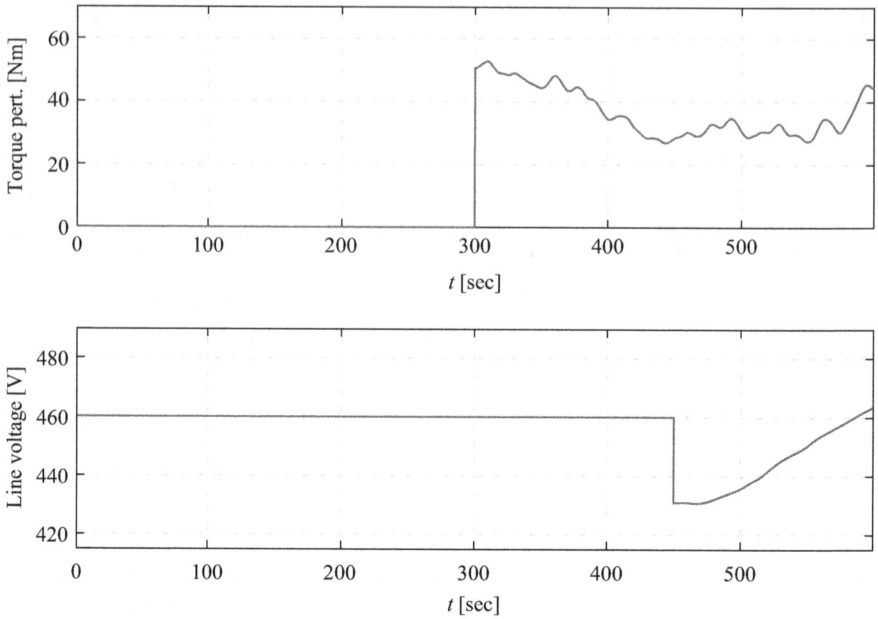

Figure 4.1.6 External disturbances: friction torque and line voltage variations

both the constant gain and adaptive versions of the control scheme were tested. The obtained results are compared and discussed in Section 4.1.5.1. Then, the adaptive suboptimal controller has been evaluated using the full-order model of the WECS (illustrated in the Appendices A.1 and A.2) under the same conditions (wind, disturbances, etc.) used with the reduced-order model. The simulation results considering the full-order model are presented and commented in the Section 4.1.5.2.

4.1.5.1 Simulations with the reduced-order model

The controller parameters V and α^* were selected on the basis of inequalities (4.1.19)–(4.1.20) using the computed uncertainty bounds (4.1.16). Particularly, since there is some degree of freedom in choosing the (α^*, V) parameter pair, a study was made aimed at identifying the minimum achievable value of V by identifying the minimum of function $\Gamma(\alpha^*) = \max\left\{\frac{1}{\alpha^*}, \frac{4G_m}{3G_m-\alpha^* G_M}\right\}$ in (4.1.20) over the admissible range $\alpha^* \in (0, 0.78)$ given in (4.1.19). It resulted (Figure 4.1.7)

$$\alpha^* = 0.383, \qquad V = \frac{F}{G_m}\Gamma(\alpha^*) = 3.4. \tag{4.1.43}$$

The parameter values in (4.1.43) were used when simulating the fixed-gain realization of the control algorithm.

For tuning the adaptive algorithm, the pair (T, N^*) should be jointly selected. This is easily done by selecting first the interval length T (the value $T = 0.668$ s was

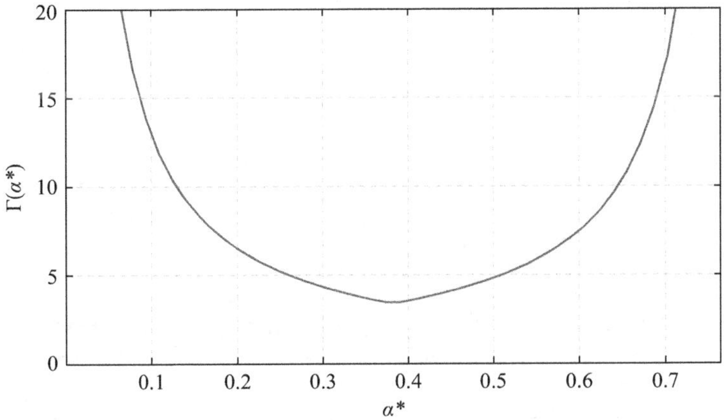

Figure 4.1.7 Plot of function $\Gamma(\alpha^*) = max\left\{\frac{1}{\alpha^*}, \frac{4G_m}{3G_m - \alpha^* G_M}\right\}$

chosen being equal to 200 times the sampling time at which the measurements are taken and the state of the converter is changed) and then making a simulation (or experimental) test using the fixed-gain controller and looking at the actual "average" number of switches of the sliding variable along the time windows of length T. Then N^* is computed as a fraction of that average (e.g., taking the 25%). The rationale for this procedure is that the average switching frequency of the sliding variable will significantly decrease when the magnitude of the discontinuous controller gain will be diminished. The T_a and Λ_1 parameters are chosen arbitrarily. It is convenient to use relatively small values of Λ_1 to avoid an excessively "nervous" adaptation of the gain. Parameter Λ_2 has to fulfill the inequality in (4.1.34), which may result in too large values since parameter P is usually highly conservative. To optimize the actual performance of the algorithm, a better choice is to select Λ_2 bigger than Λ_1 (between two and three times higher will normally suffice). The design parameters of the adaptive suboptimal controller were finally selected as follows:

$$T = 0.6 \text{ s}, \; N^* = 12, \; T_a = 10^{-3} \text{ s}, \; \Lambda_1 = 0.6, \; \Lambda_2 = 1.6. \tag{4.1.44}$$

Both the fixed-gain and adaptive controllers achieved the primary goal of tracking the optimal speed reference. This can be observed in Figure 4.1.8, where the mechanical speed profiles corresponding to the fixed-gain and adaptive controller implementations appear to be almost perfectly overlapped with the corresponding reference.

These achievements are also visible in the time evolutions of the sliding variable, shown in Figure 4.1.9, which is steered near zero and maintained there in both cases. The significant chattering reduction provided by the adaptive controller should be noticed.

The plots shown in Figure 4.1.9 include the time interval during which the torque perturbation is abruptly incorporated. Its sudden appearance causes the temporary loss of the practical-sliding condition in the adaptive SOSM controller case, whereas

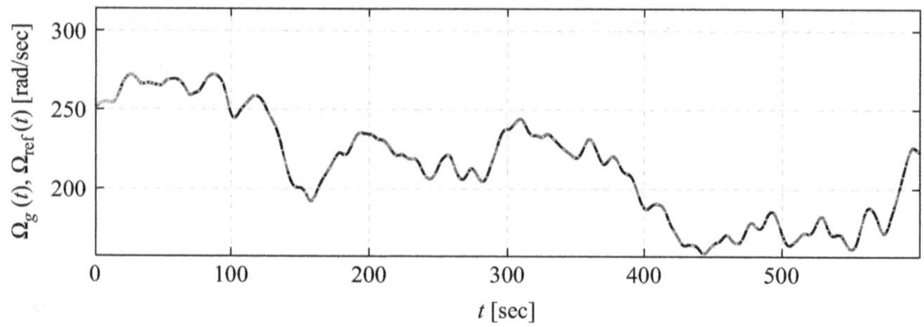

Figure 4.1.8 Mechanical speed Ω_g using the fixed-gain and adaptive SOSMC, and reference profile (reduced-order model)

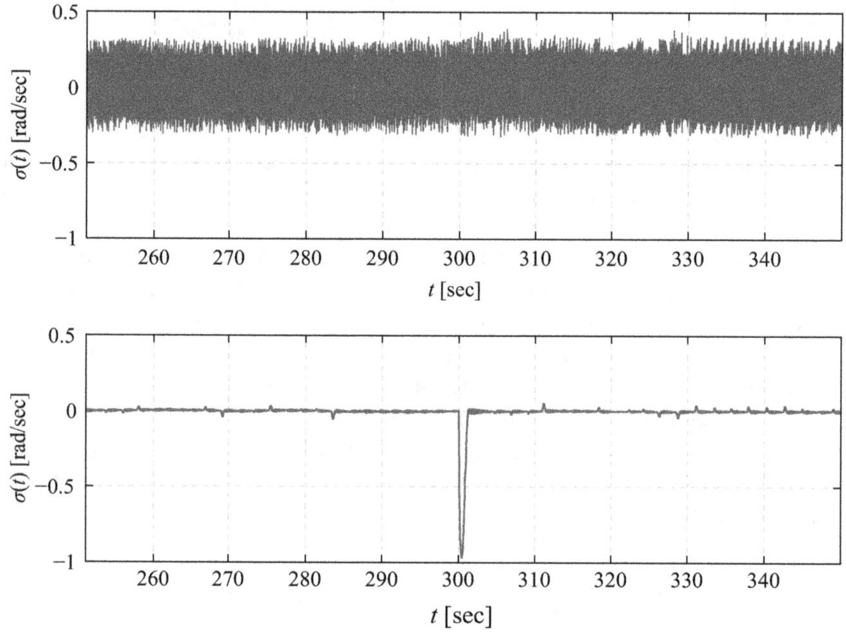

Figure 4.1.9 Sliding variable $\sigma(t)$ with the fixed-gain (upper plot) and adaptive (bottom plot) SOSM controller (reduced-order model)

the fixed-gain controller is insensitive to it, as expected. After the introduction of the disturbance, the adaptive gain of the controller then quickly increases to deal with it, as shown in Figure 4.1.10, driving back the sliding variable σ to zero in a short time. The long-term evolution of the actual control time derivative $\dot{u}(t)$ is shown for both controllers in Figure 4.1.11. The remarkable amplitude reduction achieved by means of the adaptation logic is apparent. The corresponding control actions $u(t)$, are shown in Figure 4.1.12, where the smoothest behavior obtained using the adaptive controller is noticeable.

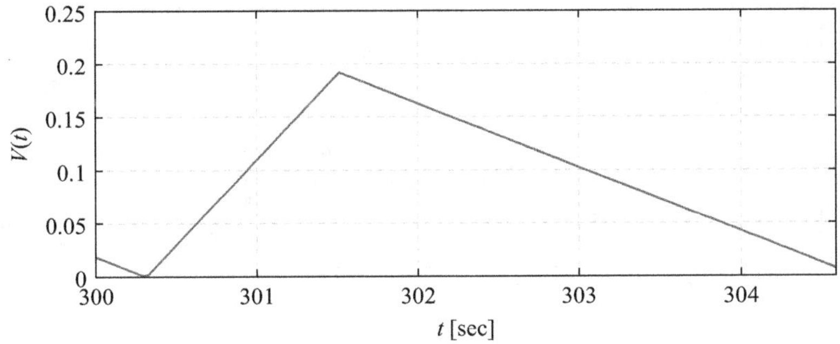

Figure 4.1.10 Adaptive SOSMC: detail of $V_M^j(t)$ evolution after the abrupt torque perturbation (reduced-order model used)

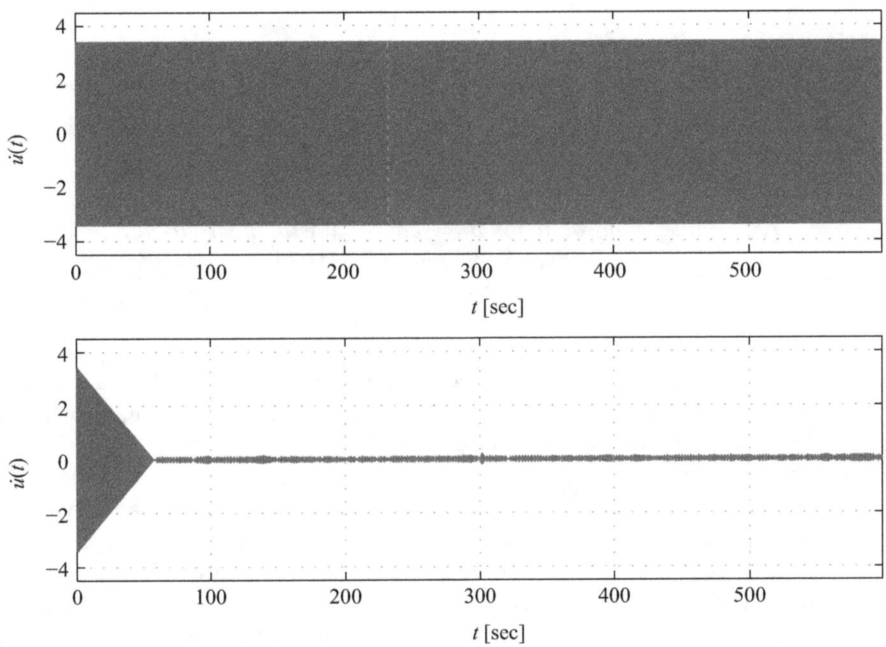

Figure 4.1.11 The actual input derivative $\dot{u}(t)$ with the fixed-gain (upper plot) and adaptive (lower plot) controller (reduced-order model)

The constant parameters of the fixed-gain algorithm have been selected large enough to address the worst-case uncertainty scenario, causing several negative effects on the system, such as chattering and the generation of strong mechanical efforts. These can be observed in Figure 4.1.13, where the two generator torques corresponding to the fixed-gain (upper plot) and adaptive (lower plot) controllers are shown.

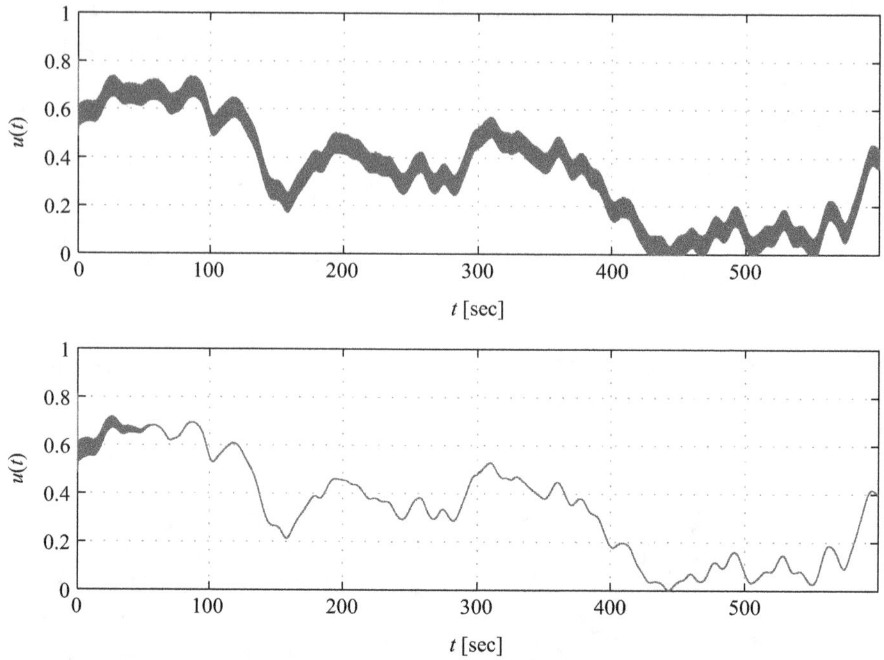

Figure 4.1.12 Applied control action $u = |\cos(\alpha_{fa})|$ with the fixed-gain (upper plot) and adaptive (bottom plot) SOSMC (reduced-order model)

Counteracting the rate of variation of the generator torque is of paramount importance to increase the lifetime of the WECS, and the adaptive SOSMC provides a remarkable attenuation of the generator torque vibrations.

In a further series of tests, the sensitivity of the controller performance against the modification of the tuning parameters is verified to show that the adaptive controller tuning is not critical and, particularly, that the algorithm provides satisfactory performance by considering a different set of tuning parameters than those previously used. The upper plot of Figure 4.1.14 presents the time evolution of the mechanical speed using different values of N^*, showing that all the curves overlap the reference speed $\Omega_{\mathrm{ref}}(t)$. In addition, the lower plot shows the sliding variable in an interval of steady state operation. It can be seen that the speed tracking accuracy remains satisfactory in all tests. Figure 4.1.15 depicts the corresponding time evolutions of the control derivative $\dot{u}(t)$, showing remarkable chattering alleviation, as compared to the fixed-gain case (shown in Figure 4.1.11-top), in all the considered scenarios.

Figures 4.1.15 and 4.1.16 present a similar sensitivity analysis made with respect to the parameter T, i.e., the length of the adaptation time interval. The satisfactory speed tracking for every tested T can be observed in Figure 4.1.16 (upper plot), while in the lower plot, it can be seen that differences in the sliding accuracy arise: the larger T is selected, the worse the accuracy, as predicted by the theoretical analysis

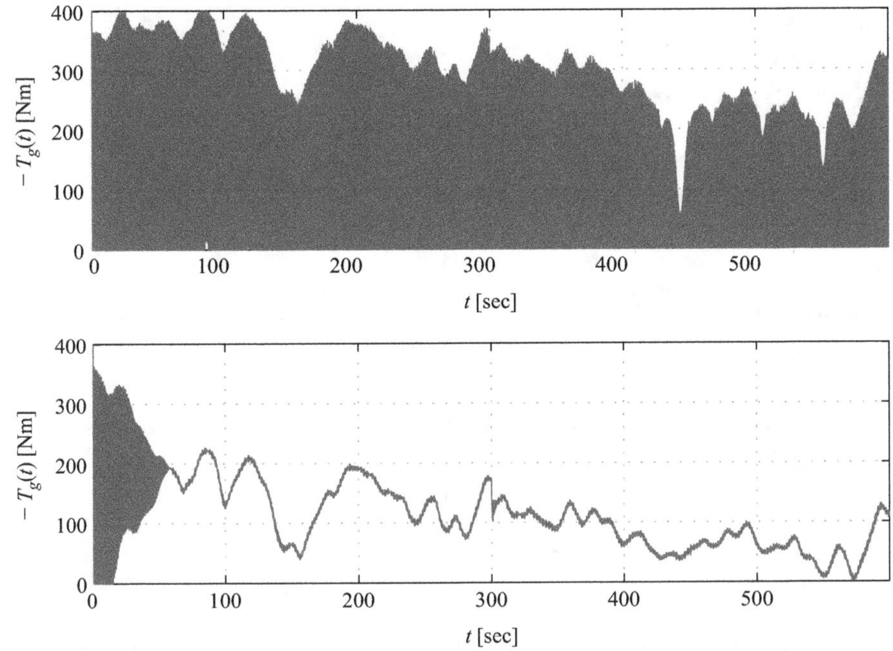

Figure 4.1.13 Electric torque of the generator, $-T_g$ with the fixed-gain (upper plot) and adaptive (bottom plot) SOSMC (reduced-order model)

of Theorem 1. However, the sliding accuracy is acceptable in all tests. Figure 4.1.17 shows that satisfactory chattering attenuation is achieved with all tested values of T.

On the basis of the reported results, the adaptation mechanism considered in this chapter reveals to be a practical solution, rather simple and easy to implement, to address the high-performance control of WECS under significant uncertainty effects and bringing considerable improvements as compared to the corresponding nonadaptive fixed-gain counterpart.

4.1.5.2 Simulations with the full-order model

To better validate the present proposal, additional simulation runs have been made by considering the full-order WECS model reported in the Appendices A.1 and A.2. The additional electrical dynamics, which was neglected in the reduced-order model previously used for analysis and design purposes will obviously cause a reduction in the switching frequency of the sliding variable; thus, the parameter N^* was reduced to the value $N^* = 4$ while keeping unchanged all other controller parameters.

For the sake of space limitation, only the adaptive SOSMC is employed in this series of tests. Tracking of the optimal speed reference is still very accurate, as shown in Figure 4.1.18.

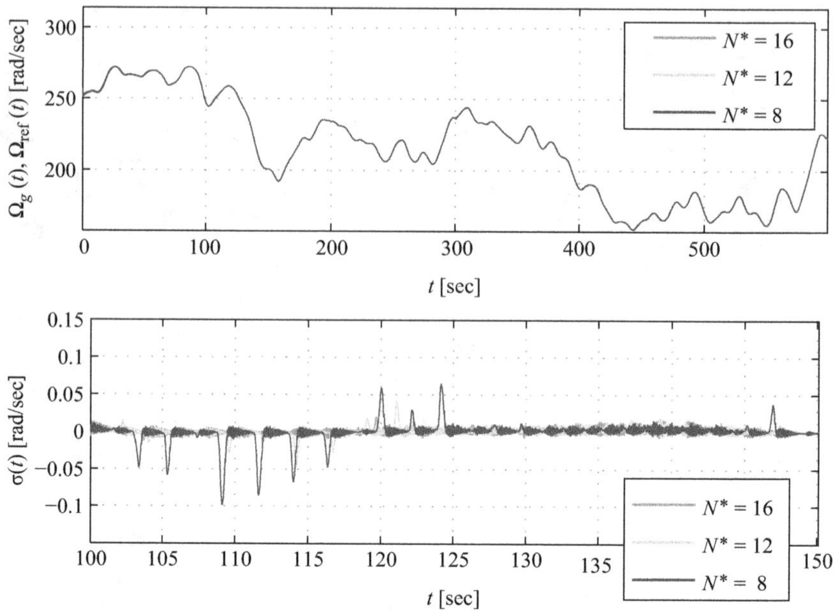

Figure 4.1.14 $\Omega_g(t)$ *and* $\Omega_{ref}(t)$ *(upper plot), and sliding variable (lower plot), for* $T = 0.668s$. *and different choices of* N^* *(reduced-order model)*

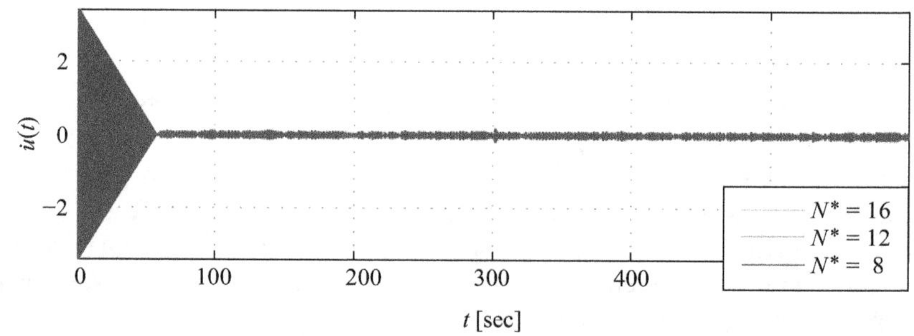

Figure 4.1.15 $\dot{u}(t)$ *for* $T = 668$ *ms and different choices of* N^* *(reduced-order model)*

The applied control input $u(t)$ becomes smoother and smoother while the adaptive SOSMC is decreasing the discontinuous gain V_M^k, and thus $|\dot{u}(t)|$, to a value comparable with the actual uncertainty to cope with (Figure 4.1.19). The corresponding behavior of $\dot{u}(t)$ is presented in Figure 4.1.20. The generator torque has a profile almost indistinguishable from the one achieved using the reduced-order model (shown in Figure 4.1.13), hence it is not reported for brevity sake.

This series of tests has shown that dealing with a realistic full-order WECS model, the adaptive SOSM controller is able to provide a significant accuracy improvement

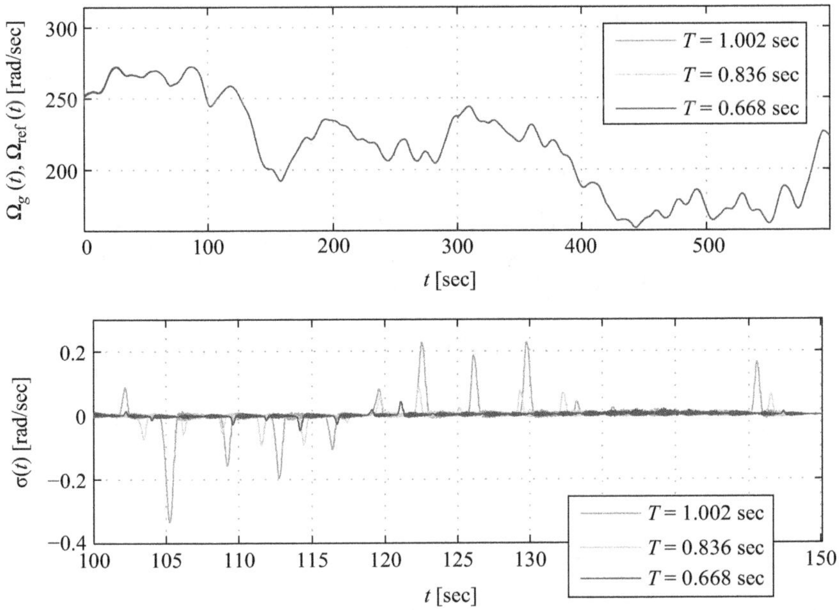

Figure 4.1.16 Mechanical speed and its reference (upper plot), and sliding variable $\sigma(t)$ (bottom), for $N^* = 12$ and different choices of T (reduced-order model)

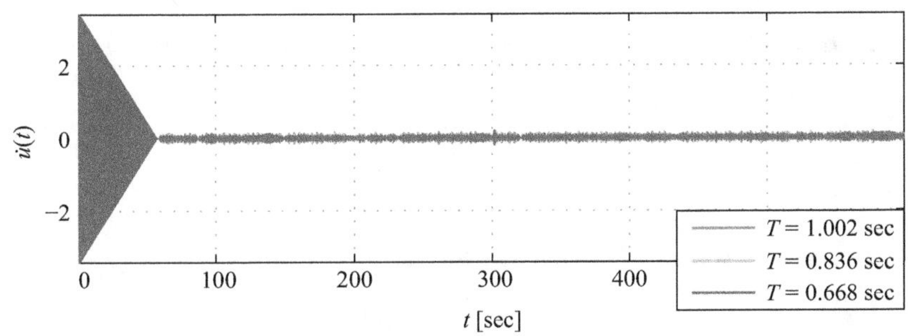

Figure 4.1.17 $\dot{u}(t)$ for different choices of T (reduced-order model)

and chattering attenuation effect. Many additional simulations using different wind profiles and uncertainty factors have been made, all giving rise to satisfactory results.

4.1.6 Conclusions

A robust controller for power conversion maximization of a wind turbine was designed. To this end, an existing adaptive SOSM control was revisited in order

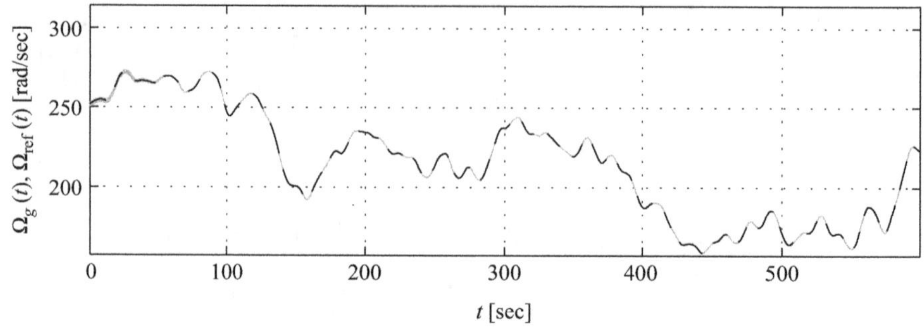

Figure 4.1.18 Mechanical speed $\Omega_g(t)$ using the adaptive SOSMC and reference profile (full-order model)

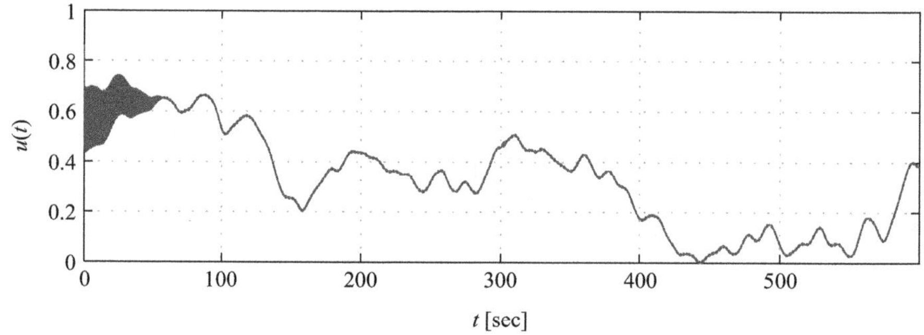

Figure 4.1.19 Applied control action $u = |\cos(\alpha_{fa})|$ (full-order model)

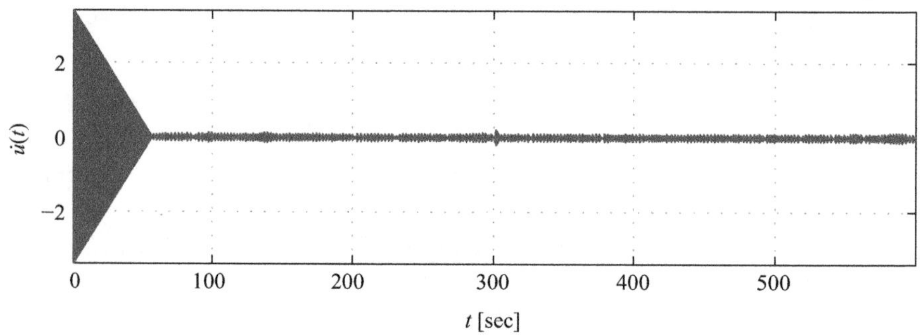

Figure 4.1.20 Control derivative $\dot{u}$ (full-order model)

to deal with quickly varying disturbances (such as those present in WECS), and its theoretical foundation was demonstrated. After thorough simulation tests, the novel adaptive WECS SOSM controller shows satisfactory performance and robustness, attesting not only the potential applicability of this combined control technique in the area of wind energy technology but also its general applicability to nonlinear systems subjected to fastly varying uncertainty sources. This research is the preliminary stage of a broader project aiming to develop and implement this class of adaptive SOSM controllers in a real small-scale wind turbine (10–60 kW) to optimize energy capture and extend its service life.

A.1 WECS full-order dynamical model

The full-order model includes the dynamics of the electrical variables and presents the correspondent expression for the generator torque. The nonlinear differential equations that describe the topology under study in a synchronously rotating direct quadrature (d-q) frame are given in compact form as follows:

$$\mathbf{V} = \begin{bmatrix} R_s + L_s p & -w_s L_s & M_s p & -\omega_s M_s \\ \omega_s & R_s + L_s p & \omega_s M_s & M_s p \\ M_s p & -s\omega_s M_s & R_b' + L_b' p & -s\omega_s L_r' \\ s\omega_s M_s & M_s p & s\omega_s L_r' & R_b' + L_b' p \end{bmatrix} \mathbf{I},$$

where $\mathbf{V} = [0;\ V_s;\ -(V_s'')'u\sin(\varphi);\ -(V_s'')'u\cos(\varphi)]^T$, $\mathbf{I} = [i_{sd};\ i_{sq};\ i_{rd}';\ i_{rq}']^T$, p is the time derivative operator, $\varphi = \tan^{-1}(\frac{i_{rd}}{i_{rq}})$, $L_b = L_r + 0.55L_f$, and $R_b = R_r + 0.55R_f$. Double quotation marks applied to the grid voltage indicate it is referred to the inverter terminals by the transformer with turns ratio n_2 (Figure 4.1.4), thus $V_s'' = V_s/n_2$. A single quotation mark applied to a variable indicates that it is referred from rotor to stator (or vice versa), through the corresponding turns ratio $n_1 = \frac{n_{\text{stator}}}{n_{\text{rotor}}}$.

The equation for the generator torque, which replaces (4.1.4), is now

$$T_g = p_p M_s (i_{sq} i_{rd}' - i_{sd} i_{rq}'). \tag{4.1.45}$$

A.2 Nominal values of the parameters

$P_{\text{rated}} = 60$ kW	$V_s = 460/\sqrt{3}$ V	$I_{\text{rated}} = 130$ A
$\omega_s = 2\pi 50$ rad/s	$\Omega_s = \omega_s/p_p$	$p_p = 2$
$n_1 = n_2 = 1.2$	$J = 7.0623$ kg m^2	$R = 6.75$ m
$R_s = 119$ mΩ	$R_r = 238$ mΩ	$R_f = 25.9$ mΩ
$L_s = L_r = 1.4$ mH	$L_f = 10.1$ mH	$M_s = 35.1$ mH
$x_s = 879.6$ mΩ	$k_{gb} = 19.85$	$\lambda_{\text{opt}} = 8.015\ k_{gb}$
$c_0 = -1.142\ 10^{-2}$	$c_1 = 2.214\ 10^{-4}$	$c_2 = -1.03\ 10^{-6}$
$c_3 = 1.191\ 10^{-9}$	$\rho = 1.2242$ kg/m^3	

Referred to fast shaft side: $T_t = T_{t\text{low}}/k_{gb}$; $\Omega_g = \Omega_{\text{low}} k_{gb}$; $J = J_t/k_{gb}^2 + J_g$. J_t and J_g are the inertia of the turbine rotor and of the generator rotating parts, respectively.

Acknowledgments

C. Evangelista and P. Puleston acknowledge the support of UNLP, CONICET, and ANPCyT (Argentina), and Marie Curie FP7-2011-IIF, ACRES (299767/911767) (EU). A. Pisano and E. Usai gratefully acknowledge the financial support from the research projects "Modeling, control and experimentation of innovative thermal storage systems," funded by Sardinia regional government, under grant agreement n. CRP-60193; and "Development, design and prototyping of an optimal control and supervision system for Micro-Grids," funded by RAS RAS-Tender (2014).

References

[1] Small wind world report 2012, World Wind Energy Association, *Tech. Rep.*, 2012. Online. Available at: http://www.wwindea.org/

[2] Renewables 2014 global status report, REN21 "Renewable Energy Policy Network for the 21st Century," Paris, *Tech. Rep.*, 2014.

[3] L. Y. Pao and K. E. Johnson, "Control of wind turbines. Approaches, challenges, and recent developments," *IEEE Contr. Syst. Magazine*, vol. 31, no. 2, pp. 44–62, 2011.

[4] I. Munteanu, A. Bratcu, N. Cutululis, and E. Ceanga, *Optimal Control of Wind Energy Systems*. Springer-Verlag, London, UK, 2007.

[5] M. Garcia-Sanz and C. Houpis, *Wind Energy Systems: Control Engineering Design*. CRC Press, Taylor & Francis Group, London, UK, 2012.

[6] S. Emelyanov, *Variable Structure Control Systems*, Nauka, Moscow, 1967.

[7] V. Utkin, "Variable structure systems with sliding modes," *IEEE Trans. Automat. Contr.*, vol. 22, no. 2, pp. 121–222, 1977.

[8] A. Sabanovic, L. Fridman, and S. Spurgeon, editors, *Variable Structure Systems: From Principles to Implementation*. IET, United Kingdom, 2004.

[9] C. Edwards, E. Fossas Colet, and L. Fridman, editors, *Advances in Variable Structure and Sliding Mode Control*. Springer, Berlin, 2006.

[10] G. Bartolini, L. Fridman, A. Pisano, and E. Usai, editors, *Modern Sliding Mode Control Theory: New Perspectives and Applications*. Lecture notes in Control and Information Sciences series, vol. 375, Springer-Verlag, Berlin, Germany, 2008.

[11] L. Fridman, J. Moreno, and R. Iriarte, editors, *Sliding Modes After the First Decade of the 21st Century*. Springer Lecture Notes in Control and Information Sciences, 2011, vol. 412.

[12] B. Bandyopadhyay, S. Janardhanan, A. Pisano, and S. Spurgeon, editors, *Advances in Sliding Mode Control—Concept, Theory and Implementation*. Springer Lecture Notes in Control and Information Sciences, 2013, vol. 440.

[13] A. Levant, "Sliding order and sliding accuracy in sliding mode control," *Int. J. Control*, vol. 58, no. 6, pp. 1247–1263, 1993.

[14] L. Fridman and A. Levant, *Robust Control Variable Structure and Lyapunov Techniques*. Springer Verlag, London, 1996, no. 217, ch. 1, pp. 106–133.

[15] G. Bartolini, A. Ferrara, and E. Usai, "Chattering avoidance by second order sliding mode control," *IEEE Trans. Automat. Contr.*, vol. 43, no. 2, pp. 241–246, 1998.

[16] A. Pisano and E. Usai, "Sliding mode control: a survey with applications in math," *Mathematics and Computers in Simulation*, vol. 81, no. 5, pp. 954–979, 2011.

[17] Y. Shtessel, C. Edwards, L. Fridman, and A. Levant, editors, *Sliding Mode Control and Observation*. Springer, New York, 2013.

[18] Y. Shtessel, J. Moreno, F. Plestan, L. Fridman, and A. Poznyak, "Super-twisting adaptive sliding mode control: a Lyapunov design," *Proc. 49th IEEE Conference On Decision and Control*, IEEE press, Piscataway, NJ, USA, December 2010, pp. 5109–5113.

[19] J. Kochalummoottil, Y. Shtessel, J. Moreno, and L. Fridman, "Output feedback adaptive twisting control: a Lyapunov design," *Proc. 2012 American Control Conference*, IEEE press, Piscataway, NJ, USA, June 2012, pp. 6172–6177.

[20] T. Gonzalez, J. Moreno, and L. Fridman, "Variable gain super-twisting sliding mode control," *IEEE Trans. Automat. Contr.*, vol. 57, no. 8, pp. 2100–2105, 2012.

[21] F. Plestan and A. Glumineau, "A new differentiator based on a second order sliding mode output feedback control," *Proc. 49th IEEE Conference on Decision and Control*, IEEE press, Piscataway, NJ, USA, December 2010, pp. 5098–5102.

[22] M. Taleb, F. Plestan, and B. Bououlid, "High order integral sliding mode control with gain adaptation," In: *European Control Conference, ECC 2013*, Zürich, Switzerland, July 2013.

[23] F. Plestan, Y. Shtessel, V. Brègeault, and A. Poznyak, "New methodologies for adaptive sliding mode control," *Int. J. Control*, vol. 3, pp. 1907–1919, 2010.

[24] G. Bartolini, A. Levant, A. Pisano, and E. Usai, "A real-sliding criterion for control adaptation," In: *Workshop on Variable Structure Systems, VSS 2002*, Sarajevo, Yugoslavia, 2002.

[25] L. Capisani, A. Ferrara, and A. Pisano, "Second-order sliding mode control with adaptive control authority for the tracking control of robotic manipulators," In: *18th IFAC World Congress*, Italy, 2011.

[26] A. Pisano, M. Tanelli, and A. Ferrara, "Time-based switched sliding mode control for yaw rate regulation in two-wheeled vehicles," In: *Decision and Control (CDC), 2012 IEEE 51st Annual Conference On*, December 2012, pp. 0743–1546.

[27] A. Pisano, M. Tanelli, and A. Ferrara, "Combined switched/time-based adaptation in second order sliding mode control," In: *Decision and Control (CDC), 2013 IEEE 52nd Annual Conference On*, 2013, pp. 4272–4277.

[28] T. Burton, D. Sharpe, N. Jenkins, and E. Bossanyi, *Wind Energy Handbook*. John Wiley and Sons, United Kingdom, 2001.

[29] P. Puleston, R. Mantz, P. Battaiotto, and F. Valenciaga, "Sliding mode control for efficiency optimization of wind energy systems with double output induction generator," *Int. J. Energ. Res.*, vol. 24, no. 1, pp. 77–92, 2000.

[30] G. Bartolini, A. Ferrara, A. Pisano, and Usai, "On the convergence properties of a 2-sliding control algorithm for non-linear systems," *Int. J. Control*, vol. 74, no. 7, pp. 718–731, 2001.

Chapter 4.2

Sliding mode control of a fuel cell–based electric power system: multiple modular configurations

Roshini S. Ashok and Y. B. Shtessel

Abstract

Control of electric power system comprised proton exchange membrane fuel cell (PEMFC), along with shared and multiple load configurations of multiple-modular boost DC–DC power converters as power conditioners, is studied using sliding mode control techniques. The nonlinear coupling of the boost converters in a shared and multiple load configurations are overcome by conventional sliding mode controllers (SMCs) that facilitate the output tracking of the load voltage in both converter configurations. The nonminimum phase output tracking is alleviated by controlling the PEMFC current using the adaptive 2-SMC. The challenge of balancing the currents in the boost converters in the shared load configuration is addressed via conventional SMC. The efficiency and robustness of the three-fold controllers for the proposed electric power system for the both configurations has been confirmed via computer simulations.

4.2.1 Introduction

Fuel cells (FCs), using external sources of hydrogen and oxygen to produce electric power, offer an alternative "green" source of electricity to power a large variety of systems, including microgrids, energy-efficient buildings, automotive applications, electric motorcycles, energy storage, and distributed systems [1, 2]. FCs can operate as long as the hydrogen and oxygen supply is maintained. FCs are used in conjunction with other power-conditioning systems like the DC–DC power converters that provide the electric power with precisely given parameters addressing the demands of a variety of loads [1–7].

Control of electric power systems comprised FC, a single DC–DC boost converter and an ultracapacitor using sliding mode control techniques, is studied in quite a few works [5, 6, 8]. Integrating FCs with DC–DC boost converters of different configurations, including the cases with individual and shared loads, allows generating electric power with certain parameters that fit the demands of a large variety of loads. Robustly controlling the FC-based electric power systems with multiple modular DC–DC boost converter of different configurations, including individual and shared loads, is the main objective of this chapter. The use of conventional SMC for both individual and shared load configuration of the multiple modular boost DC–DC converter control is justified by a high-frequency switching nature of the control function that corresponds to the opened/closed state of the converter switch that is used for charging/discharging the capacitor [2, 4, 5, 8, 9]. In this work, the proposed use of adaptive super-twisting controller [10, 11] for controlling the FC is validated by the requirement to have a continuous control function (which corresponds to a partial pressure of hydrogen and/or oxygen supply) that controls the FC current in the presence of the model uncertainties [12], whose derivative is bounded with *unknown* boundary. The proposed adaptive 2-SMC is a self-tuning controller whose adaptive gains are not overestimated [11]. The contribution of this paper is as follows

(1) The proposed decentralized three-fold controller, specifically two conventional SMCs for both individual and shared load configuration of the multiple modular DC–DC boost power converters and an adaptive-gain super-twisting controller for controlling the FC current, robustly controls the studied electric power systems.
(2) The nonminimum phase property of the DC–DC boost converters is eliminated in the studied PEMFC/multiple modular DC–DC boost converters for the both configurations by correspondingly controlling the FC current. This allows addressing the direct causal output voltage tracking in the studied electric power system.
(3) The FC current is robustly controlled via the adaptive gain 2-SM (super-twisting) controller, whose adaptive gains are not overestimated in the presence of the smooth perturbations, whose derivatives are bounded with unknown bounds.
(4) Nonlinear coupling of multiple modular DC–DC boost converters of individual load configuration through the FC is implicitly compensated by the proposed three-fold controller.
(5) The proposed three-fold controller addresses a challenge of balancing the currents of DC–DC boost converters of shared load configuration.

The structure of the paper is as follows. The direct output voltage tracking in DC–DC boost converters is discussed in Section 4.2.2. The mathematical model of the electric power system based on PEMFC/multiple modular converter configuration of individual load is presented in Section 4.2.3. Section 4.2.4 derives the mathematical model of the electric power system based on the PEMFC/multiple modular configuration of the shared load. The problem is formulated for both the individual and shared load configurations of the multiple modular DC–DC boost converters in Section 4.2.5. Section 4.2.6 outlines the three-fold controller design for both the shared and individual load configurations of the electric power system. The results of the simulation study are presented in Section 4.2.7. The conclusions are discussed in Section 4.2.8.

4.2.2 Background of direct output voltage tracking in DC–DC boost converters: nonminimum phase phenomenon

In this section, it is demonstrated that direct control of the output voltage in DC–DC boost converter is a challenging task due to the nonminimum phase property of the boost converter dynamics. A circuit diagram of a boost DC–DC converter is presented in Figure 4.2.1.

It is assumed that the voltage of *the* battery $E_0 = const$. The dynamics of the DC–DC boost power converter are well-studied and are governed by the following system of bilinear differential equations [4, 9, 13, 14]:

$$\frac{di_1}{dt} = \frac{1}{L}\left(E_0 - 0.5V_0 + V_0 v\right), \quad \frac{dV_0}{dt} = \frac{1}{C}\left(0.5i_1 - \frac{V_0}{R_0} - i_1 v\right), \tag{4.2.1}$$

where control

$$v = u - 0.5, \quad u \in [0, 1], \quad v \in [-0.5, 0.5], \tag{4.2.2}$$

and $u = 0$ corresponds to an opened switch, $u = 1$ corresponds to the closed switch in Figure 4.2.1. The output voltage tracking error dynamics are derived

$$\frac{de}{dt} = \dot{V}_0^c(t) - \frac{1}{C}\left(0.5i_1 - \frac{V_0}{R_0} - i_1 v\right), \tag{4.2.3}$$

where $e(t) = V_0^c(t) - V_0(t)$ with $V_0^c(t)$ is a given online output voltage command profile. The sliding mode control

$$v = -0.5\,\text{sign}\,(e) \tag{4.2.4}$$

is used in order to directly address the output voltage tracking problem $\lim_{t\to\infty} e(t) = 0$.

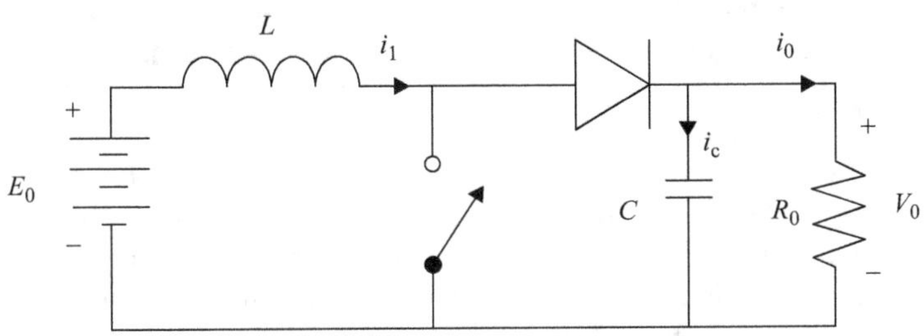

Figure 4.2.1 A circuit diagram of a boost DC–DC converter

The internal dynamics [15] of system (4.2.27) in the sliding mode is obtained by substituting equivalent control $v_{eq} = 0.5 - \frac{1}{i_1}\left(C\dot{V}_0^c + \frac{V_0}{R_0}\right)$ into the first equation of system (4.2.27). This is

$$\frac{di_1}{dt} = \frac{1}{L}\left[E_0 - \frac{V_0}{i_1}\left(\frac{V_0}{R_0} + C\dot{V}_0^c\right)\right]. \tag{4.2.5}$$

The forced zero dynamics [15] is obtained by replacing V_0 in (4.2.30) byV_0^c. Assuming for simplicity $V_0^c = const$ we obtain

$$\frac{di_1}{dt} = \frac{1}{L}\left[E_0 - \frac{\left(V_0^c\right)^2}{i_1 R_0}\right]. \tag{4.2.6}$$

It is easy to show that the equilibrium point $i_1^{eq} = \frac{(V_0^c)^2}{R_0 E_0}$ in system (4.2.31) is unstable or system (4.2.27) is nonminimum phase, while the output voltage is controlled directly.

Therefore, the sliding mode control (4.2.4) is not applicable for directly controlling the output voltage V_0 in the boost DC–DC converter, including the case with varying V_0^c given online. Using FC instead of the battery in Figure 4.2.1, the current i_1 can be easily stabilized by means of controlling the oxygen or hydrogen flow rates that are supplied to FC.

One of the main contributions of this work is eliminating the nonminimum phase property of the DC–DC boost converter in the FC-based electric power systems by means of controlling the FC current using adaptive-gain 2-SMC, specifically, adaptive twisting control [11, 16]. The use of adaptive 2-SMC is justified by the uncertain nature of FC dynamics, whose coefficients are bounded with unknown bounds.

4.2.3 Mathematical model of PEMFC/multiple modular DC–DC boost converter with an individual load configuration of an electric power system

The equivalent circuit of PEMFC/multiple modular DC–DC boost converter with an individual load configuration of electric power system is presented in Figure 4.2.2. The following notations are used in the figure: E_0 is the reversible FC voltage, i_{fc} is the FC current, V_{act} is the activation potential drop across the double-layered capacitance C_{dl}. R_{act} is the resistance that mainly causes the activation loss (V_{act}). The variable internal resistance of the FC is given to be R_{ohm}. The ohmic voltage drop is presented as V_{ohm}. The output voltages of the DC–DC boost converters 1 and 2 is represented as V_L^1, V_L^2. The corresponding output current of the DC–DC boost converters 1 and 2 is given to be i_L^1, i_L^2. The individual inductive load of the DC–DC boost converter 1 is R_L^1, L_L^1, and the corresponding inductive load of the DC–DC boost converter 2 is R_L^2, L_L^2.

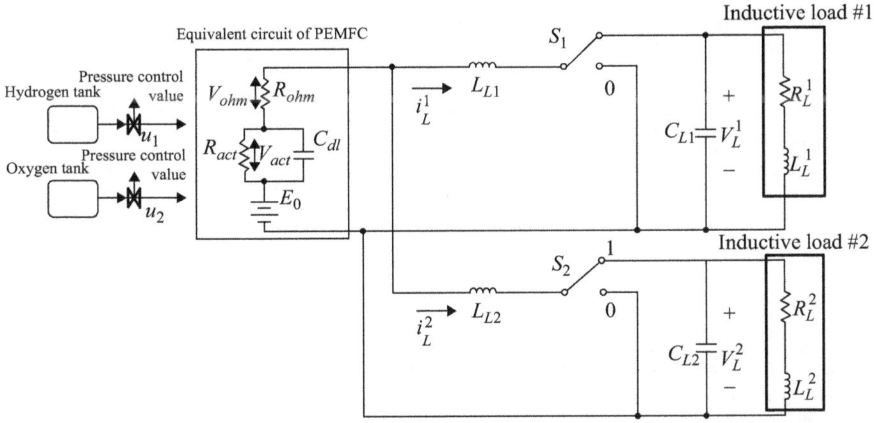

Figure 4.2.2 Equivalent circuit diagram of electric power system comprised PEMFC/multiple modular boost DC–DC converter with an individual load configuration

4.2.3.1 Mathematical model of PEMFC

Based on the equivalent circuit in Figure 4.2.2, the dynamics of activation over voltage (V_{act}) is given as,

$$\frac{dV_{act}}{dt} = \frac{i_{fc}}{C_{dl}} - \frac{V_{act}}{R_{act}C_{dl}}. \tag{4.2.7}$$

The partial pressure of hydrogen and oxygen dynamics are presented as follows [2, 7]:

$$\frac{d}{dt}P_{H_2} = -\frac{1}{\tau_{H_2}}P_{H_2} + \frac{1}{\tau_{H_2}\mathrm{k}_{H_2}}\left(q_{H_2}^{in} - 2k_\gamma i_{fc}\right), \tag{4.2.8}$$

$$\frac{d}{dt}P_{o_2} = -\frac{1}{\tau_{o_2}}P_{o_2} + \frac{1}{\tau_{o_2}\mathrm{k}_{o_2}}\left(q_{o_2}^{in} - k_\gamma i_{fc}\right), \tag{4.2.9}$$

where P_{H_2} and P_{o_2} are the partial pressures of hydrogen and oxygen, respectively. $q_{H_2}^{in}$ is the input flow rate ($kmol/s$);$\tau_{H_2} = (V_{an})/(RT)$ is the time constant associated with the hydrogen flow rate; $q_{O_2}^{in}$ is the input flow rate ($kmol/s$); $\tau_{O_2} = (V_{an})/(RT)$ is the time constant associated with the oxygen flow rate; K_γ is the consumed hydrogen flow rate, which is expressed as $K_\gamma = \frac{N_{fc}}{4F}$.

Remark 4.2.1. *Either one of the command profiles $P_{H_2}^c$ and $P_{O_2}^c$ for P_{H_2} or P_{O_2} are assumed constant. In this work, we assume that $P_{H_2}^c = const$ and is given, and the control u_1 is to be designed so that $P_{H_2}(t) \to P_{H_2}^c$ as time increases in the presence of the perturbation term $-2k_\gamma i_{fc}$.*

The actuator valve dynamics are modeled as follows [17, 18]:

$$\frac{dq_{H_2}^{in}}{dt} = \frac{1}{\tau_{qH_2}}(-q_{H_2}^{in} + \bar{q}_{H_2}^{in}), \tag{4.2.10}$$

$$\frac{dq_{O_2}^{in}}{dt} = \frac{1}{\tau_{qO_2}}(-q_{O_2}^{in} + \bar{q}_{O_2}^{in}), \tag{4.2.11}$$

where τ_{qO_2}, τ_{qH_2} are the time constants of the oxygen and hydrogen actuator flow-rate valve dynamics, respectively. The inputs of the valves $\bar{q}_{O_2}^{in}$ and $\bar{q}_{H_2}^{in}$ are considered as PEMFC control functions, i.e., $u_1 = \bar{q}_{H_2}^{in}$ and $u_2 = \bar{q}_{O_2}^{in}$.

Remark 4.2.2. *Actuator dynamics in (4.2.10) and (4.2.11) are neglected at the stage of the controller design. Equations (4.2.10) and (4.2.11) are used in the simulation study. Furthermore, the effects of the unmodeled dynamics (4.2.10) and (4.2.11) can be rigorously addressed using, for instance, the describing function techniques. The corresponding algorithms are presented in reference 19.*

The output voltage V_{tfc} of a stack of n PEMFCs defined by the following equations [20, 21]:

$$\begin{aligned} V_{tfc} &= nV_{fc}, V_{fc} = E_0 - V_{ohm} - V_{conc} - V_{act}, \\ E_0 &= \frac{\Delta G}{2F} + \frac{\Delta s}{2F}(T - T_{fc}) + \frac{RT}{2F}\left(\ln\ (P_{H_2}) + \frac{1}{2}\ln (P_{O_2})\right), \end{aligned} \tag{4.2.12}$$

$$V_{ohm} = i_{fc}R_{ohm}, \quad V_{conc} = m \cdot \exp(ni_{fc}) + b\ln\left(\frac{P_{o2}}{a}\right). \tag{4.2.13}$$

The notations in (4.2.12) and (4.2.13) are: ΔG is the Gibbs-free energy ($Jmol^{-1}$), Δs is the standard molar entropy ($Jmol^{-1}\,K^{-1}$), R = 8.314 $Jmol^{-1}\,K^{-1}$ is the universal gas constant, $F = 96485.3415\ sA/mol$ is the Faraday constant, $T_{fc} = 298.15^{\circ}K$ is the reference temperature of the FC, and $T = 353\ K$ is the stack temperature. In this work, it is assumed that the temperature $(T - T_{fc})$ is constant during the process. The variable V_{conc} depends on the changes in the concentration of reactants as they are consumed by the electrochemical reaction going in FC. Due to the small values of m and b (specifically, the value of m is typically close to $m = 3\cdot 10^{-5}V$) in (4.2.13), the term V_{conc} can be neglected in (4.2.12) as done in references 1, 3, and 8.

4.2.3.2 Mathematical model of multiple modular DC–DC boost power converters with an individual load configuration

The dynamics of the multiple modular DC–DC boost power converters with individual load configuration are governed by the following system of differential equations [13, 22]:

$$
\begin{aligned}
\frac{dV_L^i}{dt} &= \frac{1}{C_{Li}}[(1-s_i)i_L^i - i_0^i],\ \forall i = 1,2;\\
\frac{di_{fc}}{dt} &= \frac{1}{L_{L1}}[V_L^1(1-s_1)+V_{tfc}] + \frac{1}{L_{L2}}[V_L^2(1-s_2)+V_{tfc}];\\
\frac{di_0^i}{dt} &= \frac{1}{L_L^i}[-i_0^iR_L^i + V_L^i],\ \ \forall i = 1,2,
\end{aligned}
\tag{4.2.14}
$$

where $s_i \in [0,\ 1]\ \forall = 1,\ 2$ that is transformed as

$$
s_i = v^i + 0.5, v^i \in [-0.5, 0.5]; \forall i = 1,2. \tag{4.2.15}
$$

Remark 4.2.3. *It should be noted that the PEMFC current i_{fc} is equal to $i_L^1 + i_L^2$. i_L^1 is the output current of DC–DC boost converter 1. The output current of the DC–DC boost converter 2 is given as i_L^2.*

4.2.3.3 Mathematical model of electric power system comprised PEMFC/multiple modular DC–DC converter with an individual load configuration

The mathematical model of PEMFC/multiple modular DC–DC converter-based electric power system is presented in a form of input–output dynamics and internal dynamics. Considering, $y^1 = V_L^1$, $y^2 = V_L^2$, $y^3 = i_{fc}$, and $y^4 = P_{H_2}$ as the outputs and v^1, v^2, u_2, and u_1 as the control inputs, respectively, the input–output dynamics of the system are derived based on (4.2.7)–(4.2.14)

$$
\begin{bmatrix} \dfrac{dV_L^1}{dt} \\ \dfrac{dV_L^2}{dt} \\ \dfrac{d^2 i_{fc}}{dt^2} \\ \dfrac{dP_{H_2}}{dt} \end{bmatrix}
= \begin{bmatrix} B_1 \\ B_2 \\ B_3 \\ B_4 \end{bmatrix}
+ \begin{bmatrix} -\dfrac{1}{C_{L1}} i_L^1 & 0 & 0 & 0 \\ 0 & -\dfrac{1}{C_{L2}} i_L^2 & 0 & 0 \\ 0 & 0 & \left(\dfrac{1}{L_{L1}} + \dfrac{1}{L_{L2}}\right)\left(\dfrac{0.5\cdot RT}{2F}\left(\dfrac{1}{\tau_{O_2} k_{O_2} P_{O_2}}\right)\right) & 0 \\ 0 & 0 & 0 & \dfrac{1}{\tau_{H_2} k_{H_2}} \end{bmatrix}
\begin{bmatrix} v^1 \\ v^2 \\ u_2 \\ u_1 \end{bmatrix},
\tag{4.2.16}
$$

where

$$B_1 = \frac{1}{C_{L1}}\left(\frac{1}{2}i_L^1 - i_0^1\right),\ B_2 = \frac{1}{C_{L2}}\left(\frac{1}{2}i_L^2 - i_0^2\right),\ B_4 = -\frac{1}{\tau_{H_2}}P_{H_2} + \frac{1}{\tau_{H_2}k_{H_2}}\left(-2k_\gamma i_{fc}\right),$$

$$B_3 = \left(\frac{1}{L_{L1}}\left[\left(v^1 - 0.5\right)\ddot{V}_L^1 + 2\dot{v}^1\dot{V}_L^1 + V_L^1\ddot{v}^1\right] + \frac{1}{L_{L2}}\left[\left(v^2 - 0.5\right)\ddot{V}_L^2 + 2\dot{v}^2\dot{V}_L^2 + V_L^2\ddot{v}^2\right]\right)$$

$$-\left\{\left(i_{fc}\right)^{(1)} - \dot{V}_{act} + \left[\frac{RT}{2F}\left(\frac{1}{P_{H_2}}\left(-\frac{1}{\tau_{H_2}}P_{H_2} + \frac{1}{\tau_{H_2}k_{H_2}}\left(u_3 - 2k_\gamma i_{fc}\right)\right)\right)\right.\right.$$

$$\left.\left.+\frac{0.5}{P_{O_2}}\left(-\frac{1}{\tau_{O_2}}P_{O_2} - \frac{k_\gamma i_{fc}}{\tau_{O_2}k_{O_2}}\right)\right]\right\}.$$

Similarly, the internal dynamics of the FC system derived from the set of (4.2.7) and (4.2.14).

$$\frac{dV_{act}}{dt} = \frac{i_{fc}}{C_{dl}} - \frac{V_{act}}{R_{act}C_{dl}}, \tag{4.2.17}$$

$$\frac{di_0^1}{dt} = \frac{1}{L_L^1}\left(-i_0^1R_L^1 + V_L^1\right), \tag{4.2.18}$$

$$\frac{di_0^2}{dt} = \frac{1}{L_L^2}\left(-i_0^2R_L^2 + V_L^2\right). \tag{4.2.19}$$

Apparently, the internal dynamics in (4.2.17)–(4.2.19) appears to be stable, and the profiles V_{act} and i_{fc} are bounded, assuming that the forced output voltage profiles of the converters 1 and 2 (V_L^1, V_L^2) to be bounded.

Remark 4.2.4. *The nonminimum phase [14, 15, 23] in the DC–DC boost converters causes trouble in the direct control of the output voltages in the DC–DC boost converters 1 and 2, (V_L^1,V_L^2), thereby adding complexity to the controller design [16, 22–25]. In the case of the electric power system (4.2.16), controlling the current i_{fc} using the control function u_2 eliminates-the nonminimum phase dynamics. The detailed study of the controller u_2 design is presented in Section 4.2.6.*

4.2.4 Mathematical model of PEMFC/multiple modular DC–DC boost converter with shared load configuration of an electric power system

The equivalent circuit of PEMFC/multiple modular DC–DC boost converter with an individual load configuration of electric power system is presented in Figure 4.2.3 [2].

The following notations are used in Figure 4.2.3: E_0 is the reversible FC voltage, i_{fcs} is the FC current, V_{act} is the activation potential drop across the double-layered capacitance, and $C_{dl}\cdot R_{act}$ is the resistance that mainly causes the activation loss (V_{act}). The variable internal resistance of the FC is given to be R_{ohm}. The ohmic voltage drop

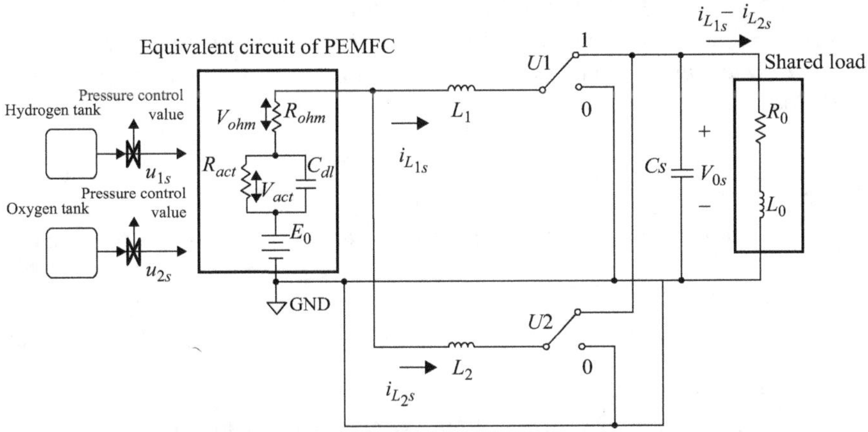

Figure 4.2.3 Equivalent circuit diagram of electric power system comprised PEMFC/multiple modular boost DC–DC converter with a shared load configuration

is presented as V_{ohm}. The output voltage of the DC–DC boost converter of a shared load configuration is represented as V_{0s}. The corresponding output current of the DC–DC boost converters 1 and 2 is given to be $i_{L_{1s}}, i_{L_{2s}}$. The shared inductive load of the DC–DC boost converter is R_0, L_0. The inputs of the valves $\bar{q}^{in}_{O_2}$ and $\bar{q}^{in}_{H_2}$ from (4.2.10) and (4.2.11) are given as, $u_{1s} = \bar{q}^{in}_{H_2}$ and $u_{2s} = \bar{q}^{in}_{O_2}$, for developing the mathematical model of the PEMFC/DC–DC boost power converter and ultracapacitor-based electric power system.

4.2.4.1 Mathematical model of multiple modular DC–DC boost power converter with shared load configuration

The dynamics of the multiple modular DC–DC boost power converters with individual load configuration are governed by the following system of differential equations [13, 22]: Bearing in mind that $i_{fcs} = i_{L_{1S}} + i_{L_{2S}}$; $i_{L_{1S}} = i_{fcs} - i_{L_{2S}}$,

$$
\begin{aligned}
\frac{dV_{0s}}{dt} &= \frac{1}{C_s}[(1 - s_{1s})(i_{fcs} - i_{L_{2S}}) - i_0 + (1 - s_{2s})\, i_{L_{2S}}],\\
\frac{di_{fcs}}{dt} &= \frac{1}{L_1}[-0.5V_{0s} + V_{0s}(1 - s_{1s}) + V_{tfc}] + \frac{1}{L_2}[-0.5V_{0s} + V_{0s}(1 - s_{2s}) + V_{tfc}],\\
\frac{di_{L_{2S}}}{dt} &= \frac{1}{L_2}[-0.5V_{0s} + V_{0s}(1 - s_{2s}) + V_{tfc}],\\
\frac{di_0}{dt} &= \frac{1}{L_0}[-i_0R_0 + V_{0s}],
\end{aligned}
\tag{4.2.20}
$$

where $s_{is} \in [0, 1]\, \forall_i = 1, 2$ that is transformed as

$$
s_{is} = v_i + 0.5, \quad v_i \in [-0.5, 0.5]; \quad \forall_i = 1, 2. \tag{4.2.21}
$$

Remark 4.2.5. *It should be noted that the PEMFC current i_{fcs} is equal to $i_{L_{1S}} + i_{L_{2S}}$, where $i_{L_{1S}}$ and $i_{L_{2S}}$ are the output currents of DC–DC boost converters 1 and 2. Next, the PEMFC current dynamics i_{fc} are rewritten in terms of the output current of the DC–DC boost converters 1 and 2 as*

$$\frac{d_{ifcs}}{dt} = \frac{di_{L_{1S}}}{dt} + \frac{di_{L_{2S}}}{dt}, \tag{4.2.22}$$

where the dynamics of the output current of DC–DC boost converter1 $i_{L_{1s}}$ are presented as,

$$\frac{di_{L_{1S}}}{dt} = \frac{1}{L_1}[-0.5V_{0s} + V_{0s}(1 - s_{1s}) + V_{tfc}]. \tag{4.2.23}$$

4.2.4.2 Mathematical model of PEMFC/multiple modular DC–DC converter with shared load configuration of an electric power system

The mathematical model of PEMFC/multiple modular DC–DC converter with shares load configuration of an electric power system is presented in a form of input-output dynamics and internal dynamics. Considering $y_{1s} = V_{0s}$, $y_{2s} = i_{L_{2s}}$, $y_{3s} = i_{fcs}$, and $y_{4s} = P_{H_2}$ as the outputs and v_1, v_2, u_{2s}, and u_{1s} as the control inputs, the input–output dynamics of the system are derived based on (4.2.17)–(4.2.21):

$$\begin{bmatrix} \frac{dV_{0s}}{dt} \\ \frac{di_{L_{2s}}}{dt} \\ \frac{d^2 i_{fcs}}{dt^2} \\ \frac{dP_{H_2}}{dt} \end{bmatrix} = \begin{bmatrix} F_1 \\ F_2 \\ F_3 \\ F_4 \end{bmatrix} + \begin{bmatrix} -\frac{(i_{fcs} - i_{L_{2S}})}{C_s} & 0 & 0 & 0 \\ 0 & \frac{V_{0s}}{L_2} & 0 & 0 \\ 0 & 0 & \left(\frac{1}{L_1} + \frac{1}{L_2}\right)\left(\frac{0.5 \cdot RT}{2F}\left(\frac{1}{\tau_{O_2} k_{O_2} P_{O_2}}\right)\right) & 0 \\ 0 & 0 & 0 & \frac{1}{\tau_{H_2} k_{H_2}} \end{bmatrix} \begin{bmatrix} v_1 \\ v_2 \\ u_{2s} \\ u_{1s} \end{bmatrix}, \tag{4.2.24}$$

where

$$F_1 = \frac{1}{C_s}\left(\frac{1}{2}(i_{fcs} - i_{L_{2S}}) - i_0 + \frac{1}{2} i_{L_{2S}} - \frac{i_{L_{2S}}}{C_s} v_2\right), \; F_2 = \frac{1}{L_2}[-0.5V_{0s} + V_{tfc}], \; F_4 = K_4$$

$$\begin{aligned} F_3 = & \left(\frac{1}{L_1}[(v_1 - 0.5)\ddot{V}_{0s} + 2\dot{v}_1 \dot{V}_{0s} + V_{0s}\ddot{v}_1] + \frac{1}{L_2}[(v_2 - 0.5)\ddot{V}_{0s} + 2\dot{v}_2 \dot{V}_{0s} + V_{0s}\ddot{v}_2]\right) \\ & - \left\{(i_{fcs})^{(1)} - \dot{V}_{act} + \left[\frac{RT}{2F}\left(\frac{1}{P_{H_2}}\left(-\frac{1}{\tau_{H_2}} P_{H_2} + \frac{1}{\tau_{H_2} k_{H_2}}(u_3 - 2k_\gamma i_{fcs})\right)\right)\right.\right. \\ & \left.\left. + \frac{0.5}{P_{O_2}}\left(-\frac{1}{\tau_{O_2}} P_{O_2} - \frac{k_\gamma i_{fcs}}{\tau_{O_2} k_{O_2}}\right)\right]\right\}. \end{aligned}$$

Similarly, the internal dynamics of the PEMFC/multiple modular DC–DC converter with shared load configuration of an electric power system are derived from (4.2.7) and (4.2.20).

$$\frac{dV_{act}}{dt} = \frac{i_{fcs}}{C_{dl}} - \frac{V_{act}}{R_{act}C_{dl}}, \tag{4.2.25}$$

$$\frac{di_0}{dt} = \frac{1}{L_0}[-i_0R_0 + V_{0s}]. \tag{4.2.26}$$

Apparently, the internal dynamics in (4.2.25) and (4.2.26) appear to be stable, and the profiles V_{act} and i_{fcs} are bounded, assuming the forced output voltage profiles of the converter with a shared load configuration (V_{0s}) to be bounded on application of feedback control law.

Remark 4.2.6. *The nonminimum phase [14, 23] in the DC–DC boost converters in the case of the electric power system (4.2.24) is eliminated by controlling the PEMFC current i_{fcs} using the control function u_{2s}. The main challenge is to balance the DC–DC boost converter currents $i_{L_{1S}}$, $i_{L_{2S}}$ by forcing the current $i_{L_{2S}}$ to follow the current $i_{L_{1S}}$, which is implemented by following the command profile $i_{L_{1Sc}}$.*

4.2.5 Problem formulation

4.2.5.1 Problem formulation of PEMFC/multiple modular DC–DC converter with an individual load configuration of an electric power system

In particular, the goal is to design the decoupled sliding mode controllers v^1, v^2, u_2, and u_1, where u_2 is the control designed to control the current i_{fc} of the PEMFC, v^1 is the control that drives the output voltage of the DC–DC boost converter 1 V_L^1 to a given reference profile $y^{1c} = V_L^{1c}(t)$. The control v^2 is designed to drive the output voltage of the converter 2 V_L^2 to a given reference profile $y^{2c} = V_L^{2c}(t)$. The control u_2 is designed to drive the output current of PEMFC i_{fc} to a corresponding current command profile $y^{3c} = i_{fc}^c(t)$ using power balance. All control laws are to be designed in the presence of unknown bounded parameters of PEMFC as well as the inductive loads (R_L^1, L_L^1 and R_L^2, L_L^2).

Using the relative degree approach [5, 9, 13] the *input–output* dynamics (4.2.23) of *vector-relative degree* $r = [1, 1, 2, 1]$ is given in a decoupled format as

$$\frac{dV_L^1}{dt} = B_1 - \frac{1}{C_{L1}} i_L^1 v^1 \tag{4.2.27}$$

for the output voltage control in the boost DC–DC converter 1,

$$\frac{dV_L^2}{dt} = B_2 - \frac{1}{C_{L2}} i_L^2 v^2 \tag{4.2.28}$$

for the output voltage control in the boost DC–DC converter 2,

$$\frac{d^2 i_{fc}}{dt^2} = B_3 + \left(\frac{1}{L_{L1}} + \frac{1}{L_{L2}}\right)\left(\frac{0.5 \cdot RT}{2F}\left(\frac{1}{\tau_{O_2} k_{O_2} P_{O_2}}\right)\right) u_2 \tag{4.2.29}$$

for the current control in PEMFC.

Finally,

$$\frac{d}{dt} P_{H_2} = B_4 + \frac{1}{\tau_{H_2} k_{H_2}} u_1 \tag{4.2.30}$$

for the hydrogen partial pressure regulation. Here $B_1, B_2, B_3,$ and B_4 are the perturbation terms.

Equations (4.2.27)–(4.2.30) are rewritten in terms of output tracking errors $e_1 = V_L^{1c} - V_L^1$, $e_2 = V_L^{2c} - V_L^2$, $e_3 = i_{fc}^c - i_{fc}$, $e_4 = P_{H_2}^c - P_{H_2}$, where $V_L^{1c}(t)$, $V_L^{2c}(t)$, $i_{fc}^c(t)$, $P_{H_2}^c$ are command profiles for $V_L^1(t)$, $V_L^2(t)$, $i_{fc}(t)$, and $P_{H_2}(t)$, respectively ($P_{H_2}^c$ is assumed to be constant):

$$\frac{de_1}{dt} = \underbrace{\dot{V}_L^{1c} - B_1}_{\psi_1} + \frac{1}{C_{L1}} i_L^1 v^1, \tag{4.2.31}$$

$$\frac{de_2}{dt} = \underbrace{\dot{V}_L^{2c} - B_2}_{\psi_2} + \frac{1}{C_{L2}} i_L^2 v^2, \tag{4.2.32}$$

$$\frac{d^2 e_3}{dt^2} = \left(i_{fc}^c\right)^{(2)} - B_3 - \left(\frac{1}{L_{L1}} + \frac{1}{L_{L2}}\right)\frac{0.5 \cdot RT}{2F}\left(\frac{1}{\tau_{O_2} k_{O_2} P_{O_2}}\right) u_2, \tag{4.2.33}$$

$$\frac{de_4}{dt} == \frac{1}{\tau_{H_2}} P_{H_2} + \frac{2k_\gamma}{\tau_{H_2} k_{H_2}} i_{fc} - \frac{1}{\tau_{H_2} k_{H_2}} u_1. \tag{4.2.34}$$

Equation (4.2.33) of relative degree 2 is reduced to relative degree 1 with respect to the variable

$$\sigma = \dot{e}_3 + c e_3, \tag{4.2.35}$$

where $c > 0$ is selected in order to provide a desired eigenvalue placement to differential equation $\dot{e}_3 + c e_3 = 0$.

Then the σ–dynamics is derived

$$\frac{d\sigma}{dt} = \underbrace{\left(i_{fc}^c\right)^{(2)} - B_3 + c e_3^{(1)}}_{\psi_3} - \underbrace{\left(\frac{1}{L_{L1}} + \frac{1}{L_{L2}}\right)\left(\frac{0.5 \cdot RT}{2F}\left(\frac{1}{\tau_{O_2} k_{O_2} P_{O_2}}\right)\right)}_{g_1} u_2. \tag{4.2.36}$$

The following assumptions about the disturbances ψ_1, ψ_2, ψ_3 are made: $|\psi_1| \leq M_1, |\psi_2| \leq M_2, |\psi_3| \leq M_3|$, respectively.

Given the output tracking error dynamics in (4.2.31), (4.2.32), (4.2.34), and (4.2.36), the problem is formulated as the control functions v^1, v^2, u_2, u_1 design so that e_1, e_2, e_3, $e_4 \to 0$ as time increases in the presence of bounded uncertainties, including unknown $R_L^1, L_L^1, R_L^2, L_L^2, C_{dl}$.

4.2.5.2 Problem formulation of PEMFC/multiple modular DC–DC converter with shared load configuration of an electric power system

The control motive is to design the decoupled sliding mode controllers v_1, v_2, u_{2s}, u_1, where u_{2s} is the control designed to control the current i_{fcs} of the PEMFC, v_1 is the control that drives the output voltage of both DC–DC boost converters V_{0s} to a given reference profile ($y_{1c} = V_{0sc}(t)$). The control v_2 is designed to drive the output current of the converter 2 $i_{L_{2S}}$ to the output current of the converter 1 reference profile ($y_{2c} = i_{L_{1cs}}(t)$). The control u_{2s} is designed to drive the output current of PEMFC to a current command profile ($y_{3c} = i_{fsc}(t)$), which is derived using power balance. The control laws are to be designed in the presence of unknown bounded parameters of PEMFC as well as the inductive load (R_0, L_0).

Using the relative degree approach [5, 9, 13, 15] the *input–output* dynamics (4.2.24) of *vector-relative degree* $r = [1, 1, 2, 1]$ is given in a decoupled format as

$$\frac{dV_{0s}}{dt} = F_1 - \frac{\left(i_{fcs} - i_{L_{2S}}\right)}{C_s} v_1, \tag{4.2.37}$$

$$\frac{di_{L_{2s}}}{dt} = F_2 + \frac{V_{0s}}{L_2} v_2, \tag{4.2.38}$$

$$\frac{d^2 i_{fcs}}{dt^2} = F_3 + \left(\frac{1}{L_{L1}} + \frac{1}{L_{L2}}\right)\left(\frac{0.5 \cdot RT}{2F}\left(\frac{1}{\tau_{o_2} k_{o_2} P_{O_2}}\right)\right) u_{2s}, \tag{4.2.39}$$

$$\frac{d}{dt} P_{H_2} = F_4 + \frac{1}{\tau_{H_2} k_{H_2}} u_{1s}. \tag{4.2.40}$$

Here F_1, F_2, F_3, F_4, are perturbation terms. Equations (4.2.37)–(4.2.40) are rewritten in terms of output tracking errors $e_{1s} = V_{0sc} - V_{0s}, e_{2s} = i_{L_{2sc}} - i_{L_{1s}}, e_{3s} = i_{fcsc} - i_{fcs}, e_{4s} = P^c_{H_2} - P_{H_2}$, where $V_{0sc}(t), i_{L_{2sc}}(t), i_{fcsc}(t), P^c_{H_2}$ are command profiles for $V_{0sc}(t)$, $i_{L_{1s}}(t)$, $i_{fcsc}(t)$, $P_{H_2}(t)$, respectively ($P^c_{H_2}$ is assumed to be constant):

$$\frac{de_{1s}}{dt} = \underbrace{\dot{V}_{0s} - F_1}_{\bar{\psi}_1} + \frac{\left(i_{fcs} - i_{L_{2S}}\right)}{C_s} v_1, \tag{4.2.41}$$

$$\frac{de_{2s}}{dt} = \underbrace{i_{L_{2sc}} - F_2}_{\bar{\psi}_2} - \frac{V_{0s}}{L_2} v_2, \tag{4.2.42}$$

$$\frac{d^2 e_{3s}}{dt^2} = \left(i_{fcsc}\right)^{(2)} - F_3 - \left(\frac{1}{L_1} + \frac{1}{L_2}\right)\frac{0.5 \cdot RT}{2F}\left(\frac{1}{\tau_{o_2} k_{o_2} P_{O_2}}\right) u_{2s}, \tag{4.2.43}$$

$$\frac{de_{4s}}{dt} == \frac{1}{\tau_{H_2}} P_{H_2} + \frac{2k_\gamma}{\tau_{H_2} k_{H_2}} i_{fc} - \frac{1}{\tau_{H_2} k_{H_2}} u_{1s}. \tag{4.2.44}$$

Equation (4.2.43) of relative degree 2 is reduced to relative degree 1 with respect to the variable

$$\sigma_s = \dot{e}_{3s} + c_s e_{3s} \tag{4.2.45}$$

where $c_s > 0$ are selected in order to provide a desired eigenvalue placement to differential equation $\dot{e}_{3s} + c_s e_{3s} = 0$. Then the σ_s – dynamics are derived

$$\frac{d\sigma_s}{dt} = \underbrace{\left(i_{fcsc}\right)^{(2)} - F_3 + c_s e_{3s}^{(1)}}_{\bar{\psi}_3} - \underbrace{\left(\frac{1}{L_1} + \frac{1}{L_2}\right)\left(\frac{0.5 \cdot RT}{2F}\left(\frac{1}{\tau_{o_2} k_{o_2} P_{O_2}}\right)\right)}_{g_2} u_{2s}. \tag{4.2.46}$$

The following assumptions about the disturbances $\bar{\psi}_1, \bar{\psi}_2, \bar{\psi}_3$ are made: $\left|\bar{\psi}_1\right| \leq H_1$, $\left|\bar{\psi}_2\right| \leq H_2$, $\left|\dot{\bar{\psi}}_3\right| \leq H_3$.

Given the output tracking error dynamics in (4.2.41), (4.2.42), (4.2.44), and (4.2.46) the problem is formulated as the control functions v_1, v_2, u_{2s}, u_{1s} design so that $e_{1s}, e_{2s}, e_{3s}, e_{4s}, \rightarrow 0$ as time increases in the presence of bounded uncertainties, including unknown R_0, L_0, C_{dl}.

Remark 4.2.7. *For simplicity the dynamics of the valves in (4.2.10) and (4.2.11) are considered as unmodeled dynamics and are neglected for the controller design in both individual and shared load configurations, but they are used in the simulations only for control validation. A rigorous analysis of SMC/2-SMC systems with unmodeled dynamics can be accomplished using the algorithms presented in [19].*

4.2.6 Controller design

4.2.6.1 Controller design for the output voltages of the DC–DC boost converters 1 and 2 in an individual load configuration

The control is decoupled. The indices are introduced in designing the dynamics of the output voltages of the DC–DC boost converters 1 and 2 tracking errors.

$$e_i = V_L^{ic} - V_L^i; \quad \forall i = 1, 2. \tag{4.2.47}$$

The dynamics of the output voltage tracking errors of DC–DC converters 1 and 2 are given in (31) and (32), where the disturbance ψ_1, ψ_2 is assumed bounded: $|\psi_1| \leq M_1$, $|\psi_2| \leq M_2$ the controller for the output voltage of the DC–DC converters 1 and 2 is to be discontinuous, since it represents a switching function $v^i \in [-0.5, 0.5]$ $\forall i = 1, 2$. Also, the controllers are to be insensitive to the bounded disturbances (ψ_1, ψ_2). A viable candidate for this controller is a conventional SMC [4, 11, 23]:

$$v^i = -0.5 sign(e_i); \quad \forall i = 1, 2. \tag{4.2.48}$$

The sliding mode existence condition [4]:

$$e_i \dot{e}_i \leq -\rho_i |e_i|, \rho_i > 0 \forall i = 1, 2, \tag{4.2.49}$$

where ρ is to be selected to provide for a given reaching time $t_r \leq e_i(0)/\rho$, is to be verified for the controller (4.2.48). Substituting (4.2.49) into (4.2.31) and (4.2.32), we obtain

$$e_1\dot{e}_1 = e_1\left(\psi_1 + \frac{i_L^1}{2C_{L1}} sign(e_1)\right) \leq |e_1|\left(M_1 + \frac{i_L^1}{2C_{L1}}\right) \leq -\rho_1|e_1|, \quad (4.2.50)$$

$$e_2\dot{e}_2 = e_2\left(\psi_2 + \frac{i_L^2}{2C_{L2}} sign(e_2)\right) \leq |e_2|\left(M_2 + \frac{i_L^2}{2C_{L2}}\right) \leq -\rho_2|e_2|. \quad (4.2.51)$$

The existence conditions (4.2.49)–(4.2.51) will be satisfied if

$$i_L^i \geq 2C_{Li}(\mathrm{M}_i + \rho_i)\forall i = 1, 2 \quad (4.2.52)$$

Equations (4.2.27) and (4.2.28) imply that the command to the inductance current command i_L^{ic} must be large enough to fulfill inequality (4.2.52).

4.2.6.2 Controller design for the output voltage of the DC–DC boost converter and output current of the DC–DC boost converter 2 in a shared load configuration

The control is decoupled. The error of the output voltage and output current of the converter 2 of the DC–DC boost converter is given as

$$e_{1s} = V_{0cs} - V_{0s}, \quad (4.2.53)$$

$$e_{2s} = i_{L_{2sc}} - i_{L_{1s}}. \quad (4.2.54)$$

The dynamics of the output voltage and output current of converter 2 tracking errors of DC–DC converters are given in (4.2.38) and (4.2.39), where the disturbance $\bar{\psi}_1, \bar{\psi}_2$ is assumed bounded: $|\bar{\psi}_1| \leq H_1, |\bar{\psi}_2| \leq H_2$. The controller for the output voltage of the DC–DC converters 1 and 2 is to be discontinuous, since it represents a switching function $v_i \in [-0.5, 0.5]\,\forall i = 1, 2$. Also, the controllers are to be insensitive to the bounded disturbances (ψ_1, ψ_2). A viable candidate for this controller is a conventional SMC [4, 11, 23]

$$v_i = -0.5 sign(e_{is}); \forall i = 1, 2. \quad (4.2.55)$$

The sliding mode existence condition [4]:

$$e_{is}\dot{e}_{is} \leq -\rho_{is}|e_{is}|, \quad \rho_{is} > 0 \quad \forall i = 1, 2, \quad (4.2.56)$$

where ρ_{is} is to be selected to provide for a given reaching time $t_{r_{is}} \leq e_{is}(0)/\rho_{is}$, is to be verified for the controller (4.2.55). Substituting (4.2.56) into (4.2.41) and (4.2.42), we obtain

$$e_{1s}\dot{e}_{1s} = e_{1s}\left(\bar{\psi}_1 + \frac{(i_{fcs} - i_{L2S})}{2C_s} sign\,(e_{1s})\right) \leq |e_{1s}|\left(H_1 + \frac{(i_{fcs} - i_{L2S})}{2C_s}\right) \leq -\rho_{1s}|e_{1s}|, \tag{4.2.57}$$

$$e_{2s}\dot{e}_{2s} = e_{2s}\left(\bar{\psi}_2 - \frac{V_{0s}}{L_2} sign\,(e_{2s})\right) \leq |e_{2s}|\left(H_2 + -\frac{V_{0s}}{L_2}\right) \leq -\rho_{2s}|e_{2s}|. \tag{4.2.58}$$

The existence conditions (4.2.56)–(4.2.58) will be satisfied if

$$(i_{fcs} - i_{L2S}) \geq 2C_s(H_1 + \rho_{1s}), \tag{4.2.59}$$

$$V_{0s} \geq L_2(H_2 + \rho_{2s}). \tag{4.2.60}$$

4.2.6.3 Adaptive gain 2-SM controller design for the PEMFC current i_{fc} and i_{fcs} for individual and shared load configurations

The error dynamics of the FC current of an individual load and shared load configurations given in (4.2.33), (4.2.34) and (4.2.45), (4.2.46) are rewritten as

$$\sigma = \dot{e}_3 + ce_3 \rightarrow \dot{\sigma} = \ddot{e}_3 + c\dot{e}_3 = \psi_3 - g_1 u_2, \tag{4.2.61}$$

$$\sigma_s = \dot{e}_{3s} + c_s e_{3s} \rightarrow \dot{\sigma}_s = \ddot{e}_{3s} + c_s\dot{e}_{3s} = \psi_{3s} - g_2 u_{2s}, \tag{4.2.62}$$

where $\sigma = \dot{e}_3 + ce_3; e_3 = i^c_{fc} - i_{fc}, \sigma_s = \dot{e}_{3s} + c_s e_{3s}; e_{3s} = i_{fcsc} - i_{fcs}$ and $\psi_3, \bar{\psi}_3$ are considered as the differentiable disturbances, whose derivative is bounded with unknown boundary, and u_2, u_{2s} are the control inputs. The controllers (u_2, u_{2s}) must be continuous. Therefore the control is designed in terms of $(\dot{u}_2, \dot{u}_{2s})$. Differentiating again, we obtain

$$\ddot{\sigma} = \dot{\psi}_3 - \underbrace{g_1\dot{u}_2}_{u_6}, \tag{4.2.63}$$

$$\ddot{\sigma}_s = \dot{\bar{\psi}}_3 - \underbrace{g_2\dot{u}_{2s}}_{u_7}. \tag{4.2.64}$$

Since the derivative of the disturbances $\dot{\psi}_3$ and $\dot{\bar{\psi}}_3$ are assumed to be bounded $|\dot{\psi}_3| \leq M_3; |\dot{\bar{\psi}}_3| \leq H_3$ with unknown boundaries M_3 and H_3, it is appropriate to apply adaptive-gain twisting controller [11, 16] that drives $e_3, \dot{e}_3$ and $e_{3s}, \dot{e}_{3s}$ to a real 2-SM in finite time. The adaptive gain twisting control is given as

$$u_6 = -\alpha_1(\text{sgn}(\sigma) + 0.5\text{sgn}(\dot{\sigma})), \tag{4.2.65}$$

$$\dot{\alpha}_1 = \left\{ \begin{array}{ll} \omega\sqrt{\frac{\gamma}{2}}\text{sgn}(\theta_1(\sigma,\dot{\sigma}) - \mu_1), & \textit{if } \alpha_1 \geq \alpha_{\min} \\ \chi_1, \quad \textit{if } \alpha_1 < \alpha_{\min} & \end{array} \right\},$$

$$u_7 = -\alpha_{1s}(\text{sgn}(\sigma_s) + 0.5\text{sgn}(\dot{\sigma}_s)), \tag{4.2.66}$$

$$\dot{\alpha}_{1s} = \left\{ \begin{array}{ll} \omega_s\sqrt{\frac{\gamma_s}{2}}\text{sgn}(\theta_{1s}(\sigma_s,\dot{\sigma}_s) - \mu_{1s}), & \textit{if } \alpha_{1s} \geq \alpha_{\min} \\ \chi_1, \quad \textit{if } \alpha_{1s} < \alpha_{\min} & \end{array} \right\},$$

where $\theta_1(\sigma,\dot{\sigma}) = \alpha_1^2\sigma^2 + \gamma|\sigma|^{3/2}\text{sgn}(\sigma)\dot{\sigma} + \alpha_1|\sigma|\dot{\sigma}^2 + \frac{1}{4}\dot{\sigma}^4, \gamma \geq \frac{1}{4}; \omega, \mu_1, \chi, \alpha_{\min}, \gamma$ are arbitrary positive constants. Finally, we have

$$u_2 = \int \frac{\alpha_1}{g_1}(\text{sgn}(\sigma) + 0.5\text{sgn}(\dot{\sigma}))dt,$$

$$u_{2s} = \int \frac{\alpha_{1s}}{g_2}(\text{sgn}(\sigma_s) + 0.5\text{sgn}(\dot{\sigma}_s))dt.$$

The problem of finding real-time robust estimation of $\dot{e}_3$ and $\dot{e}_{3s}$ is solved using arbitrary order robust exact differentiators [13]

$$\begin{aligned} \dot{z}_{0i} &= v_{0i},\ v_{0i} = -30|z_{0i} - e_3|^{1/2} sign(z_{0i} - e_3) + -z_{1i}, \\ \dot{z}_{1i} &= -100 sign(z_{1i} - v_{0i}), \end{aligned} \tag{4.2.67}$$

$$\begin{aligned} \dot{z}_{0is} &= v_{0is}, v_{0is} = -30|z_{0is} - e_{3s}|^{1/2} sign(z_{0is} - e_{3s}) + -z_{1is}, \\ \dot{z}_{1is} &= -100 sign(z_{1is} - v_{0is}), \end{aligned} \tag{4.2.68}$$

where $z_{0i} \to e_3$, $z_{1i} \to \dot{e}_3$ in finite time, and z_{1i} is used instead of $\dot{e}_3$ in $\sigma = \dot{e}_3 + ce_3$, i.e., $\sigma = z_{1i} + ce_3$. Similarly, we can write the same for $\sigma_s = \dot{e}_{3s} + c_s e_{3s}$, i.e. $\sigma_s = z_{1is} + c_s e_{3s}$.

Remark 4.2.8. *The PEMFC current command for the given electric power system is defined based on the power balance approach [5, 9] similar to the one discussed in Sections 4.2.6.2.1 and 4.2.6.2.2 for individual load and shared load configurations, respectively.*

4.2.6.4 Inductance/PEMFC current command generator for PEMFC/multiple modular DC–DC boost converter with an individual load configuration of an electric power system

The inductance current command profile $i_{fc}^c = i_L^{1c} + i_L^{2c}$ is computed based on the power balance $P_{\text{PEMFC}} = P_{Load}$[4, 5]:

$$i_L^{1c} V_{tfc} = i_0^{1c} V_L^{1c}, \tag{4.2.69}$$

where i_L^{1c} is the output current command of DC–DC boost converter 1, V_{tfc} is the measured PEMFC voltage, V_L^{1c} is the load voltage command profile, and i_0^{1c} is the inductive load (R_L^1, L_L^1) current command profile of the individual load corresponding to converter 1:

$$i_L^{1c} = i_0^{1c}\frac{V_L^{1c}}{V_{tfc}}, \frac{di_0^{1c}}{dt} = \frac{1}{L_L^1}(-i_0^{1c}R_L^1 + V_L^{1c}). \tag{4.2.70}$$

Similarly, we can derive the output current command of DC–DC boost converter 2 i_L^{2c} to be

$$i_L^{2c} = i_0^{2c}\frac{V_L^{2c}}{V_{tfc}}, \frac{di_0^{2c}}{dt} = \frac{1}{L_L^2}(-i_0^{2c}R_L^2 + V_L^{2c}). \tag{4.2.71}$$

Therefore the current command of PEMFC i_{fc}^c derived from (4.2.70) and (4.2.71) is presented as

$$i_{fc}^c = i_L^{1c} + i_L^{2c} = i_0^{1c}\frac{V_L^{1c}}{V_{tfc}} + i_0^{2c}\frac{V_L^{2c}}{V_{tfc}}. \tag{4.2.72}$$

4.2.6.5 Inductance/PEMFC current command generator for PEMFC/multiple modular DC–DC boost converter with shared load configuration of an electric power system

The PEMFC/inductance current command profile i_{fcsc} is computed based on the power balance $P_{\text{PEMFC}} = P_{Load}$ [4, 5].

$$i_{fcsc}V_{tfc} = i_0^c V_{0cs}, \tag{4.2.73}$$

where i_{fcsc} is the FC current command, V_{tfc} is the measured PEMFC voltage, $y_{1sc} = V_{0cs}$ is the load voltage command profile, and i_0^c is the load current command profile, which can be generated for the inductive R_0, L_0 load as

$$i_{fcsc} = i_0^c\frac{V_{0cs}}{V_{tfc}}, \frac{di_0^c}{dt} = \frac{1}{L_0}(-i_0^c R_0 + V_{0cs}). \tag{4.2.74}$$

Remark 4.2.9. *The PEMFC current command (i_{fcsc}) obtained from (4.2.73) based on power balance is substituted in (4.2.62) and the error $e_{3s} \to 0$ is driven to zero in finite time along with PEMFC current command for individual load configuration i_{fc}^c obtained from (4.2.72) based on power balance is substituted in (4.2.61) and the error $e_3 \to 0$ is driven to zero in finite time using 2-SM adaptive gain twisting controller without overestimating the gains.*

4.2.6.6 The control u_1 and u_{1s} design

A PI-like controller robustly addresses the set point regulation problem by driving $P_{H_2}^c - P_{H2} \to 0$ in (4.2.34) and (4.2.44) as time increases:

$$\begin{aligned} u_1 &= k_{H_2}P_{H_2} + 2k_\gamma i_{fc} + k_p e_4 + k_i \int e_4 dt, \\ u_{1s} &= k_{H_2}P_{H_2} + 2k_\gamma i_{fc} + k_p e_{4s} + k_i \int e_{4s} dt. \end{aligned} \tag{4.2.75}$$

The parameters (K_p and K_i) are tuned in the computer simulation as $K_p = 0.32$ and $K_i = 1.40$ The details of the PI controller design and tuning are straightforward [26] and, therefore, omitted for brevity.

4.2.7 Simulation study

Case 1: The simulated system of electrical energy supply consists of the multiple modular DC–DC power converters, a stack of $n = 24$ PEMFCs, and two separate inductive loads. The parameter E_0 varies accordingly to the number of PEMFC in a stack, n. The parameters of a PEMFC/multiple modular DC–DC boost converter with an individual load configuration of an electric power system are given below

$$\begin{aligned} L_{L2} &= 4.10^{-3}H,\ C_{L1} = 14.10^{-3}F,\ C_{L2} = 16.10^{-3}F, \\ R_{act} &= 0.08\Omega,\ C_{dl} = 68.10^{-3}F,\ L_{L1} = 3.10^{-3}H, \\ T &= 343^{\circ}K,\ R_{ohm} = 0.06\Omega,\ V_L^1(0) = 10V, \\ V_L^2(0) &= 7V,\ R_L^1 = 10 - 20\Omega,\ R_L^2 = 20 - 30\Omega. \end{aligned}$$

The following load voltage and current command profiles were selected in accordance with reference 3, to ensure that the proposed FC system can handle high power demands.

$$V_L^{1c}(t) = \begin{cases} 50V \quad t \in [0, 1)s, & \\ 70 + 10\sin t)V, & (1 \le t \le 3)s; \\ (70 + 10\sin t)V - (100 + 10\sin t)V, & (3 \le t \le 6)s; \\ (100 + 10\sin t)V - (70 + 10\sin t)V, & (6 \le t \le 9)\,s; \end{cases} \tag{4.2.76}$$

$$V_L^{2c}(t) = \begin{cases} 30V - (50 + 20\sin t)V, & (0 \le t \le 3)s; \\ (50 + 20\sin t)V - (80 + 20\sin t)V, & (3 \le t \le 3.5)s; \\ (80 + 20\sin t)V - (50 + 20\sin t)V, & (3.5 \le t \le 8)s. \end{cases}$$

The load resistor was changed at $t = 0.85s$:

$$\begin{aligned} R_L^1 &= \begin{cases} 20\Omega - 60\Omega, & (t \le 4)s, \\ 60\Omega - 90\Omega, & (4 \le t \le 7)s, \end{cases} \\ R_L^2 &= \begin{cases} 10\Omega - 50\Omega, & (t \le 2)s, \\ 50\Omega - 80\Omega, & (2 \le t \le 4)s. \end{cases} \end{aligned} \tag{4.2.77}$$

Figures 4.2.4 and 4.2.5 illustrate the output voltages of the DC–DC boost converters 1 and 2 (V_L^1, V_L^2) following the respective command profiles (V_L^{1c}, V_L^{2c}). Figure 4.2.6 shows that the FC's current (i_{fc}) follows the respective FC command profile (i_{fc}^c). Figure 4.2.7 shows that the adaptive gain (α_1) is not overestimated in the presence of

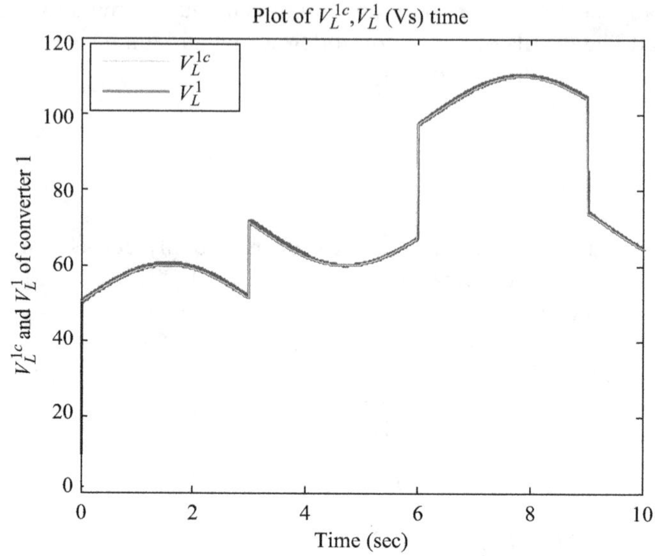

Figure 4.2.4 Plot of $V_L^{1c}(t)$ and (V_L^1)

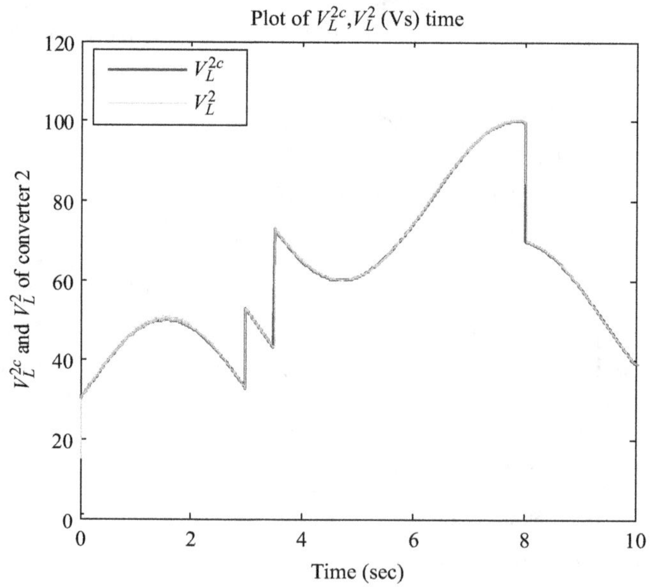

Figure 4.2.5 Plot of $V_L^{2c}(t)$ and (V_L^2)

bounded disturbance with unknown boundaries. Figure 4.2.8 indicates the FC current control function (u_2). Figure 4.2.9 shows the time histories of the controls (v^1, v^2) for the corresponding converters 1 and 2. Figure 4.2.10 shows the error corresponding to the PEMFC current and the voltages of converters 1 and 2 (e_1, e_2, e_3). Figure 4.2.11

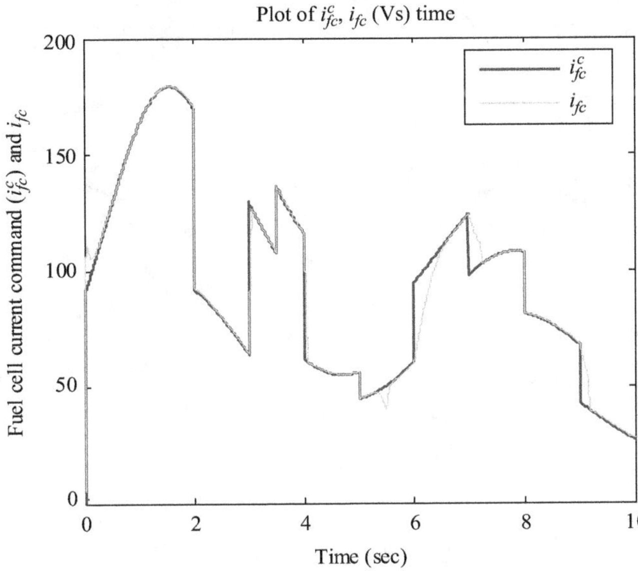

Figure 4.2.6 Plot of $i_{fc}^c(t)$ and (i_{fc})

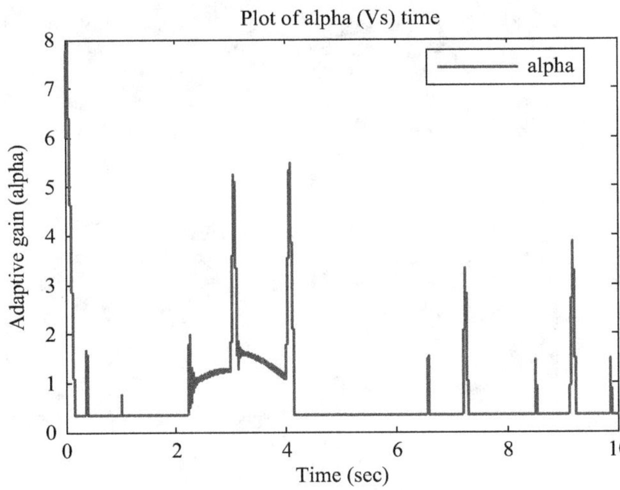

Figure 4.2.7 Plot of α_1

shows that the partial pressure of hydrogen (P_{H_2}) is maintained constant as time increases with respect to its corresponding command profile ($P_{H_2}^c$).

Case 2. The goal of this simulation experiment is to demonstrate efficacy of the output voltage tracking and balancing of the converter currents in the PEMFC/multiple

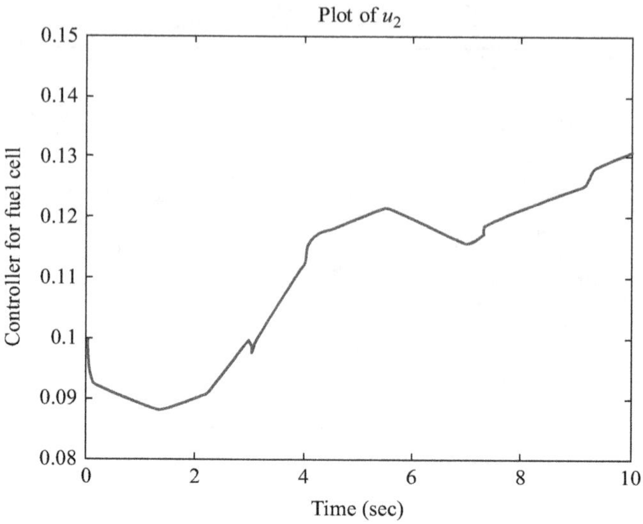

Figure 4.2.8 Plot of u_2

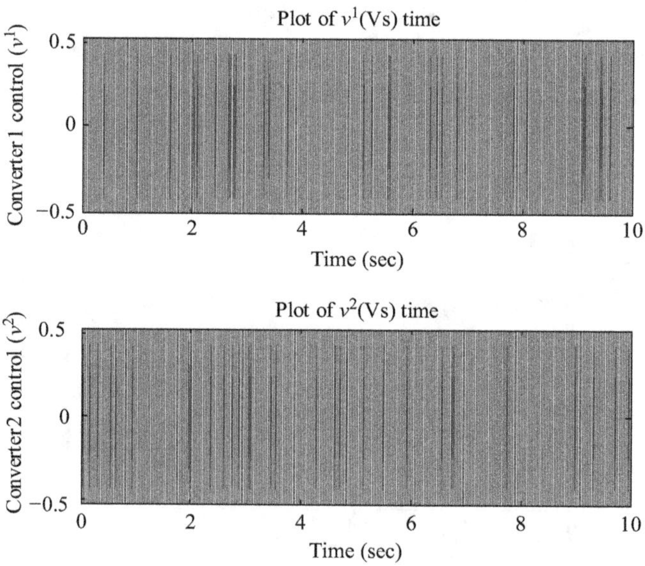

Figure 4.2.9 Plot of v^1 and v^2

modular DC–DC boost converter with shared load configuration of an electric power system with

$$L_2 = 4.10^{-3}H, C_s = 14.10^{-3}F, R_{act} = 0.08\Omega, C_{dl} = 68.10^{-3}F, L_1 = 3.10^{-3}H,$$
$$T = 343°K, R_{ohm} = 0.06\Omega, V_{0s}(0) = 10V, R_0 = 10 - 20\Omega, L_0 = 3.5 \cdot 10^{-3}H.$$

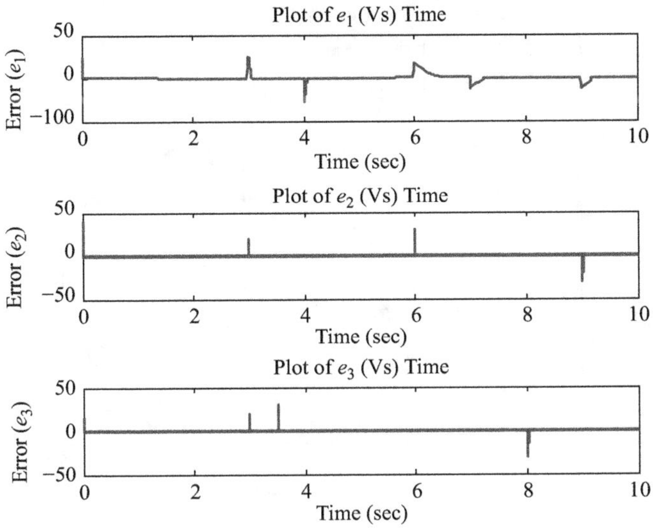

Figure 4.2.10 Plot of errors (e_1, e_2, e_3)

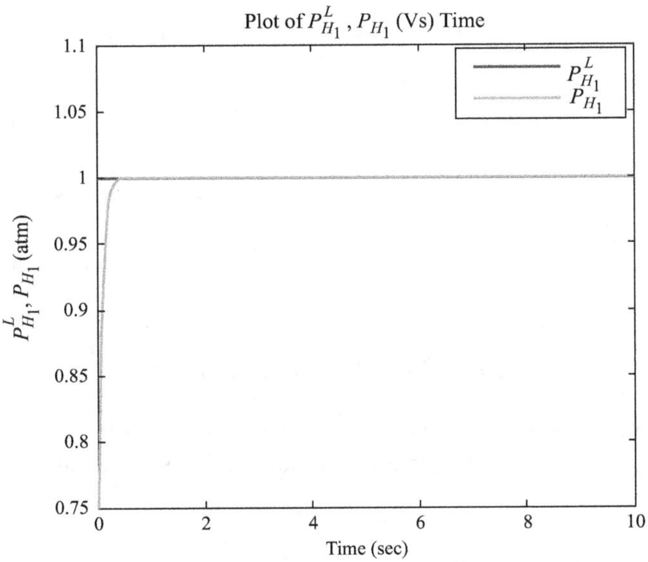

Figure 4.2.11 Plot of $P_{H_2}^c$ and P_{H_2}

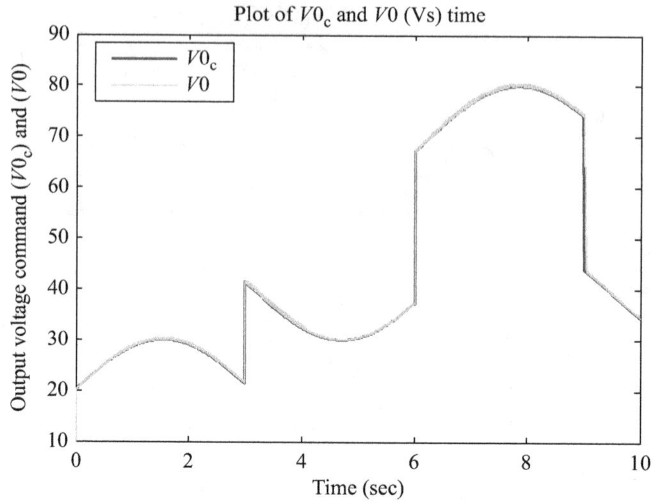

Figure 4.2.12 Plot of (V_{osc}) and (V_{os})

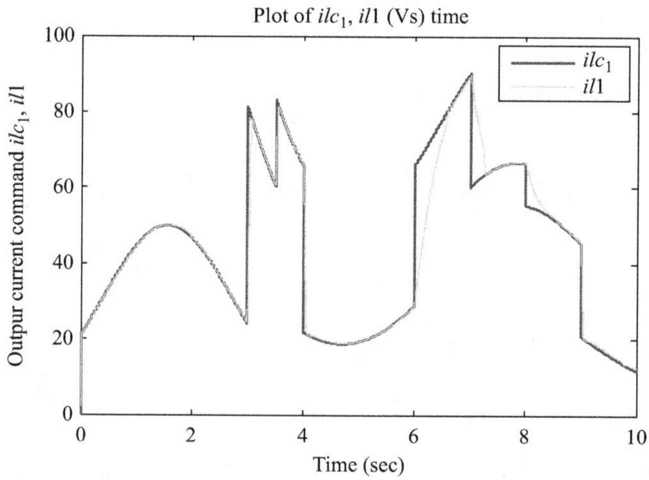

Figure 4.2.13 Plot of ($i_{L_{1Sc}}$) and ($i_{L_{1S}}$)

The following load voltage and current command profiles were selected in accordance with reference 3 to ensure that the proposed FC system can handle high power demands:

$$V_{0sc}(t) = \begin{cases} 20V \quad t \in [0,1)s, \\ (40 + 10\sin t)V, & (1 \le t \le 3)s; \\ (40 + 10\sin t)V - (70 + 10\sin t)V, & (3 \le t \le 6)s; \\ (70 + 10\sin t)V - (40 + 10\sin t)V, & (6 \le t \le 9)s; \end{cases} \tag{4.2.78}$$

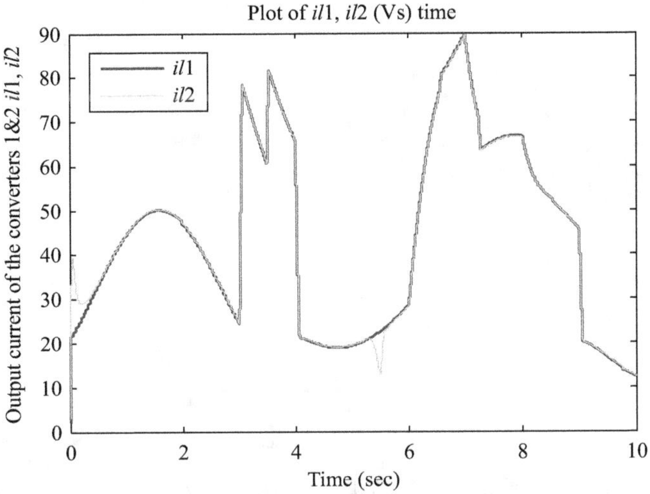

Figure 4.2.14 Plot of $(i_{L_{1S}})$ *and* $(i_{L_{2S}})$ *versus time*

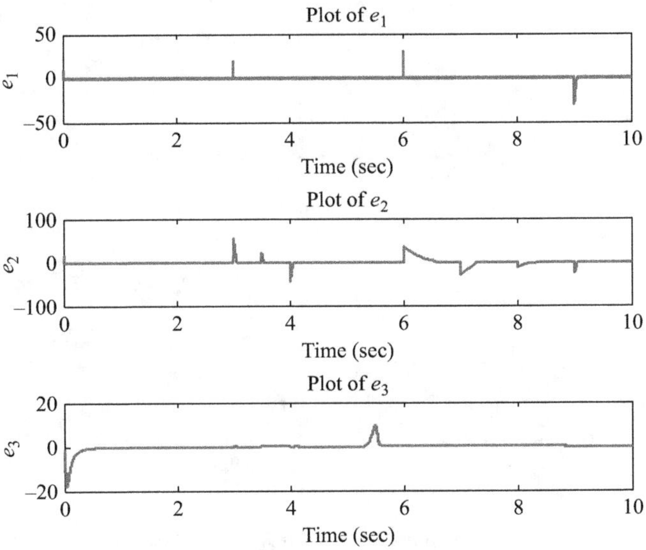

Figure 4.2.15 Plot of errors e_{1s}, e_{2s}, e_{3s}

The load resistor was changed at $t = 0.85s$:

$$R_0 = \begin{cases} 20\Omega - 60\Omega, & (t \leq 4)s, \\ 60\Omega, -90\Omega & (4 \leq t \leq 7)s. \end{cases} \tag{4.2.79}$$

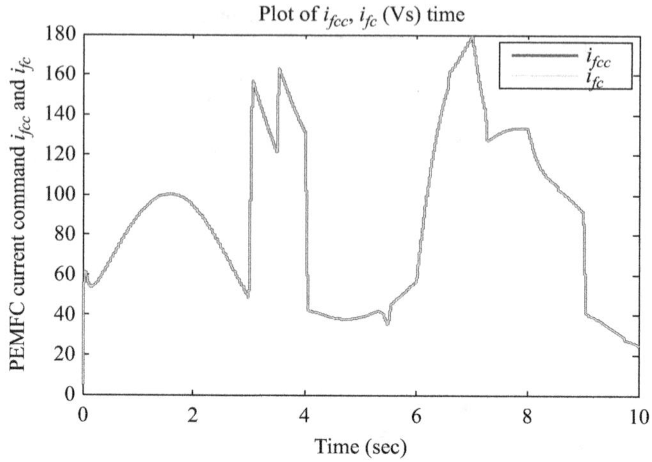

Figure 4.2.16 Plot of (i_{fcs}) and (i_{fcsc})

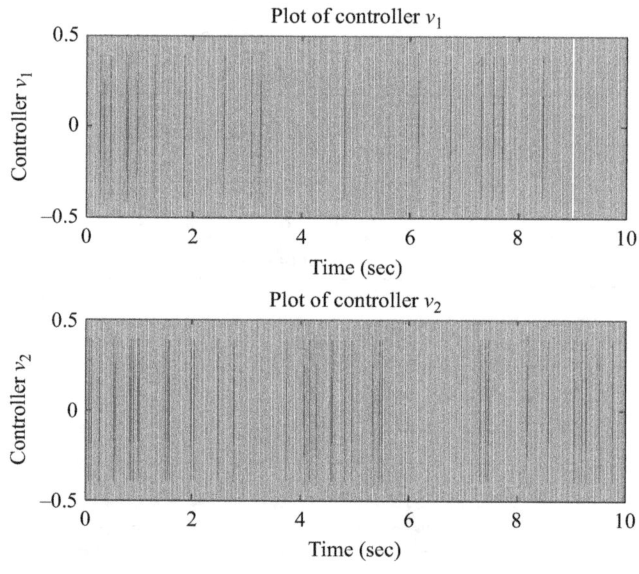

Figure 4.2.17 Plot of v_1 and v_2

The plot in Figure 4.2.12 illustrates the robust high-accuracy direct tracking of the output voltage of the DC–DC boost converter, V_{0s}, its respective command profile via SMC. The output current of converter 1 ($i_{L_{1S}}$) follows the reference profile ($i_{L_{1Sc}}$) in Figure 4.2.13. Figure 4.2.14 shows that the output current of the DC–DC boost converter 2 follows the output current of converter 1 accurately. The errors associated with the output voltage of the DC–DC boost converter, output current of DC–DC boost converter 2, and PEMFC current are illustrated in Figure 4.2.15.

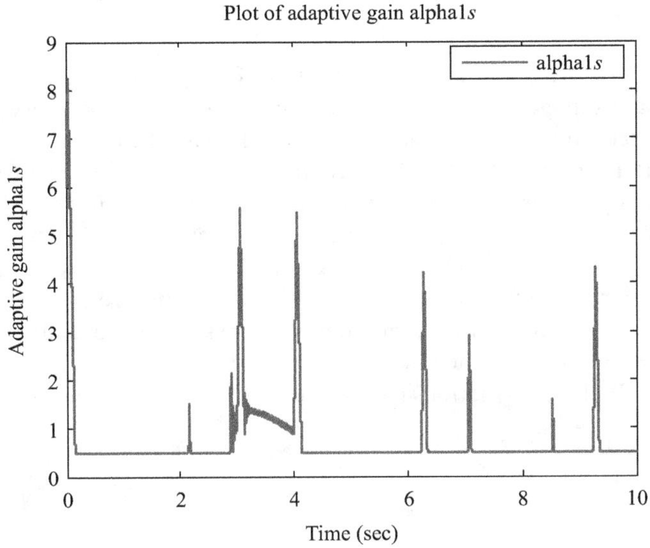

Figure 4.2.18 Plot of (α_{1s})

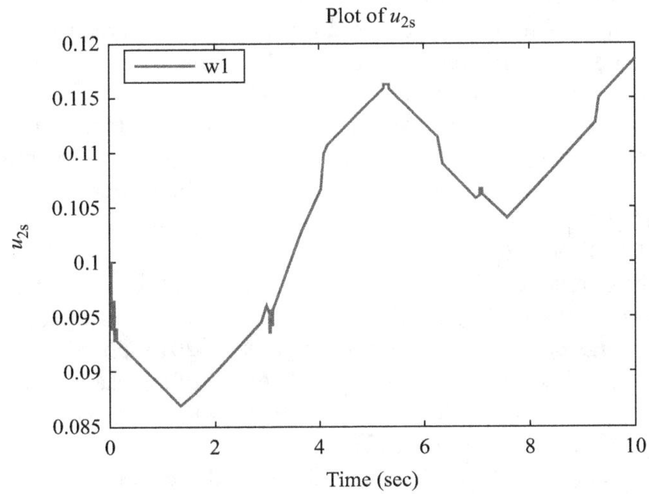

Figure 4.2.19 Plot of (u_{2s})

High-accuracy tracking of the PEMFC current, i_{fcs}, command profile is confirmed via plots in Figure 4.2.16. Figure 4.2.17 shows the controllers (v_1, v_2) for controlling the output voltage of converter 1 and balancing the currents in converter 1 and 2.The time history of adaptive control gain is demonstrated in Figure 4.2.18. The control (u_{2s}) for the PEMFC current i_{fcs} is given in Figure 4.2.19.

4.2.8 Conclusion

Control of an electric power system comprised PEMFC, along with the shared and individual load configurations of multiple modular boost DC–DC power converters, has been studied using sliding mode control techniques. The nonlinear effects of the boost converters in an individual load configuration coupled through PEMFC are overcome by conventional SMCs. The nonminimum phase output tracking is alleviated by controlling the PEMFC current using the adaptive 2-SMC. The conventional sliding mode controllers are implemented to control the output tracking of the voltage in both converter configurations. The challenge of balancing the currents in the boost converters in the shared load configuration is addressed via conventional SMC. The efficiency and robustness of the three-fold controllers for the proposed electric power system for the both configurations has been confirmed via computer simulations.

References

[1] C. Kunusch, P. F. Puleston, M. A. Mayosky, and J. Riera, "Sliding mode strategy for PEM FCS stacks breathing control using a super-twisting algorithm," *IEEE Transactions on Control System Technology*, vol. 17, no. 1, 2009, pp. 167–174.

[2] Phatiphat Thounthong and Bernard Davat, "Study of a multiphase interleaved step-up converter for fuel cell high power applications," *Energy Conversion and Management,* vol. 51, 2010, pp. 826–832.

[3] C. Kunusch, "Second order sliding mode control of a FC stack using a twisting algorithm," MS thesis, Electr. Dept., Nat. Univ. La Plata, Argentina, 2006.

[4] Y. Shtessel and R. S. Ashok, "PEM FC/DC-DC boost power converter system control via traditional and higher order sliding modes," *Proceedings of the CDC,* 2011, pp. 8261–8266.

[5] R. Ashok, Y. Shtessel, and J. Smith, "Sliding mode control of electric power system comprised of fuel cells, DC-DC boost converters and ultracapacitors," *Proceedings of American Control Conference,* 2013, pp. 5766–5771.

[6] Y. Zhang and B. Zhou, "Modeling and control of a portable proton exchange membrane fuel cell-battery power system," *Journal of Power Sources*, vol. 196, no 20, 2011, pp. 8413–8423.

[7] Xuancai Zhu, Xiao Li, Guoqiao Shen, and Dehong Xu, "Design of the dynamic power compensation for PEMFC distributed power system," *IEEE Trans. Industrial Electronics,* vol. 57, no. 6, 2010.

[8] W. Greenwell and A. Vahidi, "Predictive control of voltage and current in a fuel cell–ultracapacitor hybrid," *IEEE Trans. Industrial Electronics,* vol. 57, no. 6, 2010.

[9] R. S. Ashok and Y. B. Shtessel, "Control of fuel cell-based electric power system using adaptive sliding mode control and observation techniques," *Journal of the Franklin Institute*, doi:10.1016/j.jfranklin.2015.04.010

[10] L. Fridman and A. Levant, "Higher order sliding modes," In: *Sliding Mode Control in Engineering*, 2002, Ch. 3, pp. 53–101.

[11] Y. Shtessel, M. Taleb, and F. Plestan, "A novel adaptive-gain *super-twisting* sliding mode controller: methodology and application," *Automatica*, vol. 48, no 5, 2012, pp. 759–769.

[12] S. Spurgeon and R. Davies, "A nonlinear control strategy for robust sliding mode performance in the presence of unmatched uncertainty," *Int. J. Control*, vol. 57, no. 5, 1993, pp. 1107–1123.

[13] Y. Shtessel, C. Edwards, L. Fridman, and A. Levant, *Sliding Mode Control and Observation*, Birkhauser, New York, 2014.

[14] J. M. Olm, X. R. Oton, and Y. Shtessel, "Stable inversion-based robust tracking control in DC-DC nonminimum phase switched converters," *Automatica,* **47**, 1, 2011, pp. 221–226.

[15] A. Isidori, *Nonlinear Control Systems*, New York, 1995.

[16] J. Patterson and Y. B. Shtessel, "Sliding mode tracking control of output voltage in multiple modular boost power converters using the method of stable system center," *Proceedings of the Conference on Decision and Control,* 2005.

[17] K. Belmokhtar, M. H. Hammoudi, M. L. Doumbia, and K. Agbossou, "Modelling and fuel flow dynamic control of proton exchange membrane fuel cell," *Proceedings of 4th International Conference on Power Engineering, Energy and Electrical Drives*, 2013, pp. 415–420.

[18] J. Padullésa, G. W. Aultb, and J. R. McDonald, "An integrated SOFC plant dynamic model for power systems simulation," *Journal of Power Sources*, vol. 86, Issues 1–2, pp. 495–500.

[19] I. Boiko, *Discontinuous Control Systems. Frequency Domain Analysis and Design.* Birkhauser, 2009.

[20] R. Lindeborg, A. J. Swoboda, J. Burk, and S. Buratto, "Optimization of platinum nanoparticles for proton exchange membrane fuel cells using pulse electrochemical deposition," *The Harvard Undergraduate Research Journal*, 2013,vol. 6, Issue 1, pp. 9–15.

[21] J. Correa, F. Farret, L. Canha, and M. Simoes, "An electrochemical-based fuel- cell model suitable for electrical engineering automation approach," *IEEE Transactions on Industrial Electronics,* vol. 51, no. 5, 2004.

[22] Y. Shtessel, O. Raznopolov, and L. Ozerov, "Control of multiple modular dc-to-dc power converters in conventional and dynamic sliding surfaces," *IEEE Transactions on Circuits and Systems—I: Fundamental Theory and Applications,* vol. 45, no. 10, 1998.

[23] A. Zinober, Y. Shtessel, E. Fossas, J. Olm, and J. Patterson, "Non-minimum-phase output tracking strategies for DC-to-DC power converters control," *Advances in Variable Structure and Sliding Mode Control*, Lecture Notes in Control and Information Science vol. 334, 2006, pp. 447–482.

[24] Y. Shtessel, A. Zinober, and I. Shkolnikov, "Sliding mode control of boost and buck-boost power converters using method of stable system centre," *Automatica*, vol. 39, no. 6, 2003, pp. 1061–1067.

[25] Y. Shtessel, A. Zinober, and I. Shkolnikov, "Sliding mode control of boost and buck-boost power converters using the dynamic sliding manifold," *International Journal of Robust and Nonlinear Control,* vol. 13, no. 14, 2003, pp. 1285–1298.
[26] R. C. Dorf and R. H. Bishop, *Modern Control Systems*, 12th Edition, Prentice Hall, 2010.

Chapter 4.3
Networked model-based event-triggered sliding mode control

Antonella Ferrara and Gian Paolo Incremona

Abstract

In this chapter, the concept of model-based event-triggered control is revisited combining it with sliding mode control (SMC), so as to design robust networked SMC schemes. After some preliminaries on the conventional model-based event-triggered control, two alternative design frameworks are proposed along with the discussion of their stability properties. Finally, an example of application is presented to illustrate the two proposed strategies.

4.3.1 Introduction

The last decades progress in telecommunication has brought along the development of new control configurations, named networked control systems (NCSs), with considerable advantages in terms of remote data transmission, costs, flexibility, and modularity of the communication architectures [1–4]. However, in NCSs, the occurrence of packet loss, jitter, and delayed transmissions is quite frequent, so that the control systems performance could be deteriorated. In order to overcome the problems induced by the network presence, several control schemes designed taking into account the network presence have been proposed in the literature in recent years (see references 5–8 and the references therein). Among these, event-triggered control schemes represent an effective solution. In an event-triggered control scheme, the state of the plant is transmitted over the network only if a prespecified triggering condition holds [9–14]. A similar approach is adopted in model-based NCSs which use an explicit model of the plant asynchronously updated with the actual state of the system, when this is transmitted through the network [15, 16]. Both the approaches significantly reduce the measurement transmission rate, though guaranteeing satisfactory performance.

Sliding mode control (SMC) [17–20] may represent an alternative easy-to-implement solution able to guarantee robustness properties of the controlled system even in case of NCSs (see, for instance, reference 21). SMC and event-triggered control have been profitably coupled in some works recently appeared in the literature [22–24]. In this chapter the design of SMC schemes of model-based event-triggered type is illustrated making reference to such works. Specifically, some preliminary issues about event-triggered control and model-based event-triggered control are first briefly recalled. Then, two alternative design frameworks of model-based event-triggered SMC are discussed. Both the SMC structures are networked switched control schemes including a model-based controller, which contains the nominal model of the plant. Both rely on a suitably defined triggering condition. For the sake of simplicity, in both cases, the network presence is assumed only between the sensor and the model-based controller, since, in case of single-input plants, the control variable transmission is definitely less onerous than the state transmission. Furthermore, it is worth noting that the nominal model of the plant receives the same control variable fed into the real plant, and it is updated whenever a triggering event occurs.

The first approach uses a sliding variable determined by relying on the nominal model during a portion of the reaching phase and enables the actual state transmission only when the sliding variable is within a predefined boundary layer of the sliding manifold. As a result, the state of the nominal model is used out of the boundary layer, while the actual plant state is used inside the boundary layer. In this way, it is possible to guarantee the same robustness with respect to matched uncertainties as in conventional SMC schemes, even if the actual plant state is not always used to close the feedback.

The second approach, in contrast to the first one, out of the boundary layer uses a sliding variable defined on the basis of the actual plant state. Then, the state transmission is switched off when the sliding variable enters the boundary layer. Within this layer, the state of the nominal model is used to compute the control variable. This means that when the sliding variable belongs to the boundary layer, no state measurement transmission is required.

If, on one hand, SMC has a number of positive features [17, 18], on the other hand, the implementation of SMC laws can cause the notorious chattering phenomenon [25–28], i.e., high-frequency oscillations of the controlled variable due to the discontinuity of the control law during the sliding mode. Several methods have been developed in the literature to reduce chattering, such as Filtered Sliding Mode Control, Boundary Layer Control, and Higher Order SMC [25, 29–39]. In this chapter, making reference to the second networked SMC scheme herein presented, the chattering problem is explicitly considered. In particular, because of the model-based nature of the proposal, a chattering alleviation strategy, which appears to be intrinsic in the scheme, is designed. Specifically, when the control law relies on the model state, i.e., inside the boundary layer, a "pseudo-equivalent control" is analytically computed, and it is applied in place of the discontinuous control law. The pseudo-equivalent control is continuous, which implies a chattering alleviation effect. Note that the continuous control is called pseudo-equivalent, since it differs from the theoretical "equivalent control" [17] because of the absence of the matched uncertain term, and it is computed relying on the nominal model state.

In this chapter, the two model-based event-triggered SMC strategies are theoretically analyzed. They are also assessed in simulation, putting into evidence their capability of reducing the transmissions of the state measurement at the cost of a limited decay of the performance with respect to a genuine SMC scheme.

4.3.2 Model-based event-triggered control: preliminaries

Consider a plant, the model of which can be expressed as a perturbed chain of integrators, i.e.,

$$\begin{cases} \dot{x}_i(t) = x_{i+1}(t) \quad i = 1, \dots, n-1, \\ \dot{x}_n(t) = f(x,t) + b(x,t)u(t) + h(x,t), \\ y(t) = \sigma(x(t)), \end{cases} \tag{4.3.1}$$

where $x \in \mathbb{R}^n$ is the state, $u \in \mathbb{R}$ is the control variable, and $\sigma : \mathbb{R}^n \to \mathbb{R}$ is a smooth output function, called sliding variable. The functions $f(\cdot)$ and $b(\cdot)$ are known with $b(\cdot) \neq 0$, while $h(\cdot)$ is the uncertainty affecting the system which is assumed to be matched, i.e.,

$$h(x,t) = b(x,t)u_m(t) \tag{4.3.2}$$

with

$$|h(x,t)| \leq h_{max}, \tag{4.3.3}$$

where h_{max} is a positive constant. The nominal model of system (4.3.1), which is not affected by the uncertain term by definition, is the following:

$$\begin{cases} \dot{\hat{x}}_i(t) = \hat{x}_{i+1}(t) \qquad i = 1, \dots, n-1, \\ \dot{\hat{x}}_n(t) = f(\hat{x},t) + b(\hat{x},t)u(t), \\ \hat{y}(t) = \sigma(\hat{x}(t)), \end{cases} \tag{4.3.4}$$

where $\hat{x} \in \mathbb{R}^n$ is the state, u and $\sigma(\cdot)$ are the same as in (4.3.1).

In an event-triggered control framework, the problem to solve is to find a control law able to stabilize the closed-loop plant under reduced communication. In the considered case, the problem is further complicated by the presence of the uncertainty.

In practical implementation the state is sampled at certain time instants $t = t_0, t_1, \dots t_k, \dots, k \in \mathbb{N}$, and the control law, computed as $u(t_k)$, is held constant between two successive sampling time instants (it can be expressed as $u(t) = u(t_k)$, $\forall t \in [t_k, t_{k+1}]$, $k \in \mathbb{N}$). In traditional control, the sequence $\{t_k\}$, $k \in \mathbb{N}$, is typically periodic and the time interval $T = t_{k+1} - t_k$, $k \in \mathbb{N}$, is a priori fixed. The control approach, in that case, is classified as "time-triggered." If instead a triggering rule, which implicitly depends on the state, is specified, for instance,

$$|e| < \lambda|\sigma(x(t))| \tag{4.3.5}$$

with $0 < \lambda < 1$, and e being the measurement error defined as

$$e(t) = \sigma(x(t_k)) - \sigma(x(t)). \tag{4.3.6}$$

$\forall t \in [t_k, t_{k+1}[, \; k \in \mathbb{N}$, and the state of the controlled plant is transmitted over the network only when such a condition is violated, then the control approach can be classified as "event-triggered" [9]. In case the actuator (included in the plant) has the capability to hold the control variable computed at the last triggering time instant t_k for any $t \in [t_k, t_{k+1}[$, this approach tends to minimize the transmissions over the network path from the sensor to the controller. Moreover, by using the event-triggered control, in spite of the aperiodic transmission of the measurements, interesting stability properties can be enforced. Specifically, in reference 13, it was proved that in case of nonlinear systems, relying on a suitably defined triggering condition with a state feedback control law, the input-to-state stability of the controlled system can be guaranteed.

In a model-based control approach, an explicit model of the plant is added to the controller in order to determine the control law, whenever possible, on the basis of the state of the model rather than on the actual measurement. In this case, the measurement error depends on the actual plant state and on the state of the model, i.e.,

$$e(t) = \sigma(\hat{x}(t)) - \sigma(x(t)). \tag{4.3.7}$$

The sensor included in the scheme is assumed to have some computation capability, i.e., to be "smart." The smart sensor contains a copy of the model and the triggering condition. It continuously measures the actual plant state and computes the state error (4.3.7). Then, in case of a fixed threshold strategy, it compares the magnitude of the error to a predefined threshold λ, i.e., $|e| \leq \lambda$, and transmits the plant state to update the model state and, consequently, the control law, only if the error is greater than the threshold. Otherwise, the state of the system evolves in open loop receiving in input the control variable computed on the basis of the model state $\hat{x}$.

Another possibility to reduce the packet rate over the network is to choose a relative threshold as in (4.3.5). By virtue of this choice in reference 16, making reference to the class of linear time invariant systems, the global asymptotic stability of the system controlled via a model-based event-triggered state feedback control law was proved, also guaranteeing satisfactory results in terms of reduction of the number of transmissions of the actual plant state over the network.

In the next sections, making reference to the class of perturbed chains of integrators affected by matched uncertainty, i.e., the class of nonlinear uncertain systems in (4.3.1), two model-based event-triggered SMC schemes (Strategies 1 and 2) will be presented. They result in being satisfactory examples of networked SMC schemes. The key elements of the two proposals are a triggering condition based on the sliding variable and a boundary layer of the sliding manifold, so that the control law, at any time instant, depends on the measured state or on the state of the nominal model, as dictated by the specific adopted strategy.

4.3.3 Strategy 1: model-based event-triggered SMC

The control scheme we rely on to implement Strategy 1 is indicated in Figure 4.3.1. The plant, which is modeled by system (4.3.1), is connected to the sensor through a switch,

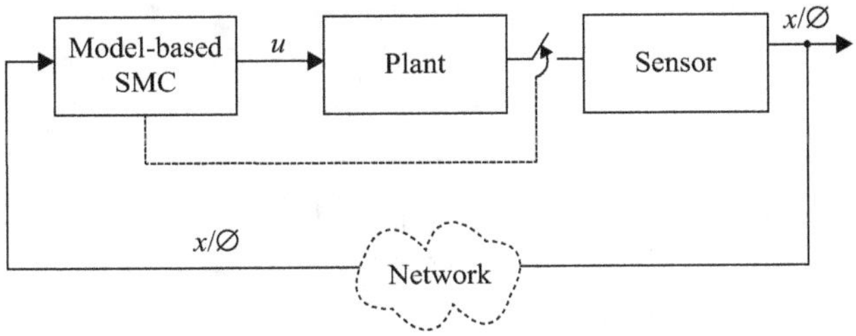

Figure 4.3.1 The model-based event-triggered sliding mode control scheme

controlled by the model-based SMC block. This block includes the sliding mode controller, the nominal model of the plant and the triggering condition. Assuming, for the time being, a generic choice of the sliding variable $\sigma(\cdot)$, the triggering condition is written in terms of the sliding variable computed relying on the nominal model state, i.e., $\hat{\sigma}(t) = \sigma(\hat{x}(t))$, as

$$|\hat{\sigma}(t)| \leq \lambda_1, \tag{4.3.8}$$

where λ_1 is a positive constant. Note that, in Figure 4.3.1, the notation $x/\varnothing$ means that the actual state or no variable is received through the network on the basis of condition (4.3.8).

Given the sliding variable and condition (4.3.8), a suitable boundary layer of the sliding manifold can be defined as

$$\Omega_{\lambda_1} := \{\hat{\sigma}(t) : \ |\hat{\sigma}(t)| \leq \lambda_1\}. \tag{4.3.9}$$

Assume, at the initial time instant t_0, that $\hat{\sigma}(t_0) = \sigma(t_0)$, consequently $sign(\hat{\sigma}(t_0)) = sign(\sigma(t_0))$, is known and that $\hat{\sigma}(t_0) \notin \Omega_{\lambda_1}$. Considering (4.3.1)–(4.3.4) with the triggering condition (4.3.8), and the control scheme in Figure 4.3.1, the problem is that of suitably stabilizing the system state, while guaranteeing robustness properties in front of matched disturbances, and ensuring a reduction of the packet rate over the network with respect to a conventional (i.e., nonnetworked) solution. Note that Strategy 1 is based on the assumption that the network is not described by a model and that it is present only between the sensor and the model-based controller. This is reasonable in a single-input SMC framework, since the control variable to be transmitted is just a real number, while the state has dimension n, with n possibly large. In fact, if the control amplitude were stored in the actuator, only a bit needs to be transmitted to indicate the sign of the control, which is clearly noncritical even in case of networked implementations.

4.3.3.1 The operative modes of strategy 1

Because of its event-triggered nature, the system controlled via Strategy 1 switches between two operative modes, as illustrated in Figure 4.3.2.

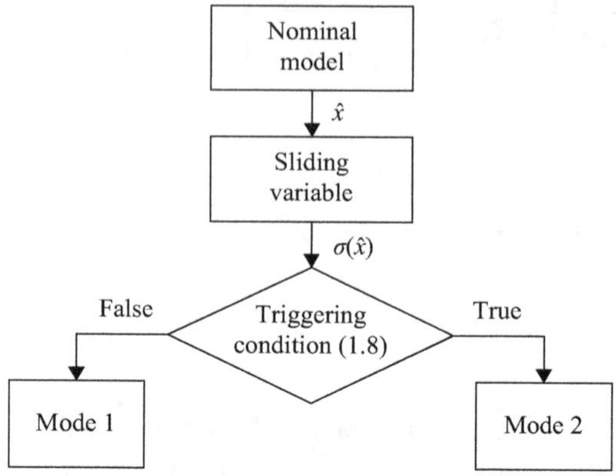

Figure 4.3.2 Schematic view of the two operative modes of Strategy 1

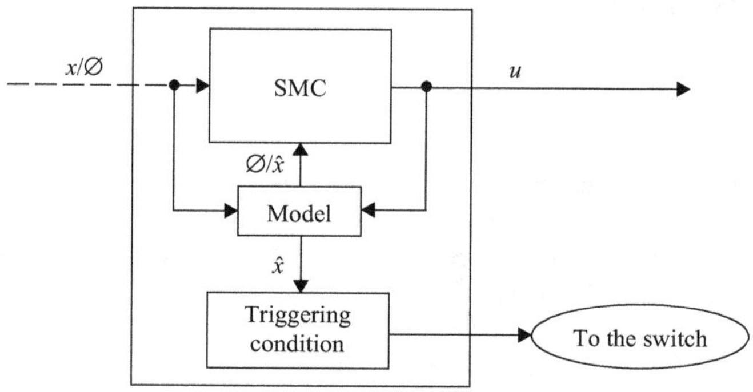

Figure 4.3.3 A representation of the model-based controller with SMC law

The state $\hat{x}$ of the nominal model is provided to the triggering condition block (Figure 4.3.3) which computes the sliding variable relying on the nominal model state. Assuming to adopt a sliding variable which is a linear combination of the considered state, one has

$$\hat{\sigma}(t) = \hat{x}_n(t) + \sum_{i=1}^{n-1} m_i \hat{x}_i(t) \tag{4.3.10}$$

with m_i being positive constants such that the characteristic equation has all roots with negative real part. Whenever condition (4.3.8) is violated, Mode 1 becomes the active operative mode. When (4.3.8) holds, Mode 2 is activated. The two operative modes are hereafter described.

Mode 1 (Condition (4.3.8) *is violated):* In this mode, the state $\hat{x}$ is used by the sliding mode controller, and the following control law

$$u(t) = -U_{\max}\operatorname{sign}(\hat{\sigma}(t)) \tag{4.3.11}$$

is generated, with $U_{\max} > 0$ being a design parameter chosen so as to satisfy, $\forall t \geq t_0$, the η-reaching conditions

$$\dot{\hat{\sigma}}(t)\hat{\sigma}(t) \leq -\hat{\eta}|\hat{\sigma}(t)|, \quad \dot{\sigma}(t)\sigma(t) \leq -\eta|\sigma(t)| \tag{4.3.12}$$

where $\hat{\eta}$ and η are positive constants, and σ has the same form indicated in (4.3.10), but is computed relying on the actual plant state, i.e.,

$$\sigma(t) = x_n(t) + \sum_{i=1}^{n-1} m_i x_i(t). \tag{4.3.13}$$

The control law (4.3.11) is sent as input to both the plant and the nominal model.

Mode 2 (Condition (4.3.8) holds): When condition (4.3.8) is verified, the switch between the plant and the sensor is closed and the measured state x is sent over the network. The nominal model is reinitialized with the plant state, and the following control variable

$$u(t) = -U_{\max} sign(\sigma(t)) \tag{4.3.14}$$

is used as input for both the plant and the nominal model. Note that in Figure 4.3.3 the notation $\varnothing/\hat{x}$ means that the state of the model is used only if the actual state has not been received through the network.

A peculiar aspect of the present networked SMC strategy with respect to a standard SMC law is the fact of using the state of the nominal model outside the boundary layer, so that, in practice, during the portion of the reaching phase in which the sliding variable approaches Ω_{λ_1}, no state transmission is needed. When $\hat{\sigma}$ enters Ω_{λ_1}, robustness issues becomes priority, and this motivates the fact of using the actual plant state as a feedback variable. Indeed, the nominal model-based control variable cannot guarantee disturbance rejection. Note that in order to avoid the Zeno phenomena over the communication network, one could consider an additional condition which allows the use of the nominal model state inside the boundary layer, when the sliding variable is "practically" zero.

4.3.3.2 Theoretical analysis

With reference to Strategy 1, some results on the stability properties of the algorithm and its robustness with respect to matched uncertainties can be proved. Let t_{λ_1} be the time instant when $\hat{\sigma}$ enters the boundary layer Ω_{λ_1} (t_{λ_1} coincides with t_0 if, at the initial time instant, $\hat{\sigma} \in \Omega_{\lambda_1}$).

Theorem 4.3.1. *Given the uncertain system (4.3.1)–(4.3.4), the control laws (4.3.11) and (4.3.14) depending on the triggering condition (4.3.8), then a sliding mode on* $\sigma(t) = 0$ *is enforced in a finite time* $t_r \geq t_{\lambda_1}$. □

Theorem 4.3.2. *Given the uncertain system (4.3.1)–(4.3.4), the control laws (4.3.11) and (4.3.14) depending on the triggering condition (4.3.8), then, $\forall\, t \geq t_r$, the origin of the model-based event-triggered controlled system results in being an asymptotically stable equilibrium point.* □

The proofs are reported in Appendices 1 and 2.

4.3.4 Strategy 2: model-based event-triggered SMC with pseudo-equivalent control

Now, consider the control scheme illustrated in Figure 4.3.4. In such a scheme, the plant is modeled by (4.3.1), and what is denoted with the term "smart sensor" is a sensor fastened to the plant, equipped with some computation capability. In particular, it measures the plant state, it determines the sliding variable σ as in (4.3.13), and then it checks if the triggering condition

$$|\sigma(t)| < \lambda_2 \tag{4.3.15}$$

with λ_2 being a small positive constant is verified. If this is the case, no state transmission is performed, so that the control law needs to be generated relying on the nominal model of the plant indicated in (4.3.4). Note that also in Figure 4.3.4, the notation $x/\varnothing$ means that the actual state or no variable is received over the network, depending on condition (4.3.15). Moreover, in analogy with Strategy 1, a boundary layer, namely,

$$\Omega_{\lambda_2} := \{\sigma(t) : |\sigma(t)| < \lambda_2\} \tag{4.3.16}$$

can be defined, which will have the role of convergence boundary layer.

Then, with reference to (4.3.1)–(4.3.4), and to the control scheme in Figure 4.3.4, the problem is that of making the sliding variable associated with the controlled system ultimately bounded, allowing the amplitude of the convergence boundary layer to be arbitrarily set. In this boundary layer an approximability property, according to which

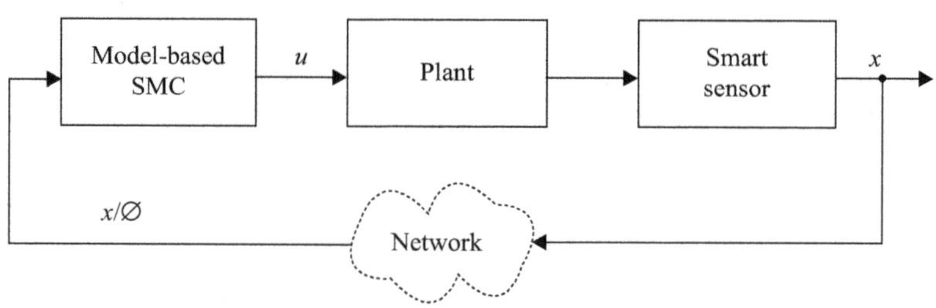

Figure 4.3.4 The model-based event-triggered SMC with pseudo-equivalent control scheme

the actual plant state evolution remains close to that of the nominal model state, has to be guaranteed. Moreover, the overall control strategy has to ensure a reduction of the plant state transmissions with respect to conventional SMC configurations.

4.3.4.1 The operative modes of Strategy 2

The smart sensor, as depicted in Figure 4.3.5, provides to the triggering condition block the actual state of the plant. This block, relying on the measured state x, computes the sliding variable σ, referring to (4.3.13), and verifies the triggering condition (4.3.15). Also, Strategy 2 switches between two different operative modes (Figure 4.3.6).

Mode 1 (Condition (4.3.15) is violated): The actual state x is sent over the network and the nominal model of the plant, contained in the model-based controller detailed in Figure (4.3.7), is reinitialized. In this case, the controller computes the SMC law in (4.3.14), which is sent to the plant and also fed into the nominal model. Note that, in Figure 4.3.7, the notation $\varnothing/\hat{x}$ means that the state of the model is used if the actual state has not been received through the network.

Mode 2 (Condition (4.3.15) holds): If condition (4.3.15) is verified, the smart sensor does not send the actual state x over the network. In this case, the state $\hat{x}$ of the nominal model is used to compute the sliding variable $\hat{\sigma}$. Note that, since the

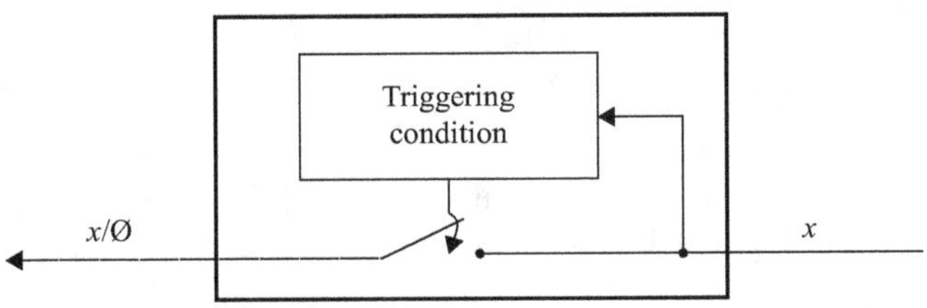

Figure 4.3.5 A representation of the smart sensor included in the process

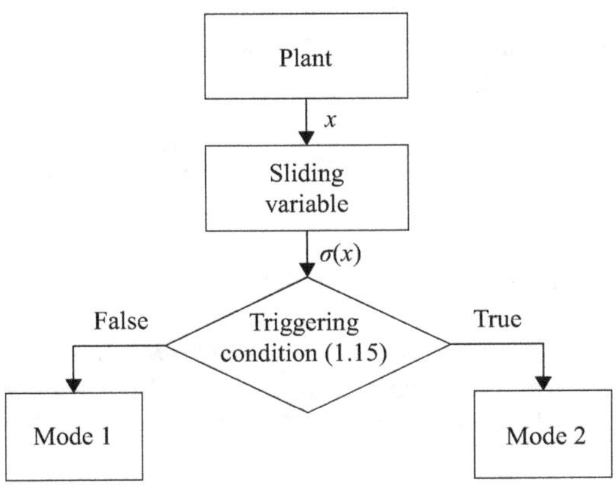

Figure 4.3.6 Schematic view of the two operative modes of Strategy 2

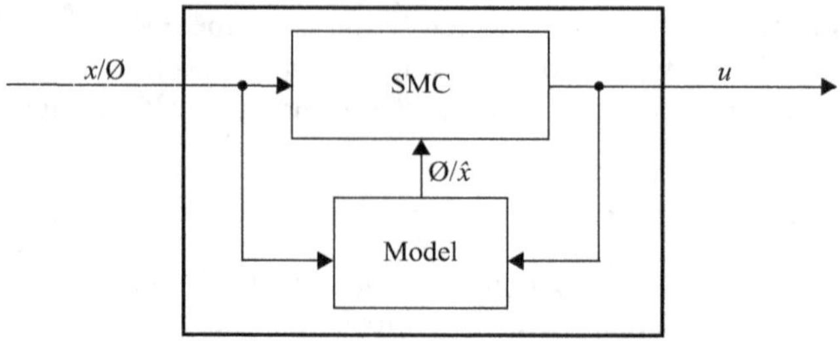

Figure 4.3.7 A representation of the model-based controller

nominal model in (4.3.4) is perfectly known, a sort of equivalent control, hereafter named "pseudo-equivalent control," can be analytically defined by posing $\dot{\hat{\sigma}} = 0$, i.e.,

$$\begin{aligned}\dot{\hat{\sigma}}(\hat{x}(t)) &= \dot{\hat{x}}_n(t) + \sum_{i=1}^{n-1} m_i \dot{\hat{x}}_i(t) = 0 \\ &= f(\hat{x}, t) + b(\hat{x}, t)u(t) + \sum_{i=1}^{n-1} m_i \dot{\hat{x}}_i(t) = 0. \qquad (4.3.17)\end{aligned}$$

Indicating with u_{pseq} the control law u solving (4.3.17), one has

$$u_{pseq} = -b^{-1}(\hat{x}, t)\left(f(\hat{x}, t) + \sum_{i=1}^{n-1} m_i \dot{\hat{x}}_i(t)\right), \qquad (4.3.18)$$

which differs from the theoretical equivalent control [17] since it does not take into account the action of the matched unknown term and is computed using the nominal model state.

4.3.4.2 Theoretical analysis

In order to address the stability properties of the system controlled via Strategy 2, the following definition is introduced:

Definition 4.3.1. *$\mathscr{C}l(\Omega_{\lambda_2})$ is the closure of Ω_{λ_2}, i.e., the set Ω_{λ_2} with all its limits points.*

Then, the following results can be proved.

Theorem 4.3.3. *Given the uncertain system (4.3.1)–(4.3.4), the control laws (4.3.14) and (4.3.18) depending on the triggering condition (4.3.15), then, the sliding variable σ in (4.3.13) is ultimately bounded in $\mathscr{C}l(\Omega_{\lambda_2})$.* □

The previous theorem is relevant to the boundedness of the sliding variable. Now, the attention is focused on the system state and on its approximability in the presence

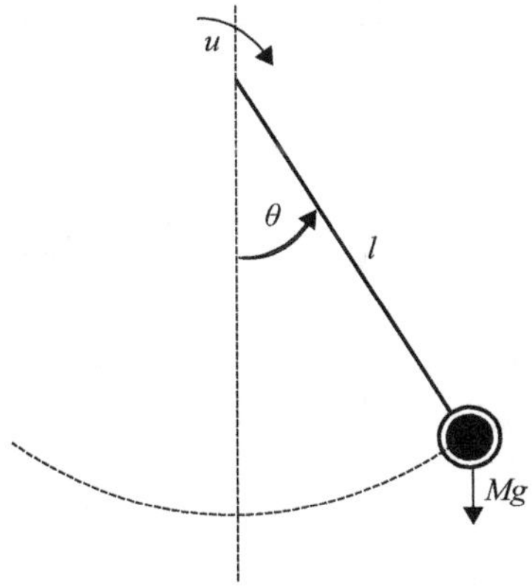

Figure 4.3.8 A schematic view of the considered illustrative example

of the considered uncertainty. More specifically, the next theorem proves that when the controlled system state evolution is such that $\sigma(t) \in \Omega_{\lambda_2}$, then its difference from the nominal model state evolution is bounded, which implies that the controlled system has an appreciable insensitiveness to the action of the uncertain term.

Theorem 4.3.4. *Given the plant (4.3.1)–(4.3.3), with $\sigma(t_0) \in \Omega_{\lambda_2}$, the nominal model (4.3.4), and the control law (4.3.18), then, there exists a positive number ε such that $\|x(t) - \hat{x}(t)\| \leq \varepsilon$, $\forall\, t \geq t_0$.* □

The proofs are reported in Appendices 3 and 4.

4.3.5 Illustrative example

In this section, the application of the proposed model-based event-triggered SMC strategies to an academic example of mechanical nature, i.e., a pendulum, illustrated in Figure 4.3.8, is presented.

Let x_1 and x_2 be the angular displacement θ and its first time derivative $\dot{\theta}$, respectively. Then the dynamics of the plant can be modeled as follows:

$$\begin{cases} \dot{x}_1(t) = x_2(t), \\ \dot{x}_2(t) = -\dfrac{g}{l}\sin(x_1(t)) + \dfrac{1}{Ml^2}(u(t) + u_m(t)), \end{cases} \tag{4.3.19}$$

where with reference to system (4.3.1),

$$f(x(t)) = -\frac{g}{l}\sin(x_1(t)), \qquad b = \frac{1}{Ml^2}. \tag{4.3.20}$$

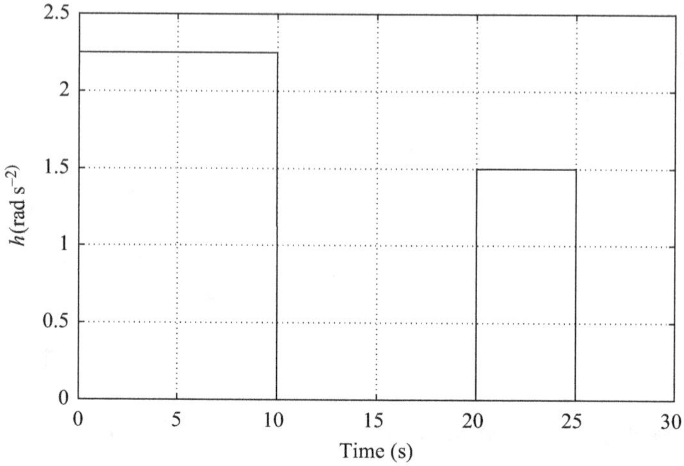

Figure 4.3.9 Matched disturbance affecting the pendulum

Moreover, let $g = 9.81$ m s^{-2} be the gravitational acceleration, $l = 0.6$ m be the length of the inextensible massless pendulum, $M = 0.1$ kg be the load mass, while let u be the control torque. The existence of an angular acceleration disturbance, which can be modelled as a matched uncertain term $h = bu_m$, is considered. The uncertain term is such that $|h| \leq 2.25$ rad s^{-2} (Figure 4.3.9). Furthermore, the associated nominal model is

$$\begin{cases} \dot{\hat{x}}_1(t) = \hat{x}_2(t), \\ \dot{\hat{x}}_2(t) = -\dfrac{g}{l}\sin(\hat{x}_1(t)) + \dfrac{1}{Ml^2}u(t). \end{cases} \tag{4.3.21}$$

The standard Euler solver, with numerical integration step $\tau = 0.001$ s and simulation time $T_s = 30$ s, has been used during the simulation tests with initial condition, both of the plant and of the nominal model, equal to $[3\ 0]^T$. The widths of the boundary layers have been set to $\lambda_1 = \lambda_2 = 0.05$. Then, the sliding variables $\hat{\sigma}$ and σ are as in (4.3.10) and (4.3.13), with $m_1 = 1$, while the control amplitude in (4.3.11) and (4.3.14) is selected as $U_{\max} = 5$. In Strategy 2, the pseudo-equivalent control results in being

$$u_{pseq} = Ml^2\left(\frac{g}{l}\sin(\hat{x}_1(t)) - m_1\hat{x}_2\right). \tag{4.3.22}$$

In order to evaluate the closed-loop performance, we have considered five indices defined as follows: (i) the root mean square (RMS) value of the plant state, x_{RMS}; (ii) the RMS value of the sliding variable, σ_{RMS}; (iii) the control effort, E_c; (iv) the percentage of updates of the actual plant state, n_{up}, which corresponds to the number

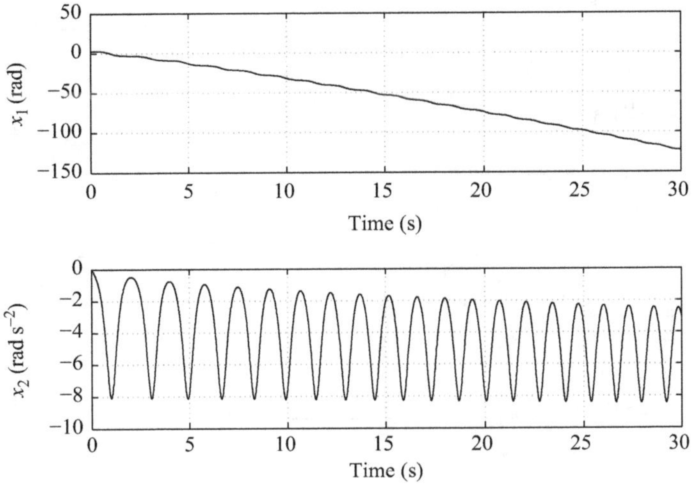

Figure 4.3.10 Open loop evolution of the plant states

of measurement transmissions; and (v) the number of sign changes of the sliding variable, n_{pn}. They are computed as follows:

$$x_{RMS} = \sqrt{\frac{\sum_{i=1}^{n_s}\sum_{j=1}^{n} x_{ji}^2}{n_s}}, \quad \sigma_{RMS} = \sqrt{\frac{\sum_{i=1}^{n_s}\sigma_i^2}{n_s}}, \tag{4.3.23}$$

$$E_c = \sqrt{\frac{\sum_{i=1}^{n_s} u_i^2}{n_s}}, \quad n_{up} = \frac{\sum_{i=1}^{n_s} f_{up}(i\tau)}{n_s}, \quad n_{pn} = \sum_{i=1}^{n_s} f_{pn}(i\tau),$$

where $f_{up}(\cdot)$ is a flag equal to 1 when the actual plant state is sent through the network, $f_{pn}(\cdot)$ is a flag equal to 1 when $[\sigma(i\tau)\sigma((i-1)\tau)] < 0$, while n_s is the number of integration steps during simulations. Moreover, x_{ji}, σ_i and u_i are the jth component of the state vector, the sliding variable, and the control variable at the ith integration step, respectively.

Figure 4.3.10 illustrates the open loop (autonomous) evolution of the plant, to be compared with the time evolution of the plant state and of the nominal model state, which are both steered to zero by using Strategy 1, in Figure 4.3.11, and Strategy 2, in Figure 4.3.12. Figures 4.3.13 and 4.3.14 report the behavior of the sliding variable of both the plant and the nominal model, as well as the flag function, representing the actual plant state transmissions, in both the considered cases. Figure 4.3.15 shows the detail of the sliding variables σ and $\hat{\sigma}$, in case of Strategy 1, when the uncertainty acts on the plant from 20 s to 25 s. Finally, Figures 4.3.16 and 4.3.17 show the control variable u in both cases. The performance indices are reported in Table 4.3.1, and graphically rendered in Figure 4.3.18, where they are normalized with respect to the worst results. Note that the values of the RMS of the states x_{RMS} and of the sliding

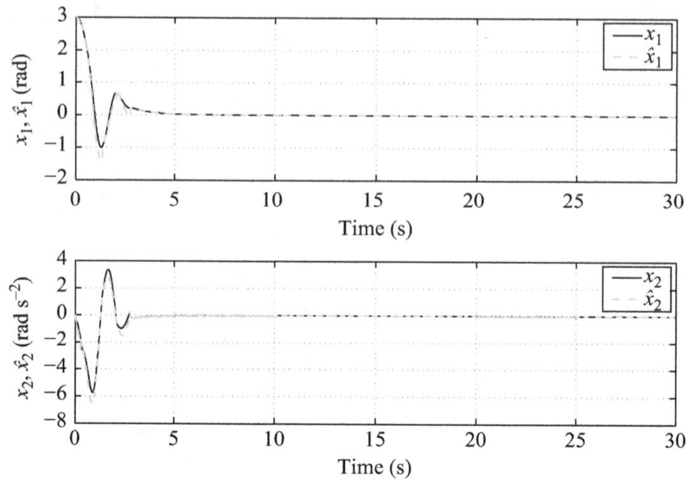

Figure 4.3.11 Time evolution of the state variables of the plant (x_1: top, x_2: bottom, solid line), and of the model ($\hat{x}_1$: top, $\hat{x}_2$: bottom, dashed line), in case of Strategy 1

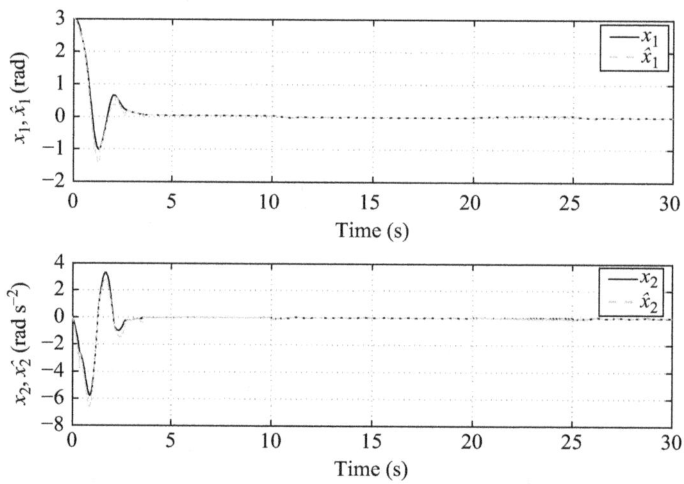

Figure 4.3.12 Time evolution of the state variables of the plant (x_1: top, x_2: bottom, solid line), and of the model ($\hat{x}_1$: top, $\hat{x}_2$: bottom, dashed line), in case of Strategy 2

variable σ_{RMS} are similar for the two strategies, while, as for the control energy E_c, the minimum value is obtained when Strategy 2 is used. Strategy 2 also outperforms the first one as far as the reduction of the measurement transmissions n_{up}, and the chattering attenuation (shown by n_{pn}) are concerned.

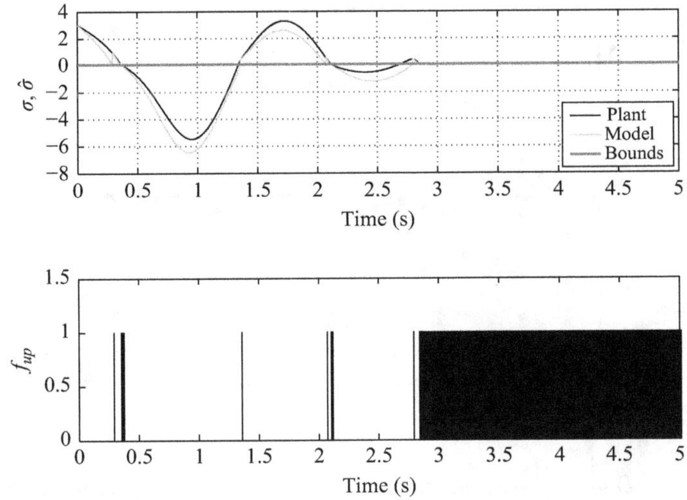

Figure 4.3.13 Time evolution of the sliding variable of the plant (solid black line), and of the model (solid gray line) with the convergence boundary layer (solid dark gray line), in case of Strategy 1

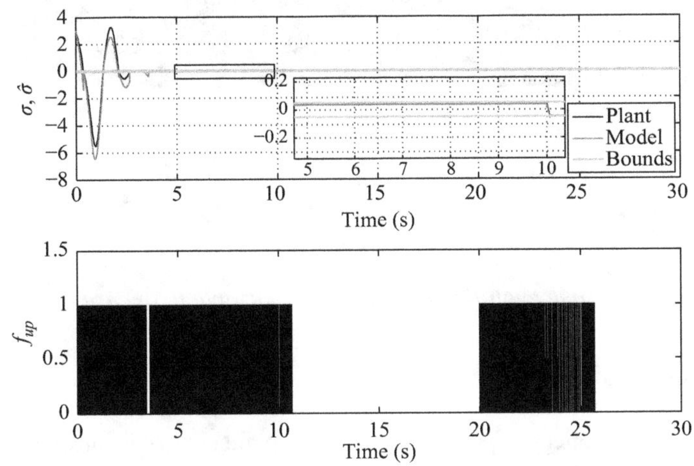

Figure 4.3.14 Time evolution of the sliding variable of the plant (solid black line), and of the model (solid dark gray line) with the convergence boundary layer (solid gray line), in case of Strategy 2

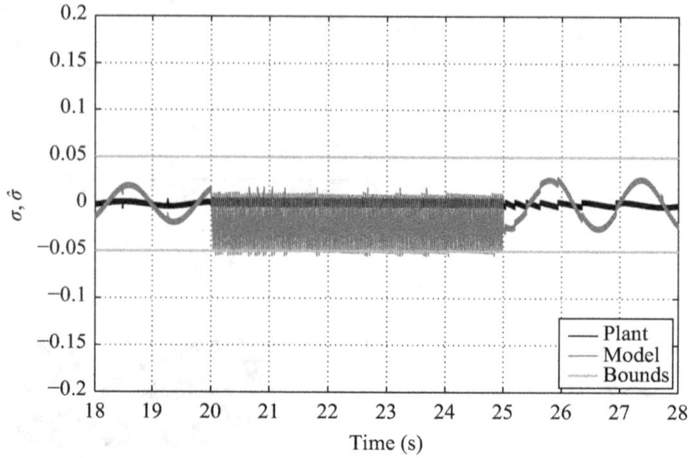

Figure 4.3.15 Detail of the time evolution of the sliding variable of the plant (solid black line), and of the model (solid dark gray line) with the convergence boundary layer (solid gray line) when the uncertainty occurs, in case of Strategy 1

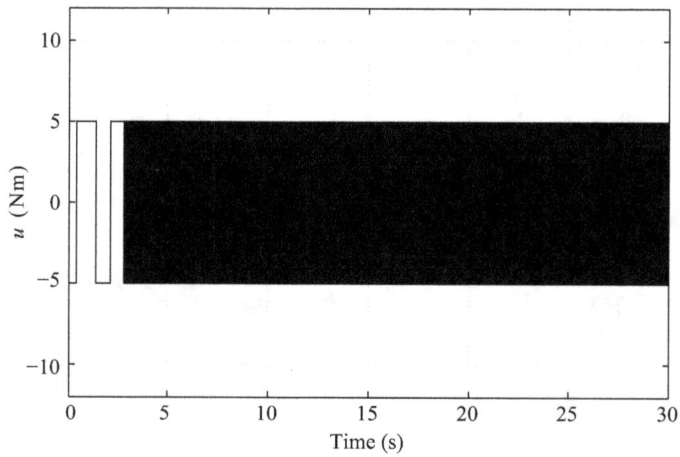

Figure 4.3.16 Time evolution of the control variable u, in case of Strategy 1

4.3.6 Conclusions

In this chapter, the concept of networked sliding mode control is introduced, and two effective model-based event-triggered SMC strategies are presented. The main design objective was to reduce the number of transmissions of the actual plant state over the network with respect to a standard SMC strategy, while guaranteeing performance analogous to that attainable in a nonnetworked implementation. The two proposals

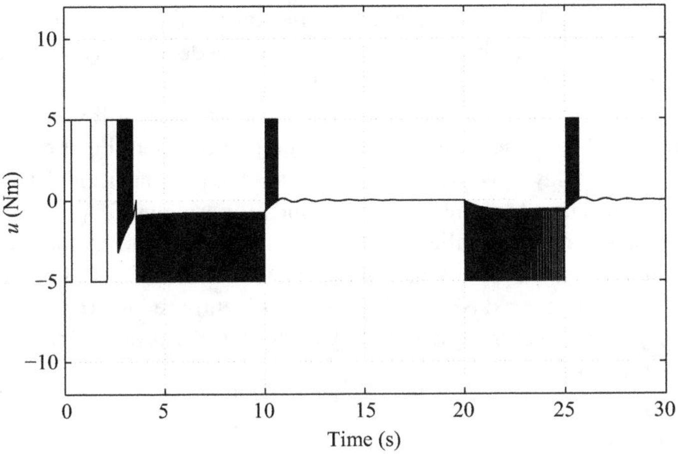

Figure 4.3.17 Time evolution of the control variable u, in case of Strategy 2

Table 4.3.1 Value of performance indices

Strategy	x_{RMS}	σ_{RMS}	E_c	n_{up}	n_{pn}
1	0.0739	0.0699	0.1697	0.8863	21979
2	0.0739	0.0667	0.1100	0.1427	121

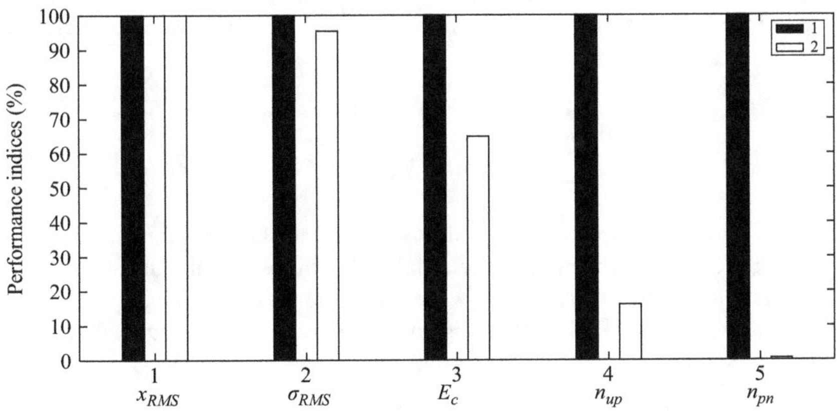

Figure 4.3.18 Relative values of the performance indices for the proposed control approach

are in some sense duals. The first one, relying on a suitable triggering condition, enables to use the actual plant state to close the feedback only when the sliding variable belongs to a prespecified boundary layer. While, outside the boundary layer, it makes use of the nominal model state. The second scheme, still of event-triggering type, enables the use of the actual state measurements when the sliding variable is outside the boundary layer, while within this latter, the nominal model state is used to compute the control law. Moreover, in the second proposal, within the boundary layer, a sort of equivalent control, called pseudo-equivalent control, is applied instead of the discontinuous control, providing a chattering alleviation effect. Both the proposed model-based event-triggered SMC schemes significantly reduce the number of state transmissions, though featuring satisfactory performance, as proved theoretically and assessed in simulation.

Appendix 1 Proof of theorem 4.3.1

Proof. By assumption, at the initial time instant t_0, one has that $\hat{\sigma}(t_0) = \sigma(t_0)$ and that $\hat{\sigma}(t_0) \notin \Omega_{\lambda_1}$.

By virtue of the fact that $\hat{\sigma}(t_0) \notin \Omega_{\lambda_1}$, the triggering condition is violated and the operative Mode 1 becomes active. Then, the state $\hat{x}$ of the nominal model is used to generate the control law (4.3.11).

Since the first of the reaching conditions (4.3.12) holds, the amplitude $U_{\max}$ of the control law (4.3.11) is adequate to dominate the uncertainty term and make the sliding variable $\hat{\sigma}$ be ultimately bounded in the boundary layer Ω_{λ_1}. The reaching time instant t_{λ_1} is the time instant when $|\hat{\sigma}(t)| = \lambda_1$. From that time instant the triggering condition (4.3.8) holds, then the actual plant state is used to generate the control variable. Since the second of the reaching conditions (4.3.12) also holds, the actual sliding manifold $\sigma(t) = 0$ is attractive, and the actual plant state is steered to it in a finite time $t_r \geq t_{\lambda_1}$, generating a sliding mode, which concludes the proof. □

Appendix 2 Proof of theorem 4.3.2

Proof. The sliding variable of the model is steered to the boundary layer Ω_{λ_1} in a finite time t_{λ_1}, and then, the model is reinitialized with the actual plant state x. Within Ω_{λ_1}, the state of the plant is eventually used, and the corresponding sliding variable is steered to zero in $t_r \geq t_{\lambda_1}$, according to Theorem 4.3.1. The equivalent system, obtained by posing $\sigma = 0$, is the following:

$$\begin{cases} \dot{x}_i(t) = x_{i+1}(t), \qquad i = 1, \ldots, n-1; \\ \dot{x}_{n-1}(t) = -\sum_{i=1}^{n-1} m_i x_i, \end{cases} \tag{4.3.24}$$

which is of reduced order with respect to system (4.3.1), and does not depend on the uncertain term. Note that system (4.3.24) can be expressed as

$$\dot{x} = Ax, \tag{4.3.25}$$

where $x \in \mathbb{R}^{n-1}$, and matrix A is

$$A = \begin{bmatrix} 0 & 1 & 0 & \cdots & 0 \\ 0 & 0 & 1 & \ddots & 0 \\ \vdots & & \ddots & & \vdots \\ -m_1 & -m_2 & -m_3 & \cdots & -m_{n-1} \end{bmatrix}. \tag{4.3.26}$$

Since $m_i > 0$ such that the characteristic equation has all roots with negative real part, then, $i = 1, \ldots, n-1$, $A \in \mathbb{R}^{(n-1)\times(n-1)}$ is Hurwitz, which proves the theorem. □

Appendix 3 Proof of theorem 4.3.3

Proof. Two different cases can be considered.

Case 1 ($\sigma(t_0) \notin \Omega_{\lambda_2}$): In this case, the system operates in Mode 1 and the control law in (4.3.14) is applied. According to the SMC theory, the convergence to the sliding manifold $\sigma = 0$ is ensured in finite time and, as a consequence, at the time instant $t_{\lambda_2} \geq t_0$ the border of Ω_{λ_2} is reached. Such a border can be regarded as an auxiliary sliding manifold expressible as

$$\sigma^*(t) = \sigma(t) - \lambda_2 sign(\sigma(t)). \tag{4.3.27}$$

The SMC law (4.3.14) forces the state into the boundary layer, so that the operative mode switches to Mode 2. Assume that the pseudo-equivalent control (4.3.18), which ignores the presence of the uncertain term, cannot maintain the sliding variable σ within Ω_{λ_2}. So, the operative mode switches back to Mode 1 and the SMC law (4.3.14) is applied again. This mechanism keeps repeating itself.

Case 2 ($\sigma(t_0) \in \Omega_{\lambda_2}$): The operative mode is Mode 2 and the control law in (4.3.18) is applied. Assume that the pseudo-equivalent control (4.3.18) cannot maintain the sliding variable σ within Ω_{λ_2}. Then, the mechanism discussed in Case 1 is produced also in this case.

This means that, in both cases, the switching between Mode 1 and Mode 2 produces a controlled system evolution which is equivalent to that attainable by applying to system (4.3.1) the discontinuous control

$$u = -U^*_{\max} sign(\sigma^*(t)) \tag{4.3.28}$$

with $U^*_{\max} < U_{\max}$. If instead, in both cases, the pseudo-equivalent control is capable of maintaining $\sigma(t)$ in Ω_{λ_2}, then, no more switching between Mode 1 and Mode 2 occurs. In both situations, σ turns out to be ultimately bounded in $\mathscr{C}l(\Omega_{\lambda_2})$, which concludes the proof. □

Appendix 4 Proof of theorem 4.3.4

Proof. Consider system (4.3.1), and, relying on (4.3.2), express it as

$$\dot{x}(t) = \mathscr{F}(x,t) + \mathscr{B}(x,t)(u + u_m) \tag{4.3.29}$$

with $\mathscr{F}^T(\cdot) = [x_2\ x_3 \ldots f(x,t)]$, and $\mathscr{B}^T(\cdot) = [0\ 0 \ldots b(x,t)]$ being $n \times 1$ vectors. Analogously, consider the nominal model (4.3.4) expressed as

$$\dot{\hat{x}}(t) = \mathscr{F}(\hat{x},t) + \mathscr{B}(\hat{x},t)u. \tag{4.3.30}$$

Within the boundary layer Ω_{λ_2} the pseudo-equivalent control u_{pseq} in (4.3.18), which allows the regularization of the problem, is sent to both the plant and the nominal model, so that

$$\begin{aligned}\dot{x}(t) &= \mathscr{F}(x,t) + \mathscr{B}(x,t)u_m \\ &\quad - \mathscr{B}(x,t)b^{-1}(\hat{x},t)\left(f(\hat{x},t) + \sum_{i=1}^{n-1} m_i\dot{\hat{x}}_i(t)\right),\end{aligned} \tag{4.3.31}$$

$$\dot{\hat{x}}(t) = \mathscr{F}(\hat{x},t) - \mathscr{B}(\hat{x},t)b^{-1}(\hat{x},t)\left(f(\hat{x},t) + \sum_{i=1}^{n-1} m_i\dot{\hat{x}}_i(t)\right). \tag{4.3.32}$$

Integrating (4.3.31) and (4.3.32), and computing the norm of the difference, it yields

$$\begin{aligned}\|x(t) - \hat{x}(t)\| &\le \|x(t_0) - \hat{x}(t_0)\| + \Bigg\| \int_{t_0}^{t} \mathscr{B}(x,\zeta)u_m d\zeta \\ &\quad + \left(\int_{t_0}^{t} \mathscr{F}(x,\zeta) - \mathscr{F}(\hat{x},\zeta)\right)d\zeta \\ &\quad - \int_{t_0}^{t} (\mathscr{B}(x,\zeta) - \mathscr{B}(\hat{x},\zeta))b^{-1}(\hat{x},\zeta)\Big(f(\hat{x},\zeta) \\ &\quad + \sum_{i=1}^{n-1} m_i\dot{\hat{x}}_i(\zeta)\Big)d\zeta \Bigg\|.\end{aligned} \tag{4.3.33}$$

Because of the smoothness assumption for the right-hand side of (4.3.32), there exists a Lipschitz constant L, and for the right-hand side of (4.3.31), there exist two positive constants M and N such that

$$\|\mathscr{F}(x,t) + \mathscr{B}(x,t)(u_{pseq} + u_m)\| \le M + N\|x\|. \tag{4.3.34}$$

Hence, according to the results proved in reference 17, the solution of (4.3.31) is bounded on a finite time interval $[t_0, T]$, and one has

$$\|x(t) - \hat{x}(t)\| \le H + L\int_{t_0}^{T} \|x(\zeta) - \hat{x}(\zeta)\|d\zeta, \tag{4.3.35}$$

where H depends on λ_2, on the right-hand side of (4.3.32), and on the initial conditions. By applying the Bellman–Gronwall Lemma to (4.3.35), it yields

$$\|x(t) - \hat{x}(t)\| \leq \varepsilon, \tag{4.3.36}$$

where $\varepsilon = He^{LT}$, which proves the theorem. □

References

[1] J. P. Hespanha, P. Naghshtabrizi, and Y. Xu, "A survey of recent results in networked control systems," *Proc. of the IEEE*, vol. 95, no. 1, pp. 138–162, 2007.

[2] F. Wang and D. Liu, *Networked Control Systems: Theory and Applications.* Springer-Verlag London Limited, Springer London, 2008.

[3] R. Gupta and M.-Y. Chow, "Networked control system: overview and research trends," *IEEE Trans. Ind. Electron.*, vol. 57, no. 7, pp. 2527–2535, 2010.

[4] J. Ackermann, A. Bartlett, D. Kaesbauer, W. Sienel, and R. Steinhauser, *Robust Control: Systems with Uncertain Physical Parameters*, Springer-Verlag: London; New York, 1993.

[5] R. Luck and A. Ray, "An observer-based compensator for distributed delays," *Automatica*, vol. 26, no. 5, pp. 903–908, 1990.

[6] J. Nilsson and B. Bernhardsson, "Analysis of real-time control systems with time delays," In: *Decision and Control, 1996., Proceedings of the 35th IEEE Conference on*, vol. 3, pp. 3173–3178, 1996.

[7] G.-P. Liu, Y. Xia, J. Chen, D. Rees, and W. Hu, "Networked predictive of systems with random network delays in both forward and feedback channels," *IEEE Trans. Ind. Electron.*, vol. 54, no. 3, pp. 1282–1297, 2007.

[8] J. Wenjuan, C. Canudas De Wit, O. Sename, and J. Dumon, "A new mathematical model for car drivers with spatial preview," In: *Proc. 18th IFAC World Cong.*, (Milan, Italy), August 2011.

[9] K. J. Aström, "Event based control," In: *Analysis and Design of Nonlinear Control Systems* (A. Astolfi and L. Marconi, eds.), Springer: Berlin, Heidelberg, pp. 127–147, 2008.

[10] W. P. M. H. Heemels, J. H. Sandee, and P. P. J. Van Den Bosch, "Analysis of event-driven controllers for linear systems," *Int. J. Control*, vol. 81, pp. 571–590, 2008.

[11] K. J. Aström and B. Bernhardsson, "Comparison of Riemann and Lebesgue sampling for first order stochastic systems," In: *Proc. 41th IEEE Conf. Decision Control*, vol. 2 (Las Vegas, NV), pp. 2011–2016, December 2002.

[12] M. Mazo and P. Tabuada, "On event-triggered and self-triggered control over sensor/actuator networks," In: *Proc. 47th IEEE Conf. Decision Control*, (Cancun, Mexico), pp. 435–440, December 2008.

[13] P. Tabuada, "Event-triggered real-time scheduling of stabilizing control tasks," *IEEE Trans. Automat. Control*, vol. 52, no. 9, pp. 1680–1685, 2007.

[14] H. Yu and P. Antsaklis, "Event-triggered real-time scheduling for stabilization of passive and output feedback passive systems," In: *Proc. American Control Conf.*, pp. 1674–1679, 2011.

[15] L. A. Montestruque and P. Antsaklis, "Stability of model-based networked control systems with time-varying transmission times," In: *Proc. American Control Conf.*, vol. 49, pp. 1562–1572, 2004.

[16] E. Garcia and P. Antsaklis, "Model-based event-triggered control with time-varying network delays," In: *Proc. 50th IEEE Conf. Decision Control*, (Orlando, FL), pp. 1650–1655, December 2011.

[17] V. I. Utkin, *Sliding Modes in Control and Optimization*. Springer-Verlag-Heidelberg, Springer Berlin Heidelberg, 1992.

[18] S. K. Spurgeon and Christopher Edwards, *Sliding Mode Control*. Taylor and Francis, London, 1998.

[19] V. I. Utkin, *Sliding Modes Control in Discrete-Time and Difference Systems*. Springer-Verlag London Limited, Springer Berlin Heidelberg, vol. 193, pp. 87–107, 1993.

[20] V. I. Utkin, J. Guldner, and J. Shi, *Sliding Mode Control in Electromechanical Systems*. Taylor and Francis, Abington, 1999.

[21] L. M. Capisani, T. Facchinetti, and A. Ferrara, "Real-time networked control of an industrial robot manipulator via discrete-time second-order sliding modes," *Int. J. Control*, vol. 83, pp. 1595–1611, 2010.

[22] P. Marti, M. Velasco, A. Camacho, E. X. Martin, and J. M. Fuertes, "Networked sliding mode control of the double integrator system using the event-driven self-triggered approach," In: *Proc. IEEE 20th Int. Symposium Ind. Electron.*, (Gdansk, Poland), 2011.

[23] A. Ferrara, G. P. Incremona, and L. Magni, "Model-based event-triggered robust MPC/ISM," In: *European Control Conference*, pp. 2931–2936, 2014.

[24] A. Ferrara, G. P. Incremona, and V. Stocchetti, "Networked sliding mode control with chattering alleviation," In: *Proc. 53th IEEE Conf. Decision Control*, (Los Angeles, CA), pp. 5542–5547, December 2014.

[25] L. Fridman, "Singularly perturbed analysis of chattering in relay control systems," *IEEE Trans. Automat. Control*, vol. 47, no. 12, pp. 2079–2084, 2002.

[26] I. Boiko, L. Fridman, A. Pisano, and E. Usai, "Analysis of chattering in systems with second-order sliding modes," *IEEE Trans. Automat. Control*, vol. 52, pp. 2085–2102, 2007.

[27] A. Levant, "Chattering analysis," *IEEE Trans. Automat. Control*, vol. 55, pp. 1380–1389, 2010.

[28] I. Boiko, "Analysis of chattering in sliding mode control systems with continuous boundary layer approximation of discontinuous control," In: *Proc. American Control Conf.*, (San Francisco, CA, USA), pp. 757–762, July 2011.

[29] G. Bartolini, A. Ferrara, and E. Usai, "Output tracking control of uncertain nonlinear second-order systems," *Automatica*, vol. 33, pp. 2203–2212, 1997.

[30] G. Bartolini, A. Ferrara, and E. Usai, "Chattering avoidance by second-order sliding mode control," *IEEE Trans. Automat. Control*, vol. 43, pp. 241–246, 1998.

[31] G. Bartolini, A. Ferrara, A. Levant, and E. Usai, "On second order sliding mode controllers," In: *Variable Structure Systems, Sliding Mode and Nonlinear Control* (K. Young and Ü. Özgüner, eds.), *Lecture Notes in Control and Information*. Springer-Verlag: London, UK, pp. 329–350, 1999.

[32] L. Fridman, "The problem of chattering: an averaging approach," In: *Variable structure systems, sliding mode and nonlinear control* (K. Young and Ü. Özgüner, eds.), vol. 247 of *Lecture Notes in Control and Information Sciences*. Springer-Verlag, Springer London, pp. 363–386, 1999.

[33] G. Bartolini, A. Ferrara, E. Usai, and V. Utkin, "On multi-input chattering-free second-order sliding mode control," *IEEE Trans. Automat. Control*, vol. 45, pp. 1711–1717, 2000.

[34] A. Levant, "Higher-order sliding modes, differentiation and output-feedback control," *Int. J. Control*, vol. 76, pp. 924–941, 2003.

[35] A. Levant, "Quasi-continuous high-order sliding-mode controllers," *IEEE Trans. Automat. Control*, vol. 50, pp. 1812–1816, 2005.

[36] I. Boiko and L. Fridman, "Frequency domain input–output analysis of sliding-mode observers," *IEEE Trans. Automat. Control*, vol. 51, no. 11, pp. 1798–1803, 2006.

[37] H. Lee and V. I. Utkin, "Chattering suppression methods in sliding mode control systems," *Annual Reviews in Control*, vol. 31, no. 2, pp. 179–188, 2007.

[38] F. Dinuzzo and A. Ferrara, "Higher order sliding mode controllers with optimal reaching," *IEEE Trans. Automat. Control*, vol. 54, pp. 2126 –2136, 2009.

[39] M. Tanelli and A. Ferrara, "Enhancing robustness and performance via switched second order sliding mode control," *IEEE Trans. Automat. Control*, vol. 58, pp. 962–974, 2013.

Chapter 4.4

Step-by-step super-twisting observer for DC series motor in the presence of magnetic saturation

L. Amet, M. Ghanes, J.-P. Barbot, and G. Santomenna

Abstract

DC series motors produce very high torque at zero or low speed. It is for this reason and its simplicity these motors can still be found in oil drilling applications, among others. In oil drilling, the environment can be very harsh, leading to violent vibrations and shocks. Position sensors, needed to perform closed-loop speed or position control, can break easily under these conditions, causing production losses. Moreover, they increase the complexity and cost of the system. Therefore, it is desirable to avoid them. In this work, we propose the use of a step-by-step super twisting observer in order to determine the speed of the motor. These observers have the property of finite time convergence. However, two inherent properties of DC series motors could make the use of observers impossible. On one hand, there exists an observability singularity at zero current. On the other hand, there is the magnetic saturation which leads to incorrect speed observation if not taken into account. To overcome these limitations, we propose the use of an observer/estimator scheme and introduce the magnetic saturation into the observer model. An industrial application is conducted to highlight the performance of the proposed solution in the context of a sensorless speed control of a DC series motor.

4.4.1 Introduction

Among the electrical machines used in industry, the DC motor is one of the simplest because it is governed by continuous voltages and currents. In addition, it presents a great flexibility since it can be configured in several different ways, depending on the connection between stator and armature windings. These configurations present different characteristics, allowing the machine to be adapted to the constraints of its specific application.

In this work, we consider the *DC series motor*, in which the field circuit is connected in series with the armature circuit. Two advantages arise from this electrical connection: on one hand, only one static converter (e.g., controlled rectifier) is needed; on the other hand, the electromagnetic torque produced is proportional to the square of the current (under linear electromagnetic flux conditions). For this reason, DC series motors are used in applications where high starting torques are needed, such as trains, elevators, hoists; or to produce high torque at slow speeds in applications such as dragline excavation or oil drilling.

The mathematical model of the DC series motor is nonlinear, which inspired the application of different control techniques, ranging from open loop to nonlinear techniques. In references 1 and 2, two open-loop strategies are presented, based on PWM and nonlinear predictive techniques, respectively; in the context of closed-loop control strategies, we found techniques such as fuzzy-tuned PI controllers [3], singular perturbation approach [4], feedback linearization design [5, 6], backstepping method [7, 8], and the application of port-controlled Hamiltonian systems equivalence [9].

In order to implement closed-loop control techniques, the speed is usually measured by means of position sensors, which implies higher economical costs and increases the complexity of the system. In addition, in some applications, mechanical components are subject to very harsh conditions, in which the position sensor is susceptible to failure. These reasons lead us to consider replacing them with some speed estimation technique.

DC series motors present an observability singularity at zero current, so special attention must be paid when estimating the speed near this condition. In [10, 11] and [12], the authors propose the application of adaptive-extended and high-gain extended Kalman filters in the observable zone, and compare their performances with those of the extended Kalman filter; nevertheless, neither solution is given to deal with the observability singularity nor sensorless control is implemented. In references 5 and 6, a nonlinear observer (with linear error dynamics) is presented, showing that sensorless control of the DC motor is feasible. However, to the best of our knowledge, no approach including the observability singularity is provided.

On the other hand, in DC series motors drives, the magnetic saturation is usually neglected and most of the existing controllers/observers consider the electromagnetic characteristic (flux versus current) linear.

In this work, a step-by-step super twisting observer is proposed in order to determine the speed of the motor by taking into account both the observability singularity and the magnetic saturation. An industrial application is presented to highlight the performances of the proposed methodology where the observability singularity and the magnetic saturation of the motor are involved.

First, a mathematical model is developed in section 4.4.2. In section 4.4.3 an observability analysis is performed, revealing a singularity at zero current operation of the DC motor. In section 4.4.4, when the DC motor is outside the observability singularity, a *step-by-step* observer [13] based on *second-order sliding modes differentiators* [14] is designed. The design offers the advantage of good differentiator properties such as finite time convergence, robustness, and design simplicity with respect to noise [15]. When the DC motor approaches the observability singularity, a simple estimator with observer/estimator swapping system is proposed in section 4.4.5. This system allows the DC motor to work properly near zero current. In section 4.4.7 the observer/estimator structure is tested experimentally in the context of a sensorless speed control by using classical proportional–integral (PI) techniques. In section 4.4.8 conclusion and future work are drawn.

4.4.2 Mathematical model

In the DC series motor, the field and armature windings are connected in series. The dynamics of the current of this circuit is given by

$$L_a \frac{di}{dt} + \frac{\partial \phi_{mf}}{\partial i} \frac{di}{dt} = -(R_f + R_a)\, i - e + v \tag{4.4.1}$$

with $e = K_m\, \phi_{mf}(i)\, \omega$, where

- L_a is the inductance of the armature winding,
- $\phi_{mf}(i)$ is the flux created by the magnetic field circuit,
- R_{mf} and R_a are the magnetic field and armature circuit resistances, respectively,
- e is the back electromotive force (back EMF),
- K_m is the back EMF constant,
- ω is the angular speed, and
- v is the voltage applied to the machine.

By setting $R = R_a + R_{mf}$ and $L = L_a + \frac{\partial \phi_{mf}}{\partial i}$, (4.4.1) can be rewritten as follows:

$$L \frac{di}{dt} = -R\, i - K_m\, \phi_{mf}(i)\, \omega + v. \tag{4.4.2}$$

On the other hand, we have the mechanical subsystem, whose dynamics is given by

$$J \frac{d\omega}{dt} = \Gamma_{em} - B\omega - \Gamma_L, \tag{4.4.3}$$

where

- J is the moment of inertia,
- B is the viscous friction,
- Γ_{em} is the electromagnetic torque developed by the machine, and
- Γ_L is the load torque.

The electromagnetic torque is given by the following equation:

$$\Gamma_{em} = K\phi_{mf}(i)\, i, \tag{4.4.4}$$

where K is the torque constant.

The electromechanical energy conversion reads as

$$\begin{aligned} ei &= \Gamma_{em}\omega, \\ K_m\, \phi_{mf}(i)\, i\, \omega &= K\, \phi_{mf}(i)\, i\, \omega; \end{aligned} \tag{4.4.5}$$

hence,

$$K = K_m. \tag{4.4.6}$$

The nonlinear model of the DC series motor is obtained rewriting (4.4.2) and (4.4.3) as follows:

$$\begin{cases} \dfrac{di}{dt} &= \dfrac{1}{L}\left(-Ri - K_m\, \phi_{mf}(i)\, \omega + v\right), \\ \dfrac{d\omega}{dt} &= \dfrac{1}{J}\left(K_m\, \phi_{mf}(i)\, i\, - B\omega - \Gamma_L\right), \end{cases} \tag{4.4.7}$$

Remark 4.4.1. *Γ_L is considered as a perturbation.*

4.4.2.1 Per-unit model

Model (4.4.7) is expressed in SI units. In order to facilitate the comparison of variables, we propose the *per-unit* model, whose quantities have no units. Furthermore, all the nominal values become 1, so the absolute value of the new variables are less than 1 most of the time. This per-unit system is given by the following scale:

- input:

$$v_{pu} = \frac{v}{V_{nom}} \tag{4.4.8}$$

- states:

$$\begin{cases} i_{pu} &= \dfrac{i}{I_{nom}} \\ \omega_{pu} &= \dfrac{\omega}{\omega_{nom}} \end{cases} \tag{4.4.9}$$

- perturbation:

$$\Gamma_{Lpu} = \frac{\Gamma_L}{\Gamma_{Lnom}} \tag{4.4.10}$$

Then, system (4.4.7) becomes

$$\begin{cases} \dfrac{di_{pu}}{dt} = \dfrac{1}{L}\left(-Ri_{pu} - K_m\phi_{mf}(i_{pu}I_{nom})\omega_{nom}\omega_{pu} + \dfrac{V_{nom}}{I_{nom}}v_{pu}\right), \\ \dfrac{d\omega_{pu}}{dt} = \dfrac{1}{J}\left(\dfrac{K_m}{\omega_{nom}}\phi_{mf}(i_{pu}I_{nom})\, i_{pu}I_{nom} - B\omega_{pu} - \dfrac{\Gamma_{nom}}{\omega_{nom}}\Gamma_{Lpu}\right). \end{cases} \tag{4.4.11}$$

Remark 4.4.2. *If the flux $\phi_{mf}(i)$ is considered to be linear, which is not the case in this work, it can be modeled as follows:*

$$\phi_{mf}(i) = L_{mf}i, \tag{4.4.12}$$

where L_{mf} is the inductance of the magnetic field circuit. Under this assumption, the nonlinear model (4.4.11) *is given by the following equations:*

$$\begin{cases} \dfrac{di_{pu}}{dt} = \dfrac{1}{L}\left(-Ri_{pu} - K_mL_{mf}i_{pu}\omega_{nom}\omega_{pu} + \dfrac{V_{nom}}{I_{nom}}v_{pu}\right), \\ \dfrac{d\omega_{pu}}{dt} = \dfrac{1}{J}\left(\dfrac{K_mL_{mf}I_{nom}^2}{\omega_{nom}}i_{pu}^2 - B\omega_{pu} - \dfrac{\Gamma_{nom}}{\omega_{nom}}\Gamma_{Lpu}\right), \end{cases} \tag{4.4.13}$$

where $L = L_a + L_{mf}$.

4.4.3 Observability analysis of the DC series motor

For a reminder of nonlinear local weakly observability in the electrical motor context see, e.g., [16], section 2. In this section, we recall the *rank criterion* given by reference 17. To do this, we first introduce the *Lie-Bäcklund* derivatives [18] of a system of order n, which are noted as

$$\begin{cases} L_fh &= \dfrac{\partial h}{\partial x}f, \\ L_f^2h &= \dfrac{\partial L_fh}{\partial x}f + \dfrac{\partial L_fh}{\partial u}\dot{u}, \\ L_f^3h &= \dfrac{\partial L_f^2h}{\partial x}f + \dfrac{\partial L_f^2h}{\partial u}\dot{u} + \dfrac{\partial L_f^2h}{\partial \dot{u}}\ddot{u}, \\ \vdots & \\ L_f^{n-1}h &= \dfrac{\partial L_f^{n-2}h}{\partial x}f + \sum_{i=1}^{n-2}\dfrac{\partial L_f^{n-2}h}{\partial u^{(i-1)}}u^{(i)}, \\ \vdots & \\ L_f^ph &= \dfrac{\partial L_f^{p-1}h}{\partial x}f + \sum_{i=1}^{p-1}\dfrac{\partial L_f^{p-1}h}{\partial u^{(i-1)}}u^{(i)}, \end{cases} \tag{4.4.14}$$

where $u^{(i)}$ is the i'th derivation of u, and p is a natural number which may be greater than n.

Criterion 4.4.1. *(Rank criterion)*
Given a system

$$\Sigma : \begin{cases} \dot{x} = f(x) + g(x)u, \\ y = h(x), \end{cases} \tag{4.4.15}$$

it is *locally regularly observable* at x_0 if

$$\operatorname{rank}(J_c) = \operatorname{rank} \left. \begin{pmatrix} dh \\ dL_f h \\ dL_f^2 h \\ \vdots \\ dL_f^i h \\ \vdots \\ dL_f^{n-1} h \end{pmatrix} \right|_{x_0} = n. \tag{4.4.16}$$

Remark 4.4.3. Regularity *implies that n derivations of the output (rows of matrix J_c) suffice for the system Σ to be observable.*

Considering the following equivalences, the model of the DC series motor (system (4.4.7)) can be written in the form (4.4.15):

$$x = \begin{bmatrix} x_1 \\ x_2 \end{bmatrix} = \begin{bmatrix} i \\ \omega \end{bmatrix}, \tag{4.4.17}$$

$$u = \begin{bmatrix} u_1 \\ u_2 \end{bmatrix} = \begin{bmatrix} v \\ \Gamma_L \end{bmatrix}, \tag{4.4.18}$$

$$f(x) = \begin{bmatrix} f_1(x) \\ f_2(x) \end{bmatrix} = \begin{bmatrix} -\dfrac{R}{L}x_1 - \dfrac{K_m}{L}\phi_{mf}(x_1)x_2 \\ \dfrac{K_m}{J}\phi_{mf}(x_1)x_1 - \dfrac{B}{J}x_2 \end{bmatrix}, \tag{4.4.19}$$

$$g(x) = \begin{bmatrix} g_1(x) \\ g_2(x) \end{bmatrix} = \begin{bmatrix} 1/L \\ 1/J \end{bmatrix}, \tag{4.4.20}$$

$$h(x) = x_1. \tag{4.4.21}$$

Remark 4.4.4. *The output $y = h(x) = x_1$ is the* measured *output (i.e., the current).*

We now determine the matrix J_c, given in (4.4.16). The first row is given by

$$dh(x) = \left[\frac{\partial h}{\partial x_1}, \frac{\partial h}{\partial x_2}\right] = [1, 0]. \tag{4.4.22}$$

To determine the second row, we need first to calculate $L_f h$:

$$L_f h = \left[\frac{\partial h}{\partial x_1}, \frac{\partial h}{\partial x_2}\right] \cdot \begin{bmatrix} f_1(x) \\ f_2(x) \end{bmatrix} = f_1(x); \tag{4.4.23}$$

therefore,

$$dL_f h = \left[\frac{\partial f_1(x)}{\partial x_1}, \frac{\partial f_1(x)}{\partial x_2}\right]. \tag{4.4.24}$$

From (4.4.22) and (4.4.24), we obtain

$$J_c = \begin{bmatrix} 1 & 0 \\ \dfrac{\partial f_1(x)}{\partial x_1} & \dfrac{\partial f_1(x)}{\partial x_2} \end{bmatrix}. \tag{4.4.25}$$

The determinant of J_c is

$$\det(J_c) = \frac{\partial f_1(x)}{\partial x_2} = -\frac{K_m}{L}\phi_{mf}(x_1), \tag{4.4.26}$$

which is different from zero *if and only if* $\phi_{mf}(x_1) = 0$, which implies that the current is exactly equal to the remanent flux current in order to demagnetize the magnetic circuit. In the considered application, the remanent flux current is neglected. Thus, according to the *rank criterion*, the DC series motor is *locally regularly observable* everywhere, except at zero current.

4.4.4 Observer design

From the previous section we conclude that an observer can be designed for the series DC machine in the case where the current is different from zero. In this section, we propose a *step-by-step super twisting algorithm* (see reference 13 for an overview of this algorithm) to estimate, in finite time, the speed (ω_{pu}) and the load torque (Γ_{Lpu}). We distinguish two stages in this design: stage 1 and stage 2, where speed and load torque observations are accomplished, respectively.

4.4.4.1 Stage 1: speed observation

This stage is associated to the current dynamics (first equation of system (4.4.11)). We propose the following change of coordinates to linearize this equation:

$$\begin{cases} z_1 = i_{pu}, \\ z_2 = -\dfrac{K_m}{L}\dfrac{\phi_{mf}(i_{pu}I_{nom})}{I_{nom}}\omega_{nom}\omega_{pu}, \\ z_3 = -\dfrac{\Gamma_{Lnom}}{J\omega_{nom}}\Gamma_{Lpu}. \end{cases} \tag{4.4.27}$$

So the dynamics of the current (z_1) is as follows:

$$\dot{z}_1 = -\frac{R}{L}z_1 + z_2 + \frac{V_{nom}}{L I_{nom}}v_{pu}, \tag{4.4.28}$$

which is now linear. We propose the first observation stage, in which z_1 (the current) is measured:

$$\begin{cases} \dot{\hat{z}}_1 = -\dfrac{R}{L}z_1 + \tilde{z}_2 + \dfrac{V_{nom}}{L I_{nom}}v_{pu} + \lambda_1|e_1|^{1/2}\operatorname{sign}(e_1), \\ \dot{\tilde{z}}_2 = \alpha_1 \operatorname{sign}(e_1), \end{cases} \tag{4.4.29}$$

where $e_1 = z_1 - \hat{z}_1$, and $\lambda_1, \alpha_1 \in \mathbb{R}$ are tuning parameters. From (4.4.28) and (4.4.29), we obtain the error dynamics:

$$\begin{cases} \dot{e}_1 = z_2 - \tilde{z}_2 - \lambda_1|e_1|^{1/2}\operatorname{sign}(e_1), \\ \dot{\tilde{z}}_2 = \alpha_1 \operatorname{sign}(e_1), \end{cases} \tag{4.4.30}$$

which is the error dynamics of a *super-twisting* algorithm [14].

Assumption 4.4.1. *The variable z_2 is a differentiable function with Lipschitz constant C_1.*

Assumption 4.4.1 and conditions (4.4.31) ensure that e_1 and $\dot{e}_1$ converge, in a finite time T_1, to zero (see reference 19 for a proof of the convergence and an estimation of T_1).

$$\begin{cases} \alpha_1 > C_1, \\ \lambda_1 > \sqrt{\dfrac{2}{\alpha_1 - C_1}}(\alpha_1 + C_1). \end{cases} \tag{4.4.31}$$

Therefore, after a finite time T_1, $\tilde{z}_2$ converges to z_2.

Remark 4.4.5. *Physically assumption 4.4.1 is not verified for magnetic component because it exists always a remanent flux. However, the remanent flux can be neglected with respect to the considered application.*

4.4.4.2 Stage 2: load torque observation

This stage is associated to the speed dynamics (second equation of system (4.4.11)). A new change of coordinates is proposed:

$$\begin{cases} \chi_1 = z_1 \\ \chi_2 = -\dfrac{L}{K_m \omega_{nom}} \dfrac{z_2}{\phi_{mf}(z_1)} \quad \text{if} \quad \phi_{mf}(z_1) \neq 0. \\ \chi_3 = z_3 \end{cases} \tag{4.4.32}$$

In this coordinate system the speed dynamics is

$$\dot{\chi}_2 = \frac{K_m}{J\omega_{nom}} \phi_{mf}(\chi_1 I_{nom})\, \chi_1 I_{nom} \ - \frac{B}{J}\chi_2 + \chi_3. \tag{4.4.33}$$

We propose the following observer:

$$\begin{cases} \dot{\hat{\chi}}_2 = E_1 \left(\dfrac{K_m}{J\omega_{nom}} \phi_{mf}(\chi_1 I_{nom})\, \chi_1 I_{nom} \ - \dfrac{B}{J}\tilde{\chi}_2 + \tilde{\chi}_3 \ + \lambda_2 |e_2|^{1/2} \operatorname{sign}(e_2) \right), \\ \dot{\tilde{\chi}}_3 = E_1 \alpha_2 \operatorname{sign}(e_2), \end{cases} \tag{4.4.34}$$

with $e_2 = \tilde{\chi}_2 - \hat{\chi}_2$, and E_1 defined as follows:

$$\begin{cases} E_1 = 1 \quad \text{if} \quad |e_1| \leq \varepsilon, \\ E_1 = 0, \quad \text{otherwise}, \end{cases} \tag{4.4.35}$$

where ε is a sufficiently small positive constant. Once the first stage has converged, $E_1 = 1$ and $\tilde{z}_2 = z_2$, which implies from (4.4.32) that $\tilde{\chi}_2 \approx \chi_2$. Under these conditions the error dynamics become

$$\begin{cases} \dot{e}_2 \approx \chi_3 - \tilde{\chi}_3 - \lambda_2 |e_2|^{1/2} \operatorname{sign}(e_2), \\ \dot{\tilde{\chi}}_3 = \alpha_2 \operatorname{sign}(e_2). \end{cases} \tag{4.4.36}$$

As in the previous subsection, (4.4.36) corresponds to the super twisting error dynamics, and $\tilde{\chi}_3$ is proven to converge to χ_3 in a finite time T_2, for χ_3 differentiable with Lipschitz constant C_2 (see assumption 4.4.1 and remark 4.4.5), under the following conditions:

$$\begin{cases} \alpha_2 > C_2, \\ \lambda_2 > \sqrt{\dfrac{2}{\alpha_2 - C_2}}\,(\alpha_2 + C_2). \end{cases} \tag{4.4.37}$$

The speed and load torque estimations are given, respectively, by

$$\hat{\omega}_{pu} = \hat{\chi}_2, \tag{4.4.38}$$

and

$$\hat{\Gamma}_{Lpu} = -\frac{J\Gamma_{nom}}{\omega_{nom}}\tilde{\chi}_3. \tag{4.4.39}$$

Remark 4.4.6. *It is important to note that the function $\phi_{mf}(\chi_1 I_{nom})$ in* (4.4.27), (4.4.29), (4.4.32), *and* (4.4.34) *of the proposed observer is obtained by a data table as it can be shown hereafter in the experimental results of section 4.4.7.1.*

4.4.5 Estimator

In order to prevent the observer to work near the observability singularity, we suggest the use of the following speed estimator:

$$\dot{\hat{\omega}}_{pu} = -\frac{1}{\tau_{vf}}\hat{\omega}_{pu} - \frac{1}{\tau_{cf}}\,\mathrm{sign}(\hat{\omega}_{pu}), \tag{4.4.40}$$

where τ_{vf} and τ_{cf} are *time constants* associated to the *viscous friction* and the *Coulomb friction*, respectively. This estimator will provide a good speed estimation only in the case where these parameters are well approximated. However, even in the case of bad parameter approximation, this allows us to 'simulate' a deceleration in the case of low currents.

4.4.5.1 Switching between estimator and observer modes

The switching between estimator and observer modes (see, e.g., for a first result [20] about observability singularity) is given by a condition over the current value. We establish a threshold I_{thr}[1] such that

- if $|i_{pu}| > I_{thr}$ the speed estimation is performed in *observer* mode,
- if $|i_{pu}| \leq I_{thr}$ the speed estimation is performed in *estimator* mode (Figure 1.1).

4.4.6 Observer and estimator discretization

In order to implement the proposed observer/estimator scheme in an actual real-time processor, their continuous models must be discretized. The derivatives are approximated by *backward differences*, and the resulting *difference equations* are then solved applying the *explicit Euler* method. To represent a signal $m(t)$, sampled at a regular sampling period T, we use the following notation:

$$m(kT) = m[k], \quad k \in \mathbb{Z}. \tag{4.4.41}$$

[1] I_{thr} is chosen with respect to the nonlinear magnetic flux characteristic according to the determinant equation given by (4.4.26).

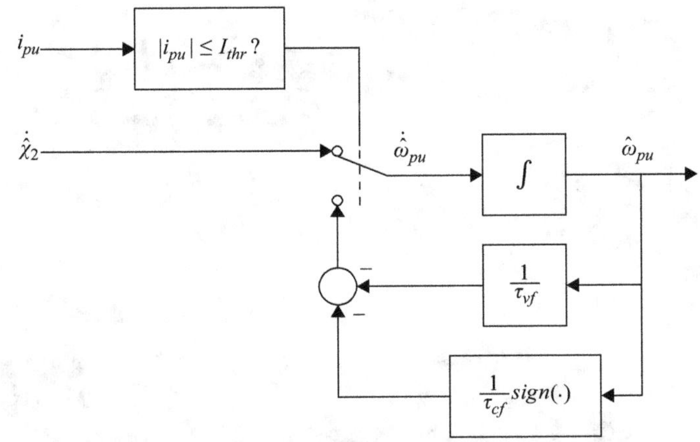

Figure 4.4.1 Switching scheme between observer and estimator modes

4.4.6.1 Discrete observer

The difference equations describing the first and second stages of the observer are, respectively,

$$\begin{cases} \hat{z}_1[k] = \hat{z}_1[k-1] + T\left(-\frac{R}{L} z_1[k-1] + \tilde{z}_2[k-1] + \frac{V_{nom}}{L I_{nom}} v[k-1] \right. \\ \qquad \left. + \lambda_1 |e_1[k-1]|^{1/2} \operatorname{sign}(e_1[k-1])\right), \\ \tilde{z}_2[k] = \tilde{z}_2[k-1] + T\alpha_1 \operatorname{sign} e_1[k-1] \end{cases} \tag{4.4.42}$$

and

$$\begin{cases} \hat{\chi}_2[k] = \hat{\chi}_2[k-1] + T E_1 \left(\frac{K_m}{J \omega_{nom}} \phi_{mf}(\chi_1 I_{nom}) I_{nom}\, \chi_1[k-1] \right. \\ \qquad \left. - \frac{f}{J} \hat{\chi}_2[k-1] + \tilde{\chi}_3[k-1] + \lambda_2 |e_2[k-1]|^{1/2} \operatorname{sign}(e_2[k-1])\right), \\ \tilde{\chi}_3[k] = \tilde{\chi}_3[k-1] + T E_1 \alpha_2 \operatorname{sign}(e_2[k-1]). \end{cases} \tag{4.4.43}$$

4.4.6.2 Discrete estimator

The difference equation that define the discrete observer is

$$\hat{\omega}[k] = \hat{\omega}[k-1] - \frac{T}{\tau_{vf}} \hat{\omega}[k-1] - \frac{T}{\tau_{cf}} \operatorname{sign}(\hat{\omega}[k-1]). \tag{4.4.44}$$

Figure 4.4.2 SCR_Flex motor control board from GS maintenance

4.4.7 Experimental results

In this section, we test the proposed observer/estimator scheme on a real motor. A sensorless closed-loop control of the motor is performed to validate the proposed observation–estimation strategy. The implemented control, in turn, consists of two classical nested PI loops. Detailed information about observer based feedback controllers for nonlinear systems can be found in reference 21. Note that all the tests presented in this section are performed without load since this is the worst working condition for a DC series connected motor.

Both the observer/estimator scheme and the controller were implanted in an SCR_Flex control board (Figure 4.4.2) from *GS Maintenance*. This board is based on a fixed point DSP from Texas Instruments, TMS320F2812, and a Spartan-3 FPGA from Xilinx. The sampling frequency is fixed to 6 kHz. The nominal characteristics and the parameters of the DC motor are presented in Table 4.4.1.

The parameters of the nested PI loops are given in Table 4.4.2.

The observer and estimator parameters are given in Tables 4.4.3 and 4.4.4, respectively.

Table 4.4.1 Nominal values and parameters of the DC motor

Variable	Nominal value
U_{nom}	220 V
I_{nom}	15 A
ω_{nom}	104.72 rad/s (1000 rpm)
Γ_{nom}	27 Nm
Parameter	**Value**
R_a	0.6 Ω
R_f	1.8 Ω
L_a	1 mH
L_f	220 mH
K_m	0.12
B	0.02 Nms
J	0.2 Nms^2

Table 4.4.2 PIs parameters

Parameter	PI speed	PI current
k_p	2	3
T_i	10	0.02

Table 4.4.3 Observer parameters

Parameter	Speed stage	Torque stage
α	10	4
λ	8	5
ε	–	0.05 (pu)

Table 4.4.4 Estimator parameters

Parameter	Value
τ_{vf}	∞
τ_{cf}	3.5 s
I_{thr}	0.2 (pu)

Table 4.4.5 Approximated inductance

Current (A)	Inductance (H)
0	0.085
0.06	0.09
0.117	0.091
0.3	0.11
0.6	0.209
1.455	0.2247
3	0.23
4.5	0.229
6	0.2239
7.5	0.2118
9	0.203
10.5	0.1913
12	0.148
13.5	0.116
15	0.1086

4.4.7.1 Motor inductance approximation

As stated in Remark 4.4.5, in this section the inductance of the motor (nonlinear flux approximation) is approximated by a data table. Let us consider the current and voltage measurements at zero speed (blocked rotor) and applying the following equation: $v = Ri + L\frac{di}{dt}$. Knowing R, L is very easily approximated for different values of the current. The result of this approximation is given in Table 4.4.5.

As we can see from this table the inductance is not constant and thus the flux is not linear as expected.

4.4.7.2 Experimental results of closed-loop sensored control with nonlinear flux approximation

We inserted some of the values given in Table 4.4.5 into the memory of the DSP in order to approximate the armature inductance. These values are given in Table 4.4.6. The intermediate values between points in the table are determined by linear interpolation. This leads to some errors in the approximation of the inductance and then, in the observed speed.

In Figures 4.4.3 and 4.4.4, we show estimated and measured speed in a sensored closed control loop context. In both accelerations, the current goes through the nonlinear flux zone. In both cases the observed speed is very close to the measured speed, which confirms that the flux is nonlinear and that the inductance approximation is good enough.

Table 4.4.6 Inductance values inserted in the DSP

Current (A)	Inductance (H)
0	0.078
0.015	0.078
1.5	0.2332
6	0.2267
7.5	0.207
15	0.146
16	0.146

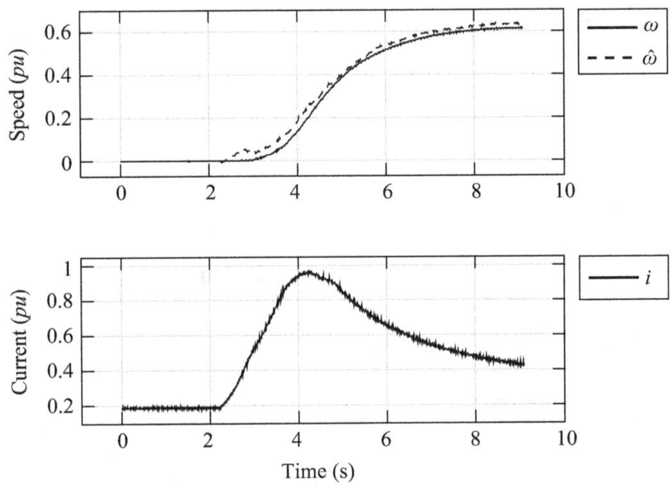

Figure 4.4.3 Sensored closed-loop control: acceleration from 0 to 0.6 units

4.4.7.3 Experimental results of closed-loop sensorless control with nonlinear flux approximation

A first experiment was run to show the switching between observer and estimator modes and, at the same time, the observer convergence (first and second stages). This is shown in Figure 4.4.5. In this case, it is not the speed the variable under control but only the current. Until 1.5 s, $\hat{\omega}$ follows ω and $i > I_{thr}$, hence the system is in observer mode. At around 1.5 s the current drops to zero, which forces the system to switch to estimator mode. The parameter τ_{cf} was set to 3.5 s, which allows $\hat{\omega}$ to converge to zero in finite time. The current is then augmented softly. When it crosses the value of 0.2 units (I_{thr}), the system switches back to observer mode and $\hat{\omega}$ converges to ω.

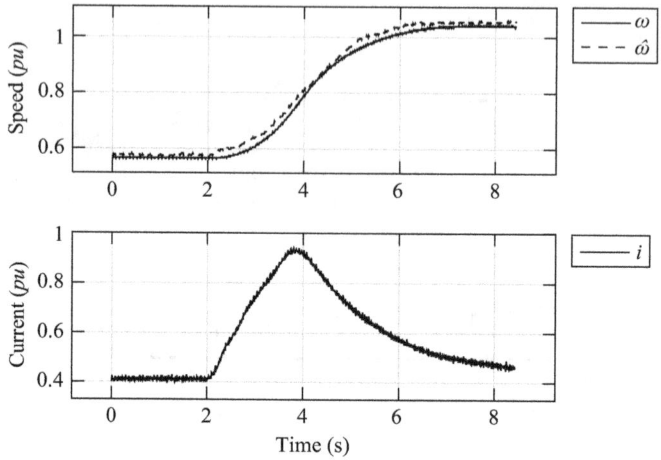

Figure 4.4.4 Sensored closed-loop control: acceleration from 0.6 to 1 units

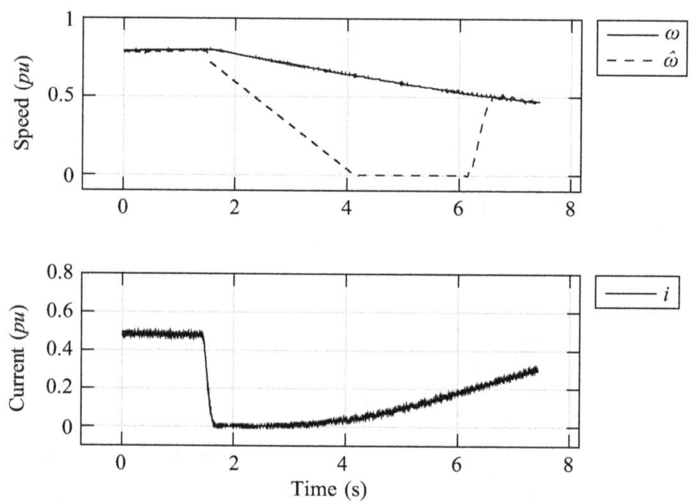

Figure 4.4.5 Switching between observer and estimator modes

As we can see, in estimator mode, $\hat{\omega}$ does not follow ω unless τ_{vf} and τ_{cf} are known, but this prevent the observer to work near the observability singularity.

Figures 4.4.6 and 4.4.7 show the sensorless control performance throughout the whole speed and current-operating ranges. Note that even if the motor is not driving a load, a current of 0.4 units is present in the steady state due to Coulomb friction. Both captures show that the system performs properly despite a small gap between $\hat{\omega}$ and ω which occurs when the current varies abruptly. This is probably a consequence of an error on L_a, which arises in the presence of large values of di/dt.

Figure 4.4.6 Sensorless closed-loop control: acceleration from 0 to 0.6 units

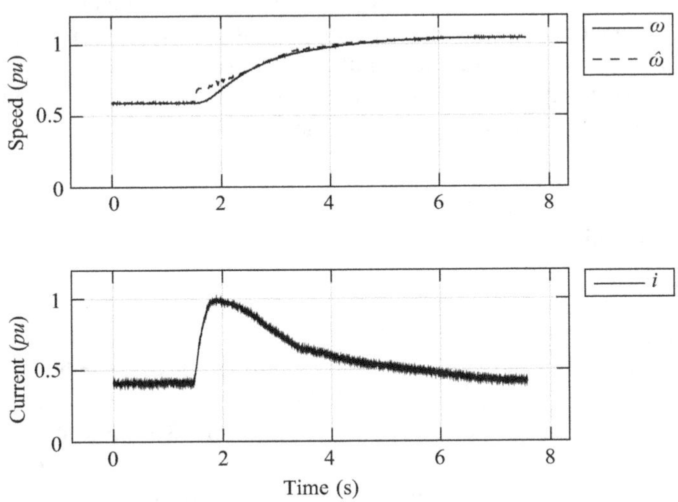

Figure 4.4.7 Sensorless closed-loop control: acceleration from 0.6 to 1 units

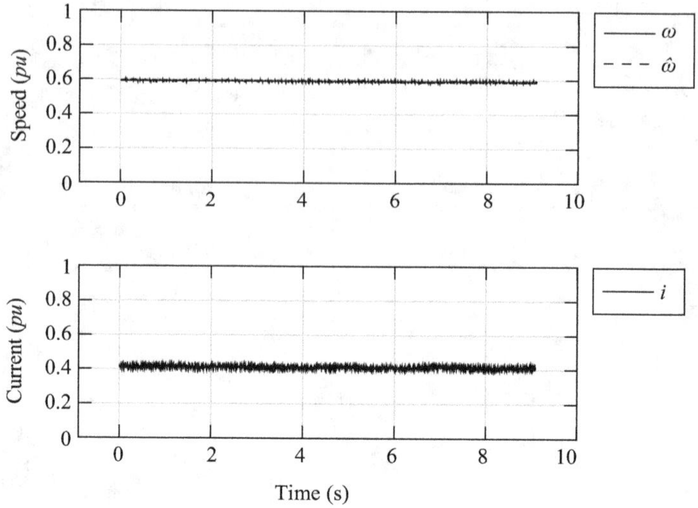

Figure 4.4.8 Sensorless control in steady state

At last, it is shown in Figure 4.4.8 that speed and current in steady state are free of oscillations.

4.4.7.4 Experimental results under linear flux assumption

In order to justify the inclusion of the magnetic saturation as proposed in this work, we have performed a sensored control of the motor under linear flux assumption. The results obtained under this assumption were not those predicted by simulation under which (4.4.13) are used and simulated. These results have showed that the approximation of linear flux is not valid, and closing the loop with the estimated speed would result in the destabilization of the system (Figure 4.4.9).

4.4.8 Conclusion

In this work a sensorless speed control for a DC series motor was proposed. An observability analysis revealed an observability singularity at zero current. This led us to design an observer/estimator approach. The proposed observer is based on second order sliding mode techniques, whose excellent properties such as finite time converge, robustness, and design simplicity with respect to noise are well known. On the other hand, the estimator operates at low currents, providing a speed estimation when the systems is near the observability singularity. Moreover, the observer/estimator was designed by considering the magnetic saturation of the motor. The whole scheme was validated by means of experimental results. The obtained results have showed that the linear approximation of the inductance is not valid, and the observer leads to

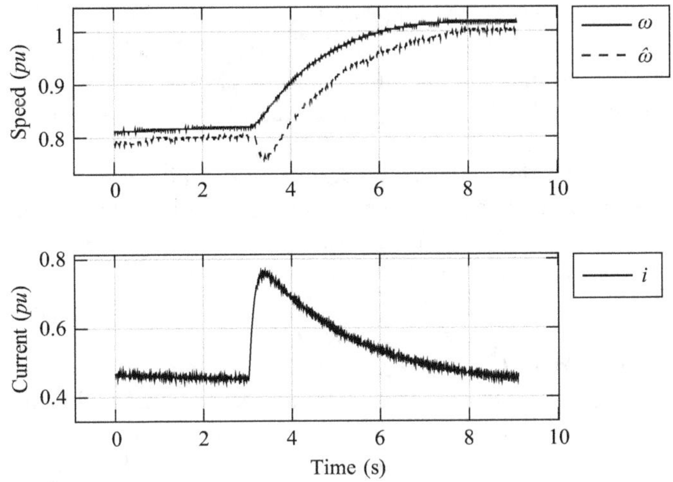

Figure 4.4.9 Experimental results under linear flux assumption

instability at zero current. That is why the magnetic saturation and the estimator are taken into account in this work in order to avoid the incorrect speed observation.

Our ongoing work focus on the stability proof of the proposed sensorless control.

References

[1] Santana J., Naredo J., Sandoval F., Grout I., Argueta O. 'Simulation and construction of a speed control for a DC series motor,' *Mechatronics*. 2002, **12**(9–10), pp. 1145–1156

[2] Siller-Alcalá I., Liceaga-Castro J. U., Alcántara-Ramírez R., Jaimes-Ponce J. 'Speed nonlinear predictive control of a series dc motor for bidirectional operation.' *Proceedings of the 13th IASME/WSEAS*;Angers, France, November 2011. Stevens Point, Wisconsin, USA: WSEAS Press; 2011, pp. 182–187

[3] Iracleous D., Alexandridis A. 'Nonlinear control of a series connected DC motor using singular perturbation and feedback linearisation techniques.' *Proceedings of the 3rd European Control conference*; Roma, Italy, September 1995. European Control Association; 1995, pp. 325–329

[4] Iracleous D., Alexandridis A. 'Fuzzy tuned PI controllers for series connected DC motor drives.' *Proceedings of the IEEE International Symposium on Industrial Electronics*; Athens, Greece. July 1995. IEEE; 1995, pp. 495–499

[5] Chiasson J. 'Nonlinear differential-geometric techniques for control of a series DC motor,' *IEEE Transactions on Control Systems Technology*. 1994, **2**(1), pp. 35–42

[6] Mehta S., Chiasson J. 'Nonlinear control of a series DC motor: theory and experiment,' *IEEE Transactions on Industrial Electronics*. 1998, **45**(1), pp. 134–141

[7] Burridge, M. J., Qu Z. 'An improved nonlinear control design for series DC motors,' *Computers & Electrical Engineering*. 2003, **29**(2), pp. 273–288

[8] Dongbo Z. 'An improved nonlinear speed controller for series DC motors.' *Proceedings of the 17th World Congress*; Seoul, Korea. July 2008. IFAC; 2008, pp. 11047–11052

[9] Iracleous D. 'Series connected DC motor tracking using port controlled Hamiltonian systems equivalence.' *Proceedings of the 13th WSEAS International Conference on SYSTEMS*; Rodos, Greece, July 2009. Stevens Point, Wisconsin, USA: WSEAS Press; 2009, pp. 591–595

[10] Boizot N., Busvelle E., Gauthier J. P., Sachau J. 'Adaptive gain extended Kalman filter: Application to a series-connected DC motor.' *Conference on Systems and Control*; Marrakech, Morocco, 2007, pp. 16–18

[11] Boizot N., Busvelle E., Sachau J. 'High-gain observers and Kalman filtering in hard real-time.' Presented at RTL 9th Workshop, 2007

[12] Boizot N. 'Adaptive high-gain extended Kalman filter and applications.' PhD dissertation, Université de Bourgogne; Université du Luxembourg, Luxembourg-France, 2010.

[13] Floquet T., Barbot J. P. 'Super twisting algorithm-based step-by-step sliding mode observers for nonlinear systems with unknown inputs,' *International Journal of Systems Science*. 2007, **38**(10). pp. 803–815

[14] Levant A. 'Robust exact differentiation via sliding mode technique,' *Automatica*. 1998, **34**(3). pp. 379–384

[15] Angulo M. T., Moreno J. A., Fridman L. 'The differentiation error of noisy signals using the generalized super-twisting differentiator.' *51st IEEE Conference on Decision and Control (CDC)*; Maui, Hawaii, USA, December 2012. IEEE; 2012, pp. 7383–7388

[16] Zaltni D., Ghanes M., Barbot J. P., Abdelkrim M. N. 'Synchronous motor observability study and an improved zero-speed position estimation design.' *49th IEEE Conference on Decision and Control (CDC)*; Atlanta, USA, December 2010. IEEE; 2010, pp. 5074–5079

[17] Hermann R., Krener A. 'Nonlinear controllability and observability,' *IEEE Transactions on Automatic Control*. 1977, **22**(5), pp. 728–740

[18] Fliess M., Lévine J., Martin P., Rouchon P. 'A Lie-Backlund approach to equivalence and flatness of nonlinear systems,' *IEEE Transactions on Automatic Control*. 1999, **44**(5), pp. 922–937

[19] Saadaoui H., Manamanni N., Djemaï M., Barbot J. P., Floquet T. 'Exact differentiation and sliding mode observers for switched Lagrangian systems,' *Nonlinear Analysis: Theory, Methods & Applications*. 2006, **65**(5), pp. 1050–1069

[20] De Leon J., Ghanes M., Glumineau A. 'On cascade and high gain observers comparison for sensorless closed loop induction motor control,' *IET Control Theory & Applications*. 2008, **2**(2), pp. 133–150

[21] Hajji S., Farza M., M'Saad M., Kamoun M. 'Observer-based output feedback controller for a class of nonlinear systems.' *Proceedings of the 17th IFAC World Congress*; Seoul, South Korea, July 2008. IFAC; 2010, pp. 4767–4772

Chapter 4.5
Sliding mode control of LCL full-bridge rectifiers

Arnau Dòria-Cerezo, Domingo Biel, and Enric Fossas

Abstract

This chapter presents a dynamic analysis and the control design for a family of unity power factor rectifiers with an inductive-capacitive-inductive filter. Both single-phase and three-phase topologies are considered, and the three-phase examples include a three-wire power converter and a four-wire configuration with neutral connection. In all cases, the control scheme consists in an inner current loop designed using the sliding mode control technique, and an external loop that regulates the output voltage. Differences among topologies are pointed out, such as the needed of using a decoupling matrix to solve the algebraic constraint on the grid currents (for the three-wire case) or the use of two independent controllers that simultaneously regulate the DC bus voltage and keep the split bus balanced (in the four-wire power converter). Numerical simulations validate the proposed control schemes and show satisfactory performances of all closed-loop behaviors.

4.5.1 Introduction

Nowadays, power rectifiers connected to the electrical power network include an inductive-capacitive-inductive (LCL) filter with the aim of improving the power grid quality. When compared to the traditional inductive (L) filters, LCL filters offer the possibility of reducing the harmonics caused by the switching of the power conversion system with low inductance values. However, the use of the LCL filter increases the dynamics of the three-phase full-bridge rectifier and complicates the control design, as well as the tuning of the control gains.

Several papers concerning the connection of LCL rectifiers to the power grid may be found in the literature. Mainly, they are focused on the analysis of the resonance effects as a consequence of the use of LCL filters and its interaction with the grid impedance. A common solution to this problem is the use of resistances, at the price of a reduction of efficiency. Also, more advanced techniques are reported, as active damping [1] which is specially attractive for large power converters, but limited by the use of more sensors (and an increasing complexity of the control gains tuning procedure).

Three-phase devices can be connected using the neutral point. In this case, three-phase power converters with four wires, they use a DC bus with two split capacitors.

Then, control algorithms are required to achieve three goals: an unitary power factor at the power grid connection point, a DC bus voltage regulation and a voltage balance of the split capacitors.

Usually, linear controllers are proposed for this kind of power converters, such as proportional-integral (PI) controllers [2, 3], deadbeat control [4, 5], space-state PI controllers [6], resonant PI control [7], adaptive feedforward cancellation [8], weighted average current control [9], etc. The main drawback for these control designs is that the nonlinear behavior of the full-bridge rectifier is disregarded. Recently, nonlinear control techniques were also applied to power converters with an LCL filter. The feedback linearization methodology is one of the proposed nonlinear approaches, and it has been applied to both three-phase inverters [10] and full-bridge rectifiers [11].

In power systems, control problems of three-phase devices are commonly translated to a new reference frame (dq or $\alpha\beta$ coordinates) that allows to decouple the state variables and to reduce the order of the dynamical system. In Reference 12, the sliding mode control (SMC) technique is applied to a current source rectifier using the stationary reference frame. However, the use of these kind of change of coordinates assumes a balanced three-phase system. An alternative is to directly control each current phase. A design for a three-phase three-wire voltage source inverter (VSI) is presented in Reference 13, where the obstacle of the algebraic currents constraint has been overcome using an extra decoupling dynamics that requires the perfect knowledge of the inverter parameters.

Linear controllers are often proposed for both single-phase (as the adaptive feedforward cancellation technique used in Reference 8) and three-phase power converters. Examples include a voltage regulator [14], and rectifiers [15] or active power filters [16] with a stationary reference frame. In Reference 17, a proportional-derivative (PD) control for an active power filter using a synchronous reference frame and a feedforward term to compensate undesired dynamics is presented. Bang–bang control actions have been also proposed, as the hysteretic control for the static compensator (STATCOM) [18].

The SMC technique allows to obtain robust controllers with a good performance, and is specially suitable for variable structure systems as power converters [19]. SMC algorithms for a single-phase full-bridge rectifier and a VSI are proposed in References 20 and 21, respectively.

4.5.2 Modeling LCL rectifiers

Full-bridge rectifiers with LCL are composed by an LCL filter, a set of switches, and a capacitor (or two in a split bus) where the voltage is regulated. In this work, the load is assumed to be a resistor in parallel with the capacitor. Figure 4.5.1 shows the scheme for the LCL power converters topologies.

In this section, the dynamical equations of both single-phase and three-phase LCL rectifiers are presented. The structure of the equations describing the power

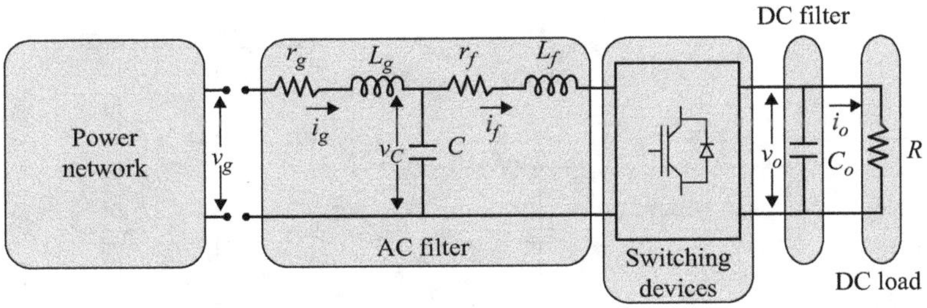

Figure 4.5.1 General scheme of LCL rectifiers

converter is similar but the differences between them make difficult to generalize a model. In Figure 4.5.1, the main parts of an LCL rectifier are shown, where

- the LCL filter is composed by the grid side and the filter side inductors, L_g and L_f, and their losses r_g, r_f, a capacitor, C, and $\boldsymbol{i}_g$, $\boldsymbol{i}_f$ are the currents of the inductors, and $\boldsymbol{v}_C$ the voltage of the capacitor;
- the switching components are controlled by a discrete signal $\boldsymbol{u}$ (or set of signals),
- the DC filter with a capacitor, C_o, with v_o as a DC voltage; and
- the load, that can be modeled as a pure resistor, R, that consumes a certain current i_o.

Notice that the voltage at the connection point, $\boldsymbol{v}_g$ is the one available for control purposes. As pointed out in the next section, $\boldsymbol{v}_g$ will be used for determining the phase and frequency for the required current.

In general, all the equations of an LCL full-bridge rectifier can be written in a compact form as

$$\frac{\mathrm{d}\boldsymbol{x}}{\mathrm{d}t} = \boldsymbol{A}(\boldsymbol{u})\boldsymbol{x} + \boldsymbol{B}\boldsymbol{v}_g,$$

where the state, $\boldsymbol{x}^T = \left(\boldsymbol{i}_g^T, \boldsymbol{v}_C^T, \boldsymbol{i}_f^T, v_o\right)$, contains the inductor currents, $\boldsymbol{i}_g$, $\boldsymbol{i}_f$, the voltage capacitor, $\boldsymbol{v}_c$, and the DC voltage, v_o. In the single-phase case, the currents and voltages are scalar, and $\boldsymbol{i}_g = i_g$, $\boldsymbol{i}_f = i_f$, $\boldsymbol{v}_C = v_C$ and $\boldsymbol{v}_g = v_g$, while, in the three-phase cases $\boldsymbol{i}_g = \left(i_{ga}, i_{gb}, i_{gc}\right)^T, \boldsymbol{i}_f = \left(i_{fa}, i_{fb}, i_{fc}\right)^T$, $\boldsymbol{v}_C = (v_{Ca}, v_{Cb}, v_{Cc})^T$, and $\boldsymbol{v}_g = \left(v_{ga}, v_{gb}, v_{gc}\right)^T$. Also, the control input for the single phase is scalar ($\boldsymbol{u} = u$) and in $\mathbb{R}^3$ for the three-phase converters, $\boldsymbol{u} = (u_a, u_b, u_c)^T$ are the control signals.

4.5.2.1 Dynamical model of a single-phase LCL rectifier

Figure 4.5.2 shows a simplified scheme of a single-phase rectifier with an LCL filter. In this case, all the inductor currents and capacitor voltages are scalar, and the dynamics of the system is described by

$$L_g \frac{\mathrm{d}i_g}{\mathrm{d}t} = -r_g i_g - v_c + v_g, \tag{4.5.1}$$

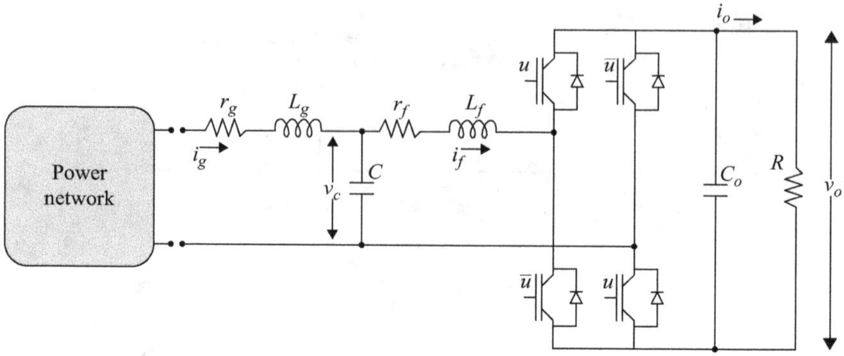

Figure 4.5.2 Simplified scheme of a single-phase rectifier with an LCL filter

$$C\frac{\mathrm{d}v_c}{\mathrm{d}t} = i_g - i_f, \tag{4.5.2}$$

$$L_f\frac{\mathrm{d}i_f}{\mathrm{d}t} = -r_f i_f + v_c - uv_o, \tag{4.5.3}$$

$$C_o\frac{\mathrm{d}v_o}{\mathrm{d}t} = ui_f - \frac{v_o}{R}, \tag{4.5.4}$$

where the grid voltage is

$$v_g = V_g \sin(\omega t), \tag{4.5.5}$$

where V_g and ω are the grid voltage amplitude and frequency, respectively, are the control signal, u, take the discrete values $u \in \{-1, 1\}$.

4.5.2.2 Dynamical model of a three-phase three-wire LCL rectifier

A three-phase rectifier with an LCL filter is shown in Figure 4.5.3. The dynamics of this system is described by

$$L_g\frac{\mathrm{d}\boldsymbol{i}_g}{\mathrm{d}t} = -r_g\boldsymbol{i}_g - \boldsymbol{M}\boldsymbol{v}_C + \boldsymbol{N}\boldsymbol{v}_g, \tag{4.5.6}$$

$$C\frac{\mathrm{d}\boldsymbol{v}_C}{\mathrm{d}t} = \boldsymbol{i}_g - \boldsymbol{i}_f, \tag{4.5.7}$$

$$L_f\frac{\mathrm{d}\boldsymbol{i}_f}{\mathrm{d}t} = -r_f\boldsymbol{i}_f + \boldsymbol{M}\boldsymbol{v}_C - \frac{1}{2}v_o\boldsymbol{M}\boldsymbol{u}, \tag{4.5.8}$$

$$C_o\frac{\mathrm{d}v_o}{\mathrm{d}t} = \frac{1}{2}\boldsymbol{i}_f^T\boldsymbol{u} - \frac{1}{R}v_o, \tag{4.5.9}$$

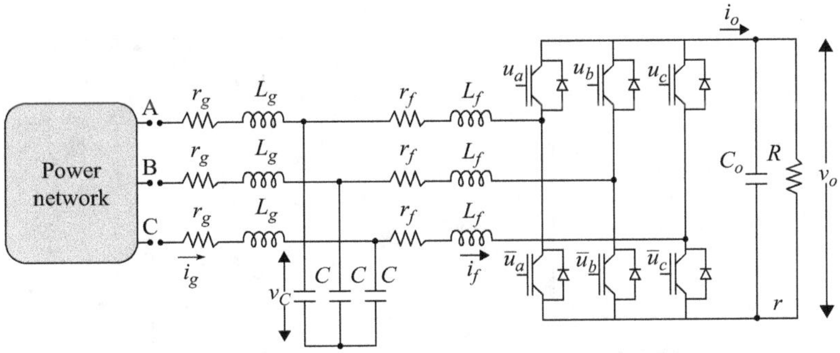

Figure 4.5.3 Simplified scheme of a three-phase three-wire rectifier with an LCL filter

where matrices $\boldsymbol{M}$ and $\boldsymbol{N}$ are

$$\boldsymbol{M} = \frac{1}{3}\begin{pmatrix} 2 & -1 & -1 \\ -1 & 2 & -1 \\ -1 & -1 & 2 \end{pmatrix}, \quad \boldsymbol{N} = \frac{1}{3}\begin{pmatrix} -2 & 1 \\ 1 & -1 \\ 1 & -2 \end{pmatrix}.$$

Notice that $\boldsymbol{M}$ is a rank-2 matrix, and $\boldsymbol{N}$ transforms the measured line-to-line voltages, $\boldsymbol{v}_g^l = \left(v_g^{ab}, v_g^{bc}\right)^T$ to the phase-to-neutral voltages $\boldsymbol{v}_g$ (which are not available for measures). The control signals, $\boldsymbol{u} = (u_a, u_b, u_c)^T$, take the discrete values $u_k \in \{-1, 1\}$, for $k = a, b, c$. Assuming a balanced grid, the grid voltage vector is

$$\boldsymbol{v}_g = \begin{pmatrix} v_{ga} \\ v_{gb} \\ v_{gc} \end{pmatrix} = V_g \begin{pmatrix} \sin(\omega t) \\ \sin\left(\omega t - \dfrac{2\pi}{3}\right) \\ \sin\left(\omega t + \dfrac{2\pi}{3}\right) \end{pmatrix}. \tag{4.5.10}$$

4.5.2.3 Dynamical model of a three-phase four-wire LCL rectifier

Finally, Figure 4.5.4 shows a simplified scheme of a three-phase four-wire rectifier with an LCL filter. In this case, the DC voltage is split in two capacitors, and then $v_o = v_{o1} + v_{o2}$. The dynamics is described by the following differential equations:

$$L_g \frac{\mathrm{d}\boldsymbol{i}_g}{\mathrm{d}t} = -r_g \boldsymbol{i}_g - \boldsymbol{v}_C + \boldsymbol{v}_g, \tag{4.5.11}$$

$$C \frac{\mathrm{d}\boldsymbol{v}_C}{\mathrm{d}t} = \boldsymbol{i}_g - \boldsymbol{i}_f, \tag{4.5.12}$$

$$L_f \frac{\mathrm{d}\boldsymbol{i}_f}{\mathrm{d}t} = -r_f \boldsymbol{i}_f + \boldsymbol{v}_C - v_{o1}\boldsymbol{u} + v_{o2}(\boldsymbol{I}_3 - \boldsymbol{u}), \tag{4.5.13}$$

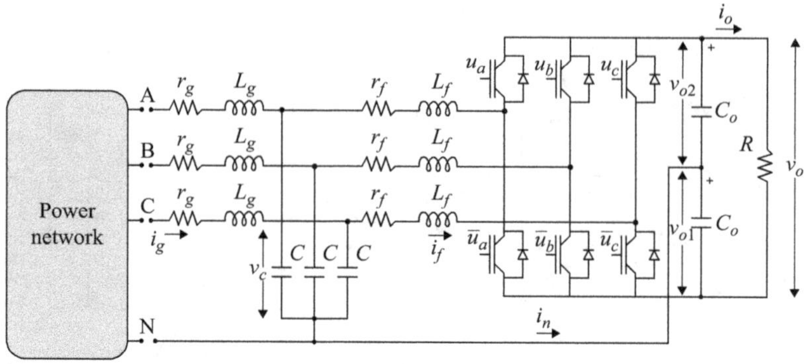

Figure 4.5.4 Simplified scheme of a three-phase four-wire rectifier with an LCL filter

$$C_o \frac{\mathrm{d}v_{o1}}{\mathrm{d}t} = \boldsymbol{i}_f^T \boldsymbol{u} - \frac{v_{o1} + v_{o2}}{R}, \tag{4.5.14}$$

$$C_o \frac{\mathrm{d}v_{o2}}{\mathrm{d}t} = \boldsymbol{i}_f^T (\boldsymbol{u} - \boldsymbol{I}_3) - \frac{v_{o1} + v_{o2}}{R}, \tag{4.5.15}$$

where $\boldsymbol{I}_3$ is a 3×1 ones vector and the control signals, $\boldsymbol{u} = (u_a, u_b, u_c)^T$, take the discrete values $u_k \in \{0, 1\}$, for $k = a, b, c$. We can also define the neutral current, i_n, as the difference of the current through each split capacitor, i.e., from (4.5.14) and (4.5.15),

$$i_n = \boldsymbol{i}_f^T \boldsymbol{I}_3$$

that is zero for a balanced $\boldsymbol{i}_f$. Unbalanced three-phase currents (in steady state, or during a transient) can vary the injected current in each capacitor and tend to unbalance the partial DC bus voltages, v_{o1} and v_{o2}. Similarly to the three-phase three-wire case, the power network is assumed balanced, and the grid voltages, $\boldsymbol{v}_g$, are in (4.5.10).

4.5.3 Overall control scheme

The control objective of the full-bridge power rectifiers is twofold: to regulate the DC link voltage, v_o, to a desired value v_o^d, and to achieve an unitary power factor. These two requirements can be reached with an input current, $\boldsymbol{i}_g$, which is in phase with the grid voltage, $\boldsymbol{v}_g$ and with an amplitude, I_g, that supplies the necessary current to keep the bus voltage at the desired value. In other words, from (4.5.5), the desired current for the single-phase case takes the form

$$i_g^d = I_g \sin(\omega t) \tag{4.5.16}$$

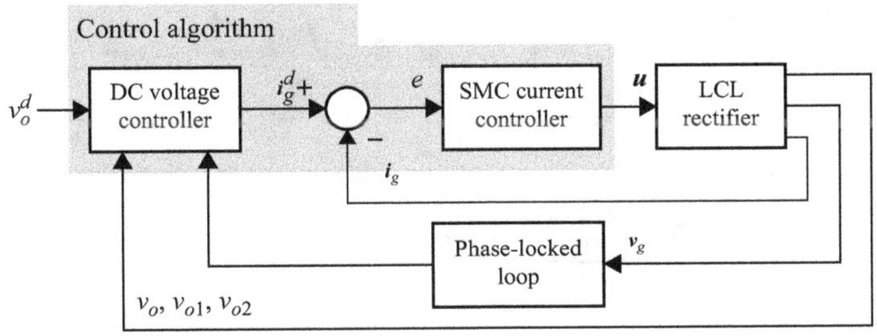

Figure 4.5.5 Overall control scheme for LCL rectifiers

and, for the three-phase rectifiers, using (4.5.10),

$$\boldsymbol{i}_g^d = I_g \begin{pmatrix} \sin(\omega t) \\ \sin\left(\omega t - \dfrac{2\pi}{3}\right) \\ \sin\left(\omega t + \dfrac{2\pi}{3}\right) \end{pmatrix}, \tag{4.5.17}$$

where I_g has to be obtained via energy balance or by a control loop. As mentioned in Section 4.5.2.3, the three-phase four-wire rectifier has a DC split bus with two capacitors. In this case, in addition, the voltage across the two capacitors must be balanced. This control specification is further commented in Section 4.5.6.

Usually, the control algorithm is split in two steps, first, an inner current loop provides the unitary power factor and then, an external loop is used to regulate the DC bus voltage assuming a perfect tracking of the grid currents. Figure 4.5.5 shows the proposed overall control scheme. The PI-based DC voltage controller provides the current references, $\boldsymbol{i}_g^d$, using the measured DC voltage and the phase and frequency of the grid at the connection point, obtained thanks to phase-locked loop (PLL). The current controller, designed using the sliding mode techniques, ensures a fast and robust tracking of the currents.

In the following sections, the control algorithms for the single-phase LCL rectifier and for the three-phase LCL rectifier (both three-wire and four-wire rectifiers) are presented. In all cases, the zero dynamic is analyzed, then the current control algorithm using sliding mode strategies is designed, and finally the outer voltage loop is presented.

4.5.4 Control design of an LCL single-phase rectifier

The proposed control scheme for the LCL single-phase rectifier is depicted in Figure 4.5.6. It contains the two loops introduced in the previous section: the inner current loop and the outer voltage loop that will be designed in the following subsections.

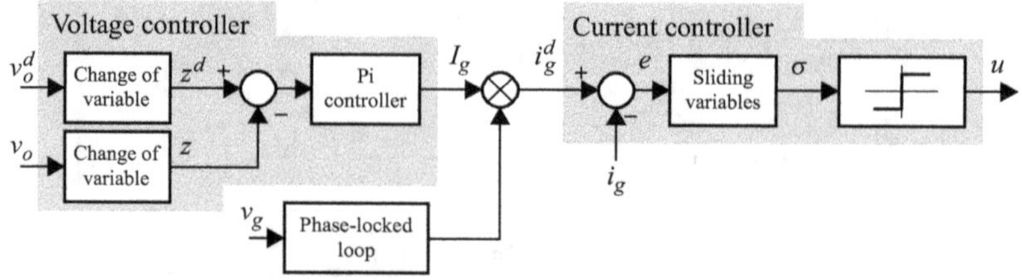

Figure 4.5.6 Proposed control scheme for a single-phase LCL rectifier

4.5.4.1 Zero-dynamics analysis

Assuming (4.5.16) in (4.5.1), (4.5.2), and (4.5.3), one gets

$$i_f = \alpha_1 \sin(\omega t) + \alpha_2 \cos(\omega t) - C\frac{\mathrm{d}v_g(t)}{\mathrm{d}t}, \quad (4.5.18)$$

$$v_o u = \beta_1 \sin(\omega t) + \beta_2 \cos(\omega t) + v_g(t) + r_f C\frac{\mathrm{d}v_g(t)}{\mathrm{d}t} + CL_f\frac{\mathrm{d}^2 v_g(t)}{\mathrm{d}t^2}, \quad (4.5.19)$$

where

$$\begin{aligned}
\alpha_1 &= I_g\left(1 - \omega^2 CL_g\right), \\
\alpha_2 &= \omega C(r_g I_g - V_g), \\
\beta_1 &= \left((r_f L_g + r_g L_f)\omega^2 C - (r_f + r_g)\right) I_g + \left(1 - \omega^2 CL_f\right) V_g, \\
\beta_2 &= \omega(r_f C(V_g - r_g I_g) - \left(L_f + L_g\left(1 - \omega^2 CL_f\right)\right) I_g).
\end{aligned}$$

Finally, replacing $v_g(t) = V_g \sin(\omega t)$, (4.5.18) and (4.5.19) in (4.5.4), the remaining dynamics of the DC voltage is

$$C_o\frac{\mathrm{d}v_o}{\mathrm{d}t} = -\frac{v_o}{R} + \frac{1}{v_o}g(t), \quad (4.5.20)$$

where

$$g(t) = (\alpha_2\beta_2 - \alpha_1\beta_1)\cos^2(\omega t) + (\alpha_1\beta_2 + \alpha_2\beta_1)\cos(\omega t)\sin(\omega t) + \alpha_1\beta_1.$$

Equation (4.5.20) is a stable Bernoulli ordinary differential equation (ODE), for $v_o > 0$, that reveals that a perfect current tracking yields oscillations at frequency 2ω in the v_o voltage, see Reference 22.

Moreover, for a general case including a current i_o flowing to the grid (bidirectional behavior), (4.5.20) transforms in

$$C_o\frac{\mathrm{d}v_o}{\mathrm{d}t} = -i_o(t) + \frac{1}{v_o}g(t),$$

which is a class A, second type, Abel ODE, and has unstable solutions for $i_o(t) < 0$, see Reference 23.

4.5.4.2 Inner-loop current control

The inner-loop current control is designed using the SMC technique. The error current is defined as

$$e_i = i_g^d - i_g$$

and the control objective is to reach $e_i = 0$. As the error, e_i, is relative degree three with respect to the control action, the following switching function is considered:

$$\sigma = \tau_0 e_i + \tau_1 \frac{\mathrm{d}e_i}{\mathrm{d}t} + \frac{\mathrm{d}^2 e_i}{\mathrm{d}t^2}, \qquad (4.5.21)$$

which is relative degree one. τ_0 and τ_1 are the control gains that can be selected to tune the error dynamics.

From (4.5.21) and using (4.5.1)–(4.5.4), one gets

$$\frac{\mathrm{d}\sigma}{\mathrm{d}t} = f(x) + \frac{1}{CL_gL_f} v_o u.$$

As $L_g, C, L_f > 0$, and assuming $v_o > 0$, the switching control action

$$u = \begin{cases} 1 & \text{if } \sigma < 0 \\ -1 & \text{if } \sigma > 0 \end{cases}$$

fulfills the reachability condition [24].

4.5.4.3 Outer-loop voltage control

The outer-loop voltage control is designed assuming a perfect tracking of the current reference, i.e., $i_g = i_g^d$. In Section 4.5.4.1, it is analyzed the zero dynamics for $i_g^d = I_g \sin(\omega t)$ and $v_g(t) = V_g \sin(\omega t)$ with I_g constant. From this analysis it can be concluded that, in steady state, there appear oscillations at 2ω frequency in the DC bus voltage, v_o. Consequently, the control objective is translated to regulate the average of the voltage v_o.

Assuming a perfect current tracking, the control law designed in the previous subsection guarantees that

$$i_g = I_g(t) \sin(\omega t), \qquad (4.5.22)$$

where $I_g(t)$ is the required amplitude to regulate the output voltage to its desired value. As proceeded in Section 4.5.4.1, replacing (4.5.22) in (4.5.1), (4.5.2), (4.5.3), and (4.5.4), one gets a complex dynamics with the predicted oscillations in (4.5.20) and the first, second and third time derivative of $I_g(t)$. Assuming $v_o(t) > 0$, the change of coordinates

$$z = \frac{1}{2} v_o^2$$

transforms (4.5.20) into a linear ODE (see Reference 22 for further details) and allows to analyze the output voltage dynamics by means of z. Finally, presuming $r_g = r_f = 0$, one gets

$$C_o \frac{\mathrm{d}z}{\mathrm{d}t} = -\frac{2}{R}z + h\left(I_g, \frac{\mathrm{d}I_g}{\mathrm{d}t}, \frac{\mathrm{d}^2 I_g}{\mathrm{d}t^2}, \frac{\mathrm{d}^3 I_g}{\mathrm{d}t^3}\right),$$

where h is a complicated function involving I_g (and its time derivatives) and sinusoidal functions. A Fourier expansion of h has been computed and the dominated coefficients[1] have been disregarded using, to this aim, the rectifier parameters used in the next subsection. Thus, to outer-loop control purposes, function h may be approximated by

$$h = (a_{01} + a_{21}\cos(2\omega t) + b_{20}\sin(2\omega t))I_g(t), \tag{4.5.23}$$

where

$$\begin{aligned} a_{01} &= \frac{1}{2}V_g, \\ a_{21} &= \frac{1}{2}\left(\omega^2 C\left(L_g + L_f - \omega^2 L_f L_g C\right) - \frac{1}{2}\right)V_g, \\ b_{20} &= \frac{1}{4}\omega C\left(\omega^2 L_f C - 1\right)V_g^2. \end{aligned}$$

Notice that averaging[2] h with a temporal window of length $\frac{2\pi}{\omega}$, the second and third terms of (4.5.23) cancel, and an (averaged) power balance equation is obtained

$$C_o \frac{\mathrm{d}\bar{z}}{\mathrm{d}t} = -\frac{2}{R}\bar{z} + \frac{1}{2}V_g I_g. \tag{4.5.24}$$

In order to regulate $\bar{z}$ to $\bar{z}^d$, where $\bar{z}^d$ is the average of $z^d = \frac{1}{2}\left(v_o^d\right)^2$, the following PI controller is considered:

$$I_g = k_p\left(\bar{z} - \bar{z}^d\right) + k_i \int \left(\bar{z} - \bar{z}^d\right)\mathrm{d}t, \tag{4.5.25}$$

where k_p and k_i are the proportional and integral control gains, respectively. Then, the closed-loop dynamics of (4.5.24) with (4.5.25) yields

$$\begin{aligned} C_o \frac{\mathrm{d}\bar{z}}{\mathrm{d}t} &= -\left(\frac{2}{R} + V_g k_p\right)\bar{z} + V_g k_i w + V_g k_p \bar{z}^d, \\ \frac{\mathrm{d}w}{\mathrm{d}t} &= \bar{z}^d - \bar{z}, \end{aligned}$$

which is asymptotically stable for $k_p > -\frac{2}{V_g R}$ and $k_i > 0$.

[1]We consider a coefficient k_1 dominant over k_2 if $\frac{k_1}{k_2} < 10^{-5}$.

[2]The averaged function of a periodic signal $f(t)$ of period T is calculated as $\bar{f}(t) = \frac{1}{T}\int_{t-T}^{t} f(\tau)\,\mathrm{d}\tau$.

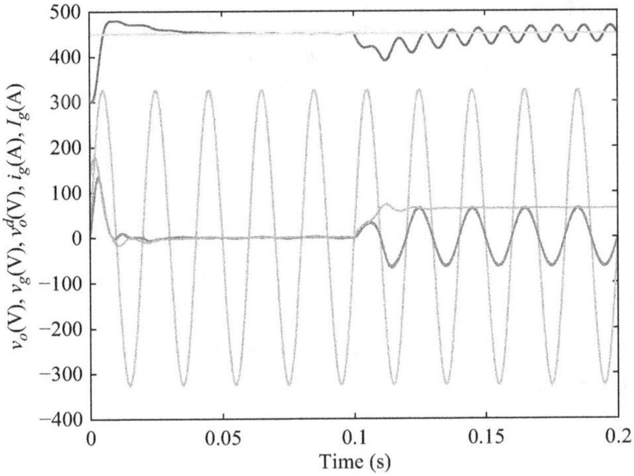

Figure 4.5.7 Simulation results: DC bus voltage, v_o, AC grid voltage, v_g, and AC grid current, i_g

4.5.4.4 Simulation results

In this section, a numerical experiment of the designed control algorithm for a single-phase rectifier with an LCL filter is presented. The parameters of the power converter are $L_g = 100$ μH, $r_g = 50$ mΩ, $C = 20$ μF, $L_f = 300$ μH, $r_f = 50$ mΩ, and $C_o =$ 2 mF. The PI control gains are $k_p = 0.8$ and $k_i = 40$. The constants of the sliding surface (4.5.21) have been fixed at $\tau_0 = 10^8$ and $\tau_1 = 2 \times 10^4$, which results in a damped dynamics with a double pole with a time constant of 100 μs in the current error. The hysteresis band, used in the SMC to implement the sign function, has been set to ensure a switching frequency between 10 and 20 kHz.

With the aim of showing the robustness in front of load variations, a sudden change from open circuit to $R = 20$ Ω at $t = 10$ ms has been done. In the simulation, the output DC voltage reference has been set at 450 V, which implies approximately 10 kW when full load and the voltage source is $v_g(t) = 220\sqrt{2}\sin(2\pi 50t)$. The simulation results are shown in Figure 4.5.7, where the following variables are plotted: the input current (i_g), the control signal given by the external controller (current reference amplitude, I_g), the voltage at the connection point (v_g), the output DC voltage (v_o), and its reference $\left(v_o^d\right)$. It can be seen that the average value of the DC voltage recovers its desired value after the load change. Notice also that the grid voltage and current are in phase, also after the load variation, which ensures an unitary power factor.

4.5.5 Control design of an LCL three-phase three-wire rectifier

The proposed control scheme is depicted in Figure 4.5.8. It follows the same structure than the single-phase LCL rectifier and only a decoupling matrix **Γ** has been introduced. See details in Section 4.5.5.2.

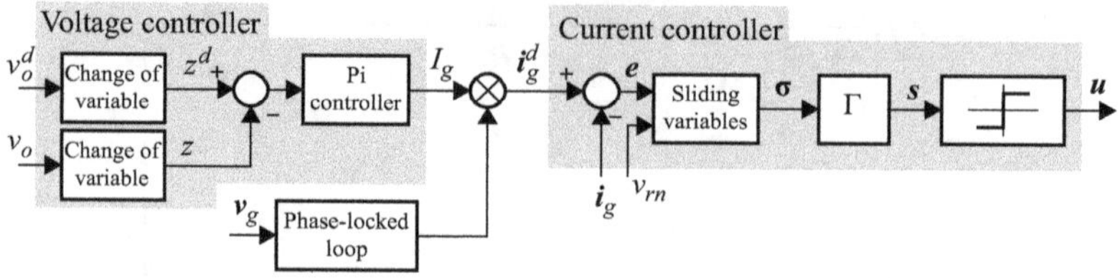

Figure 4.5.8 Proposed control scheme for a three-phase three-wire LCL rectifier

4.5.5.1 Zero-dynamics analysis

Due to the current physical constraints of the three-phase three-wire rectifier, the dynamical systems (4.5.6)–(4.5.9) can be reduced by applying the $\alpha\beta 0$ transformation[3]

$$\boldsymbol{f}^{0\alpha\beta} = \boldsymbol{T}\boldsymbol{f},$$

where $\boldsymbol{f}$ and $\boldsymbol{f}^{0\alpha\beta}$ are the original three-phase signals and the $0\alpha\beta$ transformed ones, respectively, and

$$\boldsymbol{T} = \begin{pmatrix} 1 & -\frac{1}{2} & \frac{1}{2} \\ 0 & -\frac{\sqrt{3}}{2} & \frac{\sqrt{3}}{2} \\ \frac{\sqrt{2}}{2} & \frac{\sqrt{2}}{2} & \frac{\sqrt{2}}{2} \end{pmatrix}.$$

Assuming a balanced system, all the homopolar components are null ($f^0 = 0$) and the system equation is reduced to

$$L_g \frac{\mathrm{d}\boldsymbol{i}_g^{\alpha\beta}}{\mathrm{d}t} = -r_g \boldsymbol{i}_g^{\alpha\beta} - \boldsymbol{v}_C^{\alpha\beta} + \boldsymbol{v}_g^{\alpha\beta}, \tag{4.5.26}$$

$$C \frac{\mathrm{d}\boldsymbol{v}_C^{\alpha\beta}}{\mathrm{d}t} = \boldsymbol{i}_g^{\alpha\beta} - \boldsymbol{i}_f^{\alpha\beta}, \tag{4.5.27}$$

$$L_f \frac{\mathrm{d}\boldsymbol{i}_f^{\alpha\beta}}{\mathrm{d}t} = -r_f \boldsymbol{i}_f^{\alpha\beta} + \boldsymbol{v}_C^{\alpha\beta} - \frac{1}{2} v_o \boldsymbol{u}^{\alpha\beta}, \tag{4.5.28}$$

$$C_o \frac{\mathrm{d}v_o}{\mathrm{d}t} = \frac{1}{2} \left(\boldsymbol{i}_f^{\alpha\beta}\right)^T \boldsymbol{u}^{\alpha\beta} - \frac{1}{R} v_o, \tag{4.5.29}$$

and the voltage grid and the desired currents are transformed to

$$\boldsymbol{v}_g^{\alpha\beta} = V_g \begin{pmatrix} \sin(\omega t) \\ \cos(\omega t) \end{pmatrix} \tag{4.5.30}$$

[3] In this case, the $\alpha\beta 0$ transformation is amplitude preserving.

and

$$\left(\boldsymbol{i}_g^{\alpha\beta}\right)^d = I_g \begin{pmatrix} \sin(\omega t) \\ \cos(\omega t) \end{pmatrix}.$$

Replacing (4.5.30) and forcing $\boldsymbol{i}_g^{\alpha\beta} = \left(\boldsymbol{i}_g^{\alpha\beta}\right)^d$ in (4.5.26)–(4.5.29), one gets

$$C_o \frac{\mathrm{d}v_o}{\mathrm{d}t} = -\frac{v_o}{R} + \frac{1}{v_o}\left(-a_2 I_g^2 + a_1 I_g - a_0\right), \tag{4.5.31}$$

where

$$\begin{aligned} a_2 &= \omega^2 C^2 r_g^2 r_f + r_g + r_f, \\ a_1 &= \left(2\omega^2 C^2 r_g r_f + 1\right) V_g, \\ a_0 &= C^2 \omega^2 r_f V_g^2. \end{aligned}$$

Similarly to the single-phase case, (4.5.31) is a stable Bernoulli ODE, for $v_o > 0$, that reveals that the voltage v_o can be regulated by choosing the desired current amplitude to

$$I_g = \frac{a_1 - \sqrt{a_1^2 - 4a_2\left(a_0 + \dfrac{v_o^2}{R}\right)}}{2a_2}.$$

Moreover, for the general case including a current i_o flowing to the grid (bidirectional behavior), (4.5.31) transforms in

$$C_o \frac{\mathrm{d}v_o}{\mathrm{d}t} = -i_o(t) + \frac{1}{v_o}\left(a_2 I_g^2 + a_1 I_g + a_0\right)$$

which turns to the class A, second type, Abel ODE, with unstable solutions for $i_o(t) < 0$ observed in Section 4.5.4.1.

4.5.5.2 Inner-loop current control

The objective of this controller is to track the input currents, $\boldsymbol{i}_g$, as proposed in (4.5.17). However, the $\boldsymbol{M}$ matrix is singular, implying that the control signals u_k are not independent in the current dynamics, and that $i_{ga} + i_{gb} + i_{gc} = 0$. Precisely, this fact states that if i_{ga} and i_{gb} are perfectly tracked, $i_{gc} = i_{gc}^d$. Then, the controls goals are $e_a = 0$ and $e_b = 0$, where the following error currents are defined:

$$\begin{aligned} e_a &= i_{ga} - i_{ga}^d, \\ e_b &= i_{gb} - i_{gb}^d. \end{aligned}$$

As the errors e_a, e_b are relative degree three with respect to the control action, the following switching functions are considered:

$$\sigma_a = \tau_{a0}e_a + \tau_{a1}\frac{\mathrm{d}e_a}{\mathrm{d}t} + \frac{\mathrm{d}^2 e_a}{\mathrm{d}t^2}, \tag{4.5.32}$$

$$\sigma_b = \tau_{b0}e_b + \tau_{b1}\frac{\mathrm{d}e_b}{\mathrm{d}t} + \frac{\mathrm{d}^2 e_b}{\mathrm{d}t^2}, \tag{4.5.33}$$

which are relative degree one. τ_{a0}, τ_{a1}, τ_{b0}, and τ_{b1} are the control gains that can be selected to tune the error dynamics.

The rectifier has three control actions defined by the switch position of each phase, u_k, and an extra control objective can be achieved: the floating voltage v_{rn}, which can be defined as

$$v_{rn} = -\frac{1}{3}v_o(u_a + u_b + u_c).$$

To the end of regulating v_{rn}, a third switching function is proposed

$$\sigma_{rn} = \frac{1}{L_g L_f C}\int \left(v_{rn} - v_{rn}^d\right)\mathrm{d}t.$$

The sliding control actions can be derived using the Lyapunov function candidate

$$V = \frac{1}{2}\boldsymbol{s}^T\boldsymbol{s}, \tag{4.5.34}$$

where

$$\boldsymbol{s} = \boldsymbol{\Gamma\sigma}$$

with $\boldsymbol{\sigma}^T = (\sigma_a, \sigma_b, \sigma_{rn})$ and $\boldsymbol{\Gamma}$ being a free design constant decoupling matrix. Differentiating (4.5.34) with respect to time and replacing (4.5.26)–(4.5.28),

$$\dot{V} = \boldsymbol{s}^T\boldsymbol{\Gamma F}(\boldsymbol{x}) + \frac{v_o}{6L_g L_f C}\boldsymbol{s}^T\boldsymbol{\Gamma}\hat{\boldsymbol{M}}\boldsymbol{u}, \tag{4.5.35}$$

where $\boldsymbol{F}(\boldsymbol{x})$ contains the remaining vector field depending on all the state variables, and

$$\hat{\mathbf{M}} = \begin{pmatrix} -2 & 1 & 1 \\ 1 & -2 & 1 \\ -1 & -1 & -1 \end{pmatrix}.$$

Then, a suitable choice for $\boldsymbol{\Gamma}$ is

$$\boldsymbol{\Gamma} = \begin{pmatrix} 1 & 0 & 1 \\ 0 & 1 & 1 \\ -1 & -1 & 1 \end{pmatrix},$$

which implies that $\Gamma\hat{M} = -6I_3$ (I_3 is a 3×3 identity matrix) and (4.5.35) simplifies in

$$\dot{V} = s^T\Gamma F(x) - \frac{v_o}{L_g L_f C} s^T u.$$

Finally, as all the rectifier parameters are positive, and assuming $v_o > 0$, the switching action is given by

$$u = \text{sign}(s)$$

or

$$\begin{aligned} u_a &= \text{sign}(\sigma_a + \sigma_{rn}), \\ u_b &= \text{sign}(\sigma_b + \sigma_{rn}), \\ u_c &= \text{sign}(-\sigma_a - \sigma_b + \sigma_{rn}). \end{aligned}$$

The control law summarizes in three switching control actions coming from the combination of the switching functions σ. Notice that the proposed decoupling method is considerably different than the presented one in Reference 13, where the current coupling was solved inverting the dynamics that requires the perfect knowledge of the LCL parameters, including the inductor losses.

4.5.5.3 PI outer-loop voltage control

The DC voltage, v_o, is controlled by a PI controller, with the following form:

$$I_g = -k_p\tilde{z} - k_i \int \tilde{z}\,\mathrm{d}t, \tag{4.5.36}$$

where $\tilde{z} = z - z^*$ and

$$z = \frac{1}{2}v_o^2. \tag{4.5.37}$$

Considering a perfect tracking of the current references, i.e., $i_{ga} = i_{ga}^d$ and $i_{gb} = i_{gb}^d$ and proceeding as in Section 4.5.5.1 but now assuming that the grid current amplitude is varying, $I_g(t)$, the dynamics of the DC voltage, v_o, with (4.5.37) becomes

$$C_o\frac{\mathrm{d}z}{\mathrm{d}t} = -\frac{1}{R}z + \left(V_g - L_f\frac{\mathrm{d}I_g}{\mathrm{d}t}\right)I_g, \tag{4.5.38}$$

where the inductor losses have been neglected, $r_g = r_f = 0$. The complete dynamics of the DC voltage in z for $r_g, r_f \neq 0$ can be found in Reference 25.

The stability analysis of (4.5.38) in a closed loop with (4.5.36) gives the stability conditions for the outer-loop control. Differentiating (4.5.36) with respect to time and defining (4.5.38) in terms of $\tilde{z}$

$$\begin{pmatrix} C_o & L_f I_g \\ k_p & 1 \end{pmatrix} \frac{\mathrm{d}}{\mathrm{d}t} \begin{pmatrix} \tilde{z} \\ \tilde{I}_g \end{pmatrix} = \begin{pmatrix} -\frac{1}{R} & V_g \\ -k_i & 0 \end{pmatrix} \begin{pmatrix} \tilde{z} \\ \tilde{I}_g \end{pmatrix},$$

where $\tilde{I}_g = I_g - I_g^*$ and $I_g^* = \frac{1}{V_g R} z^*$. Differentiating with respect to time the Lyapunov function candidate

$$V = \frac{1}{2}\left(\tilde{z}^2 + \tilde{I}_g^2\right),$$

one gets

$$\dot{V} = -\frac{1}{C_o - L_f k_p I_g} \begin{pmatrix} \tilde{z} & \tilde{I}_g \end{pmatrix} \begin{pmatrix} \frac{1}{R} - L_f k_i I_g & -V_g \\ -\frac{1}{R} k_p + C_o k_i & k_p V_g \end{pmatrix} \begin{pmatrix} \tilde{z} \\ \tilde{I}_g \end{pmatrix},$$

and the stability is guaranteed if the following matrix inequality is fulfilled:

$$\frac{1}{C_o - L_f k_p I_g} \begin{pmatrix} \frac{1}{R} - L_f k_i I_g & -V_g \\ -\frac{1}{R} k_p + C_o k_i & k_p V_g \end{pmatrix} > 0.$$

Considering positive tuning gain parameters, $k_p, k_i > 0$, the stability conditions yield

$$k_p < \frac{C_o}{L_f I_g},$$
$$k_i < \frac{1}{R L_f I_g}.$$

Notice that the obtained bounds for k_p and k_i also determine the region for I_g where the outer loop is designed.

4.5.5.4 Simulation results

In this section, a numerical experiment of the designed control algorithm for a three-phase three-wire rectifier with an LCL filter is presented. The voltage of the power network is 230 Vrms line-to-neutral, and the parameters of the power converter are $L_g = 100\ \mu\text{H}$, $r_g = 50\ \text{m}\Omega$, $C = 20\ \mu\text{F}$, $L_f = 300\ \mu\text{H}$, $r_f = 50\ \text{m}\Omega$, and $C_o = 2$ mF.

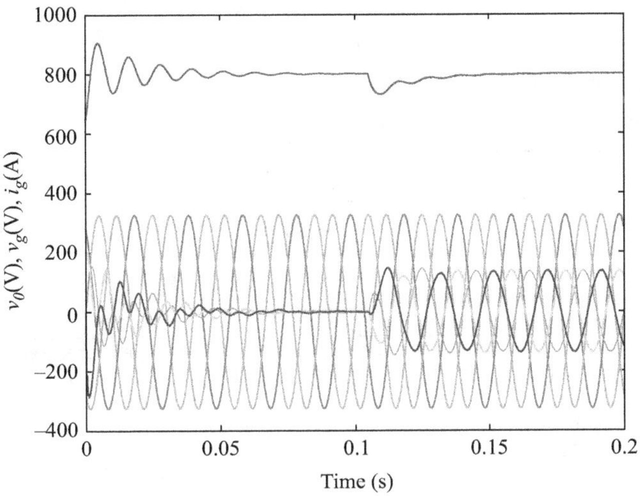

Figure 4.5.9 Simulation results: DC bus voltage, v_o, three-phase grid voltage, $\mathbf{v}_g$, and three-phase grid currents, $\boldsymbol{i}_g$

The PI control gains are $k_p = 0.75$ and $k_i = 50$. The constants of the sliding surfaces (4.5.32) and (4.5.33) have been fixed at $\tau_{a0} = \tau_{b0} = 10^8$ and $\tau_{a1} = \tau_{b1} = 2 \times 10^4$, which results in a damped dynamics with a double pole with a time constant of 100 μs in the current error. The hysteresis band has been set to ensure a switching frequency between 10 and 20 kHz. With the aim of showing the robustness in front of load variations, a sudden change from open circuit to $R = 10\ \Omega$ at $t = 105$ ms has been done. In the simulation, the output DC voltage reference has been set at 800 V, which implies approximately 64 kW when full load.

The simulation results are shown in Figure 4.5.9, where the three-phase grid currents ($\boldsymbol{i}_g$), the three-phase grid voltage ($\mathbf{v}_g$), and the output DC voltage (v_o) are plotted. From Figure 4.5.9, it can be seen that the grid voltage and current are in phase (also after the load variation) which ensures an unitary power factor, and that the DC voltage recovers its desired value even after the load change.

Figure 4.5.10 presents the steady-state behavior of the switching surfaces $\boldsymbol{s}$ and $\boldsymbol{\sigma}$. By comparing the plots, the reader can realize the decoupling effect on the surfaces.

4.5.6 Control design of an LCL three-phase four-wire rectifier

As mentioned before, in Section 4.5.3, the control objective is to regulate the total DC link voltage, $v_o = v_{o1} + v_{o2}$, to a desired value v_o^d achieving an unitary power factor. In addition, for the three-phase four-wire rectifier is required to balance the partial DC voltages, $v_{o1} = v_{o2}$. The proposed control scheme is depicted in Figure 4.5.11, with the inner current loop, that tracks the $\boldsymbol{i}_g$ three-phase currents, now with a certain

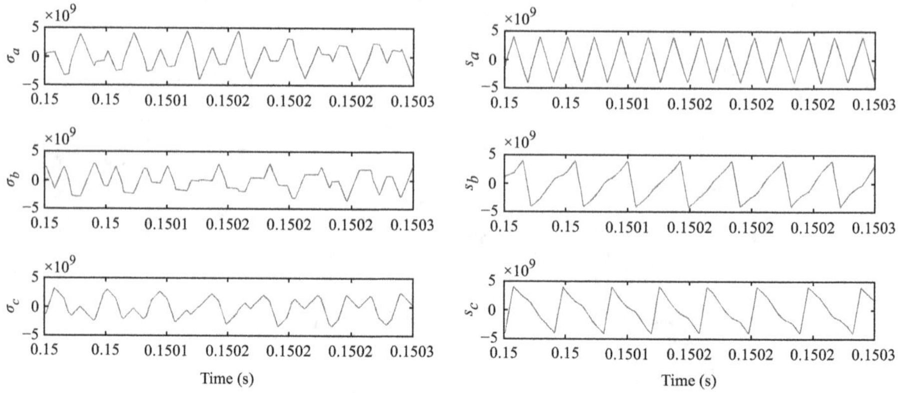

*Figure 4.5.10 Simulation results: detail of the switching functions **σ** (left) and **s** (right)*

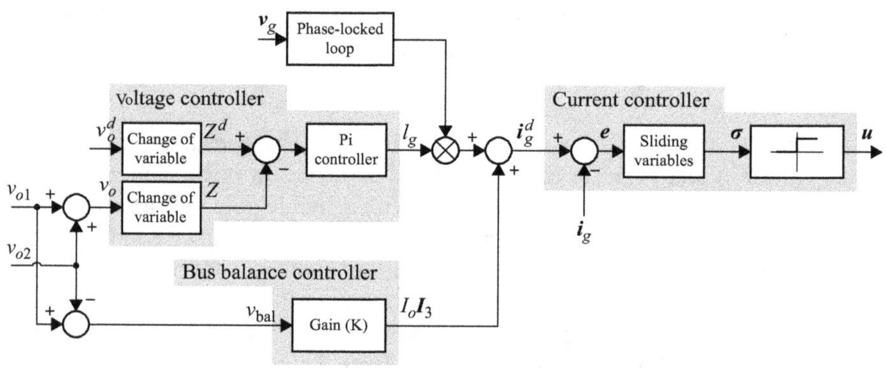

Figure 4.5.11 Proposed control scheme

amplitude, I_g and offset, I_o. The values for I_g and I_o are the outputs of the voltage controller loop and the bus balance controller, respectively.

4.5.6.1 Zero-dynamics analysis

Replacing (4.5.17) and its time derivative in (4.5.11), with (4.5.12) and (4.5.13), one gets

$$C_o \frac{dv_{o1}}{dt} = \frac{1}{v_{o1} + v_{o2}} \left(a_2 I_g^2 + a_1 I_g + a_0\right) - i_o, \tag{4.5.39}$$

$$C_o \frac{dv_{o2}}{dt} = \frac{1}{v_{o1} + v_{o2}} \left(a_2 I_g^2 + a_1 I_g + a_0\right) - i_o, \tag{4.5.40}$$

where

$$a_2 = -\frac{3}{2}(r_g + r_f) + 3\omega^2 C r_f \left(L_g - \frac{1}{2} C \left(r_g^2 + L_g^2 \omega^2 \right) \right),$$

$$a_1 = 3\omega^2 C^2 r_g r_f + \frac{3}{2} V_g,$$

$$a_0 = -\frac{3}{2}\omega^2 C^2 r_f V_g^2.$$

Following the control objectives stated in the previous section, the variables

$$v_o = v_{o1} + v_{o2}, \tag{4.5.41}$$

$$v_{\mathrm{bal}} = v_{o1} - v_{o2} \tag{4.5.42}$$

are defined. Using (4.5.39) and (4.5.40), the dynamics of v_o and v_{bal} yields

$$C_o \frac{\mathrm{d}v_o}{\mathrm{d}t} = \frac{2}{v_o}\left(a_2 I_g^2 + a_1 I_g + a_0 \right) - 2 i_o,$$

$$C_o \frac{\mathrm{d}v_{\mathrm{bal}}}{\mathrm{d}t} = 0.$$

Notice that with the proposed currents in (4.5.17) do not allow to balance the DC voltages. In order to add a new control input, the reference of the grid side currents is modified to

$$\boldsymbol{i}_g^d = I_g \begin{pmatrix} \sin(\omega t) \\ \sin\left(\omega t - \dfrac{2\pi}{3}\right) \\ \sin\left(\omega t + \dfrac{2\pi}{3}\right) \end{pmatrix} + I_o \boldsymbol{I}_3, \tag{4.5.43}$$

where I_o is a current offset included to balance the DC bus voltages. Replacing now (4.5.43) and its time derivative in (4.5.12) and (4.5.13), and using (4.5.41), (4.5.42), and $i_o = \frac{1}{R} v_o$,

$$C_o \frac{\mathrm{d}v_o}{\mathrm{d}t} = -\frac{2}{R} v_o + \frac{2}{v_o}\left(a_2 I_g^2 + a_1 I_g + a_0 \right) - \frac{3}{v_o}\left(2 b I_o^2 + v_{\mathrm{bal}} I_o \right),$$

$$C_o \frac{\mathrm{d}v_{\mathrm{bal}}}{\mathrm{d}t} = 3 I_o,$$

where

$$b = r_f + r_g.$$

4.5.6.2 Inner-loop current control

The objective of this controller is to track the input currents, $\boldsymbol{i}_g$, as proposed in (4.5.43). The control objective of the current controller is to reach $\boldsymbol{e} = 0$, where the error current is defined as $\boldsymbol{e}^T = (e_a, e_b, e_c)$ with

$$e_a = i_{ga} - i_{ga}^d,$$

$$e_b = i_{gb} - i_{gb}^d,$$

$$e_c = i_{gc} - i_{gc}^d.$$

Similarly to the single-phase case in Section 4.5.4, the error is relative degree three with respect to the control actions, and a relative degree one switching functions are considered

$$\sigma_a = \tau_{a0} e_a + \tau_{a1} \frac{\mathrm{d}e_a}{\mathrm{d}t} + \frac{\mathrm{d}^2 e_a}{\mathrm{d}t^2}, \tag{4.5.44}$$

$$\sigma_b = \tau_{b0} e_b + \tau_{b1} \frac{\mathrm{d}e_b}{\mathrm{d}t} + \frac{\mathrm{d}^2 e_b}{\mathrm{d}t^2}, \tag{4.5.45}$$

$$\sigma_c = \tau_{c0} e_c + \tau_{c1} \frac{\mathrm{d}e_c}{\mathrm{d}t} + \frac{\mathrm{d}^2 e_c}{\mathrm{d}t^2}, \tag{4.5.46}$$

where $\tau_{a0}, \tau_{a1}, \tau_{b0}, \tau_{b1}, \tau_{c0}$, and τ_{c1} are the control gains that can be selected to tune the error dynamics.

The sliding control actions can be derived using the Lyapunov function candidate

$$V = \frac{1}{2}\boldsymbol{\sigma}^T \boldsymbol{\sigma}, \tag{4.5.47}$$

where $\boldsymbol{\sigma}^T = (\sigma_a, \sigma_b, \sigma_c)$. Differentiating (4.5.47) with respect to time and replacing (4.5.11)–(4.5.13),

$$\dot{V} = \boldsymbol{\sigma}^T \boldsymbol{F}(\boldsymbol{x}) - \frac{v_o}{L_g L_f C} \boldsymbol{\sigma}^T \boldsymbol{u},$$

where $\boldsymbol{F}(\boldsymbol{x})$ contains the remaining vector field depending on all the state variables. As all the rectifier parameters are positive, and assuming $v_o > 0$, the switching action is given by

$$\boldsymbol{u} = \frac{1}{2}(\mathrm{sign}(\boldsymbol{\sigma}) + 1).$$

4.5.6.3 Outer-loop voltage control and bus balance control

The obtained dynamics assuming a perfect control tracking currents (4.5.43), simplifies in (4.5.44) and (4.5.44), i.e., the DC bus voltage balance can be independently controlled through a DC offset in the three-phase grid currents.

The proposed control for balancing the voltages is a simple gain,

$$I_o = -k_o v_{\text{bal}} \tag{4.5.48}$$

with $k_o > 0$, that transforms (4.5.44) in a linear first-order system

$$G(s) = \frac{1}{\frac{C_o}{3k_o}s + 1} \tag{4.5.49}$$

such that, k_o can be tuned to have a certain time constant.

Simultaneously, the DC bus voltage dynamics (4.5.44), with the following change of coordinates:

$$z = \frac{1}{2}v_o^2,$$

and neglecting the inductance losses $r_g = r_f = 0$ (see numerical example for a case with low losses in Reference 26) can be written as

$$C_o\frac{\mathrm{d}z}{\mathrm{d}t} = -\frac{4}{R}z + 3V_gI_g - 3v_{\text{bal}}I_o. \tag{4.5.50}$$

As $-\frac{3}{2}v_{\text{bal}}I_o$ is a vanishing term, z in (4.5.50) can be regulated by means of a PI controller, with the following form:

$$I_g = -k_p\tilde{z} - k_i\int\tilde{z}\,\mathrm{d}t,$$

where $\tilde{z} = z - z^d$ and

$$z^d = \frac{1}{2}\left(v_o^d\right)^2.$$

4.5.6.4 Simulation results

In this section, a numerical experiment of the designed control algorithm for a three-phase four-wire rectifier with an LCL filter is presented. The voltage of the power network is 230 Vrms line-to-neutral, and the parameters of the power converter are $L_g = 100\ \mu\text{H}$, $r_g = 50\ \text{m}\Omega$, $C = 20\ \mu\text{F}$, $L_f = 300\ \mu\text{H}$, $r_f = 50\ \text{m}\Omega$, and $C_o = 4$ mF.

The PI control gains of the voltage control loop are $k_p = 0.75$ and $k_i = 0.5$, and the proportional gain for balancing the DC voltages is $k_o = 0.1$. The constants of the sliding surfaces (4.5.44)–(4.5.46) have been fixed at $\tau_{a0} = \tau_{b0} = \tau_{c0} = 10^8$ and $\tau_{a1} = \tau_{b1} = \tau_{c1} = 2 \times 10^4$, which results in a damped dynamics with a double pole with a time constant of 100 μs in the current error. The hysteresis band has been set to ensure a switching frequency between 10 and 20 kHz. With the aim of showing the

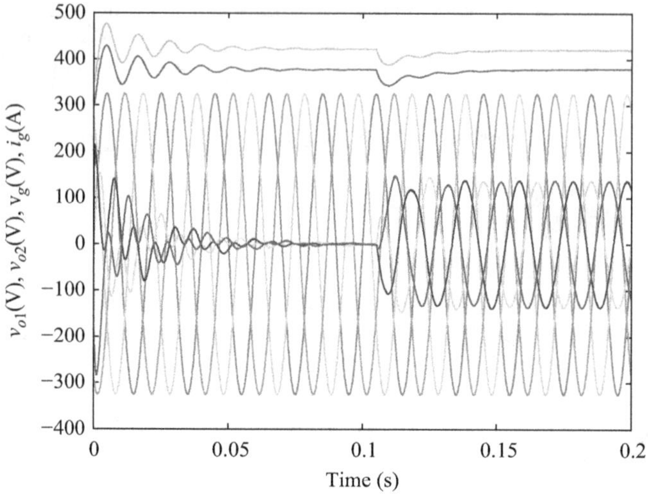

Figure 4.5.12 Simulation results: DC bus voltages, v_{o1} and v_{o2}, three-phase grid voltage, $\boldsymbol{v}_g$, and three-phase grid currents, $\boldsymbol{i}_g$, without the bus balance control

robustness in front of load variations, a sudden change from open circuit to $R = 10\ \Omega$ at $t = 105$ ms has been done. In all simulations, the output DC voltage reference has been set at 800 V, which implies approximately 64 kW when full load. In all the performed simulations the output capacitors have the following initial conditions: $v_{o1}(0) = 300$ V and $v_{o2}(0) = 350$ V.

The simulation results are shown in Figures 4.5.12–4.5.14, where the following variables are plotted: the three-phase grid current ($\boldsymbol{i}_g$), the three-phase grid voltage ($\boldsymbol{v}_g$), the three-phase filter current ($\boldsymbol{i}_f$) and the output DC voltages (v_o, v_{01}, v_{o2}).

In Figure 4.5.12, the behavior of the three-phase four-wire rectifier without the voltage balancing loop (4.5.48) is presented. It can be seen that the grid currents, $\boldsymbol{i}_g$, are in phase with the grid voltages, $\boldsymbol{v}_g$, and the total DC bus voltage is regulated at the desired value. However, after the transient, the voltages in the DC link remain unbalanced.

Figures 4.5.13 and 4.5.14 show the behavior of the closed-loop system with the DC link voltage balance control. In Figure 4.5.13, the grid currents, $\boldsymbol{i}_g$, are still in phase with the grid voltages, $\boldsymbol{v}_g$, together with a good regulation of the total DC bus voltage, but now the split voltages v_{o1} and v_{o2} are also balanced. The first-order behavior expected in (4.5.49) is shown in Figure 4.5.14.

4.5.7 Conclusions

This work presents a family of controllers for full-bridge rectifiers with an LCL filter. The control algorithms have two parts: an inner current control loop (designed using SMC techniques) which ensures an unitary power factor, and an external voltage

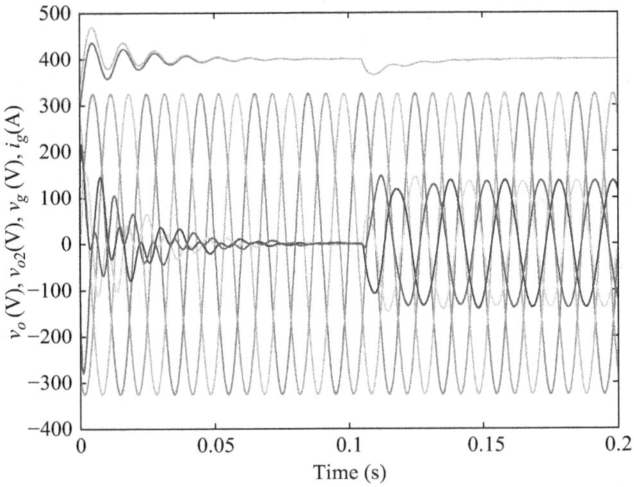

Figure 4.5.13 Simulation results: DC bus voltages, v_{o1} and v_{o2}, three-phase grid voltage, $\mathbf{v}_g$, and three-phase grid currents, $\mathbf{i}_g$, with the bus balance control

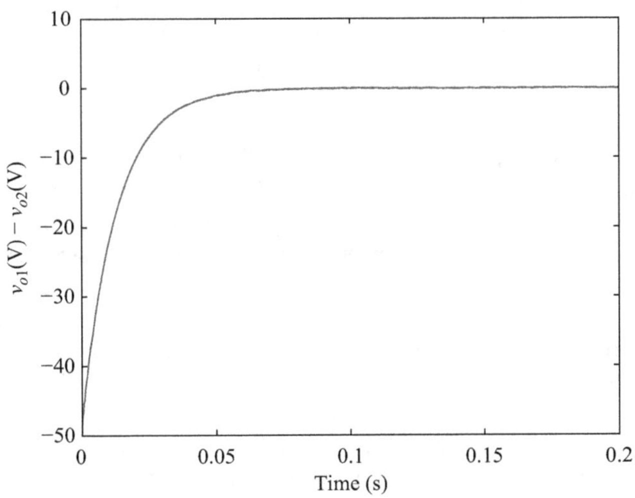

Figure 4.5.14 Simulation results: DC bus voltages, balance $v_{o1} - v_{o2}$, with the bus balance control

controller that, after an accurate dynamic analysis, is designed as a simple PI controller to regulate the DC voltage. The outer voltage loop become slightly different depending on the power converter topology; mainly, in the single-phase case, the outer loop regulates the average of DC voltage while in the four-wire case, a balancing term for the voltage of the DC split capacitors is introduced. The main feature of the three-phase three-wire rectifier controller is the use of a decoupling matrix in order to solve the current coupling of the three-phase currents. This results in a switching control action that depends on the linear combination of the "original" switching functions and the L_g, C, and L_f parameters of the LCL filter.

In all cases, the whole control law results in a very simple control algorithm (and ease of implementation), that only requires the current and voltage measures at the point of connection.

The obtained control laws have been numerically simulated. The results validate the control designs, show a good transient performances and are robust in front of load variations as well as wrong estimations on the LCL parameters. The main drawback for a future implementation could be the computation of a second time derivative in the switching functions σ_a and σ_b in (4.5.44) and (4.5.45), respectively.

References

[1] M. Malinowski and S. Bernet, "A simple voltage sensorless active damping scheme for three-phase PWM converters with an LCL filter," *IEEE Transactions on Industrial Electronics*, vol. 55, no. 4, pp. 1876–1880, 2008.

[2] J. Dannehl, C. Wessels, and F. Fuchs, "Limitations of voltage-oriented pi current control of grid-connected PWM rectifiers with LCL filters," *IEEE Transactions on Industrial Electronics*, vol. 56, no. 2, pp. 380–388, 2009.

[3] M. Liserre, F. Blaabjerg, and S. Hansen, "Design and control of an LCL-filter-based three-phase active rectifier," *IEEE Transactions on Industry Applications*, vol. 41, no. 5, pp. 1281–1291, 2005.

[4] Y.-R. Mohamed, M. Rahman, and R. Seethapathy, "Robust line-voltage sensorless control and synchronization of LCL-filtered distributed generation inverters for high power quality grid connection," *IEEE Transactions on Power Electronics*, vol. 27, no. 1, pp. 87–98, 2012.

[5] E. Wu and P. Lehn, "Digital current control of a voltage source converter with active damping of LCL resonance," *IEEE Transactions on Power Electronics*, vol. 21, no. 5, pp. 1364–1373, 2006.

[6] J. Dannehl, F. Fuchs, and P. Thogersen, "PI state space current control of grid-connected PWM converters with LCL filters," *IEEE Transactions on Power Electronics*, vol. 25, no. 9, pp. 2320–2330, 2010.

[7] Y. Tang, P. Loh, P. Wang, F. Choo, and K. Tan, "Improved one-cycle-control scheme for three-phase active rectifiers with input inductor-capacitor-inductor filters," *IET Power Electronics*, vol. 4, no. 5, pp. 603–614, 2011.

[8] M. Orellana and R. Griñó, "Digital AFC control of a three-phase three-wire unity-power-factor PWM rectifier," in *Proceedings of the Ninth Asian Control Conference*, Istanbul, Turkey, 2013.

[9] N. He, D. Xu, Y. Zhu, J. Zhang, G. Shen, Y. Zhang, J. Ma, and C. Liu, "Weighted average current control in a three-phase grid inverter with an LCL filter," *IEEE Transactions on Power Electronics*, vol. 28, no. 6, pp. 2785–2797, 2013.

[10] X. Bao, F. Zhuo, Y. Tian, and P. Tan, "Simplified feedback linearization control of three-phase photovoltaic inverter with an LCL filter," *IEEE Transactions on Power Electronics*, vol. 28, no. 6, pp. 2739–2752, 2013.

[11] A. Marzouki, M. Hamouda, and F. Fnaiech, "Nonlinear control of three-phase active rectifiers based L and LCL filters," in *Proceedings of the International Conference on Electrical Engineering and Software Applications*, Hammamet, Tunisia, 2013.

[12] X. Zhu, H. Li, Y. Peng, and X. Shi, "A control scheme of current-source PWM rectifier with sliding-mode control under unbalanced operation conditions," in *Proceedings of the Third International Conference on Electric Utility Deregulation and Restructuring and Power Technologies*, NanJing, China, 2008.

[13] E. Alarcón-Gallo, L. G. de Vicuña, M. Castilla, J. Miret, J. Matas, and A. Camacho, "Decoupled sliding mode control for three-phase LCL VSI operating at fixed switching frequency," in *Proceedings of the IEEE International Symposium on Industrial Electronics*, Hangzhou, China, 2012.

[14] S. Jiang, Y. Li, X. Lu, B. Ge, and F. Peng, "Practical control implementation for 100 kVA three-phase four-wire online voltage regulator," in *Proceedings of the 26th IEEE Applied Power Electronics Conference and Exposition*, Fort Worth, USA, 2011.

[15] S. Mohan, H. Guldner, R. Briest, and H. Wolf, "Analysis and control aspects of harmonic distortion in the front-end three-phase four-wire PWM boost rectifier," in *Proceedings of the European Conference on Power Electronics and Applications*, Dresden, Germany, 2005.

[16] O. Vodyakho and C. Mi, "Three-level inverter-based shunt active power filter in three-phase three-wire and four-wire systems," *IEEE Transactions on Power Electronics*, vol. 24, no. 5, pp. 1350–1363, 2009.

[17] S. Petterson, M. Salo, and H. Tuusa, "Applying an LCL-filter to a four-wire active power filter," in *Proceedings of the 37th IEEE Power Electronics Specialists Conference*, Jeju, South Korea, 2006.

[18] J. A. Barrado, R. Griñó, and H. Valderrama-Blavi, "Power-quality improvement of a stand-alone induction generator using a STATCOM with battery energy storage system," *IEEE Transactions on Power Delivery*, vol. 25, no. 4, pp. 2734–2741, 2010.

[19] R. Venkataramanan, Sliding mode control of power converters. Ph.D. Dissertation, California Institute of Technology, Pasadena, CA, 1986.

[20] D. Biel, A. Dòria-Cerezo, and E. Fossas, "Sliding mode control of a single-phase LCL full-bridge rectifier," in *Proceedings of the 39th Annual Conference of the IEEE Industrial Electronics Society*, Vienna, Austria, 2013.

[21] X. Hao, X. Yang, T. Liu, L. Huang, and W. Chen, "A sliding-mode controller with multiresonant sliding surface for single-phase grid-connected VSI with an LCL filter," *IEEE Transactions on Power Electronics*, vol. 28, no. 5, pp. 2259–2268, 2013.

[22] R. Griñó, E. Fossas, and D. Biel, "Sliding mode control of a full-bridge unity power factor rectifier," *Lecture Notes in Control and Information Sciences*, vol. 281, pp. 139–148, Springer, Berlin, 2002.

[23] C. Batlle, A. Dòria-Cerezo, and E. Fossas, "Bidirectional power flow control of a power converter using passive Hamiltonian techniques," *International Journal of Circuit Theory and Applications*, vol. 36, no. 7, pp. 769–788, 2008.
[24] V. Utkin, J. Guldner, and J. Shi, *Sliding Mode Control in Electromechanical Systems*. Taylor & Francis, London, 1999.
[25] D. Biel, A. Dòria-Cerezo, V. Repecho, and E. Fossas, "Sliding mode control of a three-phase three-wire LCL rectifier," in *Proceedings of the IEEE International Symposium on Industrial Electronics*, Istanbul, Turkey, 2014.
[26] D. Biel, A. Dòria-Cerezo, and E. Fossas, "Sliding mode control of a three-phase four-wire LCL rectifier," in *Proceedings of the 13th International Workshop on Variable Structure Systems*, Nantes, France, 2014.

Chapter 4.6

Adaptive solutions for robust control of electropneumatic actuators

Franck Plestan, Xinming Yan, Mohammed Taleb, and Antonio Estrada

Abstract

This chapter is focused on the design of robust controllers for electropneumatic actuators. This kind of systems is highly nonlinear, and their dynamics is uncertain due to frictions, uncertainties (especially in mass flow rate), and external perturbation. The use of robust controllers is then strongly recommended in order to get high performances for trajectories tracking or stabilization. This chapter presents very recent robust controllers based on sliding mode theory and adaptive gain. Through the application of control laws mixing high-order sliding mode and adaptive gain, the objective is to show their applicability to a real system and to compare their performances by using similar experimental benchmark.

4.6.1 Introduction

Pneumatic actuator is a system which is quite hard to accurately control, because its dynamics is a typical uncertain system. The uncertainties are caused by friction (external) perturbations and parametric uncertainties due to a very tedious identification process (e.g. the mass flow rate, which is a key-data to estimate the pressure in the actuator chamber, is very difficult to be estimated). Then, since several years, robust nonlinear controllers and (standard or high-order) sliding modes are proposed in many works [2, 3, 7, 11, 13, 14, 20] in order to get high performance. Even if they allow to get high-performances results, these works have a drawback: they require the knowledge of the uncertainties/perturbations bounds which is a very hard task given the nature of the system. These difficulties give an overestimation of these bounds, and then an overestimation of the gain inducing larger chattering phenomenon. In this case, there is a real interest to develop controllers which can be robust even if there is no information on uncertainties/perturbations. This feature can be reached by gain adaptation.

These last years, adaptive sliding mode control design has been intensively studied. In Reference 9, gain dynamics is a function of the tracking error (sliding variable): the control gain is increasing since sliding mode is not established. However, this approach induces overestimation of the gain because the gain can never be reduced. In Reference 12, the equivalent control concept is used: once sliding mode occurs, disturbance magnitude is evaluable and allows an adequate tuning of control gain. But the controller design requires the knowledge of uncertainties/perturbations bounds and the use of low-pass filter, which introduces signal magnitude attenuation, delay, and transient behavior when disturbances are acting. In Reference 8, it is proposed to use a sliding mode disturbance observer in the gain-adaptation algorithm; however, the knowledge of uncertainties bounds is required to design observer-based controller.

The main previously mentioned drawbacks have been removed in Reference 17 (and its MIMO version in Reference 18): the designed control requires no information on the uncertainties/perturbations, and the gain dynamics induces no overestimation of it. Furthermore, this approach has shown its applicability through its experimental application on electropneumatic actuator. However, accuracy and chattering limitation can be improved thanks to the use of high- (at least second-)order sliding mode controls.

This chapter is focused on the design of second-order sliding mode controllers with gain adaptation. These controllers have been proposed in very recent works, in which they have been applied to electropneumatic system. These three controllers are adaptive version of twisting [21] and super-twisting [19] algorithms and an adaptive second-order sliding mode output feedback controller [6]. Then, this chapter gives a state of the art of adaptive second-order sliding mode controllers, the novelty being that the experimentations have been on a new single benchmark (original desired position trajectory and external perturbations) in order to compare the performances of the three controllers.

This chapter is organized as following. Section 4.6.2 describes the electropneumatic system and the model. In Section 4.6.3, a second-order sliding mode controller with adaptive gains based on twisting control is presented, a sliding variable with relative degree equal to two being defined. In Section 4.6.4, with the same sliding variable, an adaptive second-order sliding mode output feedback controller is also applied on the electropneumatic system, which does not use time derivative of the sliding variable. In Section 4.6.5, an adaptive version of super-twisting algorithm is introduced. This last method is applicable to the system with relative degree equals to one; so a new sliding variable is defined. At last, the performances of these three controllers on the electropneumatic system are compared in Section 4.6.6.

4.6.2 Electropneumatic system

All the experimental tests presented in this chapter are realized on the electropneumatic system [15] set of *IRCCyN* lab, Nantes, France. The electropneumatic system (see Figures 4.6.1 and 4.6.2) has a software architecture based on a dSpace board. It is composed of two actuators. The first one, named the "main actuator," is a double acting pneumatic actuator controlled by two servodistributors and is composed of two chambers denoted P and N. The piston diameter is 80 mm, and the rod diameter

Figure 4.6.1 *Photo of electropneumatic system. On the left hand side is the "main" actuator whose position and pressure in a chamber can be controlled. On the right hand side is the "perturbation" actuator which provides varying load force*

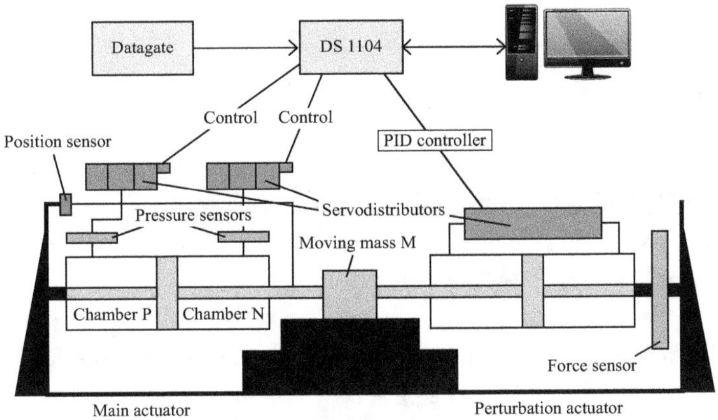

Figure 4.6.2 *Scheme of electropneumatic system*

is 25 mm. The source pressure is 7 bar, and the maximum force developed by the actuator is 2720 N. The air mass flow rates entering in the chambers are modulated by two three-way servodistributors. The pneumatic jack horizontally moves a load carriage of mass M. This carriage is coupled with the second pneumatic actuator, referred as the "perturbation actuator." This latter produces a dynamical load force on the main actuator. Pressures of both chambers and carriage position are measured, whereas velocity is derived by a differentiator. In the sequel, only the control of the main actuator position is considered.

The pneumatic plant model is obtained from three physical laws: the mass flow rate under a restriction, the pressure behavior in a chamber with variable volume, and the fundamental mechanical equation. The complete model of the experimental set-up reads as [21]

$$
\begin{aligned}
\dot{p}_P &= \frac{krT}{V_P(y)}\left[\alpha_P + \beta_P \cdot w - \frac{S}{rT} p_P v\right], \\
\dot{p}_N &= \frac{krT}{V_N(y)}\left[\alpha_N - \beta_N \cdot w + \frac{S}{rT} p_N v\right], \\
\dot{v} &= \frac{1}{M}\left[S(p_P - p_N) - b_v v - F_{ext}\right], \\
\dot{y} &= v
\end{aligned}
\tag{4.6.1}
$$

with y the main actuator piston position, v its velocity, p_P, and p_N the pressures in the both chambers (respectively, P and N chambers). F_{ext} is the external force produced by the "perturbation" actuator, and w is the control input of the system. The volume of each chamber is defined as

$$
\begin{aligned}
V_P(y) &= V_0 + S \cdot y, \\
V_N(y) &= V_0 - S \cdot y.
\end{aligned}
\tag{4.6.2}
$$

Functions α_* and β_* ($*$ being N or P) are defined as fifth-order polynomials with respect to p_X [1] such that the mass flow rate q_{m*} reads as

$$
q_{m*}(w, p_X) = \alpha_* \left(p_*\right) + \beta_* \left(p_*, \text{sign}\,(w)\right) w. \tag{4.6.3}
$$

Table 4.6.1 displays the values of the physical parameters of the experimental set-up. By denoting $x = [p_P \; p_N \; v \; y]^T$, one gets a system in the form of

$$
\dot{x} = f(x) + g(x)w \tag{4.6.4}
$$

Table 4.6.1 Physical parameters of the experimental set-up

Parameters	**Notation**	**Value**
Mass	M	3.4 kg
Piston surface	S	0.0045 m^2
Half-cylinder volume	V_0	3.40 10^{-4} m^3
Perfect gas constant	r	287 J kg^{-1} K^{-1}
Temperature	T	20°C
Polytropic constant	k	1.2
Viscous friction coefficient	b_v	50

with

$$f(x) = \begin{bmatrix} \frac{krT}{V_P(y)}\left[\alpha_P - \frac{S}{rT}p_P v\right] \\ \frac{krT}{V_N(y)}\left[\alpha_N + \frac{S}{rT}p_N v\right] \\ \frac{1}{M}\left[S(p_P - p_N) - b_v v - F_{ext}\right] \\ v \end{bmatrix}, g(x) = \begin{bmatrix} \frac{krT}{V_P(y)}\beta_P \\ -\frac{krT}{V_N(y)}\beta_N \\ 0 \\ 0 \end{bmatrix}.$$

The working domain denoted $\mathscr{X}$ is such that

$$|y| \leq 72\text{ mm},\ |v| \leq 1\text{ m s}^{-1},\ 1\text{ bar } \leq p_P, p_N \leq 7\text{ bar},\ |u| \leq 10\text{ V}.$$

The aim of the control law is to get a high accuracy in terms of position tracking of the desired trajectory, in spite of the perturbation force. In the following sections, three second-order sliding mode controllers based in different strategies and with adaptive gains will be presented.

4.6.3 Adaptive twisting controller

The so-called "Twisting" algorithm [10] is well-known as a second-order sliding mode controller, which is applicable to nonlinear systems with relative degree equal to 1 or 2. For its standard version, the gain should be tuned based on the bounds of the uncertainties and perturbations. Then, an overestimation is unavoidable, which may increase the chattering problem. In Reference 21, an adaptive twisting control (denoted "ATW" in the sequel) is proposed. The idea is to dynamically adjust the control gain by checking online whether a second-order sliding mode is established or not. The adaptation law is based on a discrete time criterion which evaluates the occurrence of a real second-order sliding mode.

4.6.3.1 Sliding variable definition

Consider the nonlinear uncertain system (4.6.4). The sliding variable $\sigma(x,t)$ is defined as

$$\sigma(x,t) = v - \dot{y}_{ref}(t) + \lambda\left(y - y_{ref}(t)\right) \tag{4.6.5}$$

with $\lambda > 0$; the relative degree of (4.6.4) with respect to σ equals 2. From (4.6.5), one gets

$$\ddot{\sigma} = \Psi(t,x) + \Phi(x,t)w \tag{4.6.6}$$

where

$$\begin{aligned} \Psi &= \Psi_{Nom} + \Delta_\Psi \\ \Phi &= \Phi_{Nom} + \Delta_\Phi \end{aligned} \tag{4.6.7}$$

$$\begin{aligned}\Psi &= \frac{krTS}{M}\left(\frac{\alpha_P}{V_P}-\frac{\alpha_N}{V_N}\right)-\frac{kS^2}{M}\left(\frac{p_P}{V_P}+\frac{p_N}{V_N}\right)v\\ &\quad+\frac{1}{M}\left(-b_v\dot{v}-\dot{F}_{ext}-y_{ref}^{(3)}\right)+\lambda\left(\dot{v}-\ddot{y}_{ref}\right)\\ \Phi &= \frac{krTS}{M}\left(\frac{\beta_P}{V_P}-\frac{\beta_N}{V_N}\right).\end{aligned} \tag{4.6.8}$$

In Reference 21, it has been numerically shown that, under current operating conditions, the bounded functions Ψ_{Nom} and Φ_{Nom} only depend on the measured variables, and $\Phi_{Nom} > 0$. Then, consider the following control law:

$$w = \frac{1}{\Phi_{Nom}}(-\Psi_{Nom}+u) \tag{4.6.9}$$

with u being the "new" control input of σ-dynamics. Applying (4.6.9) to (4.6.6)–(4.6.7), one gets

$$\ddot{\sigma} = \underbrace{\Delta_\Psi - \Delta_\Phi\Phi_{Nom}^{-1}\Psi_{Nom}}_{a} + \underbrace{\left(1+\Delta_\Phi\Phi_{Nom}^{-1}\right)}_{b}\cdot u. \tag{4.6.10}$$

The functions a and b fulfill the following assumptions for $x \in \mathscr{X}$.

A1. The function a is a bounded perturbation such that $|a| < a_M$.

A2. The function b reading as $b = b_N + \Delta b$ (nominal part plus uncertainties) is an uncertain function and $0 < b_m < \Delta b < b_M$ with b_m and b_M unknown.

A3. The time derivatives of a and b are bounded.

Then, the objective of the control law u (and then w) is to establish a real second-order sliding mode (defined as follows) with respect to σ in a finite time in spite of the perturbations and uncertainties.

Definition 4.6.1. *[10] A real second-order sliding mode is established with respect to the sliding variable σ if, after some finite time*

$$|\sigma| < \mu_0 T_e^2, \quad |\dot{\sigma}| < \mu_1 T_e$$

for some positive μ_0, μ_1, and T_e the sampling period.

4.6.3.2 Adaptive control law design [21]

Consider system (4.6.10) and an adaptive version of the twisting algorithm reading as

$$u = -K(t)\left(\text{sign}(\sigma)+\gamma\,\text{sign}(\dot{\sigma})\right),\ 0.5 \le \gamma \le 1 \tag{4.6.11}$$

with γ a constant control parameter.

The idea of the gain adaptation consists in increasing the gain K until a real second-order sliding mode with respect to σ is detected. Then, the gain K is gradually reduced following the idea that it is enough large, viewed that the sliding mode is established. At a moment, the gain K could become too much small with respect to perturbations and uncertainties: in this case, the second-order sliding mode is lost. Then, the gain K is force to increase until the second-order sliding mode is detected. From that, the gain adaptation law design is made in two steps. The first one consists in designing the second-order sliding mode detector, whereas the second one concerns the definition of the gain dynamics.

Sliding mode detector. The idea is to define a criterion for the detection of the real second-order sliding mode with respect to σ. Consider a natural number N_T and some $\mu > 0$. Define

$$\delta(t) = \begin{cases} 1 & \text{if } \forall t_j \in [t - N_T T_e, t] : |\sigma(t_j)| \leq \mu K(t_j) T_e^2, \\ -1 & \text{if } \exists t_j \in [t - N_T T_e, t] : |\sigma(t_j)| > \mu K(t_j) T_e^2, \end{cases} \tag{4.6.12}$$

where t_j are the sampling instants. The second-order sliding mode criterion is satisfied if $\delta = 1$. Note that this detector is based on Definition 4.6.1: one checks that the sliding variable is kept in a vicinity of 0 depending explicitly on T_e^2.

Gain adaptation law. The adaptation law is composed of two components. The first one is based on a continuous differential equation allowing to increase/reduce the gain, whereas the second component is based on a discrete equation allowing to make an instantaneous increment of the gain when the second-order sliding mode is lost. This latter component allows to accelerate the increase of the gain, the objective being to reduce the time taken in order to recover as soon as possible the second-order sliding mode.

Let the gradual adaptation law be

$$\dot{K} = \begin{cases} -\delta\lambda K & \text{if } K > K_m; \\ -\delta\lambda_m & \text{if } K_{mm} < K \leq K_m,\ K(0) \geq K_{mm}; \\ \lambda_m & \text{if } K \leq K_{mm}, \end{cases} \tag{4.6.13}$$

where λ and λ_m are positive adaptation parameters. The two first equations allow to increase/decrease (it depends on the state of the detector δ) the gain by two different manners: if K is larger than K_m (fixed by the user), the increase/decrease dynamics is exponential: if the sliding mode is established, the idea is to reduce as fast as possible the gain $K(t)$ in order to reduce the magnitude and the chattering of the control. If the sliding mode is not established, the exponential convergence allows to quickly increase the gain in order to counteract as soon as possible perturbations and uncertainties. If K is smaller than K_m, the increase/decrease dynamics is linear, the idea being to evolve by a smoother way. The third equation is used in order to ensure that the gain never becomes smaller than K_{mm} (fixed by the user). This latter parameter can be chosen very small.

In addition, an instant increment is implemented at each sampling instant t_i at which the second-order sliding mode criterion is violated, i.e.,

$$K(t_i) = qK(t_i^-) \quad \text{if } \delta(t_i^-) = 1 \ \&\ \delta(t_i) = -1 \tag{4.6.14}$$

with

$$K(t_i^-) = \lim_{t \to t_i} K(t), \quad t < t_i.$$

In other words, the gain $K(t_i^-)$ is the limit of $K(t)$ when t tends to t_i from the left. Finally, one of the main results of Reference 21 which concerns an adaptive version of the twisting algorithm is given through the following theorem.

Theorem 4.6.1 ([21]). *Consider system* (4.6.10) *with Assumptions A1–A3 fulfilled and controlled by* (4.6.11)–(4.6.14).

Then, for any sufficiently large λ *and* μ *and sufficiently large* $N_T \geq 4$ *(chosen after* μ*) with sufficiently small* T_e*, the parameter* $K(t)$ *features local maxima which do not exceed* $q|a|/b$*. Furthermore, the controller ensures the establishment of a real second-order sliding mode with respect to* σ *in a finite time.* □

4.6.4 Adaptive output feedback controller

As described in the previous section, the ATW controller requires both the sliding variable and its time derivative. In practice, due to noisy measurements, the use of differentiators can lead to a degradation of the controller performances. Then, in order to reduce this problem, a second-order sliding mode output feedback controller is proposed in References 4–6. With this method, the use of first-order time derivative of the sliding variable in the controller can be removed. Note that this class of controllers can be applied to uncertain nonlinear systems with a relative degree equal to 1 or 2 with respect to the sliding variable. Furthermore, an adaptation law is used to dynamically tune the gain, which drives to an adaptive output feedback controller (denoted "AOF" in the sequel) based on second-order sliding mode concept.

4.6.4.1 Sliding variable definition

As previously denoted, this class of controllers is applicable to uncertain nonlinear systems with relative degree equal to 2 or 1 with respect to the sliding variable. Then, in the sequel, a similar definition of sliding variable (4.6.5) is used. Consider also that Assumptions A1–A3. The objective of the proposed AOF controller is to establish a real second-order sliding mode with respect to σ in a finite time, without using any information of $\dot{\sigma}$, which is equivalent to stabilize the system (4.6.10) with a controller depending only on σ, i.e., stabilization of the system

$$\ddot{\sigma} = a(\cdot) + b(\cdot)u(\sigma). \tag{4.6.15}$$

4.6.4.2 Adaptive output feedback control law design [6]

The proposed controller displayed in the sequel is proposed in a sampled context in order to be realistic for applications, i.e., the control input is updated with a period T_e, and the value of the control input is constant over each sampling period. By a similar way, the measurement of σ is also synchronized on the sampling period. It means that, for example, the change of sign of σ can be detected with a delay lower or equal to T_e; it has been shown in Reference 16 that, this delay can induce divergence: a second-order sliding mode is never obtained. In Reference 16, a very simple control methodology has been proposed by using the knowledge of σ-sign, and a switching gain strategy: the control gain, tuned at a nominal value, is switching during a single sampling period, to a larger value at an adequate time, and then is evolving between two values, a nominal one and a large one. The commutation of the gain allows to "compensate" the potential delay due to sampling period and then ensures the convergence of the system. However, whereas the tuning of the nominal gain is made from the bounds of the uncertainties, a value for the large gain guaranteeing the convergence of the closed-loop system, is hardly determined. Then, to overcome this latter drawback, an adaptive approach has been improved by [6] which is detailed in the sequel.

Consider the following control law defined as

$$u = -K(t) \cdot \text{sign}(\sigma(kT_e)) \tag{4.6.16}$$

with the gain switching between two values

$$K(t) = \begin{cases} K_m & \text{if} \quad t \notin \mathcal{T} \\ \gamma K_m & \text{if} \quad t \in \mathcal{T} \end{cases} \tag{4.6.17}$$

with $K_m > 0$, $\gamma > 1$, and

$$\begin{aligned} \mathcal{T} = \{ & t \mid (k-1)T_e \leq t < kT_e,\ k \in N,\ k \geq 2 \\ & \text{and } \text{sign}(\sigma((k-2)T_e)) \neq \text{sign}(\sigma((k-1)T_e))\}. \end{aligned} \tag{4.6.18}$$

It means that the gain K is such that it takes a larger value γK_m during the sampling period following the detected commutation of the σ-sign. In the rest of time, the gain equals K_m. In the sequel, an adaptive law concerning K_m is described which limits the gain in case of reduced perturbations or uncertainties and also reduces the required knowledge of uncertainties/perturbations bounds. Let the gain $K_m(t)$ fulfill the following rules

- $K_m(t)$ is constant over each sampling period T_e;
- $T_s^0 = 0$, $K_m(0) > 0$;
- the adaptation law of K_m reads as

  $$\dot{K}_m = \Gamma \cdot \text{sign}(t - T_s^i - \eta T_e) \tag{4.6.19}$$

  with $\Gamma > 0$, $\eta > 1$, and T_s^i ($i \in N$, $i > 0$) being the ith sampled time at which a sign commutation of σ is detected.

Theorem 4.6.2 ([6]). *Consider system* (4.6.15) *with Assumptions A1–A3 fulfilled and controlled by* (4.6.16)–(4.6.19) *with γ large enough. Then, the controller ensures the establishment of a real second-order sliding mode with respect to σ in a finite time for a bounded gain $K_m(t)$.* □

4.6.5 Adaptive super-twisting controller

An other popular second-order sliding mode control law is the super-twisting algorithm [10] which is applicable to uncertain nonlinear systems with relative degree equals to one with respect to the sliding variable. In spite of this limit, this algorithm appears very interesting because, as the control law presented in the previous section, it requires only the sliding mode information. However, the main disadvantage of the standard version of the super-twisting control algorithm is that it requires the knowledge of the perturbations/uncertainties bounds. As previous controllers, this constraint implies that the tuning is made in the worst case, and then, in many cases and for the most part of the time, the gain is overestimated. In order to attenuate this drawback, in Reference 19, the super-twisting algorithm and gain adaptation are brought together. In this section, an adaptive version of super-twisting controller (denoted "ASTW") is presented for the control of the electropneumatic actuator position.

4.6.5.1 Sliding variable definition

The super-twisting control is applicable for uncertain nonlinear systems with a relative degree equals to one. So, the sliding variable $\sigma(x,t)$ has to be different than the one used for ATW and AOF controllers. Define the sliding variable σ as

$$\sigma = \omega_n^2(y - y_{ref}(t)) + 2\zeta\omega_n(\dot{y} - \dot{y}_{ref}(t)) + (\ddot{y} - \ddot{y}_{ref}(t)) \tag{4.6.20}$$

with ζ and ω_n the parameters of the sliding variable, with ζ the damping coefficient and ω_n the proper pulsation of the second-order system describing the behavior of the closed-loop system once $\sigma = 0$. From (4.6.4)–(4.6.20), one gets

$$\dot{\sigma} = \bar{\Psi}(t,x) + \bar{\Phi}(x,t)w \tag{4.6.21}$$

with

$$\begin{aligned} \bar{\Psi} &= \bar{\Psi}_{Nom} + \bar{\Delta}_{\Psi}, \\ \bar{\Phi} &= \bar{\Phi}_{Nom} + \bar{\Delta}_{\Phi} \end{aligned} \tag{4.6.22}$$

and

$$\begin{aligned} \bar{\Psi} &= \frac{krTS}{M}\left(\frac{\alpha_P}{V_P} - \frac{\alpha_N}{V_N}\right) - \frac{kS^2}{M}\left(\frac{p_P}{V_P} + \frac{p_N}{V_N}\right)v \\ &\quad + \frac{1}{M}\left(-b_v\dot{v} - \dot{F}_{ext} - y_{ref}^{(3)}\right) + \omega_n^2\left(\dot{v} - \ddot{y}_{ref}\right) + 2\zeta\omega_n\left(\dot{a} - y_{ref}^{(3)}\right), \\ \bar{\Phi} &= \frac{krTS}{M}\left(\frac{\beta_P}{V_P} - \frac{\beta_N}{V_N}\right). \end{aligned} \tag{4.6.23}$$

The bounded functions $\bar{\Psi}_{Nom}$ and $\bar{\Phi}_{Nom}$ only depend on measured variables, and $\bar{\Phi}_{Nom} > 0$. Then, by a similar way as previous sections, consider the following control law

$$w = \frac{1}{\bar{\Phi}_{Nom}} \left(-\bar{\Psi}_{Nom} + u\right) \tag{4.6.24}$$

with u being the "new" control input of σ-dynamics. Applying (4.6.24) to (4.6.21), one gets

$$\dot{\sigma} = \underbrace{\bar{\Delta}_{\Psi} - \bar{\Delta}_{\Phi}\bar{\Phi}^{-1}_{Nom}\bar{\Psi}_{Nom}}_{\bar{a}} + \underbrace{\left(1 + \bar{\Delta}_{\Phi}\bar{\Phi}^{-1}_{Nom}\right)}_{\bar{b}} \cdot u. \tag{4.6.25}$$

The functions $\bar{a}$ and $\bar{b}$ fulfill the following assumptions for $x \in \mathscr{X}$.

A4. The function $\bar{a}$ is a bounded perturbation such that $|\bar{a}| < \bar{a}_M$.

A5. The function $\bar{b}$ reading as $\bar{b} = \bar{b}_N + \bar{\Delta}b$ (nominal part plus uncertainties) is an uncertain function and $0 < \bar{b}_m < \bar{\Delta}b < \bar{b}_M$ with $\bar{b}_m$ and $\bar{b}_M$ unknown.

A6. The time derivatives of $\bar{a}$ and $\bar{b}$ are bounded.

4.6.5.2 Control law design

Define the control input u as

$$\begin{aligned} u &= -K(t)|\sigma|^{1/2}\text{sign}(\sigma) + v, \\ \dot{v} &= -\frac{k(t)}{2}\text{sign}(\sigma), \end{aligned} \tag{4.6.26}$$

the gains $K(t)$ and $k(t)$ being defined by

$$\dot{K} = \begin{cases} \omega_1\sqrt{\frac{\gamma_1}{2}}\text{sign}(|\sigma| - \mu), & \text{if } K(t) > K_{min}; \\ \eta, & \text{if } K(t) \le K_{min} \end{cases} \tag{4.6.27}$$

and

$$k(t) = 2\varepsilon K(t). \tag{4.6.28}$$

Note that the first equation of (4.6.27) consists in increasing (resp. decreasing) the gain if the sliding variable is in (resp. out) a vicinity μ of the origin. In fact, the dynamics of the gain $K(t)$ follows a similar idea than previous controllers: the quantity $[\text{sign}(|\sigma| - \mu)]$ can be viewed as a sliding mode detector, and the behavior of the gain is forced by the detector value. The second equation of (4.6.27) is used to ensure that the gain $K(t)$ never becomes negative.

Theorem 4.6.3. *[19] Consider system* (4.6.25) *with Assumptions A3–A6 fulfilled and controlled by* (4.6.26)–(4.6.28) *with* ε, ω_1, γ_1, η *positive constants, and the parameter* K_{min} *an arbitrary small positive constant. Then, the controller ensures the convergence of* σ *to a vicinity of the origin in a finite time.* □

4.6.6 Experimental comparisons

In this section, experimental results of the three previous controllers are displayed. The performances of these control laws are compared, for the first time, on the same benchmark. The experimental tests are made with a sampling time $T_e = 1$ ms. The initial conditions of the electropneumatic system are set as

$$y(0) = 0.07 \text{ m}, \quad v(0) = 0 \text{ m s}^{-1}, \quad p_P(0) = 1 \text{ bar}, \quad p_N(0) = 1 \text{ bar}. \tag{4.6.29}$$

As shown in Figure 4.6.3, the main actuator is forced by the control law to track a reference signal $y_{ref}(t)$ under the perturbation force $F_{ext}(t)$. The objective of the benchmark is to evaluate the performances of the control laws in case of trajectory tracking (the desired trajectories being more or less fast) and in presence of time varying perturbations. The position trajectory reference $y_{ref}(t)$ is composed of two sinusoidal signals with different frequencies and is formally defined as

$$\begin{aligned} y_{ref}(t) &= 0.04\cos(0.2\pi t) && \text{for} \quad 0 \geq \text{t} \geq 20 \text{ s}, \\ y_{ref}(t) &= 0 && \text{for} \quad 20 < \text{t} < 30 \text{ s}, \\ y_{ref}(t) &= 0.04\cos(0.4\pi t) && \text{for} \quad 30 \geq \text{t} \geq 50 \text{ s}. \end{aligned} \tag{4.6.30}$$

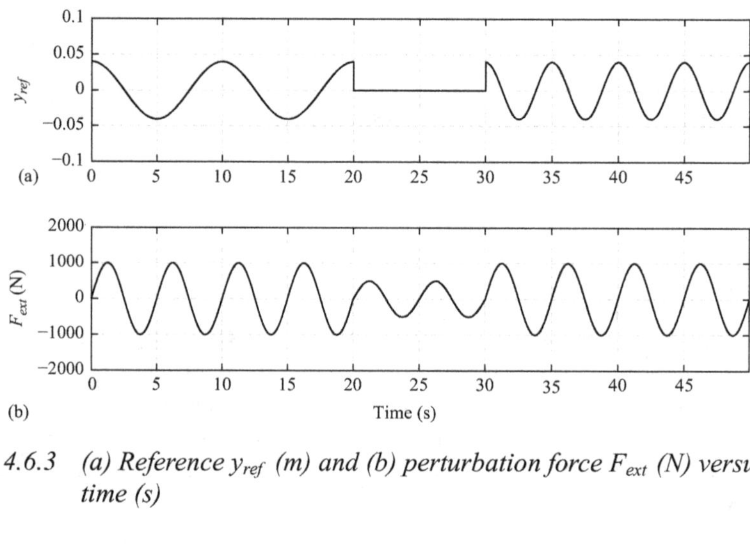

Figure 4.6.3 (a) Reference y_{ref} *(m) and (b) perturbation force* F_{ext} *(N) versus time (s)*

The perturbation force $F_{ext}(t)$ is generated by the perturbation actuator and is defined as

$$\begin{aligned} F_{ext}(t) &= 1000\sin(0.4\pi t) \quad \text{for} \quad 0 \geq \text{t} \geq 20 \text{ s}, \\ F_{ext}(t) &= 500\sin(0.4\pi t) \quad \text{for} \quad 20 < \text{t} < 30 \text{ s}, \\ F_{ext}(t) &= 1000\sin(0.4\pi t) \quad \text{for} \quad 30 \geq \text{t} \geq 50 \text{ s}. \end{aligned} \tag{4.6.31}$$

Figure 4.6.3 displays the reference trajectory and the perturbation force applied to the main actuator. The parameters setting for the three controllers are detailed in Table 4.6.2. The tuning of these parameters has been made in order to get the best performances with each controller.

The results obtained by each controller are represented in Figures 4.6.4–4.6.6 (resp. for ATW, AOF, and ASTW controllers): the reference trajectory and the measured position (subplot a), the perturbation force (subplot b), the control input (subplot c), the position tracking error (subplot d), the adaptive gains (subplot e), and the sliding variables (subplot f) are presented.

Due to a tracking error at the initial condition, the gains for all the controllers are initially increasing. After a finite time, the gains are sufficient large in order to ensure the establishment of a real second-order sliding mode: this latter occurs which induces the decrease of the gains. After that, increase and decrease of the gains alternately appear, because of loss (too small gains) and establishment (enough large gains) of the sliding mode.

Note that the frequencies of the reference signals are different for the first 20 s and the last 20 s. For a tracking reference with higher frequency, there is no doubt that larger gains are required. Comparing average gains for $t \in [10,\ 20]$ and for $t \in [40,\ 50]$, it is clear that the ATW control average gain grows from 1400 to 1800 whereas it grows from 2400 to 2900 for AOF controller, and from 1000 to 1500 for ASTW controller. A conclusion is that the gains magnitude is adjusted to the different working conditions.

Define the "energy" required by the controller as

$$E = \int_{t=0}^{t=50} |w|\mathrm{d}t. \tag{4.6.32}$$

Table 4.6.2 Parameters for three controllers

ATW	AOF	ASTW
$\alpha = 40 \quad \mu = 100$ $\lambda = 20 \quad q = 5.1$ $K_m = 50 \quad K_{mm} = 1$ $\gamma = 2/3$	$\alpha = 40 \quad \gamma = 5$ $\Gamma = 1500 \quad \beta = 2.1$ $K_m \in [500,\ 3000]$	$\omega_n = 50 \quad \zeta = 0.7$ $\varepsilon = 1 \quad \gamma_1 = 2$ $\omega_1 = 200 \quad \mu = 2$ $K_{min} = 0.01 \quad \eta = K_{min}$

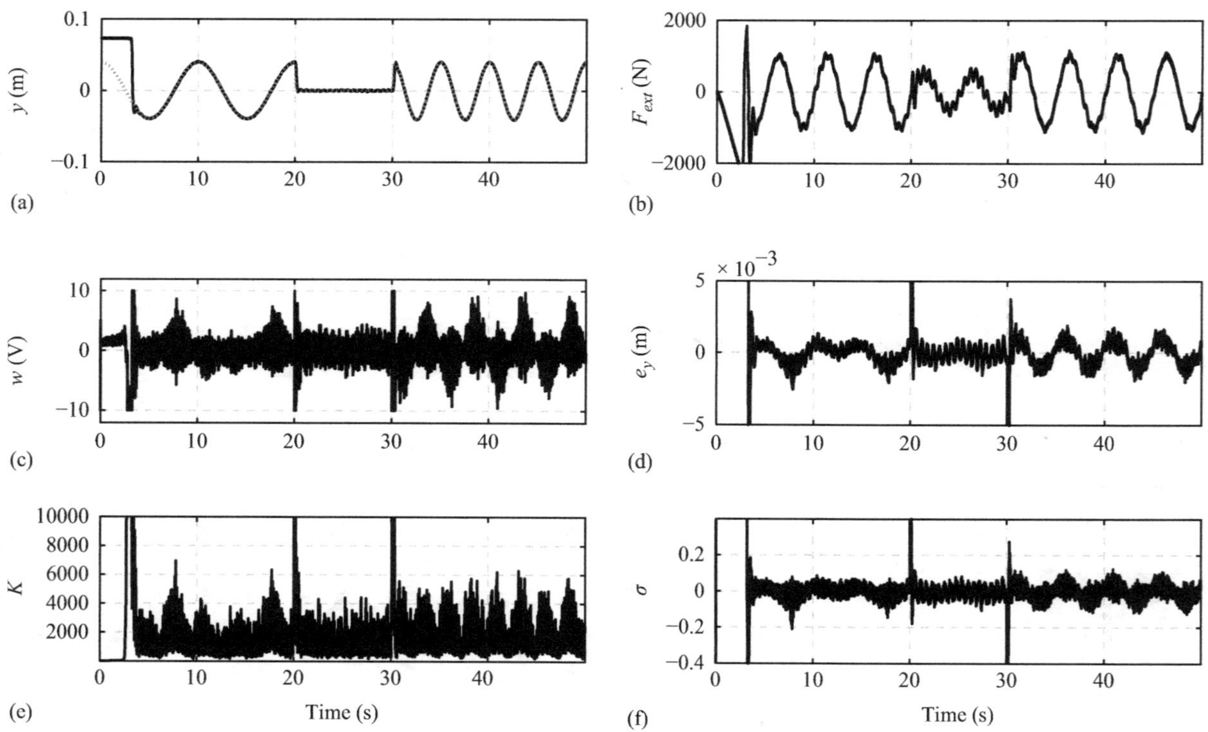

Figure 4.6.4 ATW. (a) Reference and y (m), (b) perturbation force F_{ext} (N), (c) control input u (V), (d) tracking error e_y (m), (e) gain K, and (f) sliding variable σ versus time (s)

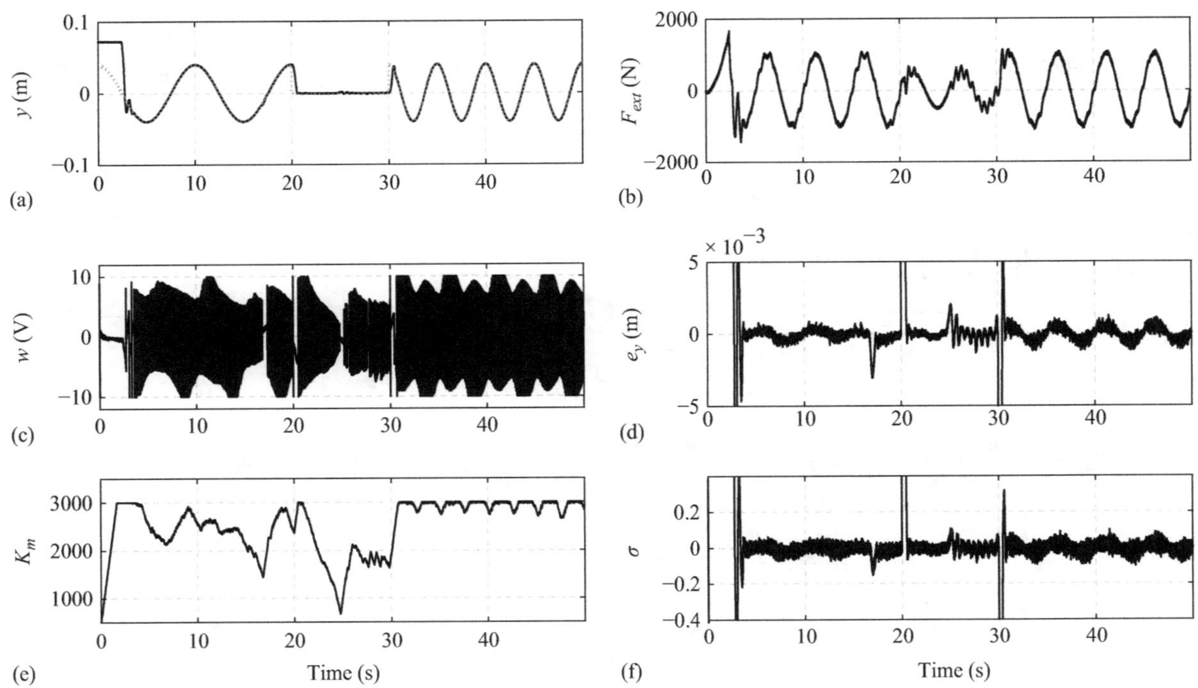

Figure 4.6.5 *AOF. (a) Reference and y (m), (b) perturbation force F_{ext} (N), (c) control input u (V), (d) tracking error e_y (m), (e) gain Km, and (f) sliding variable σ versus time (s)*

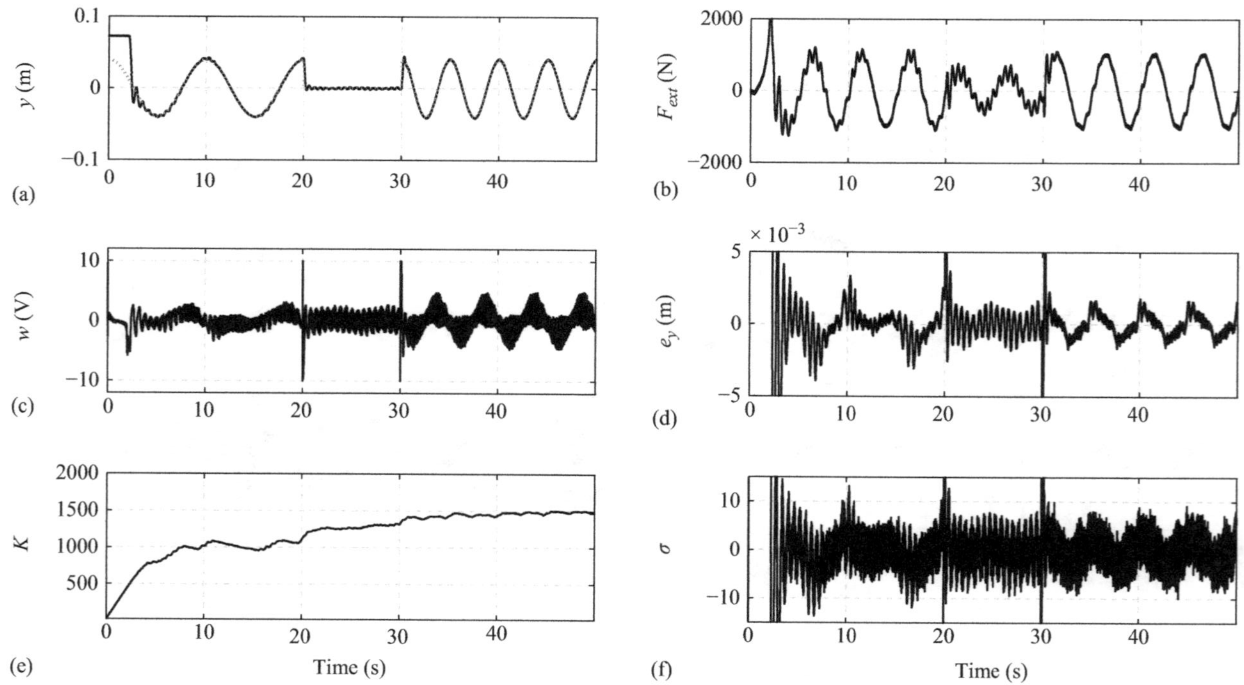

Figure 4.6.6 *ASTW. (a) Reference and y (m), (b) perturbation force F_{ext} (N), (c) control input u (V), (d) tracking error e_y (m), (e) gain K, and (f) sliding variable σ versus time (s)*

Table 4.6.3 Parameters for three controllers

	ATW	AOF	ASTW
Number of parameters	7	6	8
Differentiation order	2	1	2
Mean of $\|e_y\|$	4.30×10^{-3}	3.50×10^{-3}	3.00×10^{-3}
Std. dev. of e_y	1.42×10^{-2}	1.18×10^{-2}	9.7×10^{-3}
Mean of $\|w\|$	1.58	3.44	1.35
Std. dev. of w	2.26	4.42	1.78
Energy of w	3.70	3.67	2.14

Table 4.6.3 displays different variables allowing to compare the performances of the controllers

- *Number of parameters*. This information is crucial in terms of tuning difficulty. In fact, more this number is large, more difficult the tuning of the controller is;
- *Order of differentiation*. The use of measurement differentiation is an important point in order to reduce the quality of the control input. This order has to be the lowest;
- *Mean/standard deviation of the tracking error e_y*. This information allows to estimate the accuracy induced by each controller;
- *Mean/standard deviation of the control input w*. This information allows to estimate the behavior of the control input;
- *Energy of the control w*. This information allows to evaluate the amount of energy required by the control input.

The experimentation results show that, in the frame of this benchmark, the ASTW gets the best performances in terms of accuracy, chattering free, input quality, and energy consumption. However, the number of parameters seems important. Note also that the AOF and ATW control algorithms allow to get accurate results in spite of lower performances indicators.

4.6.7 Conclusions

This chapter has detailed experimental applications of three adaptive controllers based on second-order sliding mode concept. These controllers have been evaluated on the same benchmark consisting in tracking a desired position trajectory for an electropneumatic actuator in spite of external perturbation. The interest of such controllers is their capability to maintain a high level of performances (in terms of accuracy, reduction of chattering) with a very reduced amount of knowledge/information concerning the system.

References

[1] M. Belgharbi, D. Thomasset, S. Scavarda, and S. Sesmat, "Analytical model of the flow stage of a pneumatic servo-distributor for simulation and nonlinear control," *Scandinavian International Conference on Fluid Power SICFP'99*, Tampere, Finland, 1999, pp. 847–860.

[2] M. Bouri and D. Thomasset, "Sliding control of an electropneumatic actuator using an integral switching surface," *IEEE Transactions on Control Systems Technology*, vol. 2, no. 2, pp. 368–375, 2001.

[3] M.-K. Chang, J.-J. Liou, and M.-L. Chen, "T-S fuzzy model-based tracking control of a one-dimensional manipulator actuated by pneumatic artificial muscles," *Control Engineering Practice*, vol. 19, no. 12, pp. 1442–1449, 2011.

[4] A. Estrada and F. Plestan, "Second order sliding mode output feedback control: impulsive gain and extension with adaptation," *IEEE Conference on Decision and Control*, Maui, HI, USA, 2012.

[5] A. Estrada, F. Plestan, and B. Allouche, "An adaptive version of a second order sliding mode output feedback controller," *European Control Conference*, Zurich, Switzerland, 2013.

[6] A. Estrada and F. Plestan, "Second order sliding mode output feedback control with switching gains—application to the control of a pneumatic actuator," *Journal of the Franklin Institute*, vol. 351, no. 4, pp. 2335–2355, 2014.

[7] A. Girin, F. Plestan, X. Brun, and A. Glumineau, "Robust control of an electropneumatic actuator: application to an aeronautical benchmark," *IEEE Transactions on Control Systems Technology*, vol. 17, no. 3, pp. 633–645, 2009.

[8] C.E. Hall and Y.B. Shtessel, "Sliding mode disturbance observer-based control for a reusable launch vehicle," *AIAA Journal of Guidance, Control, and Dynamics*, vol. 29, no. 6, pp. 1315–1328, 2006.

[9] Y.-J. Huang, T.-C. Kuo, and S.-H. Chang, "Adaptive sliding-mode control for nonlinear systems with uncertain parameters," *IEEE Transactions on Systems, Man, and Cybernetics – Part B: Cybernetics*, vol. 38, no. 2, pp. 534–539, 2008.

[10] A. Levant, "Sliding order and sliding accuracy in sliding mode control," *International Journal of Control*, vol. 58, no. 6, pp. 1247–1263, 1993.

[11] S. Laghrouche, M. Smaoui, F. Plestan, and X. Brun, "Higher order sliding mode control based on optimal approach of an electropneumatic actuator," *International Journal of Control*, vol. 79, no. 2, pp. 119–131, 2006.

[12] H. Lee and V.I. Utkin, "Chattering suppression methods in sliding mode control systems," *Annual Reviews in Control*, vol. 31, pp. 179–188, 2007.

[13] R. Moreau, M.T. Pham, M. Tavakoli, M.Q. Le, and T. Redarcel, "Sliding-mode bilateral teleoperation control design for master-slave pneumatic servo systems," *Control Engineering Practice*, vol. 20, pp. 584–597, 2012.

[14] A.K. Paul, J.K. Mishra, and M.G. Radke, "Reduced order sliding mode control for pneumatic actuator," *IEEE Transactions on Control Systems Technology*, vol. 2, no. 3, pp. 271–276, 1994.

[15] F. Plestan and A. Girin, "A new experimental setup for a high performance double electropneumatic actuators system," *American Control Conference*, Saint Louis, USA, 2009.

[16] F. Plestan, E. Moulay, A. Glumineau, and T. Cheviron, "Robust output feedback sampling control based on second-order sliding mode," *Automatica*, vol. 46, no. 6, pp. 1096–1100, 2010.

[17] F. Plestan, Y. Shtessel, V. Brégeault, and A. Poznyak, "New methodologies for adaptive sliding mode control," *International Journal of Control*, vol. 83, no. 9, pp. 1907–1919, 2010.

[18] F. Plestan, Y. Shtessel, V. Brégeault, and A. Poznyak, "Sliding mode control with gain adaptation—application to an electropneumatic actuator," *Control Engineering Practice*, vol. 21, no. 5, pp. 679–688, 2013

[19] Y. Shtessel, M. Taleb, and F. Plestan, "A novel adaptive-gain supertwisting sliding mode controller: methodology and application," *Automatica*, vol. 48, no. 5, pp. 759–769, 2012.

[20] M. Smaoui, X. Brun, and D. Thomasset, "A combined first and second order sliding mode approach for position and pressure control of an electropneumatic system," *American Control Conference ACC'05*, Portland, Oregon, USA, 2005.

[21] M. Taleb, A. Levant, and F. Plestan, "Pneumatic actuator control: solution based on adaptive twisting and experimentation," *Control Engineering Practice*, vol. 21, no. 5, pp. 727–736, 2013.

Index

Index compiled by MPS Limited

www.ingramcontent.com/pod-product-compliance
Lightning Source LLC
LaVergne TN
LVHW011903200926
848484LV00002B/516